W9-DEE-481

Intraseasonal Variability in the Atmosphere–Ocean Climate System

William K. M. Lau and Duane E. Waliser

Intraseasonal Variability in the Atmosphere–Ocean Climate System

Dr. William K. M. Lau
Chief Laboratory for Atmospheres
NASA/Goddard Space Flight Center
Greenbelt
Maryland
USA

Dr Duane E. Waliser
Jet Propulsion Laboratory
California Institute of Technology
Pasadena
California
USA

SPRINGER–PRAXIS BOOKS IN ENVIRONMENTAL SCIENCES
SUBJECT *ADVISORY EDITOR*: John Mason B.Sc., M.Sc., Ph.D.

ISBN 3-540-22276-6 Springer-Verlag Berlin Heidelberg New York

Springer is part of Springer-Science + Business Media (springeronline.com)

Bibliographic information published by Die Deutsche Bibliothek

Die Deutsche Bibliothek lists this publication in the Deutsche Nationalbibliografie; detailed bibliographic data are available from the Internet at http://dnb.ddb.de

Library of Congress Control Number: 2004113457

Printed in Germany

Cover design: Jim Wilkie
Project Management: Originator Publishing Services, Gt Yarmouth, Norfolk, UK

Printed on acid-free paper

Contents

Preface

On the subject of extended-range weather forecasts, one of the pioneers of numerical weather forecasts, John von Neumann (1955) wrote:

> "The approach is to try first short-range forecasts, then long-range forecasts of those properties of the circulation that can perpetuate themselves over arbitrarily long periods of time ... and only *finally* to attempt forecast for medium-long time periods which are too long to treat by simple hydrodynamic theory and too short to treat by the general principle of equilibrium theory"

In modern phraseology, von Neuman's short-range forecasts would mean weather forecasts extending out to about five days, long-range forecasts would be equivalent to climate predictions extending out to a season or longer, and medium-long-range forecast would refer to intraseasonal predictions having lead times of the order of two to eight weeks. Numerical weather forecasting has seen tremendous improvement since its inception in the 1950s. Today, human activities are often so dependent on skillful short-term weather forecasts, that many have come to the unrealistic hope, and even expectation, that weather forecasts should be accurate all the time. However, any basic textbook on weather forecasting will point out that there exists a natural limit on *deterministic* weather forecasts of about two weeks, which is strongly dependent on initial conditions and atmospheric flow regimes.

Recently, the public has been made aware of high-impact climate phenomena such as El Niño and La Niña, which can affect weather patterns all over the world. Thanks in large part to the international climate research program, Tropical Ocean Global Atmosphere (TOGA), scientists now have the observational resources, the knowledge, and the models to make useful (deterministic) predictions of El Niño and La Niña with lead times up to 9–12 months. These predictions in turn have been helpful in making *probabilistic* seasonal-to-interannual forecasts of weather patterns (not deterministic forecasts of individual weather events) more skillful over certain space–time domains (e.g. wintertime temperature over North America and summer rainfall over the Asian monsoon region, and South America). Because the lead time for climate prediction is typically a season or longer – a time long enough for the atmosphere to lose memory of its initial state – the skill of prediction is no longer dependent on the initial conditions of the atmosphere. In contrast to weather forecasts, seasonal-to-interannual climate predictions owe their skill to a dependence on slowly changing boundary conditions at the Earth's surface, such as sea surface temperature, snow cover, and soil moisture, and the considerable impact these boundary conditions have on determining the statistics of the observed weather patterns. In the forecasting community, it is often said that weather forecasting is an initial value problem and climate prediction is more akin to a boundary value problem. What about the timescales in-between, for example lead times between about 2 weeks and 2 months? Are there atmosphere–ocean phenomena with these timescales that are predictable, and how do these phenomena and their predictability respond to the changing boundary conditions at the Earth's surface? These are among some of the issues to be addressed in this book.

Given the progress in weather forecasting and seasonal-to-interannual climate prediction, it is apparent that we are ready to more formally and thoroughly address forecasting of, in von Neuman's words, the "medium-long time periods". Improving extended-range (i.e., intraseasonal) forecasts requires fundamental knowledge built on sound research, realistic models of the atmosphere, ocean and land components of the climate system, and the training of a new generation of scientists and forecasters. Today, we have many textbooks and research reference books on weather and climate variability, and prediction, but there has been none focused specifically on intraseasonal variability (ISV). There has been a large body of scientific studies showing that ISV is far from a simple interpolation between weather and climate scales/processes, and is not just a red-noise extension of weather variability. Indeed, there are specific and unique modes of ISV that are ubiquitous and can be found in the atmosphere, the ocean, and the solid Earth, as well as in the tropics and the extra-tropics.

To improve prediction in the intermediary timescale (2 weeks to less than a season) of the atmosphere–ocean, it is vital to improve our understanding of the phenomena that are inherently intraseasonal, and the manner in which they interact with both shorter (weather) and longer (climate) timescales. Thus one of the overarching goals of this book is to summarize our current understanding of IV and its interactions with other weather and climate processes. However, in developing the framework for this book, we found that including all aspects of ISV would require too much material for one book. Thus, in order to limit the scope of this book, we have chosen to focus primarily on ISV in the tropical ocean and atmosphere, including its interactions with the extra-tropics whenever appropriate. Using this guideline, topics directly related to mid-latitude atmospheric blocking or extra-tropic annular modes, for example, will not be treated in their own right in this book, but rather discussed in the context of their interaction with tropical ISV (TISV).

Central to TISV is the Madden–Julian Oscillation (MJO) phenomenon, known also as the 40–50-day or 30–60-day oscillation. However, TISV in general refers to a broad spectrum of phenomena; some quasi-periodic, some non-periodic, some with global reach, and others with regional manifestations. To avoid possible confusion in this book with the various terminologies used in the literature, we refer throughout this book to all variability longer than synoptic timescales (~2 weeks) and shorter than a season (90 days), as ISV. The MJO is specifically referred to as the atmosphere–ocean entity that exhibits a coherent eastward propagation along the equator with quasi-periodicity of 30–60 days. In the general case, when a quasi-periodic oscillation can be identified, the term intraseasonal oscillation (ISO) will be used. When specially referring to ISV or ISO in the tropics, the acronyms TISV or TISO will be used as appropriate. In this nomenclature, MJO is a special case of a TISO.

This book is intended to be a one-stop reference book for researchers interested in ISV as well as a textbook for senior undergraduate and graduate students in Earth science disciplines. The book contains 12 chapters, each with a comprehensive bibliography. Chapter 1 provides a historical account of the detection of the MJO by R. Madden and P. Julian, who discovered the phenomena. The regional characteristics of TISV on South Asia, East Asia, the Americas, and Australia/Indonesia are covered in Chapters 2–5, respectively. Air–sea interactions and oceanic ISV are discussed in Chapters 6 and 7. Chapter 8 discusses atmospheric and solid Earth angular momentum and Earth rotation associated with ISV. Chapter 9 is on El Niño Southern Oscillation (ENSO) connections to ISV. Chapters 10, 11, and 12 are devoted to the theory, numerical modeling, and predictability of ISV, respectively. The chapters are written with self-contained material, and frequent cross-referencing to other chapters, so that they need not be read in sequence. Readers are encouraged to jump to their chapters of interest if they so desire. However, we strongly recommend everyone to read the Preface and Chapter 1 first to obtain the proper perspective of the subject matter and objectives of the book.

This book could not have been possible without the support and the dedicated efforts of the contributing authors, who provided excellent write-ups for the chapters in a timely manner. Everyone we contacted regarding this book was very enthusiastic and supportive. In addition, we thank Drs H. Annamalai, Charles Jones, Huug van den Dool, T. C. (Mike) Chen, Klaus Weickmann, Chidong Zhang, Ragu Murtugudde, William Stern, George Kiladis, and Steve Marcus, and one anonymous reviewer for providing very constructive comments in reviewing various chapters of this book. The co-chief editors will also like to thank the Earth Science Enterprise of the National Aeronautics and Space Administration, the Office of Global Programs of the National Oceanographic and Atmospheric Administration, and the Climate Dynamics and Large-Scale Dynamic Meteorology Programs of the Atmospheric Sciences Division of the National Science Foundations for providing support over the years for research on ISV.

REFERENCE

von Neumann, J. (1955) Some remarks on the problem of forecasting climate fluctuations. "Dynamics of Climate": The Proceedings of a Conference on the Application of Numerical Integration Techniques to the Problem of the General Circulation. Pergamon Press, 137.

William K. M. Lau and Duane E. Waliser
Goddard Space Flight Center, Greenbelt, Maryland
October, 2004

Figures

Abbreviations

AAM	atmospheric angular momentum
AAMWG	Asian–Australian Monsoon Working Group
ACC	Antarctic Circumpolar Current
ADCP	acoustic doppler current profiler
AMEX	Australian Monsoon Experiment
AMIP	Atmospheric Model Itercomparison Project
AO	Arctic Oscillation
AS	Arabian Sea
AVHRR	Advanced Very-High Resolution Radiometer
BoB	Bay of Bengal
BSISV	boreal summer ISV
CAPE	Convective Available Potential Energy
CC	cloud clusters
CCA	canonical correlation analysis
CID	convective interaction with dynamics
CISK	Conditional Instability of the Second Kind
CISO	climatological intraseasonal oscillation
CLIVAR	Climate Variability
CM	center of mass
CMAP	CPC Merged Analysis of Precipitation
COARE	Coupled Ocean–Atmosphere Response Experiment
CRM	cloud-resolving model
d.o.f.	degrees of freedom
DERF	Dynamical Extended Range Forecast
DOD	Department of Ocean Development (India)
DORIS	Doppler Orbitography and Radio positioning Integrated by Satellite
EA/WNP	East Asian and western North Pacific

ECHAM	European Centre for Medium-range Weather Forecast – Hamburg atmospheric model
ECMWF	European Centre for Medium-range Weather Forecasts
EEOF	extended EOF
EMEX	Equatorial Mesoscale Experiment
ENSO	El Niño Southern Oscillation
EOF	empirical orthogonal functions
EPM	eastward propagating mode
ER	equatorial Rossby
EUC	Equatorial Undercurrent
EWP	empirical wave propagation
FFT	Fast Fourier Transform
FGGF	First GARP Global Experiment
FSU	Florida State University
GARP	Global Atmospheric Research Program
GCM	global circulation model
GEWEX	Global Energy and Water Experiment
GFDL	Geophysical Fluid Dynamics Laboratory
GLA	Goddard Laboratory for Atmosphere
GLAS	Goddard Laboratory for Atmospheric Services
GMS	Geostationary Meteorological Satellite
GPS	Global Positioning System
GTS	Global Telecommunication System
IAV	interannual variability
IB	inverted-barometer
IFA	Intensive Flux Array
IMET	Improved Meteorology
IO	Indian Ocean
ISO	intraseasonal oscillation
ISV	intraseasonal variability
ITCZ	Intertropical Convergence Zone
ITF	Indonessian Throughflow
K.E.	kinetic energy
LIM	Linear Inverse Model
LLJ	low-level jet
LOD	length of day
LPS	low-pressure systems
M-SSA	multi-channel Singular Decomposition
MBF	Meiyu/Baiu front
MEM	maximum entropy method
MISI	monsoon ISO index
MJO	Madden–Julian Oscillation
MLO	mixed layer depth
MONEX	Monsoon Experiment
MRF	medium-range forecast

MRG-TD	Mixed Rossby–Gravity wave-Tropical Disturbance
MT	monsoon trough
NAME	North American Monsoon Experiment
NCAR	National Center for Atmospheric Research
NCEP	U.S. National Centers for Environmental Prediction
NEC	North Equatorial Current
NECC	North Equatorial Counter Current
NMC	National Meteorological Center (U.S.A.)
NOAA	National Oceanic and Atmospheric Administration
OLR	outgoing long-wave radiation
OLRA	OLR anomaly
PC	principal components
PNA	Pacific–North American
POPs	Principal Oscillating Patterns
PSA	Pacific–South American
PV	potential vorticity
QBM	quasi-biweekly mode
QBO	Quasi-biennial Oscillation
QSM	quasi-stationary mode
RAAM	relative atmospheric angular momentum
RHc	Relative Humidity criterion
RMM	real-time Multivariate MJO
SACZ	South Atlantic Convergence Zone
SALLJEX	South American Low-Level Jet Experiment
SCC	Supercloud dusters
SEC	South Equatorial Current
SLR	satellite laser ranging
SOI	Southern Oscillation Index
SPCZ	South Pacific Convergence Zone
SSA	single spectrum analysis
SSH	sea surface height
SST	sea surface temperature
SSTA	SST anomaly
SSWJ	subsurface westward jet
STCC	Subtropical Countercurrent
TAO	Tropical Atmosphere Ocean
TC	tropical cyclones
TCZ	Tropical Convergence Zone
TISO	tropical intraseasonal oscillation
TISV	tropical intraseasonal variability
TIW	Tropical Instability Waves
TMI	TRMM Microwave Imager
TOGA	Tropical Ocean Global Atmosphere
TRMM	Tropical Rain Measuring Mission (satellite)
VAMOS	Variability of the American Monsoon Systems

VLBI	very-long-baseline interferometry
WISHE	wind-induced surface heat exchange
WNP	western North Pacific
WWB	westerly wind burst
WWE	westerly wind event
WWW	World Weather Watch

1

Historical perspective

Roland A. Madden and Paul R. Julian

1.1 INTRODUCTION

The 1960s was a remarkable decade for research in tropical meteorology. Tropical climatology was already reasonably understood, but little was known of its variability or that of daily tropical weather. Regularly sampled data and access to computers to process data became more readily available. The excitement of looking at these data, which no one else had studied before, must have been something like that of polar explorers in the early part of the century who made their way to places no one had ever been before. The decade opened with descriptions of the remarkable Quasibiennial Oscillation (QBO) showing that neither the formally identified "Krakatoa Easterlies" nor the "Berson Westerlies" were steady features of the equatorial stratosphere (Ebdon, 1963). By the mid-1960s a theory taylored specifically to waves in the equatorial region was published, and soon after some of them were observed. These were, arguably, the first identifications of large-scale waves in the atmosphere predicted by theory. By the end of the decade the tropical atmosphere was a topic of research given similar attention to that of the mid-latitudes.

The surprising discovery of the QBO (Ebdon, 1960; Reed *et al.*, 1961) kindled new interest in the meteorology of the tropics leading eventually to the discovery, at the beginning of the next decade, of an equally surprising tropical oscillation with an intraseasonal timescale. That feature is often referred to as the Madden–Julian Oscillation (MJO) after papers appearing in the *Journal of the Atmospheric Sciences* (Madden and Julian, 1971; 1972). The discovery of the MJO resulted from the serendipitous convergence of this new interest in the tropics, new tropical data, new computers, and the increasing application of spectrum analysis. Basic features of the MJO are described here, but first, the research environment that led to its discovery is outlined.

Graystone (1959) showed the zonal wind, u, or wind blowing from west to east, in a time–height section of the lower stratosphere from October 1956 through to

W. K. M. Lau and D. E. Waliser (eds), *Intraseasonal Variability in the Atmosphere–Ocean Climate System.*

August 1958. With only 23 months of data, Graystone could not have recognized the very regular change in the *u*-wind from westerlies to easterlies and back with an approximate 26-month period, even though, in retrospect, it is clearly evident. He did "note the interesting lack of an annual variation in the data." It took a longer time series to bring out the QBO (Reed *et al.*, 1961). The QBO, a phenomenon so unexpected yet so unmistakable and so amazing, proved that the tropics was not, excepting an occasional tropical storm, a dull and uninteresting place meteorologically.

The QBO begged for an explanation and tropical meteorologists, most notably at the Universities of Tokyo and Washington, began searching for one. Yanai and Maruyama (1966) reported on wave-like disturbances in the meridional wind, *v*, or south to north component, in the tropical lower stratosphere with a timescale of five days, a horizontal length scale of 10,000 km, and a westward phase speed of about $23\,\mathrm{m\,s^{-1}}$. Motivation for their work was a search for large-scale waves in the equatorial stratosphere. It was thought that such waves might play a role in the momentum convergence needed to explain the QBO. It is now understood that the waves Yanai and Maruyama discovered do play a minor role, but more importantly their discovery marked one of the first unambiguous identifications of theoretically predicted, large-scale atmospheric waves. A theory of equatorial waves had been layed out in that same year by Matsuno (1966). It is interesting to note that Matsuno submitted his manuscript in November 1965, in which he thanks Yanai for reading it; yet Yanai and Maruyama did not immediately recognize the connection between their observational paper submitted in July 1966 and Matsuno's theory – it did not take long though. In the following year Maruyama published a second paper on the waves and identified them as mixed-Rossby gravity waves, predicted by the theory (Maruyama, 1967).

At almost the same time, Wallace and Kousky (1968) were studying the *u*-wind in the tropical stratosphere. There motivation was similarily related to the QBO. They stated: "This study of synoptic-scale disturbances in the tropical stratosphere was originally motivated by certain unsolved problems relating to the momentum budget of the quasi-biennial oscillation." They found waves with 15-day periods, length scales of 20,000–40,000 km (zonal wavenumber 2 and 1), and 6–10 km vertical scales. They identified them as Kelvin waves predicted by Matsuno's theory.

In the above and subsequent papers, Yanai *et al.* (1968), Wallace and Chang (1969), and colleagues showed the power of spectrum analysis in extracting the most out of widely scattered tropical observations and how to interpret results in the context of theoretical predictions. Their work provided a vantage for the analysis of tropical data beginning at the National Center for Atmospheric Research (NCAR) in Boulder, Colorado. The Center was collecting longer time series than had been available to the research community earlier, and it had the fastest computers devoted to meteorological studies: a Control Data Corporation (CDC) 6600, and in 1971 a CDC 7600. The computers were advanced for the time, but clock speeds and memories were only 10 mHz and 64 kb, and 36 mHz and 65.5 kb, respectively. Today a typical laptop computer (e.g., Dell 8600) has a clock speed of 1,400 mHz and 262,144 kb of random access memory.

Another fortuitous development at the time was that of the fast Fourier transform, or FFT (Cooley and Tukey, 1965), which made it feasible to perform spectrum analysis of long time series on these machines. Before the development of the FFT, a Fourier transform of an N-member time series required $N \times N$ complex multiplies. The FFT reduced that requirement to $N \times \log(N)$ which for a 10-year record of daily values reduced multiplications by a factor of 100.

In the fall of 1970, we embarked on a study "designed to provide analysis over a broader frequency range and to study the non-stationary aspects of the aforementioned wave modes" (Madden and Julian, 1971). The aforementioned wave modes were those discovered by Yanai and Maruyama and Wallace and Kousky. What we found was a variation with a timescale longer than these waves and shorter than any component one could attribute to seasonal variations, and one not predicted by any theory.

1.2 THE INTRASEASONAL, TROPOSPHERIC OSCILLATION

For our initial analysis, rawinsonde data from Kanton Island (3°S, 172°W) for the period June 1957 to March 1967 were available with only about 2.5% of observations below 500 hPa missing, and approximately 5% missing above that level. Data were one sample per day, usually 00 GMT, and they included surface pressure, winds, temperatures, and humidity often extending to pressure levels higher than the tropopause which is at about 100 hPa near the equator. Though difficult to appreciate now, analyses of 10 years of daily observations pushed computer memory to its limit.

The rawinsonde data were contained on magnetic tapes: one station per tape. Magnetic tapes required an operator to physically mount them on a tape drive. Cross-spectra among data from two stations would require two tape mounts. Since the computers served all NCAR scientists and visitors, the need to mount tapes often meant long delays. To avoid these delays, initial analyses involved reading the data from the tapes and putting out values on punched cards. Data on cards were read in with the program and there was no need for further tape mounts. Of course this meant that decks of as many as 2,000 cards containing fortran routines and data had to be fed through the card reader, and, in any case, "turn around" was still slow relative to today. This was not all bad since it allowed plenty of time to digest the results of one run before the return of a second.

Resulting cross-spectra between the u-wind in the lower troposphere below about 500 hPa and that in the upper troposphere above 500 hPa showed negative values with large magnitudes in the cospectra occurring at approximately 50-day periods. A cross-spectrum of two time series gives information on the covariability between them as a function of frequency or period. This covariability is contained in the co-spectrum and the quadrature spectrum. The co-spectrum is that part of the co-variability at some frequency that is either exactly in phase or exactly out of phase. It turned out that, near 50-day periods, upper tropospheric u-winds were out-of-phase with lower tropospheric u-winds, or, put another way,

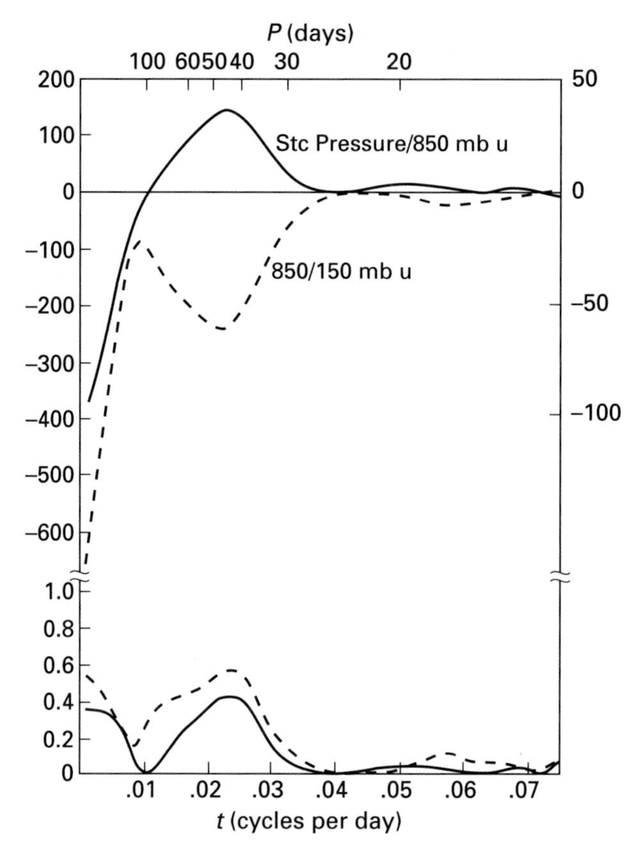

Figure 1.1. (*Top*) the co-spectrum of the 850 and 150-hPa *u*-wind (dashed line and left ordinate values), together with the co-spectrum of the station (sfc) pressure and the 850-hPa *u*-wind (solid line and right ordinate values) for Kanton Island. (*Bottom*) the coherence-squared statistic for the 850 and 150-hPa *u*-wind (dashed line) and for the station pressure and 850-hPa *u*-wind (solid line). The 0.1% prior (6% a posteriori) confidence level on the null hypothesis of no coherence is 0.25.
From Madden and Julian (1971).

the phase shift between the two was 180 degrees. If the phase shift had been different, then some, all in the case of an exactly 90-degree phase shift, of the covariablilty would have been found in the quadrature spectrum.

Figure 1.1 presents an example of those negative co-spectra along with a positive one between the 850 hPa *u* and the surface pressure. Corresponding coherence squares, a correlation as a function of frequency which includes both the co-spectrum and the quadrature spectrum parts of the covariability, are plotted at the bottom of the figure. All results have relative extrema in the 40–50-day period range. Madden and Julian (1971) were able to see these low frequency maxima where others had not by virtue of the relatively long record from Kanton that we could analyze. Phase angles (not shown) indicated that the 850-hPa u was out-of-phase with u at tropospheric levels above 600–500 hPa. Surface pressure was very nearly in-phase with the 850-hPa u. Coherence squares of 0.25 demark the 0.1% prior confidence level for a null hypothesis of zero coherence. A null hypothesis is a hypothesis about the data that can be tested statistically and accepted or rejected. In this case the null, zero coherence, is that the upper and lower tropospheric u-winds are not related.

A short digression is in order to explore statistical confidence levels. A 5% prior confidence level is one that 5% of estimates are likely to exceed due to sampling variability even if the null hypothesis were true. The kind of study we were doing might be termed "exploratory data analysis" because we had no "a priori" reason, that is no reason before we looked at the data to expect anything unusual at 50-day periods. When doing exploratory data analysis with no prior reason to expect any difference from a null, prior confidence levels are not particularily discriminatory. A "significant peak" at an arbitrary frequency may well be one of the 5% of such peaks expected. A stronger test is in order. The bandwidth of the analysis whose results are shown in Figure 1.1 is 0.008 1 cycles/day. As a result there are just over 60 (0.5/0.008 1) non-overlapping, independent estimates. Even if the null hypothesis of zero coherence were true, the 5% prior significance level would have, on average, three values (60×0.05) in a single sample spectrum exceeding it. The 0.1% prior significance level would have, on average, 0.06 values exceeding it in a single sample spectrum, or six values in 100 such sample spectra. In this case the 0.1% "prior confidence level" can be thought of as the 6% "posterior confidence level". A posterior confidence level is considerably more stringent than a prior one. Most of the coherence-squared values and spectral peaks reported in Madden and Julian (1971) exceeded zero coherence and smooth background spectral null hypotheses by the stringent 6% posterior confidence levels.

Madden and Julian (1971) concluded that at Kanton the oscillation was a relatively broadband phenomenon with maxima in coherence and power typically in the 41–53-day period range. The u-wind and pressure oscillations were in-phase with each other at a given level, but out-of-phase between lower and upper troposphere. There was a nodal surface in the 600–500-hPa levels. The v-wind did not appear to be involved. This last conclusion proved wrong, and resulted from not distinguishing results by season (see Section 1.7).

1.3 THE ELEMENTARY 4-D STRUCTURE

Cross-spectra between locations and the technique of compositing indicated that the pressure disturbance probably began in the Indian Ocean and propagated eastward moving at more than 30 m s^{-1} from Singapore (1°N, 104°E) to the Balboa Canal Zone (9°N, 80°W). Pressure oscillations were largest within 10 degrees of the Equator and from at least Singapore to Curacao (12°N, 69°W) in the Carribean. Data from six widely spaced rawinsonde stations indicated that the oscillation extended all the way around the world in the upper equatorial troposphere. In the lower troposphere the oscillations in the u-wind appeared to be limited to the Indian and western Pacific Oceans.

Troposheric temperature variations supported the out-of-phase nature of the vertical structure of the u-wind. Low pressures at the surface of the Central Pacific are accompanied by high tropospheric temperatures. This nearly out-of-phase relationship through the troposphere changes to a nearly in-phase relationship at 100 hPa. Low 100-hPa temperatures, possibly indicating a higher tropopause, are associated with low surface pressures. The temperature amplitude was of the order of 0.5°C in the troposphere and about twice that at 100 hPa. At least at Singapore and Chuuk (7°N, 152°E), high water vapor mixing ratios also accompanied low surface pressures.

Convergence of the lower level u-wind, divergence of the upper level u-wind, higher tropospheric temperatures and mixing ratios, and possibly higher tropopause were circumstantial evidence that deep convection accompanied low surface pressures. Figure 1.2 summarizes the evidence about the oscillation. Relative dates in the figure are indicated symbolically by letters in the left of each panel, and they relate to the surface pressure oscillation at Kanton. Date "A" is the time when pressure is low and "E" when it is high at Kanton. Other "dates" are intermediate ones. For a 48-day period there would be 6 days between each panel with time increasing from top to bottom. The pressure oscillation is indicated at the bottom of each panel with negative anomalies shaded. The streamlines reflect u-wind anomalies. Assumed associated convection is indicated by the cumulus and cumulonimbus clouds. The tropopause height (relatively high above surface low pressure and convective regions) is depicted by the wavey line at the top of each panel.

The behavior of the oscillation as indicated in Figure 1.2 is as follows: a negative pressure anomaly is present over East Africa and the Indian Ocean, and large-scale convection begins over the Indian Ocean (F); the pressure anomally propagates eastward past the Date Line as does the eastern edge of the zonal circulation cell (G); by the time of lowest pressure at Kanton the zonal circulation cells have a zonal wave number 1 character and the convection has moved across Indonesia (A); pressures begin to rise over the Indian Ocean and convection weakens over and east of the Date Line (B–C); finally there is highest pressure over Kanton, subsiding motion there and possibly weak rising motions over the Atlantic Ocean (E).

Figure 1.2 provides a simplified 3-D picture of the oscillation. The fourth, south–north dimension is, to first approximation, characterized by a simple weakening of the signal as one looks further from the equator.

1.4 OTHER EARLY STUDIES OF THE OSCILLATION

To our knowledge, the oscillation was not reported before 1971. Aspects were discussed in the later 1970s. Evidence of eastward propagating clouds speculated

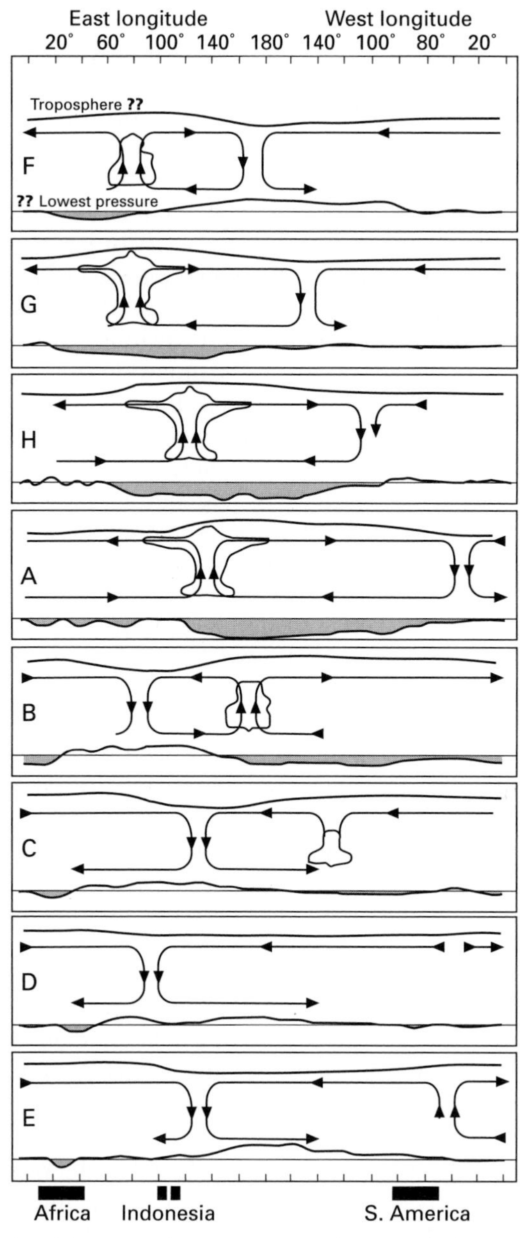

Figure 1.2. A schematic of the approximate structure of the oscillation in the equatorial plane. The situations summarized in each panel are about 4 to 8 days apart with time increasing downward. Cartoon clouds indicate large regions of increased convection. Streamlines show the east–west circulation with convergence into, and divergence out of, the convective areas in the lower and upper troposphere respectively. The wavey line at the top represents the tropopause and that at the bottom changing sea level pressure.
From Madden and Julian (1972).

by Madden and Julian (1972) was presented by Gruber (1974) who found eastward, zonal wave number 1 variance near 50 days in a space–time spectrum of cloud brightness data near the equator. Zangvil (1975) similarly reported eastward, zonal wave number 1 and wave number 2 variance near 40 days in equatorial cloud data. In addition to this eastward movement, a northward propagation of cloud zones over India with 30–40-day timescales was suggested by spatial correlations computed by Murakami (1976). A related paper published in the same year contained Dakshinamurti and Keshavamurty's (1976) spectral analyses of winds over India. They showed relative maxima in variance near 30-day periods that were associated with south-to-north movements of the monsoon trough. Later, Yasunari (1979) argued that the eastward propagating equatorial clouds and the northward movement of cloud systems in the monsoon trough were related.

Parker (1973) found the oscillation in the 100-hPa u-winds and temperatures over the equator. Parker considered the oscillation to be sufficiently like a Kelvin wave to be considered as such. Like the Madden and Julian references, Parker concluded that the oscillation affected u and not v-winds and, at least at Gan Island, u and pressure were about in phase. These are basic characteristics of the Kelvin wave. Equatorial Kelvin waves move eastward and the disturbance associated with the oscillations in variables moved eastward. The disturbance was symetric about the equator, fell off in amplitude away from the equator, and 100-hPa temperature variations tended to lead u-wind variations by 0.25 of a cycle – all properties of the Kelvin wave. At about the same time, Holton (1973) and Lindzen (1974) presented modeling and theoretical evidence that the oscillation could be the manifestation of an atmospheric Kelvin wave. Later it was suggested that the oscillation resulted from an eastward moving, forced Kelvin–Rossby wave pair as contained in Matsuno (1966) and others' work and studied by Gill (1980) (Yamagata and Hayashi, 1984; Madden, 1986).

We cannot pretend to provide an adequate summary of the considerable related theoretical work that has followed. That is found in Chapter 10. However, this section provides the opportunity to give a brief overview of its development. Theoretical work began with the aforementioned studies of Kelvin waves, but because of the large vertical scale of the oscillation, this theory typically predicted phase speeds that were much faster than observed. Adding a linear damping to the equations resulted in modes with more realistic eastward speeds (Chang, 1977). Convection was always recognized as an essential part of the oscillation, but early Kelvin wave theories did not explain what caused the convection. In addition, the recognition that convective heating near the equator forced a Kelvin–Rossby wave pair made unclear why eastward propagation associated with the Kelvin wave was selected over the westward propagation of the Rossby waves.

Wave-CISK (Conditional Instability of the Second Kind), in which the low-level moisture convergence of an existing wave produces convection and warming that acts, in turn, to reinforce the wave, could explain the convection. A problem was that most CISK formulations favor small-scale convection, not the very large scale that is observed. In addition eastward propagation is not a necessary consequence. Lau and Peng (1987) introduced "mobile wave-CISK" which included "positive only heating" unlike the wavey heating of traditional approaches. Their model favored more realistic large-scale convection and the eastward propagating Kelvin wave part of the response. To get the relatively slow propagation, the heating profile they used had what is likely an unrealistic maximum at low tropospheric levels.

A different mechanism that could explain the formation of convection and the eastward movement was "wind-induced surface heat exchange" (WISHE) (Emanual, 1987; Neelin *et al.*, 1987). Here, surface winds of the large-scale wave affect fluxes of latent heat from the ocean to preferably support new convection to the east of old. Unfortunately, WISHE favors smaller scale convection just as traditional CISK does.

None of these theories explain all of the complex features of the MJO. Undoubtedly many of the physical process that they describe are important as well as boundary layer friction (Wang, 1988; Hendon and Salby, 1994) and radiative effects (Hu and Randall, 1994). Chapter 10 adds details of the theories, and describes a model that incorporates many of their influences. Here, we continue to describe the Oscillation from an observational point of view.

1.5 THE OSCILLATION IN 1979

We separate out 1979 as a key time in the history of research into intraseasonal variations in the tropics because the meteorological research community invested considerable effort in the First Global Atmospheric Research Program (GARP) Global Experiment (FGGE) during that year. There were several MJOs in 1979. Two particularly strong oscillations passed over the region of the Monsoon Experiment (MONEX), May to August, and interest in them was renewed.

Lorenc (1984) computed the empirical orthogonal functions (EOFs) of daily values of the 200-hPa velocity potential during FGGE and found that the two leading EOFs (explaining most of the variance after the annual cycle had been removed) represented a zonal wave number 1, eastward propagating pattern, that circled the equator in 30–50 days, much like the upper tropospheric divergent circulations indicated in Figure 1.2. Lorenc noted negative velocity potential (upper level divergence) near India during the 1979 monsoon onset in mid-June and monsoon revival, or active period, in late July. In contrast, positive velocity potential (upper level convergence) ruled during a break in mid-July and during withdrawal in mid-August.

The importance of the wax (active periods) and wane (break periods) in the monsoon is illustrated in Figure 1.3. There are very large variations in Indian rainfall with roughly 40 days between maxima. These variations are associated with northward movement of the Monsoon Trough and accompanying clouds. The cloud systems also moved eastward along the equator. This particular northward and eastward moving event is clearly documented in Lau and Chan (1986b). We note that sometimes the term MJO is limited to systems whose dominant propagation is eastward, and those with considerable latitudinal

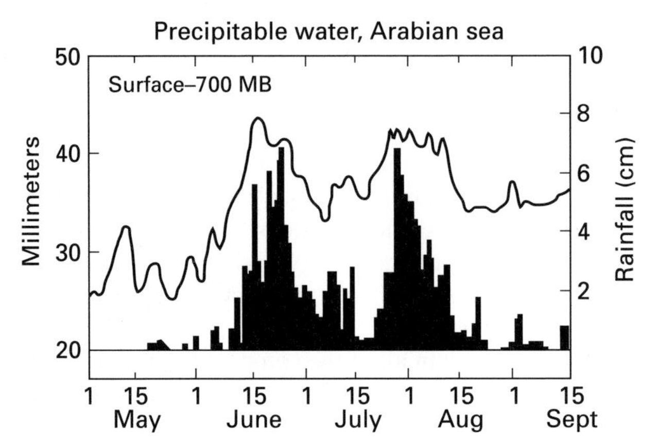

Figure 1.3. Time series of precipitable water from the surface to 700 hPa over the Arabian Sea (thin line) from TIROS-N, and the precipitation along the west coast of India during MONEX.
Adapted from Cadet (1986).

movement are considered as members of a broader class of intraseasonal oscillations (ISOs; see Chapter 10).

The MONEX period also revealed related variations in latent heat flux over the Bay of Bengal with amplitudes of $40\,\mathrm{W\,m^{-2}}$ about an average of about $200\,\mathrm{W\,m^{-2}}$ (Krishnamurti *et al.*, 1988). There were strong winds over the Bay of Bengal and accompanying positive anomalies in latent heat flux in mid-June and in late July coincident with the heavy rains over India (Figure 1.3). At those same times strong easterlies were found over the entire tropical Pacific (Madden, 1988). The resulting varying surface friction and exchange of angular momentum between the atmosphere and ocean–Earth system played an important role in angular momentum changes discussed in Section 1.8.

1.6 COMPLEXITY OF CLOUD MOVEMENT AND STRUCTURE

Although Figure 1.2 captures the essential nature of the oscillation in convection, it is a simplified picture. Wang and Rui (1990) stratified movement of cloud complexes related to the Oscillation into three categories: (1) strictly eastward near the equator from Africa to the Central Pacific; (2) similarily eastward over the Indian Ocean and then northward or southward over the western Pacific; and (3) eastward as above but connected to cloud systems that moved northward into southern Asia or into the North Pacific. Three out of four of the strictly eastward moving cloud complexes occurred from September to May. Complexes that moved eastward and then southward into the South Pacific also tended to occur during the northern winter half of the year, while those that moved into the North Pacific did so from April to December. This points to a tendency for the meridional movement to be into the

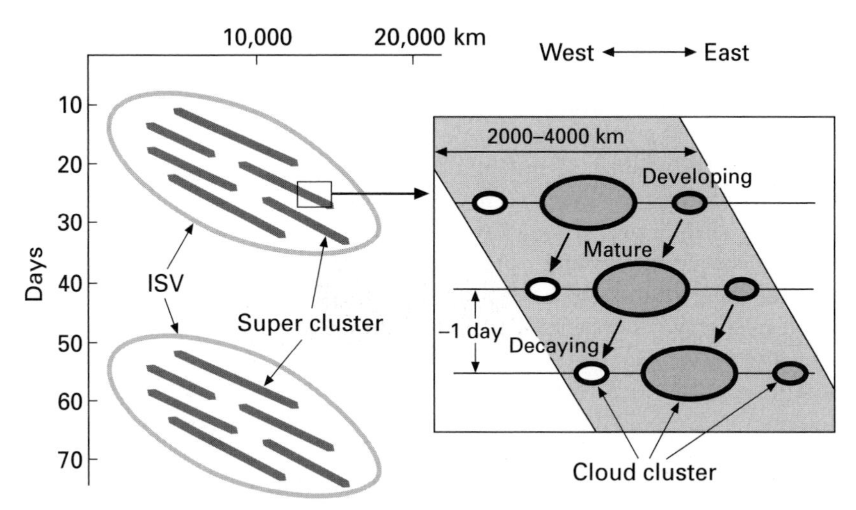

Figure 1.4. Schematic describing the details of the large-scale eastward-propagating cloud complexes (slanting ellipses marked ISV on the left-hand side). Slanting emboldened lines represent super cloud clusters (SCC) within the larger ISV. The right-hand side illustrates the fine structure of the SCC with smaller westward-moving cloud clusters (CC) that develop, grow to maturity, and decay in a few days.
From Nakazawa (1988).

summer hemisphere as was observed during MONEX. Other seasonal variations are described in Section 1.7.

High temporal and spatial resolution satellite-measured cloud data reveal added complexity in the make up of the individual cloud complexes. Figure 1.4 is a schematic that summarizes some of this complexity. The intraseasonal variability (ISV) depicted by Nakazawa (1988) corresponds to the large-scale cloud complexes indicated in Figure 1.2. Their east–west scale is of the order of 5,000–10,000 km and they move eastward half way around the Earth in about 20 days. Finer spatial resolution reveals that these large-scale complexes are made up of eastward propagating supercloud clusters (SCC) having 2,000–4,000-km horizontal scales. The SCC, in turn, are composed of westward-moving cloud clusters (CCs) which continually develop to the east and decay to the west. Life times of the CCs are only 1–2 days.

1.7 SEASONAL VARIATIONS IN THE OSCILLATION

Some interesting seasonal variations are revealed when one isolates spectral and cross-spectral quantities as a function of the time of year (Madden, 1986). The out-of-phase relationship between lower and upper troposphere evident in the Indian and western Pacific equatorial regions is strongest at stations in the summer hemisphere. This is consistent with a close connection with the divergent motions of the Intertropical Convergence Zone (ITCZ). In this regard, the Oscillation often modulates the large monsoons of the summer hemisphere. Examples for the northern summer are those during MONEX discussed above. Chapters 2, 3, 4, and 5 of this book cover variations in the monsoons in detail.

Another seasonal result reveals a role of the v-wind and illustrates the danger in interpreting spectral results of non-stationary time series. Esimates of seasonally varying coherence and phase between u and v-winds in a frequency band centered on 1/47 day at 150 hPa over Chuuk reveal that the coherence is large twice a year. In northern winter (summer) u and v are out-of-phase (in-phase) manifesting surges in the climatological south-easterlies (north-easterlies) in that season. This reflects a connection to changing upper-level outflow from the ITCZ. Coherence between time series of u and v-wind that are not stratified by season are small because the negative relation during northern winter cancels the positive one during northern summer.

Variance in the 1/47-day frequency band generally exceeds that in adjacent bands by the largest amount during December, January, and February so, by that measure, the oscillation is strongest in those months. There is no obvious change in the period of the oscillation with season. It averages from 45 to 48 days but individual oscillations range from a few weeks to more than 60 days. There is subjectivity in deciding if an MJO is present, but two methods of identification, one using winds (Madden, 1986) and one using May to October clouds (Knutsen *et al.*, 1986), suggest that they are active between 50–75% of the time. Considering a 45-day period, we might expect that there would be between four and six oscillations in a typical year. They are slightly more likely to occur during the northern winter half of the year since on average there are fewer occurrences during June, July, and August than during other seasons (Madden, 1986; Wang and Rui, 1990).

1.8 THE OSCILLATION IN THE ZONAL AVERAGE

Figure 1.2 shows that the surface pressure anomally is not a simple sinusoid but reflects changes in the zonal average as well. Another striking example of a zonally averaged component in the oscillation is in the relative atmospheric angular momentum (RAAM). The RAAM is a mass weighted integration of the u-winds over the entire globe. Figure 1.5 shows the RAAM during MONEX. There are two marked relative maxima about 45 days apart: one at the end of June and a second in mid-August. Because the angular momentum of the atmosphere–ocean–Earth system remains nearly constant, a change in RAAM can be reflected in changes in the momentum of the ocean or solid Earth. Feissel and Gambis (1980) reported on a 50-day oscillation in the angular momentum of the solid Earth as reflected in measurements of the length of day (LOD) during the MONEX period of about $0.35\,\mathrm{m\,s^{-1}}$ (10^{-3} s) peak-to-trough amplitude. The LOD change corresponding to a change in RAAM is indicated in Figure 1.5, and reveals that peak-to-trough amplitudes of about $0.2\,\mathrm{m\,s^{-1}}$ (June) to more than $0.3\,\mathrm{m\,s^{-1}}$ (August) would result if all changes in RAAM went to increasing the angular momentum of the solid Earth. LOD is longest (solid Earth momentum smallest) when RAAM is greatest. This is but one example of the consistency between RAAM and LOD that occurs on all timescales less than a

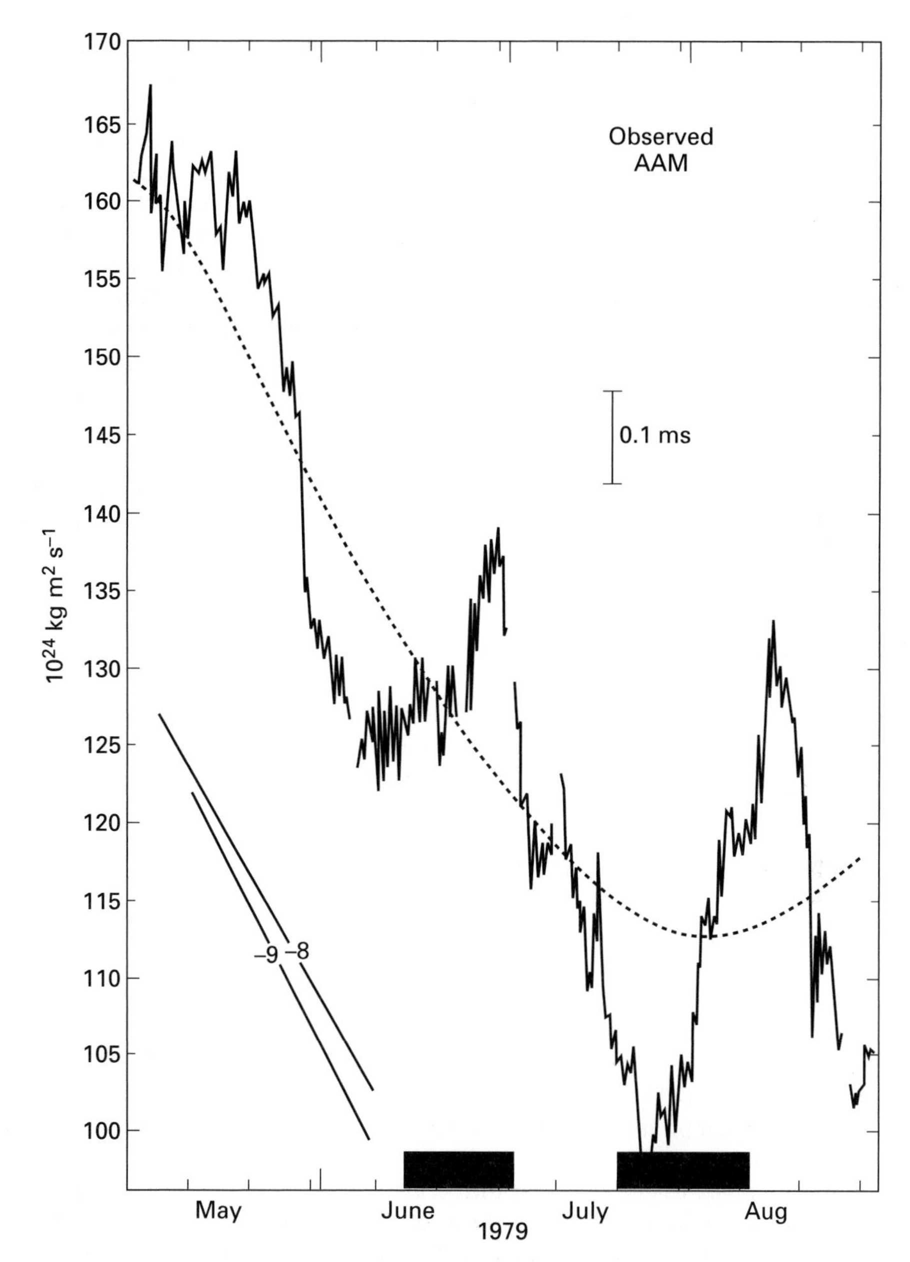

Figure 1.5. Observed relative atmospheric angular momentum (RAAM) during MONEX (thin irregular line). The dotted curved line represents approximate seasonal variation. Slanting lines on the lower left are expected seasonal variations based on two estimates of northern springtime climatalogical torques in units of $10^{18}\,\mathrm{kg\,m^2\,s^{-2}}$ (Newton, 1971; Wahr and Oort, 1984). The amplitude of a corresponding 0.1 ms change in length of day (LOD) is indicated. Thick lines at the bottom mark times of heavy monsoon rains from Figure 1.3. From Madden (1988).

few years. It is a credit to our observing systems that these two disparate time series, irregularly measured winds averaged over the Earth and estimates of tiny changes in the LOD, are so well related.

The evolving surface wind and pressure distributions during an oscillation result in changing frictional and mountain torques that vary the exchange of momentum between the atmosphere–ocean–Earth. As the cloud complex moves east the torques combine to increase the RAAM. First, frictional torques increase reflecting stronger trades over the tropical Pacific. Then mountain torques reach relative maxima, sometimes with relatively high pressure to the east of the Rocky Mountains, while relatively low pressure systems approach from the west. This results in an anomaly surface pressure gradient directed from west to east across the mountains and a positive anomaly in the mountain torque. Anomaly pressure gradients across the Himilayas (Weickmann and Sardeshmukh, 1994) and the Andes (Salstein and Rosen, 1994) are sometimes also important. The RAAM tends to reach a relative maximum shortly after the propagating clouds reach the Central Pacific (Madden and Speth, 1995). Chapter 8 in this book considers the intraseasonal exchange of momentum among the atmosphere–ocean–Earth more thoroughly.

1.9 OTHER EFFECTS OF THE OSCILLATION

A tropical phenomenon as large as the MJO is certain to affect mid-latitude weather. Weickmann *et al.* (1985) described the reach of the oscillation into mid-latitudes and Lau and Phillips (1986) linked it to wave trains propagating across the Pacific to North America. Propagation to mid-latitudes is now a research area of intense interest. Wave train propagation is dependent on the background flow and its ever changing character results in widely differing mid-latitude responses to very similar MJOs. With the growing bank of observations and improved modeling it is likely that MJOs will contribute to added skill in mid-latitude weather forecasts in the 5-day to 3-week range (e.g., Ferranti *et al.*, 1990; Jones *et al.*, 2004).

We have seen evidence of the oscillation in many aspects of the tropical atmosphere. In addition, the development of tropical cyclones is favored in regions of the upper-level, negative velocity potential of the MJO as well (Nakazawa, 1986; Liebmann *et al.*, 1994; Maloney and Hartman, 2000; Mo, 2000; Hall *et al.*, 2001). At this point, though promising, it is not clear how this relation might aid in tropical storm prediction.

Section 1.5 contains evidence that the oscillation affects the Indian Monsoon. It also influences the Australian Summer Monsoon. Holland (1986) found an average of 40 days between its active bursts. More recently, Wheeler and Hendon (2004) found a tripling of the probability of extreme (highest quintile) monsoon rainfall between the wet and dry phases of the oscillation. The oscillation has also often been implicated in the special case of the annual onset of the monsoon (Hendon and Liebmann, 1990a; Hung and Yanai, 2004; Wheeler and Hendon, 2004). It should be noted that none of these results translate into spectral evidence for a favored 40-day period in monsoon rainfall (Hendon and Liebmann, 1990b; Drosdowsky, 1996).

While the effect of the oscillation is unmistakable, the discussion of Chapter 5 will show that it is only one part of a myriad of intraseasonal monsoon phenomena.

The underlying ocean plays a part in the oscillation as well. Related ocean current variations are apparent. McPhaden (1982) found that low-level winds at Gan Island (1°S, 73°E) and 100 m depth currents, were coherent on 30–60-day timecales. Similarly, Mysak and Mertz (1984) concluded that variations in the wind stress or the wind-stress curl drove 40–60-day oscillations that they found in in the Somali Current. They found that during 1979, both u and v wind stresses in the region have spectral peaks in the 40–50 day range.

Another response to the surface wind oscillation is the excitation of ocean Kelvin waves along the equator. They move eastward and then north and south along the west coast of the Americas. There is a clear 40–60-day variation in sea level height from at least the Peruvian coast northward to northern California (Enfield, 1987). Luther (1980) had already reported 35–80-day spectral peaks in sea level height from Kanton to the Galapagos (1°S, 91°W). These variations result from Kelvin waves that are excited in the far western Pacific by surface winds of MJOs.

The oscillation affects the underlying ocean, and it can be assumed that the ocean affects the oscillation. For example, there are changes in the oscillation that appear to be driven by the El Niño/La Niña cycle (e.g., Lau and Chan, 1986a). On the other hand, evidence is growing that MJOs can play an important role in the timing of the ocean cycle itself. Lau and Chan (1986a) were the first to propose a link between MJOs and the onset of El Niño. It may also be important in the demise of El Niño (Takayabu *et al.*, 1999). The physical mechanism may be that both the anomalous surface westerlies and easterlies of the MJO can excite downwelling and upwelling ocean Kelvin waves, respectively, that then influence the sea surface temperatures (McPhaden, 1999). The El Niño/La Niña cycle is important for global climate, and the possiblity that MJOs influence its timing has kindled additional interest in them since the late 1980s. More on this important topic follows in Chapters 6 and 7.

1.10 SUMMARY

The discovery of the QBO was important for subsequent discoveries of mixed-Rossby gravity waves and Kelvin waves, and they, in turn, were the motivation for work that led to the discovery of the MJO. A basic description of the MJO is presented here. The Oscillation affects tropical clouds and precipitation, planetary-scale divergence patterns, the Asian and Australian Monsoons, zonally averaged pressures, atmospheric angular momentum and the LOD, mid-latitude weather, and the ocean beneath it. These features introduced are brought up to date in subsequent chapters based on the burgeoning research that has taken place in the last 20 years.

1.11 REFERENCES

Cadet, D. L. (1986) Fluctuations of precipitable water over the Indian Ocean during the 1979 summer monsoon. *Tellus*, **38A**, 170–177.

Chang, C.-P. (1977) Viscous internal gravity waves and low-frequency oscillations in the tropics. *J. Atmos. Sci.*, **34**, 901–910.

Cooley, J. W. and J. W. Tukey (1965) An algorithm for the machine calculation of Fourier series. *Math. Comput.*, **19**, 297–301.

Dakshinamurti, J. and R. N. Keshavamurty (1976) On oscillations of period around one month in the Indian summer monsoon. *Indian J. Meteor. Hydrol. Geophys.*, **27**, 201–203.

Drosdowsky, W. (1996) Variability of the Australian summer monsoon at Darwin: 1957–1992. *J. Climate*, **9**, 85–96.

Ebdon, R. A. (1960) Notes on wind flow at 50 mb in tropical and sub-tropical regions in January 1957 and January 1958. *Quart. J. Roy. Meteor. Soc.*, **86**, 540–542.

Ebdon, R. A. (1963) The tropical stratospheric wind fluctuation. *Weather*, **18**, 2–7.

Emanuel, K. A. (1987) Air–sea interaction model of intraseasonal oscillations in the tropics. *J. Atmos. Sci.*, **44**, 2324–2340.

Enfield, D. B. (1987) The intraseasonal oscillation in eastern Pacific sea levels – How is it forced? *J. Phys. Oceanogr.*, **17**, 1860–1876.

Feissel, M. and D. Gambis (1980) La mise en evidence de variations rapides de la Duree de Jour. *C. R. Hebd. Seances Acad. Sci.*, **Ser. B**(291), 271–273.

Ferranti, L., T. N. Palmer, F. Molteni, and E. Klinker (1990) Tropical–extratropical interaction associated with the 30–60 day oscillation and its impact on medium and extended range prediction. *J. Atmos. Sci.*, **47**, 2177–2199.

Gill, A. E. (1980) Some simple solutions for heat-induced tropical circulation. *Quart. J. Roy. Meteor. Soc.*, **106**, 447–463.

Graystone, P. (1959) Meteorological Office discussion – Tropical meteorology. *Met. Mag.*, **88**, 113–119.

Gruber, A. (1974) Wavenumber–frequency spectra of satellite measured brightness in tropics. *J. Atmos. Sci.*, **31**, 1675–1680.

Hall, J. D., A. J. Matthews, and D. J. Karoly (2001) The modulation of tropical cyclone activity in the Australian region by the Madden–Julian Oscillation. *Mon. Wea. Rev.*, **129**, 2970–2982.

Hendon, H. H. and B. Liebmann (1990a) A composite study of onset of the Australian summer monsoon. *J. Atmos. Sci.*, **47**, 2227–2240.

Hendon, H. H. and B. Liebmann (1990b) The intraseasonal (30–50 day) oscillation of the Australian summer monsoon. *J. Atmos. Sci.*, **47**, 2909–2923.

Hendon, H. H. and M. L. Salby (1994) The life cycle of the Madden–Julian Oscillation. *J. Atmos. Sci.*, **51**, 2225–2231.

Holland, G. J. (1986) Interannual variability of the Australian summer monsoon at Darwin: 1952–82. *Mon. Wea. Rev.*, **114**, 594–604.

Holton, J. R. (1973) On the frequency distribution of atmospheric Kelvin waves. *J. Atmos. Sci.*, **30**, 499–501.

Hu, Q. and D. A. Randall (1994) Low-frequency oscillations in radiative-convective systems. *J. Atmos. Sci.*, **51**, 1089–1099.

Hung, C.-W. and M. Yanai (2004) Factors contributing to the onset of the Australian summer monsoon. *Quart. J. Roy. Meteor. Soc.*, **130**, 739–758.

Jones, C., D. E. Waliser, K. M. Lau, and W. Stern (2004) The Madden–Julian Oscillation and its impact on Northern Hemisphere weather predictability. *Mon. Wea. Rev.*, **132**, 1462–1471.

Knutsen, T. R., K. M. Weickmann, and J. E. Kutzbach (1986) Global-scale intraseasonal oscillations of outgoing longwave radiation and 250 mb zonal wind during northern hemisphere summer. *Mon. Wea. Rev.*, **114**, 605–623.

Krishnamurti, T. N., D. K. Oosterhof, and A. V. Mehta (1988) Air–sea interaction on the time scale of 30 to 50 days. *J. Atmos. Sci.*, **45**, 1304–1322.

Lau, K.-M. and T. J. Phillips (1986) Coherent fluctuations of extratropical geopotential height and tropical convection in intraseasonal time scales. *J. Atmos. Sci.*, **43**, 1164–1181.

Lau, K.-M. and P. H. Chan (1986a) The 40–50 day oscillation and the El Niño/Southern Oscillation – A new perspective. *Bull. Amer. Meteor. Soc.*, **67**, 533–534.

Lau, K.-M. and P. H. Chan (1986b) Aspects of the 40–50 day Oscillation during northern summer as inferred from outgoing longwave radiation. *Mon. Wea. Rev.*, **114**, 1354–1367.

Lau, K.-M. and L. Peng (1987) Origin of low-frequency (intraseasonal) oscillations in the tropical atmosphere. Part 1: Basic theory. *J. Atmos. Sci.*, **44**, 950–972.

Liebmann, B., H. H. Hendon, J. D. Glick (1994) The relationship between tropical cyclones of the western Pacific and Indian Oceans and the Madden–Julian Oscillation. *J. Meteor. Soc. Jap.*, **72**, 401–412.

Lindzen, R. S. (1974) Wave-CISK and tropical spectra. *J. Atmos. Sci.*, **31**, 1447–1449.

Lorenc, A. C. (1984) The evolution of planetary-scale 200-mb divergent flow during the FGGE year. *Quart. J. Roy. Meteor. Soc.*, **110**, 427–441.

Luther, D. S. (1980) *Observations of Long Period Waves in the Tropical Oceans and Atmosphere*. PhD. thesis, Massachusetts Institute of Technology–Woods Hole Oceanographic Institution, 210pp.

McPhaden, M. J. (1982) Variability in the central equatorial Indian Ocean: Ocean dynamics. *J. Mar. Res.*, **40**, 157–176.

McPhaden, M. J. (1999) Genesis and evolution of the 1997–1998 El Niño. *Science*, **283**, 950–954.

Madden, R. A. (1986) Seasonal variations of the 40–50 day oscillation in the Tropics. *J. Atmos. Sci.*, **43**, 3138–3158.

Madden, R. A. (1988) Large intraseasonal variations in wind stress over the tropical Pacific. *J. Geophys. Res.*, **93**, 5333–5340.

Madden, R. A. and P. R. Julian (1971) Description of a 40–50 day oscillation in the zonal wind in the tropical Pacific. *J. Atmos. Sci.*, **28**, 702–708.

Madden, R. A. and P. R. Julian (1972) Description of global-scale circulation cells in the tropics with a 40–50 day period. *J. Atmos. Sci.*, **29**, 1109–1123.

Madden, R. A. and P. Speth (1995) Estimates of atmospheric angular momentum, friction, and mountain torque during 1987–1988. *J. Atmos. Sci.*, **52**, 3681–3694.

Maloney, E. D. and D. L. Hartmann (2000) Modulation of eastern North Pacific hurricanes by the Madden–Julian Oscillation. *J. Climate*, **13**, 1451–1460.

Maruyama, T. (1967) Large-scale disturbances in the equatorial lower stratosphere. *J. Meteor. Soc. Jap.*, **45**, 391–408.

Matsuno, T. (1966) Quasi-geostrophic motions in the equatorial area. *J. Meteor. Soc. Jap.*, **44**, 25–43.

Mo, K. C. (2000) The association between intraseasonal oscillations and tropical storms in the Atlantic basin. *Mon. Wea. Rev.*, **128**, 4097–4107.

Murakami, T. (1976) Cloudiness fluctuations during the summer monsoon. *J. Meteor. Soc. Jap.*, **54**, 175–181.

Mysak, L. A. and G. J. Mertz (1984) A 40-day to 60-day oscillation in the source region of the Somali Current during 1976. *J. Geophys. Res.*, **89**, 711–715.

Nakazawa, T. (1986) Intraseasonal variations of OLR in the tropics during the FGGE year. *J. Meteor. Soc. Jap.*, **64**, 17–34.

Nakazawa, T. (1988) Tropical super clusters within intraseasonal variations over the western Pacific. *J. Meteor. Soc. Jap.*, **66**, 823–8.

Neelin, J. D., I. M. Held, and K. H. Cook (1987) Evaporation–wind feedback and low-frequency variability in the tropical atmosphere. *J. Atmos. Sci.*, **44**, 2341–2348.

Newton, C. W. (1971) Global angular momentum balance: Earth torques and atmospheric fluxes. *J. Atmos. Sci.*, **28**, 1329–1341.

Parker, D. E. (1973) Equatorial Kelvin waves at 100 millibars. *Quart. J. Roy. Meteor. Soc.*, **99**, 116–129.

Reed, R. J., W. J. Campbell, L. A. Rasmussen, and D. G. Rogers (1961) Evidence of a downward-propagating, annual wind reversal in the equatorial stratosphere. *J. Geophys. Res.*, **66**, 813–818.

Salstein, D. A. and R. D. Rosen (1994) Topographical forcing of the atmosphere and a rapid change in the length of day. *Science*, **264**, 407–409.

Takayabu, Y. N., T. Iguchi, M. Kachi, A. Shibata, and H. Kanzawa (1999) Abrupt termination of the 1997–98 El Niño in response to a Madden–Julian Oscillation. *Nature*, **402**, 279–282.

Wahr, J. M. and A. H. Oort (1984) Friction and mountain torques and atmospheric fluxes. *J. Atmos. Sci.*, **41**, 190–204.

Wallace, J. M. and V. E. Kousky (1968) Observational evidence of Kelvin waves in the tropical stratosphere. *J. Atmos. Sci.*, **25**, 900–907.

Wallace, J. M., and C.-P. Chang (1969) Spectrum analysis of large-scale wave disturbances in the tropical lower troposphere. *J. Atmos. Sci.*, **26**, 1010–1025.

Wang, B. (1988) Dynamics of tropical low-frequency waves: An analysis of the moist Kelvin wave. *J. Atmos. Sci.*, **45**, 2051–2065.

Wang, B. and H. Rui (1990) Synoptic climatology of transient tropical intraseasonal convection anomalies. *Meteor. Atmos. Phys.*, **44**, 43–61.

Weickmann, K. M., G. R. Lussky, and J. E. Kutzbach (1985) Intraseasonal (30–60 day) fluctuations of outgoing longwave radiation and 250 mb stream function during northern winter. *Mon. Wea. Rev.*, **113**, 941–961.

Weickmann, K. M. and P. D. Sardeshmukh (1994) The atmospheric angular momentum cycle associated with the Madden–Julian Oscillation. *J. Atmos. Sci.*, **51**, 3194–3208.

Wheeler, M. C. and H. H. Hendon (2004) An all-season real-time multivarite MJO index: Development of an index for monitoring and prediction. *Mon. Wea. Rev.*, **132**, 1917–1932.

Yamagata, T. and Y. Hayashi (1984) A simple diagnostic model for the 30–50 day oscillation in the tropics. *J. Meteor. Soc. Jap.*, **62**, 709–717.

Yanai, M. and T. Maruyama (1966) Stratospheric wave disturbances propagating over the equatorial Pacific. *J. Meteor. Soc. Jap.*, **44**, 291–294.

Yanai, M., T. Maruyama, T. Nitta, and Y. Hayashi (1968) Power spectra of large-scale disturbances over the tropical Pacific. *J. Meteor. Soc. Jap.*, **46**, 308–323.

Yasunari, T. (1979) Cloudiness fluctuations associated with the Northern Hemisphere summer monsoon. *J. Meteor. Soc. Jap.*, **57**, 227–242.

Zangvil, A. (1975) Temporal and spatial behavior of large-scale disturbances in tropical cloudiness deduced from satellite brightness data. *Mon. Wea. Rev.*, **103**, 904–920.

2

South Asian monsoon

B. N. Goswami

2.1 INTRODUCTION

2.1.1 South Asian summer monsoon and active/break cycles

As the word "monsoon" (derived from an Arabic word meaning seasons) indicates, the South Asian (SA) summer monsoon is part of an annually reversing wind system (Figure 2.1(b, e) (Ramage, 1971; Rao, 1976)). The winds at low levels during the summer monsoon season are characterized by the strongest westerlies anywhere at 850 hPa over the Arabian Sea, known as the low-level westerly jet (LLJ) (Figure 2.1e), and a large-scale cyclonic vorticity extending from the north Bay of Bengal (BoB) to western India known as the "monsoon trough" (Figure 2.1(e) (Rao, 1976)). The easterly jet (Figure 2.1(f)) centered around 5°N and the Tibetan anticyclone centered around 30°N are important features of upper level winds over the monsoon region during northern summer. Millions of inhabitants of the region, however, attach much greater importance to the associated seasonal changes of rainfall. Wet summers and dry winters (Figure 2.1(a, d)) associated with the seasonal changes of low-level winds are crucial for agricultural production and the economy of the region. The monsoon, or the seasonal changes of winds and rainfall, in the region could be interpreted as a result of northward seasonal migration of the east–west oriented precipitation belt (Tropical Convergence Zone, TCZ) from southern hemisphere in winter to northern hemisphere in summer (Gadgil, 2003). The largest northward excursion of the rain belt takes place over the Indian monsoon region where it moves from a mean position of about 5°S in winter (Figure 2.1(a)) to about 20°N in northern summer (Figure 2.1(d)) (Waliser and Gautier, 1993). In the upper atmosphere (200 hPa), the equatorial easterlies are weak and confined between 5°N and 10°S while the subtropical westerlies intrude all the way to 10°N during northern winter (Figure 2.1(c)). The subtropical westerlies recede to north of 30°N during northern summer and a strong easterly jet characterizes the equatorial upper

W. K. M. Lau and D. E. Waliser (eds), *Intraseasonal Variability in the Atmosphere–Ocean Climate System.*

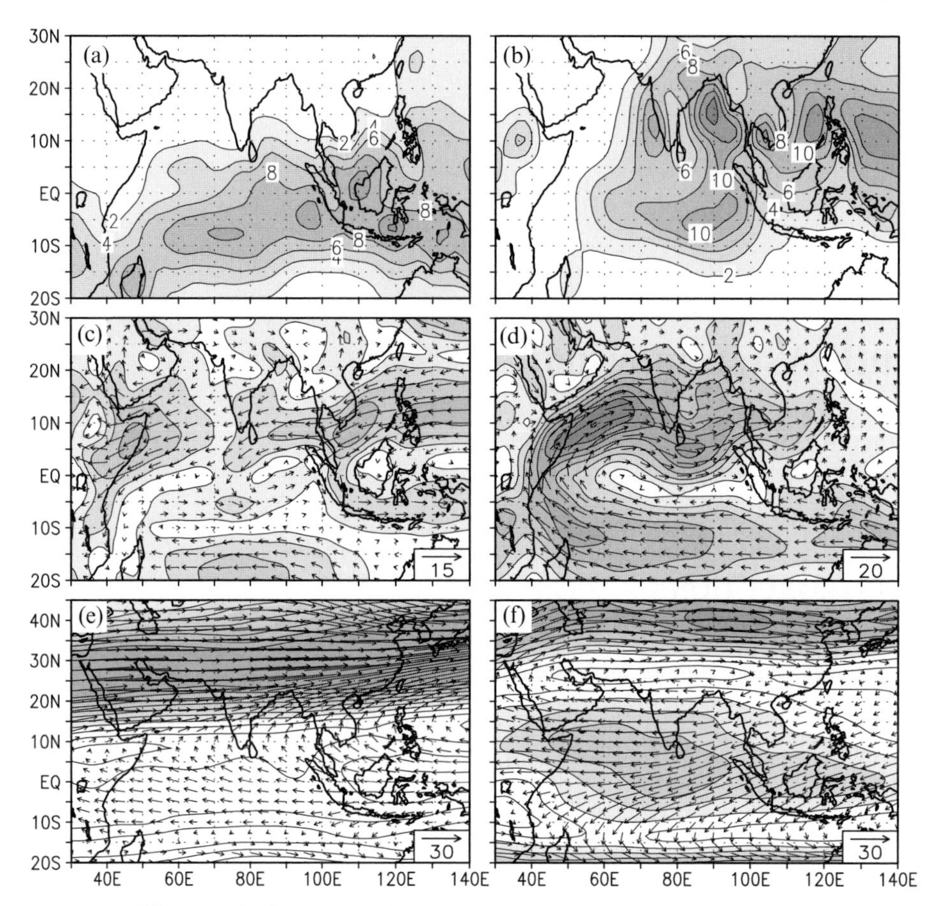

Figure 2.1. Climatological mean precipitation (mm/day) based on CMAP during (a) boreal winter (DJF) and (d) summer (JJAS). (b) and (e) are the same as (a) and (d) but for winds ($\mathrm{m\,s^{-1}}$) at 850 hPa based on NCEP Reanalysis. The contour interval for isotachs is $2\,\mathrm{m\,s^{-1}}$ with the minimum contour being 2. (c) and (f) are similar to (b) and (e) but for winds at 200 hPa. The contour interval for isotachs is $5\mathrm{m\,s^{-1}}$ with the minimum contour being 5. For better depiction of the subtropical westerly jet stream in winter and the Tibetan anticyclone in summer, a larger meridional domain is used for the 200-hPa winds (c and f).

atmosphere in the region (Figure 2.1(f)). The year-to-year variations of the long-term seasonal mean precipitation over the Indian region is strongly correlated with food production in the region (Parthasarathy *et al.*, 1988; Webster *et al.*, 1998; Abrol and Gadgil, 1999). The extremes in year-to-year variations of the long-term mean precipitation manifest in the form of large-scale floods and droughts (Parthasarathy and Mooley, 1978; Shukla, 1987; Mooley and Shukla, 1987) and cause devastating human and economic loss.

The seasonal mean rainfall of approximately $8\,\mathrm{mm\,day^{-1}}$ does not pour as a continuous deluge but is punctuated by considerable variations within the season. In addition to the day-to-day fluctuations of weather (e.g., lows and depressions) with

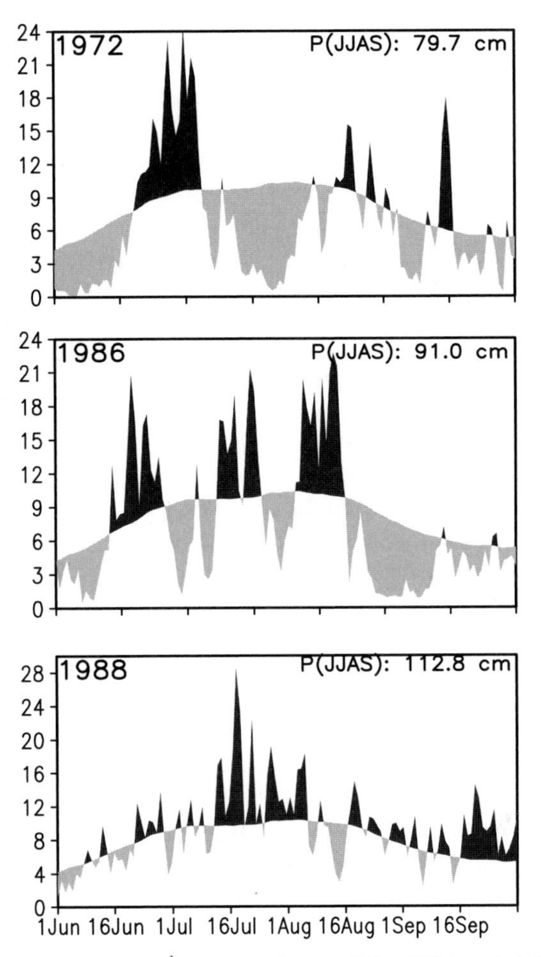

Figure 2.2. Daily rainfall (mm day^{-1}) averaged over 72°E–87°E and 10°N–25°N based on station data over the Indian continent during the summer monsoon season for three years, 1972, 1986, and 1988. Departure from the mean annual cycle (shown as the envelope) are shaded. Seasonal mean rainfall for each year are also shown in the top-right corners.

timescales of 5–7 days (also known as synoptic disturbances), a characteristic feature of monsoon rainfall is the prolonged spells of dry and wet conditions often lasting for 2–3 weeks. Examples of such spells can be seen in the time series of rainfall averaged over central India between 1 June and 30 September, 1972, 1986, and 1988 (Figure 2.2). During 1972 and 1986, the evolution of rainfall over the season went through extended periods of above-normal conditions (wet spells) followed by extended periods of below-normal conditions (dry spells). The extended above-normal rain spells can been seen to represent epochs when the monsoon was vigorous or "active" while the dry spells represent periods when the monsoon took a "break" from its activity (Ramamurthy, 1969; Raghavan, 1973) and hence are known as "active" and "break" conditions respectively. Frequent or prolonged breaks within the monsoon season, as in the case of 1972 and 1986 (Figure 2.2), leads to drought conditions and adversely affects agricultural production (Gadgil, 1995; Webster *et al.*, 1998). Similarly, the above normal seasonal rainfall in 1988 was a result of the occurrence of more active spells and an absence of extended break spells within the season. Thus, frequency of occurrence of active and break spells influence the seasonal mean rainfall and hence agricultural production. For example, long breaks in critical growth periods of agricultural crops also lead to substantially reduced yields (Gadgil and Rao, 2000). As a consequence of their influence in agricultural production and water resources, considerable attention has been paid toward understanding the nature of monsoon breaks and the possible mechanism responsible for them. In fact, the earliest reference to monsoon breaks was made more than a century ago by Blanford (1886), where he referred to the periods between two active spells as "intervals of droughts". With upper air data over the Indian continent and its neighborhood, large-scale circulation changes associated with the active and break conditions have been identified (Ramamurthy, 1969; Raghavan, 1973; Krishnamurti and Bhalme, 1976; Sikka, 1980; Alexander *et al.*, 1978). The active (break) condition is generally associated with an increase (decrease) of cyclonic vorticity and decrease (increase) of surface pressure over the central Indian monsoon trough region and strengthening (weakening) of the LLJ. Movement of the low-level trough (monsoon trough) to foothills of the Himalayas during break conditions have been known (Ramamurthy, 1969; Raghavan, 1973; Krishnamurti and Bhalme, 1976; Sikka, 1980; Alexander *et al.*, 1978). Weakening of the Tibetan anticyclone in the upper atmosphere and extension of a large amplitude trough in the mid-latitude westerlies up to northern Indian latitudes are also associated with monsoon breaks (Ramaswamy, 1962).

The dry and wet spells of the the active and break conditions represent sub-seasonal or intraseasonal variations (ISV) of the monsoon with timescales longer than synoptic variability (1–10 days) but shorter than a season. Studies have also shown (Dakshinamurthy and Keshavamurthy, 1976; Alexander *et al.*, 1978) certain preferred periodicities are associated with the monsoon ISV indicating that certain oscillations (intraseasonal oscillations (ISOs)) are involved in generating the ISV. Early studies on monsoon ISV that manifest in active and break cycles were based on station rainfall data and soundings from a few upper air stations. Availability of daily satellite cloudiness data and operational analysis in the mid-1970s with global coverage, brought new insight regarding the large-scale spatial structure and relationship between convection and circulation of monsoon ISV. Progress in modeling during the past three decades has also provided new insight regarding the origin of the monsoon ISV. During this period, we have also learned how the monsoon ISOs interact with different scales of motion. At one end of the spectrum, it interacts with the annual cycle influencing the seasonal mean, its interannual variability (IAV), and limiting the predictability of the seasonal mean, while at the other end it modulates synoptic activity and causes spatial and temporal clustering of lows and depressions. In this chapter, we attempt to provide a synthesis of the observed spatial and temporal scale of the monsoon ISOs, their regional propagation characteristics, relationships with large-scale regional and global circulation, together with a review of theories for their scale selection. The mechanism through which monsoon ISOs influence the seasonal mean and its IAV will also be highlighted. The variety of observations utilized and analysis methodology employed to highlight these different aspects of the summer monsoon ISOs are described in the Appendix (p. 56).

2.1.2 Temporal and spatial scales

Distinct from the synoptic disturbances (lows and depressions), the ISOs of monsoons essentially have timescales between 10 and 90 days. Insight regarding the spatial structure of ISOs and coupling between different variables may be obtained by constructing an index of monsoon ISOs. We construct such an index based on 10–90-day band-pass filtered pentad CPC Merged Analysis of Precipitation (CMAP) (Xie and Arkin, 1996), precipitation interpolated to daily values and averaged over the box between 70°E–90°E and 15°N–25°N during 1 June and 30 September of each year. The time series normalized by its own standard deviation (2.35 mm day^{-1}) (hereafter referred to as ISO index) $>+1$ (<-1) represents active (break) conditions as seen from Figure 2.3, where a sample of the ISO index for ten summer seasons (122 days in each season) is shown (Goswami *et al.*, 2003). Composites of ISO filtered CMAP anomalies for active and break conditions for 20 summer seasons (1979–1998) were constructed and the difference between active and break composites is shown in Figure 2.4(a) while a similar difference between active and break composites from station rainfall for 11 years (1979–1989) is shown in Figure 2.4(b). The spatial pattern of active minus break composites over the Indian continent corresponds well with active (break) patterns described in other studies (Singh *et al.*, 1992; Krishnamurthy and Shukla, 2000). The composite from CMAP within the Indian continent is very similar to that obtained from station data. It also illustrates that the monsoon ISV with active–break phases is not confined to the

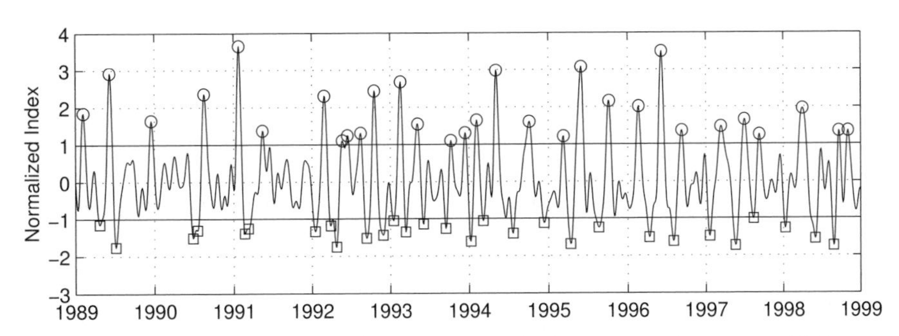

Figure 2.3. Time series of normalized monsoon ISO index between 1 June and 30 September (122 days) for a sample of ten (1989–1998) summer seasons. The ISO index is defined as 10–90-day filtered CMAP rainfall anomaly averaged between 70°E–90°E and 15°N–25°N. The time series is normalized by its own standard deviation (2.35 mm day^{-1}). Open circles and squares indicate peaks of active and break conditions respectively.

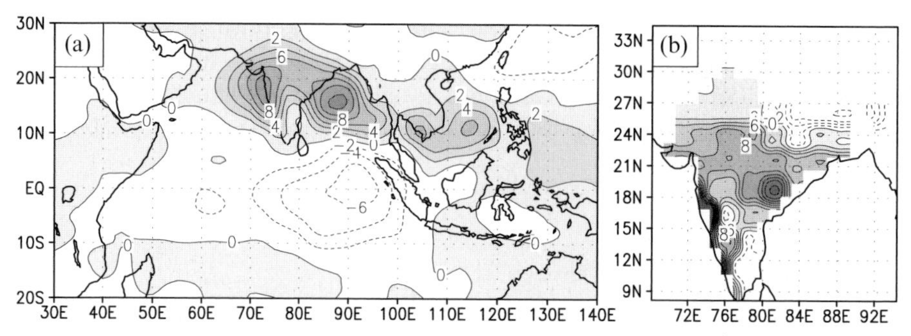

Figure 2.4. Active minus break composites of precipitation (mm day^{-1}) constructed from active (ISO index $>+1$) and break days (ISO index <-1) as defined by the ISO index (Figure 2.3) from 23 summers (1979–2001) of 10–90-day filtered CMAP anomalies (a), and 11 summers (1979–1989) of 10–90-day filtered station precipitation data (b).

Indian continent alone but has a much larger spatial scale and is associated with enhanced (decreased) rainfall extending from the western Pacific to the north BoB and central Indian continent. This observation also highlights that the ISV during northern summer over the SA monsoon region and those over the East Asian and western North Pacific (EA/WNP) monsoon region are interlinked (also see Chapter 3). One important characteristic of SA monsoon ISV is the north–south dipole in precipitation with active (break) condition being associated with enhanced (decreased) precipitation over the monsoon trough region and decreased (enhanced) precipitation over the eastern equatorial Indian Ocean (IO) (Goswami and Ajayamohan, 2001). Another aspect of the spatial structure of the dominant ISV is a dipole-like structure of opposite sign over the western equatorial Pacific and western north Pacific (Annamalai and Slingo, 2001).

A lag regression analysis of 10–90-day filtered winds from the US National Center for Environmental Prediction–National Center for Atmospheric Research (NCEP–NCAR) Reanalysis (Kalnay *et al.*, 1996; Kistler *et al.*, 2001) at a number of vertical levels and 10–90-day filtered CMAP precipitation with respect to the ISO index brings out the vertical structure and relationship between convection and circulation of the ISV. Simultaneous regressions of CMAP rainfall and 850-hPa winds and those with a lag of 14 days are shown in Figure 2.5(a, b). Anomalous meridional circulation associated with active (0 lag) and break (14-day lag) phases are shown in Figure 2.5(c, d) based on regressions of meridional and vertical velocities with respect to the ISO index averaged between 70°E–90°E. The low-level wind anomalies associated with the ISO (Figure 2.5(a, b)) are consistent with a linear response to correponding precipitation anomalies, indicating that the monsoon ISV and the active and break conditions are opposite phases of a large-scale convectively-coupled oscillation. Anomalous Hadley circulation of opposite sign associated with the active and break phases shows that the regional monsoon Hadley circulation is significantly strengthened (weakened) during the active (break) phase. The anomalous Hadley circulation also indicates a baroclinic vertical structure for the monsoon ISV.

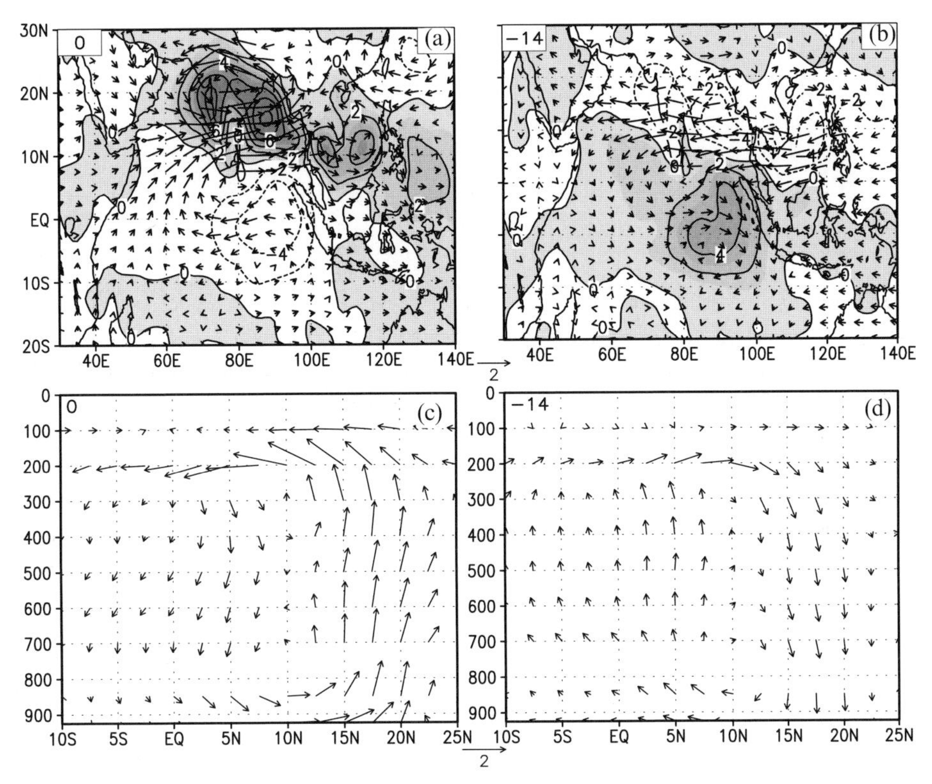

Figure 2.5. Horizontal and vertical structure of the dominant ISV. (*Top*) regressed 10–90-day filtered CMAP (contours, mm day^{-1}) and zonal and meridional wind anomalies at 850 hPa (vectors, m s^{-1}) with respect to the ISO index (Figure 2.3) at 0 lag (active condition) and 14 day lag (break condition). (*Bottom*) anomalous regional Hadley circulation associated with active and break conditions. Regressed meridional and vertical wind anomalies at a number of vertical levels averaged over 70°E–90°E. Vertical wind anomalies (hPa s^{-1}) have been scaled up by a factor of 50.

Within the broad range of 10–90-day periods, two period ranges, with periodicities between 10 and 20 days and 30 and 60 days, respectively, are particularly prominent. Several early studies (Murakami, 1976; Krishnamurti and Bhalme, 1976) showed the existence of a 10–20-day oscillation in a number of monsoon parameters. Later studies (Krishnamurti and Ardunay, 1980; Chen and Chen, 1993) show that the 10–20-day oscillation is a westward propagating mode closely related to monsoon active/break conditions. In addition to the 10–20-day oscillation, a prominent oscillation with a 30–60-day period is seen in monsoon circulation (Dakshinamurthy and Keshavamurthy, 1976), cloudiness and precipitation (Yasunari, 1979, 1980, 1981; Sikka and Gadgil, 1980). Most of these early studies estimated the spectral peaks based on limited data and hence it was not possible to establish the statistical significance of the peaks. Existence of significant power in the

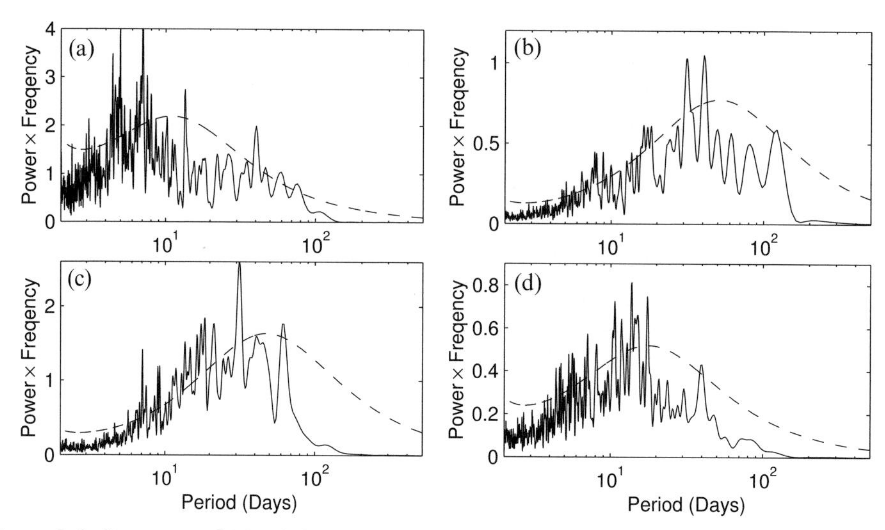

Figure 2.6. Spectrum of (a) rainfall anomalies for 20 (1971–1990) summer seasons (1 June–30 September) from station data averaged over 75°E–85°E and 15°N–25°N, and (b) zonal wind anomalies at 850 hPa for 20 (1979–1998) summer seasons from NCEP Reanalysis averaged over 55°E–65°E and 5°N–15°N (Arabian Sea). (c) same as (b) but averaged over 85°E–90°E and 10°N–15°N (Bay of Bengal). (d) Same as (b) but for meridional wind anomalies averaged over 80°E–85°E and EQ–5°N. Spectra are calculated using the Tukey-window method and the dotted lines represent a 90% confidence level with respect to a red noise null hypothesis.

two frequency ranges is illustrated in Figure 2.6 where the power spectra of four representative time series are shown. One (Figure 2.6(a)) is of daily precipitation anomalies from the raingauge data (Singh *et al.*, 1992) averaged over 75°E–85°E and 15°N–25°N (central India) for 20 (1971–1990) summer seasons (1 June–30 September) while two others (Figure 2.6(b, c)) are of daily zonal wind anomalies at 850 hPa from NCEP Reanalysis averaged over 55°E–65°E and 5°N–15°N (Arabian Sea (AS)) and over 85°E–90°E and 10°N–15°N (BoB) also for 20 summer seasons respectively. The last one (Figure 2.6(d)) is of meridional wind anomalies averaged over 80°E–85°E and EQ–5°N. A strong quasi-biweekly period is seen in the precipitation time series (Figure 2.6(a)) clearly separated from the synoptic variability (period <10 days) and the lower frequency with approximately 40-day periodicity. Significant power at the 10–20-day range is also noted in the two zonal wind time series. It is also noted that the zonal winds at low levels over the BoB (Figure 2.6(c)) have higher power at this frequency range than those over the AS (Figure 2.6(b)). The meridional wind over the central equatorial IO (Figure 2.6(d)) shows prominant power at 10–20-day timescales well separated from synoptic disturbances and the low-frequency 40-day mode. There is also significant power at the 30–60-day range in all the four time series. To put the role of these two oscillations in the context of variability of full daily anomalies, zonal wind anomalies at 850 hPa between 1 June and 30 September for 20 years were band-pass filtered to retain periods between 10–20 days and 30–60 days using a Lanczos filter. The ratio

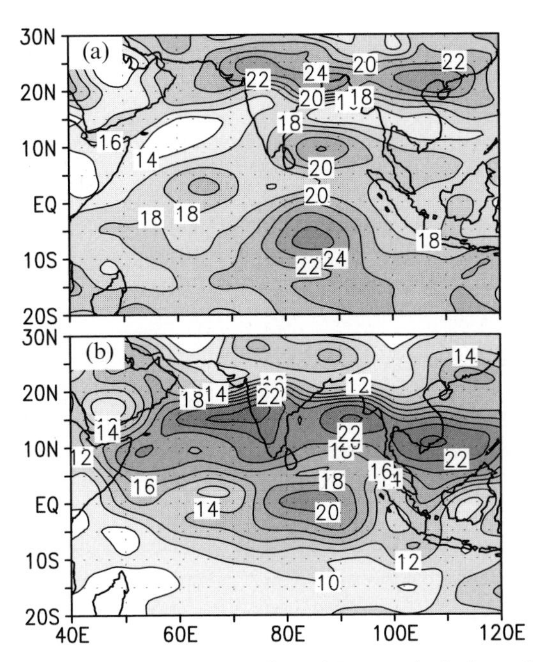

Figure 2.7. Percentage of total daily variance of 850-hPa zonal winds explained by (a) 10–20-day mode and (b) 30–60-day mode during summer monsoon season (1 June–30 September) for 20 years (1979–1998).

between variances of the 10–20-day mode and of the 30–60-day mode and that of total daily anomalies are shown in Figure 2.7. It may be noted that variance of zonal winds at low level contributed by each of these modes is considerable, ranging between 15% and 25% of the total daily variance. However, the importance of the two ISO modes go far beyond these numbers of percentage of total daily variance explained, as the ISO strongly modulates the synoptic activity (see Section 2.4) responsible for a large fraction of the total daily variability. The global structure of the 30–60-day oscillation has been explored in a number of studies (Krishnamurti *et al.*, 1985; Knutson *et al.*, 1986; Lau and Chen, 1986; Murakami *et al.*, 1986; Knutson and Weickmann, 1987; Nakazawa, 1986). In contrast, the spatial structure and propagation characteristics of the 10–20-day mode has been addressed only by a limited number of studies (Krishnamurti and Ardunay, 1980; Chen and Chen, 1993; Goswami and Ajayamohan, 2001; Chatterjee and Goswami, 2004). It may be noted that the 10–20-day and the 30–60-day ISV is not unique to the SA monsoon alone. The EA/WNP monsoon also exhibits 12–24-day and 30–60-day varibility during boreal summer (see Chapter 3).

The primary features of horizontal and vertical structures of the two modes are summarized here. The horizontal and vertical structure of the 10–20-day mode is illustrated in Figure 2.8. A reference time series is constructed by averaging 10–20-day filtered zonal winds at 850 hPa over a box between 85°E–90°E and 5°N–10°N

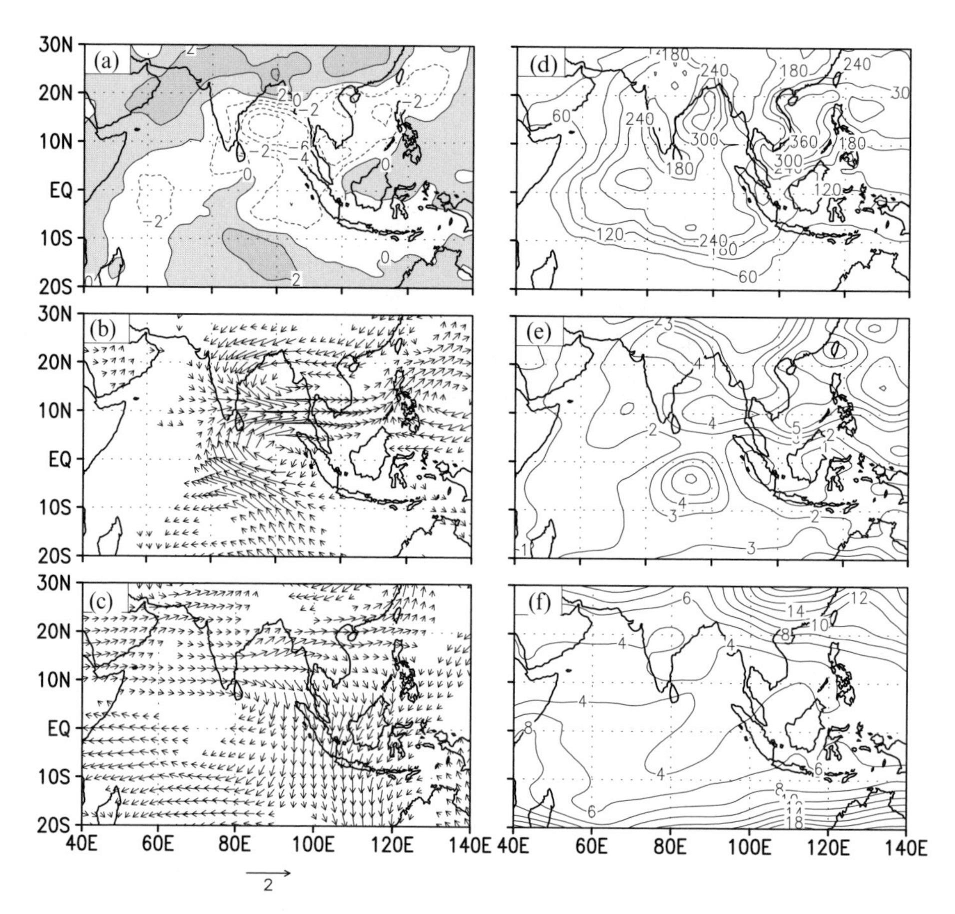

Figure 2.8. Spatial structure and amplitude of the 10–20-day mode. Regressed 10–20-day filtered anomalies of (a) OLR (in $\mathrm{W\,m^{-2}}$), (b) 850-hPa winds, and (c) 200-hPa winds (in $\mathrm{m\,s^{-1}}$) with respect to a reference time series of 10–20-day filtered zonal winds averaged over 85°E–90°E and 5°N–10°N with 0 lag. Only regressed wind anomalies significant at 95% confidence level are plotted, with a mean variance of 10–20-day filtered (d) OLR (in $\mathrm{W^2\,m^{-4}}$), (e) 850-hPa, and (f) 200-hPa zonal winds (in $\mathrm{m^2\,s^{-2}}$) based on 20 (1979–1998) summers (1 June–30 September).

during the summer season (1 June–30 September) for 20 years (1979–1998). The reference box is selected to be in a region of high variance of the 10–20-day filtered zonal winds at 850 hPa. Lag regressions with the reference time series of 10–20-day filtered zonal and meridional winds at 850 hPa and 200 hPa, together with that of OLR at all grid points, are constructed. Simultaneous regressed OLR and wind vector anomalies at 850 hPa and 200 hPa are shown in Figure 2.8(a, b, c) respectively. The low-level wind structure of the mode is characterized by two vortices, one centered around 18°N while the other has its center close to the equator around 3°S. It may be recalled that the gravest meridional mode ($n = 1$)

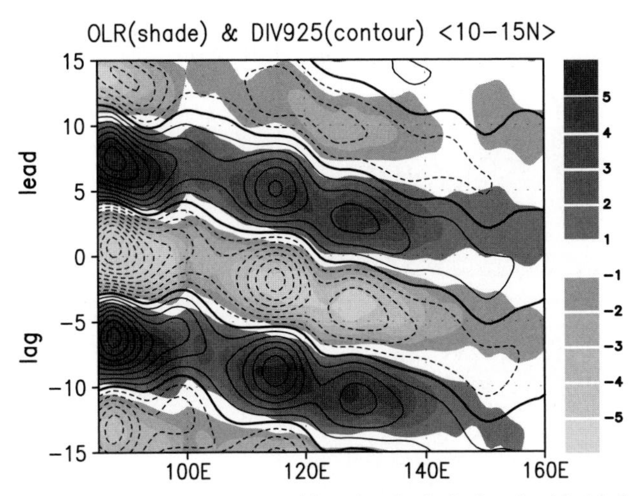

Figure 2.9. Coupling between convection and low-level winds for the 10–20-day mode. Lag–longitude plot of regressed 10–20-day filtered anomalies of OLR (in $W m^{-2}$; shaded) and divergence of 925-hPa winds (contour) with respect to the same reference time series described in Figure 2.8 averaged over 10°N–15°N. Solid (dashed) lines indicate positive (negative) divergence, with a contour interval of $0.1 \times 10^{-6} s^{-1}$, and with thick lines showing the zero contour.

equatorial Rossby wave is also characterized by two vortices similar to those in Figure 2.8(b) but centered around the equator (Matsuno, 1966; Gill, 1982). It has been recently shown (Chatterjee and Goswami, 2004) that the low-level spatial structure of the mode may be interpreted as the gravest meridional mode ($n = 1$) equatorial Rossby wave with a wavelength of about 6,000 km but shifted to the north by about 5 degrees by the background summer mean flow. The phase of the vortices in the vertical remain the same from the surface up to 200 hPa and change sign around 150 hPa, indicating its vertical structure to have a significant barotropic component (also see Chen and Chen, 1993) together with a baroclinic component. The figure also contains average variance associated with 10–20-day filtered OLR zonal winds at 850 hPa and 200 hPa (Figure 2.8(d, e, f). Significant fluctuations of OLR (standard deviation of 15–20 $W m^{-2}$) and zonal winds at both lower and upper levels ($2 m s^{-1}$) are associated with this oscillation. Coherent evolution of the OLR and the circulation anomalies throughout the oscillation indicate that convective coupling is involved with the genesis and propagation of the mode. This is further illustrated in Figure 2.9 where regressed OLR and divergence at 925 hPa averaged over 5°N–15°N are shown as a function of longitudes and lags. Close association between boundary layer convergence (divergence) and negative (positive) OLR anomalies is apparent. Also the convergence center being slightly west of the OLR center seems to be responsible for the westward propagation of the mode.

The horizontal structure of the 30–60-day mode is studied in a similar manner by constructing a reference time series of 30–60-day filtered zonal winds at 850 hPa averaged over 80°E–90°E and 10°N–15°N and calculating lag regressions of

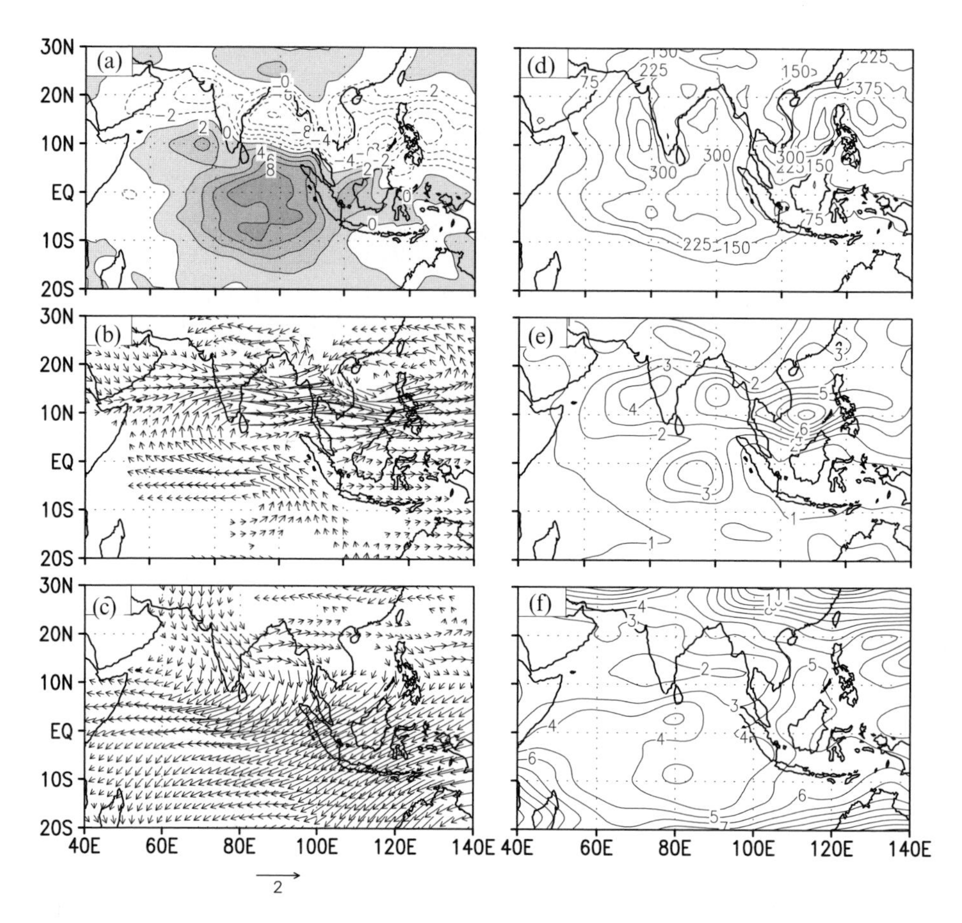

Figure 2.10. Spatial structure and amplitude of the 30–60-day mode. Regressed 30–60-day filtered anomalies of (a) OLR (in $W m^{-2}$), (b) 850-hPa winds, and (c) 200-hPa winds (in $m s^{-1}$) with respect to a reference time series of 30–60-day filtered zonal winds averaged over 85°E–90°E and 5°N–10°N with 0 lag. Only regressed wind anomalies significant at 95% confidence level are plotted, with a mean variance of 30–60-day filtered (d) OLR (in $W^2 m^{-4}$), (e) 850-hPa, and (f) 200-hPa zonal winds (in $m^2 s^{-2}$), based on 20 (1979–1998) summers (1 June–30 September).

30–60-day filtered zonal and meridional winds and OLR everywhere. The simultaneous OLR and vector wind anomalies at 850 hPa and 200 hPa associated with the mode are shown in Figure 2.10(a, b, c) respectively. The mean variance associated with the 30–60-day filtered OLR and zonal winds at 850 hPa and 200 hPa are shown in Figure 2.10(d, e, f) respectively. The spatial structure of OLR and low-level winds associated with the mode are similar to those found in various studies (e.g., Goswami and Ajayamohan, 2001; Annamalai and Slingo, 2001; Webster *et al.*, 1998). The horizontal scale of the 30–60-day mode (half wavelength of about 10,000 km) is much larger than that of the 10–20-day mode (Figure 2.8) which is rather regional in character. The other interesting point to note is that the low-level wind anomalies associated with the 30–60-day mode (Figure 2.10(b)) has structure similar to that of the seasonal mean (Figure 2.1(e)) strengthening (weakening) the seasonal mean in its active (break) phases. It may also be noted that the horizontal structure of the 30–60-day mode around the Indian longitudes is characterized by two vortices of opposite sign flanked on either side of a vortex centered close to the equator, similar to the spatial structure of $n = 2$ equatorial Rossby mode. The difference between the two (e.g., asymmetry in the strength of the northern and the southern vortices, a shift of the vortices to the north) are likely to be due to the modification of the Rossby wave by the summer mean background flow. A comparison of Figure 2.5 and Figure 2.10 indicates that the large-scale structure of the active/break conditions seems to come largely from the 30–60-day mode. The 200-hPa anomalies associated with the 30–60-day mode are opposite to those at low levels but with a tilt to the west. The phase transition takes place (not shown) at around 500 hPa. Therefore, a first baroclinic mode vertical structure emerges for the mode.

Coherent evolution of the OLR and relative vorticity anomalies at 850 hPa shown in Figure 2.11 over a cycle of the oscillation indicates strong convective coupling for the 30–60-day mode also. The evolution of the OLR anomalies over a cycle of the 30–60-day mode is similar to that found in other studies (e.g., Annamalai and Slingo, 2001) using other methods. One interesting point that emerges from this figure is that the 30–60-day variability over the SA monsoon region and that over the EA/WNP region (see Chapter 3) during boreal summer is governed by the same 30–60-day mode of variability. Main difference being that the phase of the northward propagation of the mode in the EA/WNP is shifted with respect to that over the SA monsoon region by about 10 days. The convection first starts at the equatorial IO (day −20) and moves northward to about 10°N (day −10) when convection starts in the south China Sea region. When the convection reaches 25°N in the SA region by day 0, it progresses to about 15°N in the EA/WNP region. Another important point to note is that bands of cyclonic (anticyclonic) relative vorticity at 850 hPa move coherently northward with bands of negative (positive) OLR anomalies with relative vorticity maxima being about 3°N of the convection maxima. We shall show later (see Section 2.2.1.1) that this phase relationship between convection and low-level relative vorticity is important for understanding the mechanism of northward propagation of the mode.

2.1.3 Regional propagation characteristics

The seminal work of Yasunari (1979) and Sikka and Gadgil (1980) led the discovery that the zonally oriented cloud band Tropical Convergence Zone (TCZ) repeatedly propagates northward starting from south of the equator to the foothills of the Himalayas during the summer monsoon season and that the propagation of this cloud band is intimately associated with the active and break cycles of the monsoon ISV (Krishnamurti and Subrahmanyam, 1982; Murakami *et al.*, 1984; Webster *et al.*, 1998). In terms of the two modes, the northward propagation is primarily associated with the 30–60-day mode. Regressed anomalies of relative vorticity based on

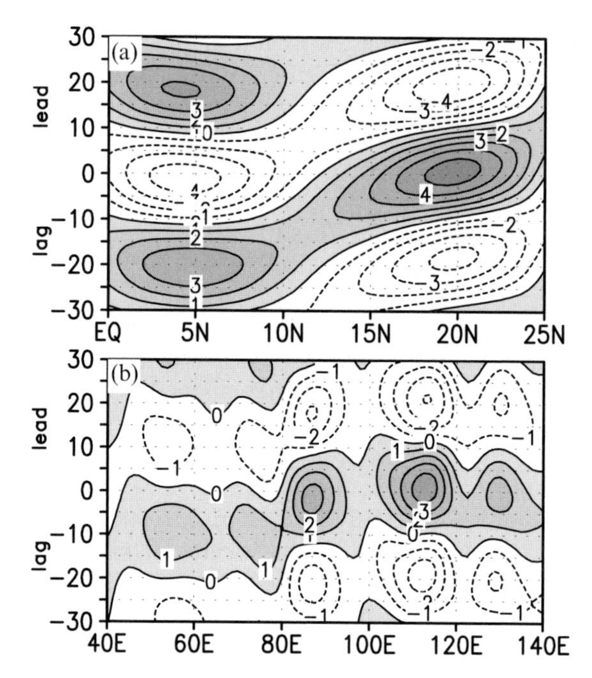

Figure 2.12. (a) Lag–latitude section of regressed anomalies of 30–60-day filtered 850-hPa relative vorticity ($10^{-6}\,s^{-1}$) with respect to the same reference time series described in Figure 2.10 and averaged over 80°E–90°E. (b) Lag–longitude section of regressed anomalies of 30–60-day filtered 850-hPa relative vorticity ($10^{-6}\,s^{-1}$) with respect to the same reference time series described in Figure 2.10 and averaged over 10°N–20°N.

30–60-day filtered data for 20 summer seasons averaged over 80°E–90°E (Figure 2.12(a)) illustrate that, on the average, the mode propagates northward north of the equator up to about 25°N and southward south of the equator to about 10°S (not shown). The phase speed of northward propagation is about 1° latitude per day. Similar regressed anomalies averaged between 10°N–20°N as a function of longitude and lag (Figure 2.12(b)) indicate a rapid eastward phase propagation at the rate of about $10\,m\,s^{-1}$ for the mode at this latitude belt between 40°E and 80°E and almost stationary in character between 80°E and 140°E. Although, the northward propagation characteristics of the mode over the Indian longitudes is rather robust, there is considerable event-to-event and year-to-year variability (not shown). In some years (e.g., 1979) there were several regular northward propagating periods extending up to 25°N while in some other years (e.g., 1981), the northward propagating periods seem to be terminated at about 15°N. In other years still (e.g., 1986) there were very few clear northward propagating periods. While the averaged northward propagation (Figure 2.12(a)) is a measure of regularity and predictability, it coexists with a certain amount of irregularity that limits the predictability of the mode.

A plot similar to Figure 2.12(a) for the 10–20-day mode (Figure 2.13(a)) shows that this mode is not associated with any significant northward propagation

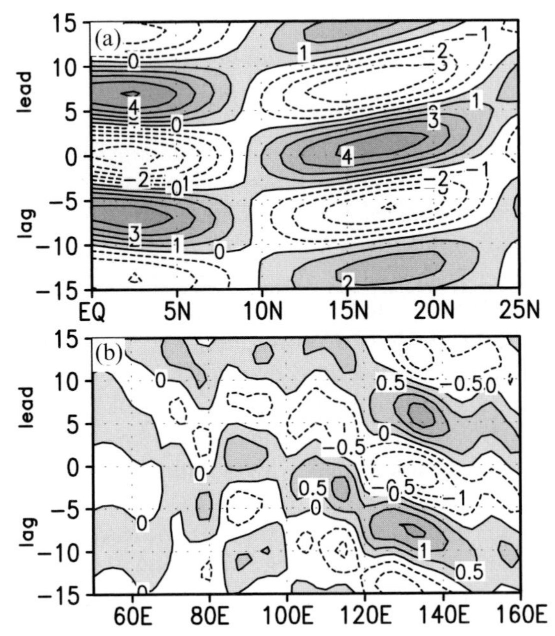

Figure 2.13. (a) Same as Figure 2.12(a), but for the 10–20-day mode. (b) Same as Figure 2.12(b), but for the 10–20-day mode of regressed anomalies averaged over 5°N–15°N. The reference time series used is same as the one described in Figure 2.8.

(Yasunari, 1981; Krishnamurti and Ardunay, 1980; Chen and Chen, 1993). However, the mode is clearly associated with a westward propagation from the western Pacific to about 70°E (Figure 2.13(b)). Both Figure 2.9 and Figure 2.13(b) indicate that the phase speed of westward propagation is about $4.5\,\mathrm{m\,s^{-1}}$ (Krishnamurti and Ardunay, 1980; Chen and Chen, 1993; Chatterjee and Goswami, 2004). Another point to note from Figure 2.9 and Figure 2.13(b) is that the 10–20-day mode over the SA monsoon region is closely linked with that over the EA/WNP monsoon region during northern summer and that westward propagation from the EA/WNP monsoon region influence the 10–20-day variability over the SA monsoon region (Fukutomi and Yasunari, 1999; Annamalai and Slingo, 2001).

2.1.4 Relationship between poleward propagating ISO and monsoon onset

The SA summer monsoon ISOs are strongly tied up with the annual evolution of the mean monsoon. A notable event in the seasonal evolution of the Indian summer monsoon is the "onset", representing a sudden transition from dry to wet conditions of the annual cycle. Although no objective criterion exists for fixing the date of onset of the SA monsoon, the primary indicator from early days of Indian meteorology has been a sharp and sustained increase in rainfall at a group of stations in Kerala in the southern tip of the Indian continent (Ananthakrishnan *et al.*, 1967). Dramatic changes in some of the regional circulation features are known to occur around the time of the onset (Pearce and Mohanty, 1984; Krishnamurti, 1985; Ananthakrishnan and Soman, 1988a, b; Soman and KrishnaKumar, 1993; Joseph *et al.*, 1994). The dramatic feature of the onset is the sudden increase in precipitation over the monsoon region (70°E–110°E, 10°N–30°N) followed by a sudden increase of the kinetic energy (K.E.) over the low-level jet (LLJ) region (55°E–65°E, 5°N–15°N) by a factor of 5–10. This is illustrated in Figure 2.14 (see color section)- where the northward propagation of precipitation anomalies (from CMAP) averaged over 70°E–90°E together with the K.E. of winds at 850 hPa averaged over 55°E–65°E and 5°N–15°N for three years are shown. Almost invariably, onset of the monsoon is triggered by a poleward propagating monsoon ISO. It may also be noted that a northward propagating ISO pulse in April or early May does not lead to an onset. A northward propagating pulse of convection and precipitation in early May, as in 1979 (also in 1995 and 2002) is often called a *bogus onset* (Flatau *et al.*, 2001, 2003) and sometimes could be confused as the real onset. The *bogus onset* is usually associated with a bifurcation of the Madden–Julian Oscillation (MJO) over the BoB. While the dynamic and thermodynamic conditions over the Indian region is not yet ready for the onset, it leads several days later to the onset of the east Asian or south China Sea monsoon (Lau *et al.*, 2002) (see also Chapter 3). The real onset is substantially delayed during years with a bogus onset. A delayed onset (as in 1979) or an early onset (as in 1984), therefore, depends on the phase of the first pulse of the northward propagating monsoon ISO. The interannual variability of the phase of the first monsoon ISO and hence that of the onset seems to be related to, among others, sea surface temperatures (SST) over south tropical IO and western equatorial Pacific (Joseph *et al.*, Flatau *et al.*, 2003).

While the high level of mean K.E. of the LLJ is maintained by the large-scale non-adiabatic tropospheric heat source that is set up by the Tibetan plateau heating and deep convection over India and the BoB (Li and Yanai, 1996), the fluctuations of total K.E. of the LLJ within the season are closely related with the intraseasonal fluctuations of precipitation (non-adiabatic heating) over India and BoB. The northward propagating pulse of precipitation ISO over India and the BoB in September, however, does not accelerate the LLJ over the AS. This may be understood if we keep in mind that the wind response depends on the spatial structure of the non-adiabatic heat source. The center of the non-adiabatic heating moves eastward with the season and is centered around 10°N and 100°E in September due to a decrease of sensible heating over the Tibetan plateau (Yanai and Tomita, 1998). As a result, the cross equatorial flow shifts to the central and eastern IO during September and does not reflect in the K.E. of the western IO winds.

2.1.5 Relationship with the MJO

An oscillation with period between 30–60 days trapped around the equator, generally known as an MJO was discovered in the early 1970s and has since been studied extensively (see Chapters 1 and 7–12 for details). The MJO is strongest during the boreal winter and spring seasons when it appears as an eastward propagating large-scale system in convection, zonal winds, and upper level velocity potential (Hendon and Salby, 1994). During the boreal summer, while the MJO is typically weaker and more complex in character (Madden, 1986; Madden and Julian, 1994), the northward propagating monsoon ISO is vigorous. The similarity in temporal character of the two ISOs raises a natural question: are the two in any way related? This question led some authors (Yasunari, 1979; Lau and Chen, 1986; Madden and Julian, 1994) to suggest that the northward propagating convection over the Indian monsoon region is related to eastward propagating clouds along the equator. Tracing all the 122 intraseasonal systems in pentad mean intraseasonal OLR (equivalent to 10–90-day filtered) data between 1975 and 1985, Wang and Rui (1990) concluded that almost half of the northward propagating events during boreal summer were not associated with eastward propagating equatorial convection (MJOs). Building on the study of Wang and Rui (1990), a comprehensive study based on 24 years of convection data was carried out by Jones *et al.* (2004) using a more objective tracking of convection anomalies. They arrived at conclusions similar to those of Wang and Rui (1990) for the summer ISOs. With the higher sample size compared to that of Wang and Rui (1990), they could also study some IAV of the summer ISOs. Using an index of ISV based on the first two empirical orthogonal functions (EOFs) of 25–80-day filtered OLR anomalies between 1975 and 1999 (excluding 1978), Lawrence and Webster (2001) concluded that about 78% of the northward moving convection is associated with eastward moving convection at the equator, although they also found some independent northward moving events. The discrepancy between their study and that of Wang and Rui (1990) on the fraction of independent northward propagating events appear to be partly due to the fact that only a small fraction (about 20%) of the total ISO variance is represented by the ISO index of Lawrence and Webster (2001), and partly due to a different criterion being used in the latter study to define independent northward moving events. While a fraction of boreal summer events are associated with an eastward propagating MJO, a significant fraction (up to 50%) of them are independent northward moving events. The northward propagating character, larger meridional scale, and weaker eastward penetration make the summer ISOs distinct from the winter MJO (Jones *et al.*, 2004). As a result, MJO theories may not be sufficient to explain genesis or northward movement full of the spectrum of summer monsoon ISOs.

2.2 MECHANISM FOR TEMPORAL SCALE SELECTION AND PROPAGATION

As noted in Figure 2.7, both the ISOs are significant, each explaining 15–25% of the daily variability. Although, they are not single frequency sinusoidal oscillations, peak power around the broadband spectrum indicates preferential excitation and amplification around these frequencies. A question naturally arises: what is responsible for the preferential excitation and amplification around these frequencies? Such a mechanism must also explain the very large horizontal scale and northward movement of the 30–60-day mode and smaller horizontal scale and westward propagation of the 10–20-day mode. We also noted that evolution of circulation is strongly coupled with that of convection and precipitation for both the modes. Interaction between organized convection and flow (circulation) must, therefore, be at the heart of explaining not only the basic preferred periodicities but also their propagation characteristics.

2.2.1 30–60-day mode

The seasonal mean summer monsoon is characterized by two maxima in the rainfall, one over the monsoon trough region between 15°N and 25°N and another in the IO between the equator and 10°S. Both the locations are associated with low-level cyclonic vorticity and represent two preferred locations of the TCZ. The oceanic preferred location is associated with a SST maximum (Hastenrath and Lamb, 1979) in the region and the oceanic TCZ is maintained by the meridional gradient of SST and zonally symmetric dynamics discussed in several studies (Schneider and Lindzen, 1977; Held and Hou, 1980; Goswami *et al.*, 1984). The off-equatorial location is along the monsoon trough (MT), a low-level quasi-stationary cyclonic vorticity arising from interaction of cross-equatorial flow and Himalayan topography. Cross-equatorial flow is set up by a large-scale pressure gradient due to large-scale non-adiabatic heating gradient. Tomas and Webster (1997) studied the role of inertial instability in maintaining such off-equatorial precipitation when the zero potential vorticity line is located off the equator due to the large-scale north–south surface pressure gradient. Since the horizontal scale of the 30–60-day mode is large and similar to that of the seasonal mean (Figure 2.10), the mode appears as fluctuations of the TCZ between the two preferred locations (Sikka and Gadgil, 1980; Krishnamurti and Subrahmanyam, 1982; Goswami, 1994). Dynamics of zonally symmetric TCZ, therefore, provide a paradigm for understanding genesis and northward propagation of the 30–60-day mode (Webster, 1983; Goswami and Shukla, 1984; Nanjundiah *et al.*, 1992; Srinivasan *et al.*, 1993). Webster (1983) simulated northward propagating ISOs in a zonally symmetric model with land north of 14°N and an advective mixed layer ocean south of it (hereafter referred to as the W-model). The simulated ISOs, however, had period of about 15 days and unlike observed ISOs started from the land–ocean boundary. Using a zonally symmetric version of the GLAS (Goddard Laboratory for Atmospheric Sciences) global circulation model (GCM), Goswami and Shukla (1984) simulated the ISO of the TCZ during northern summer having period and northward propagation similar to those observed. The simulated unfiltered precipitation band (figure 4 in Goswami and Shukla, 1984) starts from about 5°S and has longer residence time over the northern location of the TCZ, similar to the characteristics of the observed cloud band (Sikka and Gadgil, 1980). They further showed that the simulated ISO with a period similar to that of the observed arises due to an interaction between organized convection and large-scale Hadley circulation. According to their *convection–thermal relaxation feedback* mechanism, convective activity results in an increase of static stability which depresses convection itself. Meanwhile, dynamical processes and radiative relaxation decreases moist static stability and brings the atmosphere to a new convectively unstable state.

Gadgil and Srinivasan (1990) and Srinivasan *et al.* (1993) derived additional insight regarding factors responsible for determining the period of the oscillations by experimenting with the W-model. They showed that poor simulation of the period of the ISO by the original W-model was due to lack of thermal inertia of land and low moisture holding capacity. With the W-model modified to include increased thermal inertia of the land and increased moisture holding capacity (from 10 cm to 30 cm), Srinivasan *et al.* (1993) were able to simulate the observed periodicity of the ISOs. The zonally symmetric mixed layer in the W-model had a large systematic error in simulating the meridional structure of the SST over the IO region and prescription of observed SST, instead of simulating it with the mixed layer model, allowed simulation of propagation of the oscillations from near the equator.

Very large zonal scale compared to the meridional scale of the dominant summer monsoon ISO invited the zonally symmetric dynamics to be invoked for an explanation of the genesis and temporal scale selection. However, these theories can not shed light on the observed zonal phase propagation of a large fraction of the summer ISO. Therefore, wave dynamics need to be involved for a fuller explanation of the genesis and phase propagation of the ISO. Krishnan *et al.* (2000) indicated that westward propagating Rossby waves are important in causing the transition from an active to a break phase of the summer ISO. As the theory for genesis and eastward propagating MJO is much better developed (see Chapter 10) in terms of an equatorial convectively coupled Kelvin–Rossby couplet, some attempts have been made (Lau and Peng, 1990; Wang and Xie, 1997) to seek an explanation for scale selection and northward propagation of summer monsoon ISOs within the same framework. Lau and Peng (1990) show that interaction of the eastward propagating equatorial ISOs with the mean monsoon flow can lead to an unstable quasi-geostrophic baroclinic Rossby wave with a wavelength of 3,000–4,000 km over the Indian monsoon region between 15°–20°N. As these waves grow, low-level air is drawn northward resulting in a rapid northward propagation of the area of deep convection. Based on results of a modeling study, Wang and Xie (1997) described the summer ISOs as a convection front formed by continuous emanation of equatorial Rossby waves from convection in the equatorial western Pacific. An initial convection cell over the equatorial IO moves eastward as a Kelvin wave and forms Rossby wave cells on either side with typical scale of 2,000–4000 km. Feedback with convection and westward propagation of the emanated Rossby waves makes the convection front tilt north-westward from the equator to 20°N, resulting in an apparent northward movement. The westward propagating Rossby waves eventually decay over the AS due to a sinking of dry air mass from Africa while a southern cell initiates an equatorial perturbation and starts the next cycle. The life cycle takes about 4 weeks.

Thus, convectively coupled wave dynamics involving a Kelvin and Rossby wave pair can give rise to an ISO with a timescale of 30–40 days. This mechanism is probably responsible for the summer monsoon ISOs that are associated with eastward propagation at the equator. A significant fraction of the summer ISOs have northward moving events not associated with eastward moving equatorial disturbances. Zonally symmetric dynamics are required to explain these summer ISOs as wave dynamics can not explain the genesis of this fraction of the summer ISOs. Thus, both wave dynamics and zonally symmetric physics are required for the explanation of genesis of totality of the summer monsoon 30–60-day mode.

2.2.1.1 *Mechanism for poleward propagation of the 30–60-day mode*

As described in Section 1.3, a major characteristic of the 30–60-day mode is its poleward propagation. Hence, a considerable amount of literature has been devoted toward understanding the poleward propagation of this mode. However, the temporal scale selection and the northward propagation are intimately related. Therefore, here we describe how different mechanisms for temporal scale selection, discussed above, address the question of observed northward propagation of the mode. Webster (1983) proposed that the northward gradient of sensible heat flux is responsible for the poleward propagation of the cloud band. Goswami and Shukla (1984) also indicated that ground hydrology feedback may be responsible for the northward propagation. Later, Gadgil and Srinivasan (1990) and Nanjundiah *et al.* (1992) showed that the northward gradient of moist static stability (poleward side being more unstable than the equatorward side) is responsible for the poleward propagation of the cloud band. The existence of a meridional gradient of the near-surface background humidity in this region contributes to the meridional gradient of the moist static stability. While these studies followed a zonally symmetric paradigm and identified some processes that are important for the northward propagation, a few other studies (Lau and Peng, 1990; Wang and Xie, 1997) attempted to explain the basic genesis and the northward propagation resulting from unstable Rossby waves interacting with the basic monsoon mean flow. However, the physical mechanism that led to the northward component of the Rossby wave propagation was not clearly identified in these studies. Keshavamurthy *et al.* (1986) also indicate that the northward propagation may arise from a zonally symmetric component of an eastward propagating equatorial oscillation interacting with the mean flow over the Asian monsoon region.

To gain some insight regarding the mechanism responsible for the poleward propagation of the cloud band, a diagnostic analysis is presented here. Lag regression of 30–60-day filtered OLR, relative vorticity at 850 hPa, and divergence at 925 hPa with respect to the same reference time series used in Figure 2.10 were carried out. Lag–latitude plot of regression of OLR and 850-hPa relative vorticity averaged over 80°E–90°E are shown in Figure 2.15(a) (see color section), while that for 850-hPa relative vorticity and divergence at 925 hPa are shown in Figure 2.15(b). It is seen from Figure 2.15(a) that, at any given time, cyclonic (anticyclonic) vorticity at 850 hPa is to the north of the negative (positive) OLR anomalies representing cloudy (clear) belts. This is consistent with observations made from Figure 2.11 (see color section). The cyclonic (anticyclonic) vorticity at 850 hPa is associated with convergence (divergence) of moisture in the boundary layer as seen by the in-phase relationship between the two in Figure 2.15(b) (see color section). The atmospheric circulation driven by the diabatic heating (Gill, 1980) associated with the zonally oriented cloud band in the presence of background mean flow with easterly vertical shear produces a cyclonic vorticity with a maximum about 3°N of the center of the cloud band. Cyclonic vorticity drives frictional convergence in the planetary boundary layer and leads to higher moisture convergence north of the cloud band. The meridional gradient of the mean boundary layer moisture also helps in making the moisture convergence larger to the north of the cloud band. This leads the convection center to move northward. Kemball-Cook and Wang (2001) and Lawrence and Webster (2001) also indicated that the planetary boundary layer convergence maximum, being north of the convection maximum, is responsible for the northward propagation of the summer monsoon 30–60-day oscillation. Hsu and Weng (2001) also suggested that a similar frictional convergence driven by low-level vorticity is responsible for the north-westward movement of the EA/WNP monsoon 30–60-day oscillation.

The crucial question for the northward propagation of the 30–60-day mode is to understand what is responsible for producing the low-level vorticity maximum north of the convection maximum. Also, the northward propagation takes place from about 5°S, but the frictional convergence is ineffective near the equator. Therefore, an alternative mechanism is required for explaining the northward propagation near the equator. Although several earlier studies (Goswami and Shukla, 1984; Gadgil and Srinivasan, 1990; Nanjundiah *et al.*, 1992), using zonally symmetric dynamics, identified some processes to be important for the northward propagation of the mode, a clear physical picture had not emerged. Using a simple zonally symmetric model to interpret results from a GCM simulation and observations, Jiang *et al.* (2004) provide a clearer picture of the physical processes responsible for the poleward propagation. They propose that a combination of a "vertical wind-shear mechanism" and a "moisture–convection feedback mechanism" is responsible for the poleward propagation of the convection band. They show that the easterly mean vertical shear in the region gives rise to the generation of barotropic vorticity to the north of the convection center, which in turn generates barotropic divergence in the free atmosphere north of the convection center. This leads to boundary layer convergence north of the convection maximum. The mean flow and mean boundary layer humidity during summer monsoon season also allow perturbation moisture convergence to be maximum north of the convection center. Both these processes contribute to the poleward propagation of the mode. Near the equator, the moisture–convection feedback mechanism is the dominant mechanism producing the poleward propagation. Although, this mechanism is essentially based on zonally symmetric dynamics, a similar mechanism may be at work in producing the northward component of the Rossby wave motion in the presence of the background summer mean flow (see Chapter 10).

Based on the understanding of genesis of the large-scale (zonally symmetric) component of the 30–60-day mode from modeling studies (e.g., Goswami and Shukla, 1984), and understanding gained on the poleward propagation from diagnostic studies (Figure 2.15, see color section) and theoretical analysis (e.g., Jian *et al.*, 2004), the evolution of the mode in the meridional plane over the SA monsoon region is schematically shown in Figure 2.16. Low-level convergence of moisture associated with the SST gradient in a conditionally unstable atmosphere produces

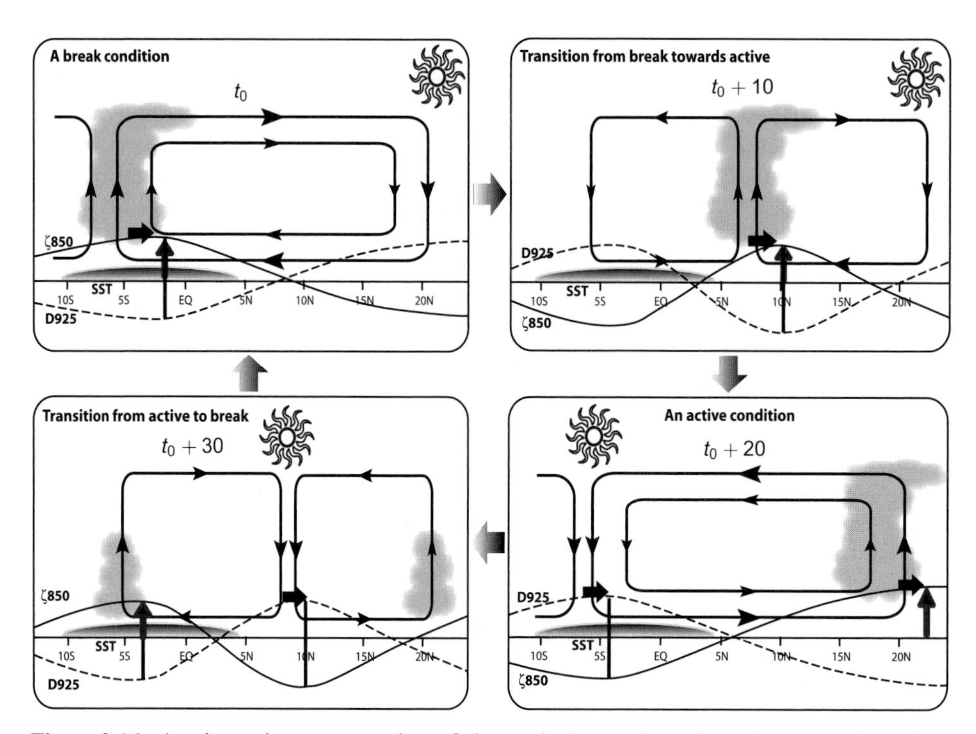

Figure 2.16. A schematic representation of the evolution and northward propagation of the meridional circulation associated with the 30–60-day mode in the meridional plane. The thin arrows indicate the anomalous Hadley circulation. The thick vertical arrow indicates the location of the center of the boundary layer moisture convergence, while the thick horizontal arrow indicates the direction of poleward motion of the cloud band. The thin solid (dotted) line indicates the phase of the relative vorticity at 850 hPa (divergence at 925 hPa) with positive (negative) phase being above (below) the base line. The location of clear sky conditions is shown by the sun-like symbol.

organized convection and intensifies the TCZ over the SST maximum over the equatorial IO with subsidence and clear conditions over the MT region (Figure 2.16(a)). The anomalous Hadley circulation has ascending motion over the oceanic position and descending motion over the MT region. This situation corresponds to a break condition. It is also associated with anomalous cyclonic vorticity over the oceanic location and anticyclonic vorticity over the MT. The cyclonic vorticity at low levels and associated boundary layer moisture convergence is maximum north of the zone of maximum convection and makes it move northward (also see Figure 2.15, color section). After about 10 days ($t_0 + 20$, Figure 2.16(b)) the convection zone reaches about 10°N, with both MT and oceanic region being under subsidence and clear sky. After another 10 days ($t_0 + 20$, Figure 2.16(c)), the convection zone reaches the MT zone around 25°N with divergence maximum slightly north of the anticyclonic vorticity maximum over the oceanic zone (SST maximum). The anomalous Hadley circulation has descending motion over the oceanic position and ascending motion over the MT region. This situation corresponds to an active monsoon condition. During the next 10 days ($t_0 + 30$), the convection over the MT moves further northward to the foothills of the Himalayas and the clear condition over the equatorial zone also moves northward from 5°S. Decrease in subsidence over the oceanic zone is associated with weak winds and the continued clear condition raises the surface temperature as the net heat flux at the surface becomes positive (also see Section 2.3). This acts against the subsidence and eventually makes the convection to break out. The convection builds up to become maximum over the oceanic zone in about 10 more days, repeating the cycle.

2.2.2 10–20-day mode

In contrast to the considerable attention received by the genesis and scale selection of the summer monsoon 30–60-day mode, the 10–20-day mode has received very little attention. Having observed that both cloudiness and the moist static stability have a quasi-biweekly oscillation, Krishnamurti and Bhalme (1976) proposed a mechanism in terms of cloud–radiation–convection feedback. According to them, the net radiative effect warms up the surface over the MT region under normal summer monsoon conditions and builds up dry and moist static stability of the lower atmosphere. This leads to moist convection, increase in cloud cover, and reduction of solar radiation at the surface. Resultant surface cooling and warming of the middle layer of the atmosphere due to latent heat release associated with the convection stabilizes the temperature profile and cuts down convection. With dissipation of the clouds and increase in solar radiation at the surface, moist instability again builds up and the process can repeat itself. While this is a plausible scenario, they did not demonstrate that the mechanism would preferentially select the quasi-biweekly period. In fact, modeling studies (Goswami and Shukla, 1984; Gadgil and Srinivasan, 1990) indicate that a very similar mechanism leads to the 30–60-day oscillation. Chen and Chen (1993) made a comprehensive study of the structure and propagation characteristics of the quasi-biweekly mode (QBM) but did not address the question of scale selection. Using a shallow water model with fixed vertical structure and a relaxation timescale for moisture variable, Goswami and Mathew (1994) proposed that the QBM is an unstable mode of the tropical atmosphere driven by evaporation–wind feedback in the presence of background westerlies. However, the exact nature of the unstable mode is unclear as the meridional structure is not shown. Also, the zonal wavelength of their most unstable mode is between 9,000–12,000 km, much larger than observed. Recently, Chatterjee and Goswami (2004) (hereafter referred to as CG) provided a mechanism for scale selection of the QBM. As shown in Figure 2.8(b, c) they noted that the horizontal structure of the QBM at low level resembles that of a first meridional mode ($n = 1$) equatorial Rossby wave with wavelength of about 6,000 km, but shifted to north by about 5°. What makes the Rossby wave unstable and affects the selection of observed horizontal scale? A pure $n = 1$ equatorial Rossby wave has the two vortices centered around the equator. What is responsible for the shift of the structure of the QBM in the meridional direction (Figure 2.8(b))? According to CG, the QBM is an equatorial Rossby wave driven unstable by convective feedback in the presence of a frictional boundary layer and modified by the summer mean flow.

Using a two-layer atmosphere and a steady Ekman boundary layer with simple parameterization for convective heating, they showed that convective feedback forced by boundary layer convergence drives an $n = 1$ Rossby wave unstable with maximum growth corresponding to a period of 16 days and a wavelength of 5,800 km. The westward phase speed of the unstable Rossby mode corresponding to this period and wavelength is close to the observed westward phase speed of about $4.5\,\mathrm{m\,s^{-1}}$. The positive feedback arises from the fact that the baroclinic component of low-level vorticity for the $n = 1$ Rossby wave drives an in-phase boundary layer Ekman convergence. As vorticity is in phase with the baroclinic component of potential temperature, latent heat release from the boundary layer moisture convergence increases the potential temperature perturbation leading to a positive feedback. They noted that turbulent entrainment in the planetary boundary layer is important in achieving the in-phase relationship between the convergence in the boundary layer and the potential temperature perturbation at the top of the boundary layer. They also demonstrated that inclusion of realistic summer time mean flow in their model results in a shift of the unstable Rossby wave to the north by about 5° without seriously affecting the preferred periodicity and wavelength. The shift of the spatial structure to the north by about 5° can be interpreted as a "dynamic equator" effect. As a free symmetric Rossby wave has the tendency to be centered around the background zero absolute vorticity (sum of relative vorticity and Earth's vorticity), the zero background absolute vorticity line may be called the "dynamic equator". The vorticity of the background mean flow can make the dynamic equator shift from the geographical equator. It may be noted that the low-level summer mean flow makes the zero absolute vorticity line move to about 5°N in the IO/Indonesian region (Tomas and Webster, 1997). The observed shift of the structure of the QBM appears to be due to the shift of the dynamic equator by the background mean flow. Thus, this model explains both the scale selection and propagation characteristics of the QBM. It is also shown that the same model may explain the genesis and scale selection of the northern winter QBM observed in the western Pacific.

2.3 AIR–SEA INTERACTIONS

As noted in Section 2.1.2, the monsoon ISOs are associated with significant fluctuations of cloudiness and surface winds, and hence one may expect considerable fluctuations in the net heat flux at the surface. What is the amplitude of the net heat flux (Q_{net}) associated with the monsoon ISOs? What is the amplitude of intraseasonal SST fluctuations? Do the intraseasonal SST fluctuations have a large spatial scale similar to that of the atmospheric parameters and Q_{net}? What is the role of the SST fluctuations in determining the periodicities and northward movement of the monsoon ISOs? We shall attempt to get some answers to these questions in this section.

The first evidence that the ISOs involve significant fluctuations of SST and latent heat flux at the interface over the BoB and western Pacific came from Krishnamurti *et al.* (1988) who examined data during 1979 from FGGE (First GARP Global Experiment). While the tropical atmosphere and ocean (TAO) moorings in the Pacific have given good data on SST fluctuations associated with the MJO, very little high-resolution direct measurement of SST is available in the IO region. Recent measurements on a few moored buoys by the Department of Ocean Development (DOD) of India (Sengupta and Ravichandran, 2001) have shown large intraseasonal fluctuations of SST and net heat flux over the BoB. Measurements during special field experiments such as BOBMEX (Bhat *et al.*, 2001) and JASMINE (Webster *et al.*, 2002) also show large intraseasonal fluctuations of net heat flux. However, the large-scale structure of ISV of SST was not available as the IR-based satellite which derived weekly SST (Reynolds and Smith, 1994) was shown to be sufficiently inadequate to represent the observed intraseasonal SST variability (Sengupta and Ravichandran, 2001). With the availability of reliable high-resolution SST measurements using microwave sensors, TMI (TRMM Microwave Imager) on board the TRMM (Tropical Rain Measuring Mission) satellite, it became possible to describe the spatial distribution of ISV of SST over the IO region. Using TMI SST, wind speed, and NOAA OLR data, Sengupta *et al.* (2001) showed that the intraseasonal SST oscillations with an amplitude of 0.6–0.8°C has a large horizontal scale similar to that of atmospheric ISO (Figure 2.5) and possesses northward movement in the region coherent with OLR, surface wind speed, and Q_{net} during the 1998, 1999, and 2000 summer monsoon seasons. Such coherent evolution of 10–90-day filtered anomalies of Q_{net}, surface wind speed, precipitation, and SST for the summer season of 2002 is shown in Figure 2.17. The components of the net heat flux are estimated using the same formulation used in Sengupta *et al.* (2001). It is noteworthy that the amplitude of interseasonal Q_{net} fluctuations is quite large (70–90 $W\,m^{-2}$). The phase lag between the SST and the precipitation bands (Figure 2.17, lower panels) is another important aspect of the observed summer monsoon ISOs. The warm (cold) SST band follows the dry (rainy) band with a time lag of 7–10 days. Vecchi and Harrison (2002) indicate that this phase relation could be exploited to predict the monsoon breaks. The speed of northward movement of SST is the same as that for the atmospheric fields but lags behind the Q_{net} by about 7 days. The quadrature relationship between SST and Q_{net} indicates that the intraseasonal SST fluctuations are essentially being driven by the atmospheric ISV through Q_{net}. The quadrature relationship between SST and Q_{net} is not found to hold good over the equatorial IO and the western IO coastal region (Sengupta *et al.*, 2001) where the ocean dynamics also seem to play a role in determining the SST. Waliser *et al.* (2003b) studied an ocean model with a mixed layer with daily wind stresses and heat fluxes associated with the atmospheric ISOs. While the net heat flux associated with ISOs is a major forcing, they found that the mixed layer depth (MLD) variations tended to contribute positively to the SST variations. Further, contributions from advection and/or entrainment within the Somali Current region and equatorial IO is significant.

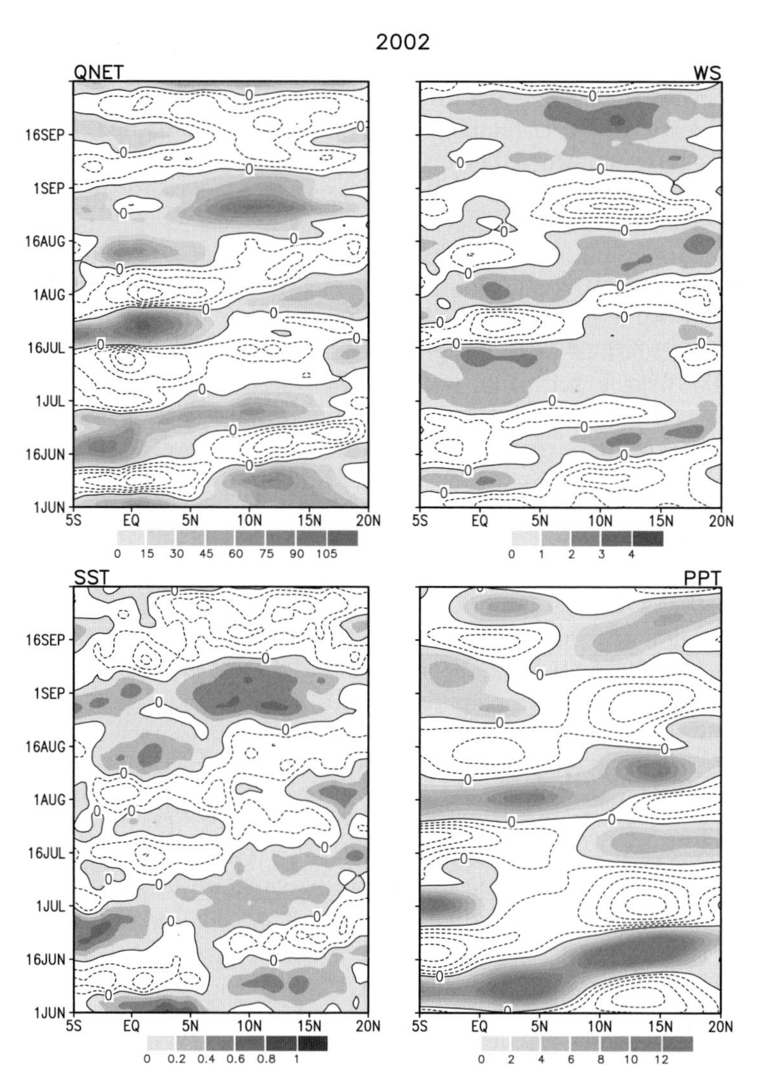

Figure 2.17. Time–latitude section of 10–90-day filtered net heat flux (Q_{net}; W m^{-2}), surface wind speed (WS; m s^{-1}), SST (°C), and CMAP precipitation (PPT; mm day^{-1}) for the 2002 summer season averaged over 85°E–90°E. Positive (negative) anomalies are shown with shading (contours). Contour intervals are the same for both positive and negative anomalies.

From the above discussion, it is clear that atmospheric ISOs do force ISO in SST over the IO. However, the role of the intraseasonal fluctuations of SST in determining either the scale selection or northward movement of the Asian monsoon ISOs is unclear. Is air–sea coupling crucial for existence of the *observed* ISOs? Air–sea coupling is certainly important for the *observed* summer ISOs as the quadrature phase relationship between atmospheric convection (or precipitation) and SST can not be simulated by an AGCM forced by observed daily SST (Wu *et al.*, 2002; Fu *et al.*, 2003). Although intraseasonal variations of SST in this region may act almost like a slave to the atmospheric ISV, coupling between ocean and atmosphere is essential to simulate the correct phase relationship between convection and SST. This was demonstrated by Fu *et al.* (2003) where they coupled an AGCM to a relatively simple ocean model without flux corrections, and were able to simulate the observed period, northward movement, and phase relationship between precipitation and SST quite well. When the AGCM is forced by daily SST from the coupled run, the AGCM simulated precipitation band is in phase with the SST. Zheng *et al.* (2004) and Rajendran *et al.* (2004) carried out studies similar to that of Fu *et al.* (2003) where they compared simulation of a coupled GCM (CGCM, an AGCM coupled to an OGCM) with simulations of the AGCM forced with CGCM SSTs, and came to very similar conclusions. In their study, Zheng *et al.* (2004) find that coupling also improves the spatial structure of the summer ISO variability over the IO. One aspect of the air–sea interaction that is not clear even from these studies is how the ISV of SST feeds back to the convective activity of the atmosphere and modifies the ISO characteristics.

Is air–sea interaction crucial for the *existence* of the summer ISOs? As discussed in Sections 2.1 and 2.2, the theories of atmospheric ISOs indicate that the basic temporal scale selection and northward propagation of the atmospheric summer ISOs arise from internal feedback within the atmosphere. Several AGCMs forced with prescribed SST exhibit ISV levels at or above that found in observations with spatial patterns that resemble the observed pattern (Waliser *et al.*, 2003b; Kemball-Cook and Wang, 2001; Jiang *et al.*, 2004) and include some form of northward propagation. This also supports the hypothesis that the basic oscillation and the northward propagation may be of internal atmospheric origin. The air–sea interactions can modify these internally triggered oscillations in amplitude, frequency domain, as well as in northward propagation characteristics. Kemball-Cook and Wang (2001) find that the air–sea coupling enhances northward propagation of the summer ISOs significantly. Waliser *et al.* (2004) find that air–sea coupling also improves the space–time characteristics of the summer ISO over the IO. Fu *et al.* (2002) also find that ocean–atmosphere coupling improves simulation of the amplitude and northward propagation characteristics of the ISOs. However, some of these results could be influenced by the bias of the mean states of the individual components of the coupled model. Therefore, quantitative estimate of modification of the summer ISO by air–sea coupling is still not well established. Modeling of air–sea interactions associated with northern summer monsoon ISOs is still in its infancy. More studies would be required to unravel the quantitative role played by the SST on determining the amplitude and phase propagation of the monsoon ISO.

2.4 CLUSTERING OF SYNOPTIC EVENTS BY THE ISOs

The modulation of synoptic activity by the large-scale circulation anomalies associated with the MJO has been demonstrated by Liebmann *et al.* (1994) and Maloney and Hartmann (2000). The horizontal structure of low-level winds associated with the summer monsoon 30–60-day mode (Figure 2.10(b)) also have large scales and are similar to that of the seasonal mean (Figure 2.1(e)). Therefore, the meridional shear of the low-level zonal winds and cyclonic vorticity at 850 hPa are significantly enhanced (weakened) during an active (break) phase of the ISO. Hence, conditions for cyclogenesis are much more favorable during an active phase compared to a break phase. Similar to MJO, do the monsoon ISOs modulate the synoptic activity in the region during northern summer? Using genesis and track data for low-pressure systems (LPSs) for 40 years (1954–1993), Goswami *et al.* (2003) show that genesis of an LPS is nearly 3.5 times more favorable during an active condition (147 events corresponding to normalized index $> +1$) compared to a break condition (47 events corresponding to normalized index < -1) of the monsoon ISO. They also show that the LPSs are spatially strongly clustered to be along the MT region during an active condition (Figure 2.18). Day-to-day fluctuations of precipitation are essentially governed by the synoptic activity. As the synoptic activity is clustered in time and space by the ISO, a prediction of ISO phases about three weeks in advance may allow one to also predict the probability of high (low) rainfall activity with such a lead time.

Due to the much larger horizontal scale of the monsoon ISO as well as the MJO compared to that of the synoptic disturbances, all these studies (Liebmann *et al.*, 1994; Maloney and Hartmann, 2000; Goswami *et al.*, 2003) argue that collective effect of randomly occurring synoptic disturbances could not influence the structure of the ISO significantly. However, a recent study by Straub and Kiladis (2003) indicates that the westward propagating Mixed Rossby–Gravity wave-Tropical Disturbance (MRG-TD)-type synoptic disturbances may have some influence on the structure of the summer ISO.

2.5 MONSOON ISO AND PREDICTABILITY OF THE SEASONAL MEAN

The prediction of summer monsoon rainfall at least one season in advance is of great importance for the agro-based economy of the region. For over a century, attempts have been made to predict seasonal mean monsoon rainfall using statistical methods involving local and global antecedent parameters that correlate with the monsoon rainfall (e.g., Blanford, 1884; Walker, 1923, 1924; Gowarikar *et al.*, 1989; Sahai *et al.*, 2003; Rajeevan *et al.*, 2004). The linear or non-linear regression models as well as neural network models (Goswami and Srividya, 1996) indicate some skill when the monsoon is close to normal (about 70% of the years over the past 130-year period) but fails to predict the extremes with useful skill. Almost all statistical models failed to predict the drought of 2002. Delsole *et al.* (2002) argue that regression models with many predictors (e.g., the 16 parameter model of Indian Meteorological Department (Gowarikar *et al.*, 1989)) may possess some artificial skill and often regression models with two or three parameters produce better forecasts on average than regression models with multiple predictors. Thus, the usefulness of

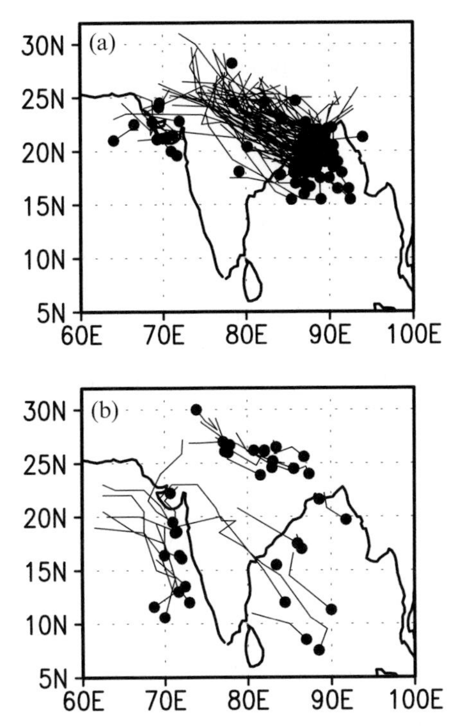

Figure 2.18. Tracks of LPSs for the period 1954–1983 during extreme phases of monsoon ISO. (a) Active ISO phase (monsoon ISO index (MISI) $> +1$) and (b) break ISO phase (MISI < -1). The MISI used here is 10–90-day filtered relative vorticity during the summer monsoon season (1 June–30 September) averaged over 80°E–95°E and 12°N–22°N. Dark dots represent the genesis point and the lines show their tracks. Large number of LPSs during the active phase are strongly clustered along the MT. The few LPSs that form during breaks clearly avoid the MT region and form either near the foothills of the Himalayas or off the western coast and move westward.
After Goswami *et al.* (2003). © American Geophysical Union.

the statistical models is limited. A series of sensitivity studies (Charney and Shukla, 1981; Shukla, 1981, 1988; Lau, 1985) have shown that the tropical climate is, in general, much less sensitive to initial conditions and hence more predictable than the extra-tropical climate. These studies lay the foundation for deterministic climate prediction in the tropics, and dynamical prediction of seasonal mean monsoon using state-of-the-art climate models appears to be a logical alternative to statistical prediction. Although, the climate models have improved significantly over the years in simulating the mean climate, they still do not have skill better than the statistical models in predicting the seasonal mean monsoon (Wang *et al.*, 2004; Kang *et al.*, 2002b). Almost all present-day climate models have serious difficulty in simulating the seasonal mean monsoon climate and its interannual variations (Sperber and Palmer, 1996; Gadgil and Sajani, 1998; Kang *et al.*, 2002a, b; Wang *et al.*, 2004). Even though climates of certain tropical regions show very little sensitivity to initial conditions (e.g., Shukla, 1998), the Indian summer monsoon appears to be an exeption within the tropics and appears to be quite sensitive to initial conditions (Sperber and Palmer, 1996; Sperber *et al.*, 2001; Krishnamurthy and Shukla, 2001), making it probably the most difficult climate system to simulate and predict.

What makes the Indian monsoon such a difficult system to simulate and predict? The sensitivity of monsoon climate to initial conditions indicates existence of significant "internal" low frequency (LF) variability in the monsoon region (Goswami, 1998). The predictability of the monsoon is going to be determined by the extent to which internal LF variability govern IAV of the monsoon. What is responsible for such internal LF variability in the monsoon region? We recall (see Section 2.2) that monsoon ISOs arise due to internal dynamical feedback between organized convection and large scale circulation with the possibility of SST coupling playing a role. Could monsoon ISOs lead to any significant LF internal variability? If they do, that part of monsoon IAV would be unpredictable. Ajaya Mohan and Goswami (2003) made estimates of internal IAV of circulation based on daily data from NCEP–NCAR Reanalysis for more than 40 years and convection data for more than 20 years, and showed that almost all internal IAV in the tropics arises from the ISOs.

How do the ISOs influence the seasonal mean and its IAV? We noted in Section 2.1.2 that the spatial structure of the 30–60-day mode (Figure 2.10) is similar to that of the seasonal mean (Figure 2.1), strengthening (weakening) the seasonal mean in its active (break) phases. As shown in Figure 2.19, the ISV and IAV of the Asian monsoon are, in fact, governed by a common spatial mode of variability (Fennessy and Shukla, 1994; Ferranti *et al.*, 1997; Goswami *et al.*, 1988; Goswami and Ajaya Mohan, 2001). Horizontal structures of the ISO and the seasonal mean being similar, higher probability of occurrence of active (break) conditions within a season could result in a stronger (weaker) than normal monsoon. If monsoon ISOs were a single-frequency sinusoidal oscillation, this could not happen. However, due to the existence of a band of frequencies, the ISOs are rather quasi-periodic and hence higher probability of occurrence of active (break) phases within a season could take place. Goswami and Ajaya Mohan (2001) have indeed shown that a strong (weak) Indian monsoon is associated with higher probability of occurrence of active (break) conditions. Sperber *et al.* (2000) also show that ISV and IAV of the Asian monsoon are governed by a common mode of spatial variability and that strong (weak) monsoons are associated with higher probability of occurrence of active (break) conditions. The fact that ISV influences the seasonal mean and its predictability was also concluded by Waliser *et al.* (2003b), who compared simulation of the seasonal mean and ISV by a number of AGCMs and found that higher ISV is associated with higher intra-ensemble variance (internal variability) and poorer predictability of the seasonal mean.

The predictability of the seasonal mean monsoon is governed by a relative contribution of slowly varying external component of forcing (such as that associated with the ENSO) and internal variability to the observed IAV of the monsoon. High (low) predictability is associated with a higher (lower) contribution of the external forcing to the IAV compared to that from the internal variability. How much of the total IAV of the Asian monsoon is actually governed by the LF internal

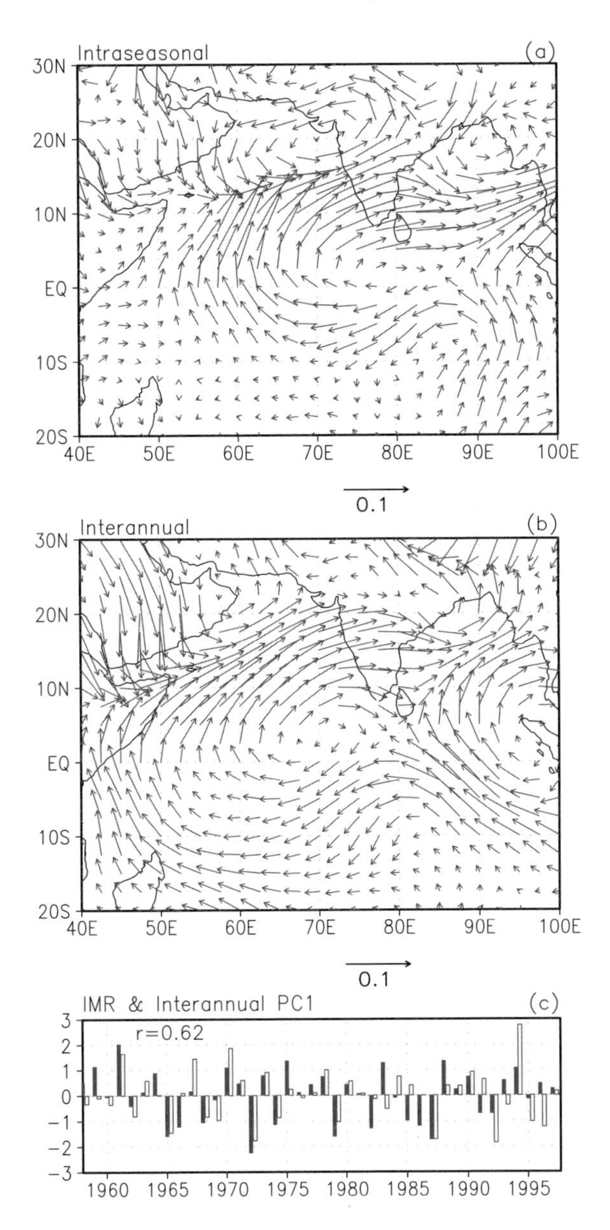

Figure 2.19. First EOF of the intraseasonal and interannual 850-hPa winds. (a) Intraseasonal EOFs are calculated with ISO filtered winds for the summer months (1 June–30 September) for a period of 20 years (1978–1997). (b) Interannual EOFs are calculated with the seasonal mean (JJAS) winds for a 40-year period (1958–1997). Units of vector loading are arbitrary. (c) Relation between all India monsoon rainfall (IMR; unfilled bars) and interannual PC 1 (filled bar). Both time series are normalized by their own standard deviation. Correlation between the two time series is shown.

(a, b) After Goswami and Ajaya Mohan (2001), (c) American Meteorological Society.

variability? Estimates made using an AGCM (Goswami, 1998) and using long observations (AjayaMohan and Goswami, 2003) indicate that about 50% of total IAV of the Asian monsoon is governed by the internal component coming primarily from the ISOs. Thus, ISOs make the Asian monsoon a difficult system to predict by making the unpredictable "noise" comparable to the externally forced predictable "signal". For many years, a consensus on the fraction of total IAV of the Indian monsoon governed by the ISOs was lacking. However, a consensus towards what is concluded here is slowly evolving. Therefore, the seasonal mean summer monsoon will remain a difficult system to predict. Clever methods will have to be devised to simulate and identify the weak signal from a background of noise of comparable amplitude. The prospect of predicting the seasonal mean monsoon would have improved if the statistics of the summer ISOs were strongly modulated (or constrained) by the slowly varying forcing (such as that associated with the ENSO). Modeling studies so far, however, indicate that the summer monsoon ISOs over the Asian monsoon region are not sufficiantly influenced by slowly varying SST changes associated with the ENSO.

2.6 PREDICTABILITY AND PREDICTION OF THE MONSOON ISOs

While the monsoon ISOs make the seasonal mean monsoon difficult to predict, they themselves may possess predictability beyond the current skill of medium-range prediction by virtue of their quasi-periodic nature. Prediction of long dry spells two to three weeks in advance are important to farmers in planning for sowing, harvesting, and water management. What is the limit on prediction of dry and wet spells or break and active phases of the monsoon ISO? These and other issues related to predictability and extended range prediction of the summer monsoon ISOs are discussed at length in Chapter 12 and hence will not be repeated here. Two important findings are summarized here. The first finding is that the potential predictability of the monsoon breaks is much higher than that for the monsoon active conditions (Goswami and Xavier, 2003; Waliser *et al.*, 2003a). Secondly, simple empirical models developed during the last couple of years (Goswami and Xavier, 2003; Webster and Hoyos, 2004) demonstrate a potential for predicting the summer monsoon ISOs up to three weeks in advance.

2.7 SUMMARY AND DISCUSSION

A synthesis of large-scale spatial and temporal structures and regional propagation characteristics of the Asian summer monsoon ISV is presented in this chapter, based on advances made in global observations. Such observations have revealed that the active and break SA monsoon or the wet and dry spells over the Indian continent, are manifestation of a superposition of 10–20-day and 30–60-day oscillations. Both the 10–20-day oscillation and the 30–60-day oscillation contribute roughly equally to the total ISV in the SA monsoon region. While 30–60-day oscillation has a very large zonal scale encompassing both the SA and the EA/WNP monsoon regions, the

10–20-day oscillation has a smaller zonal scale and is regional in character. The 30–60-day mode is characterized by a northward propagation while the 10–20-day mode is characterized by a westward propagation. ISV on a 30–60-day timescale over the EA/WNP region (see Chapter 3) and that over the SA region are closely related through the evolution (northward propagation) of the large spatial structure associated with the 30–60-day mode. Also the 10–20-day variability over the SA region are associated with 10–20-day oscillation propagating from the western Pacific and amplification over the BoB. Thus, the ISV over the SA and EA/WNP monsoon regions are intimately related.

Advances made in understanding the scale selection for the 30–60-day mode and its northward propagation as well as that for the 10–20-day mode and its westward propagation are reviewed based on analysis of observations and a hierarchy of modeling studies. Two mechanisms seem to contribute to the temporal scale selection of the 30–60-day mode. One is a "convection–thermal relaxation feedback mechanism", according to which convective activity results in an increase of static stability which depresses convection itself. As convection dies, dynamical processes and radiative relaxation decreases moist static stability and brings the atmosphere to a new convectively unstable state. This mechanism does not involve wave dynamics and may be responsible for the northward propagating 30–60-day oscillations not associated with eastward propagation of convection in the equatorial region. The other mechanism involves eastward propagation of convection over the equatorial IO in the form of a Kelvin wave and west-north-west propagation of Rossby waves emanated over the western Pacific. The timescale is determined in this case by propagation time of the moist Kelvin wave from the eastern IO to the western Pacific and the moist Rossby waves from the western Pacific to the AS where they decay and a new equatorial perturbation is generated.

An important advance has also been made in understanding the poleward propagation of the 30–60-day mode. Several modeling studies did indicate that ground hydrology and meridional gradient of the moist static stability were important for the northward propagation of the mode. However, a clear physical picture had not emerged. Some diagnostic studies then showed that the low-level relative vorticity drives a boundary layer moisture convergence that is maximum about 3°N of the convection maximum. What is responsible for the low-level vorticity to be maximum about 3°N of the convection maximum has been elucidated in some recent modeling and theoretical studies. The easterly vertical shear of the summer mean flow couples the barotropic and the baroclinic components of response to the convective heating and generates a barotropic vorticity maximum north of the convection maximum. The barotropic vorticity maximum forces boundary layer moisture convergence to be maximum to the north of the heating maximum. Thus, a better understanding is emerging for the mean northward propagation. However, we recall that the northward propagation is rather intermittent within a summer season and varies from year to year. For predictability of the ISO phases, we need to understand the cause of variability of the northward propagation of the 30–60-day mode. This is still an important outstanding problem and more theoretical and modeling work is required in this direction.

Another major advance has been made in understanding the genesis and scale selection of the 10–20-day mode. Until recently, no clear physical mechanism for the selection of the 10–20-day mode period, wavelength, and westward phase propagation was known. A unified model now explains the spatial structure (wavelength), period, and westward phase speed of both summer and winter 10–20-day oscillations or the QBM. It is demonstrated that the QBM is an $n = 1$ equatorial Rossby wave with about 6,000 km wavelength and a period of 14–16 days, that is shifted to the north (south) of the equator by about 5 degrees by the summer (winter) background mean flow (Chatterjee and Goswami, 2004). For some time, the driving mechanism for the observation of equatorial Rossby waves with a 10–20-day timescale was a puzzle, as some theoretical studies indicated that convective feedback could not make the $n = 1$ equatorial Rossby mode unstable. A recent study (Chatterjee and Goswami, 2004) shows that inclusion of a proper boundary layer (inclusion of turbulent entrainment) allowed the $n = 1$ equatorial Rossby mode to become unstable with a maximum growth rate corresponding to observed period and wavelength.

Interaction between the ocean and the atmosphere on intraseasonal timescales during the northern summer and its role in the scale selection and northward propagation of monsoon ISO are also reviewed. This is an area where our knowledge just began to grow. Two things became clear during the last couple of years. A reasonable estimate of ISV of net heat flux at the surface (made possible by the availability of reliable SST from TMI, surface winds from QuikSCAT and NOAA OLR on daily timescales) showed that ISV of heat flux is a major driving force for the ISV of SST over most of the tropical IO, although advection and entrainment play roles in the equatorial IO and the Somali Current region. It is also noted that there exists a quadrature phase relationship between northward propagation of SST and precipitation on the 30–60-day timescale and that air–sea coupling is crucial for this observed phase relationship between SST and precipitation. Thus, atmospheric ISOs seem to lead to the ISOs in SST (largely through heat fluxes) and the air–sea coupling is certainly required for the observed phase relationship between SST and precipitation. However, it is still unclear how the ISOs in SST feedback to the ISOs in convection, and modify them. It appears that the basic genesis, temporal scale selection, and northward phase propagation may arise from atmospheric internal dynamics and the air–sea coupling modify the spatio-temporal character in some way. However, quantitative contribution of air–sea coupling to the space–time spectra and the northward propagation is not well settled at this time. Much more theoretical and modeling work is required to resolve these issues.

Developing coupled ocean–atmosphere GCMs for the Asian monsoon region is a challenging task, although some initial work is being made. This is because, almost all AGCMs have large systematic errors in simulating the mean SA monsoon and most OGCMs have more than 1°C error in simulating the mean SST over the IO. These systematic errors of the component models lead to drift of the coupled model climate that may influence the quantitative estimates sought above. However, the following observation provides a silver lining. We note that the signal in SST fluctuations on the intraseasonal timescale ($\sim$1°C) is larger than that on the interannual timescale ($\sim$ 0.5°C) during northern summer over the IO. Also, the amplitude of the dominant forcing, namely the net surface heat flux, on the intraseasonal timescale ($\sim$60–80 W m^{-2}) is much larger than that on the interannual timescale ($\sim$15–20 W m^{-2}). This provides hope that modeling of air–sea interactions on the intraseasonal timescale over the Asian monsoon region may still be easier than modeling the same on the interannual time scale.

Interactions between the summer monsoon ISOs and various scales of motion are also summarized. What started in the mid-1970s (Dakshinamurthy and Keshavamurthy, 1976; Alexander *et al.*, 1978) and early 1980s (Yasunari, 1979; Sikka and Gadgil, 1980) as innocuous quasi-periodic oscillations that contribute to the active and break spells within the monsoon season, the ISOs of the SA monsoon, have emerged as a major building block of the SA monsoon itself. On the one hand, they produce space–time clustering of the synoptic disturbances (lows and depressions) and control the day-to-day fluctuations of precipitation, while on the other hand, they influence the seasonal mean and contribute significantly to the IAV of the seasonal mean precipitation. It is estimated that the modulation of the seasonal mean monsoon by slowly varying external forcing is rather weak and up to 50% of the total IAV may be contributed by internal variability arising from the monsoon ISOs. This leads to poor predictability of the seasonal mean SA monsoon, as the ISOs are primarily of internal atmospheric origin and the component of IAV of the seasonal mean contributed by them may be unpredictable. The potential predictability of the monsoon would have been enhanced if the statistics of the ISOs were also modulated by the slowly varying external forcing. Currenly available studies indicate that ISO statistics over the Asian monsoon region is only weakly modulated by slowly varying SST forcing. However, now we know that air–sea coupling is involved with the ISV of the SA monsoon. Coupled evolution of SST and circulation and precipitation on the intraseasonal timescale may introduce certain constraints on the internal variability generated by the ISOs. However, this question is just being raised and no study has addressed it so far. In the coming years, CGCMs should investigate whether predictability of the seasonal mean monsoon is enhanced by air–sea coupling of the summer monsoon ISOs. Monsoon predictability could also be influenced by interdecadal variability of the external forcing and interdecadal variability of ISO statistics. The role of ISOs on interdecadal variability of predictability of the SA monsoon needs to be studied using long observations and coupled models.

Exploiting the quasi-periodic nature of the ISOs, it has been shown (Goswami and Xavier, 2003; Webster and Hoyos, 2004) that the phases of the ISOs could be predictable up to three weeks in advance. This knowledge is likely to be put to the practical use of extended range forecasting of dry and wet spells of the monsoon and flood forecasting. Improvement in the extended range forecasting of the ISO phases, however, may come only from better understanding and simulation of the within-the-season and year-to-year variability of the northward propagating events. Fundamental work is required to advance understanding of this aspect of the ISOs.

For better long-range prediction of the seasonal mean as well as for better extended range prediction of the ISOs themselves, it is apparent that the CGCMs must simulate the climatology of the ISOs correctly. However, ISOs do influence the annual cycle and ISO activity is related to the internal variability of the seasonal mean. Waliser *et al.* (2003b) find that AGCMs with higher (lower) internal variability are associated with a stronger (weaker) annual cycle. Thus, the ISO activity may be indirectly tied up with the hydrological cycle of a model. The hydrological cycle of a model, in turn, depends on various parameterizations of the model, such as the cumulus scheme, the land surface processes, etc. The model-to-model variability in simulation of the statistics of the ISOs may be related to the differences in these parameterization schemes. Correct simulation of the climatology of the observed ISOs, therefore, remains a challenging task and continued focussed efforts must be made for improving the summer ISO simulations in GCMs.

2.8 ACKNOWLEDGMENTS

This work is partially supported by the Department of Ocean Development, Government of India and Indian National Centre for Ocean Information Services (INCOIS), Hyderabad. I thank D. Sengupta for his comments on the draft of this manuscript. I am grateful to Duane, Bill, and an anonymous reviewer for detailed and constructive comments that significantly improved the presentation of the chapter. Some of the results presented in the chapter grew out of work done in collaboration with my colleague D. Sengupta and students R. S. Ajaya Mohan, Retish Senan, and Prince Xavier. I am thankful to Prince Xavier and R. Vinay for help in preparing the manuscript.

2.9 APPENDIX

Several data sets have been utilized in preparing the figures presented in the article. Primary among them are, the daily circulation data from NCEP–NCAR Reanalysis (Kalnay *et al.*, 1996; Kistler *et al.*, 2001), daily interpolated outgoing long-wave radiation (OLR) data (Liebmann and Smith, 1996), and pentad precipitation estimates from CMAP (Xie and Arkin, 1996) produced by merging of raingauge data over land and five different satellite estimates of precipitation. The pentad CMAP data was linearly interpolated to daily values. We have also used gridded precipitation over the Indian continent for 20 years (1971–1990) originally compiled by Singh *et al.* (1992) based on 306 raingauge stations well distributed within the continent. The NCEP–NCAR Reanalysis, the interpolated OLR, as well as CMAP precipitation are available at $2.5^\circ \times 2.5^\circ$ horizontal resolution. Daily anomalies are constructed as deviations of the daily observations from an annual cycle defined as a sum of the annual mean and first three harmonics. To extract band-pass filtered data we have generally used a Lanczos filter (Duchon, 1979). The surface winds and SST obtained from Microwave Imager on board the TRMM satellite (Wentz *et al.*, 2000) and are not affected by clouds, aerosols, and atmospheric water vapor. A three-day running mean provided data at $0.25^\circ \times 0.25^\circ$ horizontal resolution.

2.10 REFERENCES

Abrol, Y. P. and S. Gadgil (eds) (1999) *Rice in a Variable Climate*. APC Publications Pvt. Ltd., New Delhi, 243 pp.

Ajaya Mohan, R. S., and B. N. Goswami (2003) Potential predictability of the Asian summer monsoon on monthly and seasonal time scales. *Met. Atmos. Phys.*, doi: 10.1007/s00703-002-0576-4.

Alexander, G., R. Keshavamurty, U. De, R. Chellapa, S. Das, and P. Pillai (1978) Fluctuations of monsoon activity. *J. Met. Hydrol. Geophys.*, **29**, 76–87.

Ananthakrishnan, R. and M. Soman (1988a) The onset of the south-west monsoon over Kerala: 1901–1980. *J. Climatol.*, **8**, 283–296.

Ananthakrishnan, R. and M. Soman (1988b) Onset dates of the south-west monsoon over Kerala for the period 1870–1900. *Int. J. Climatol.*, **9**, 321–322.

Ananthakrishnan, R., U. R. Acharya, and A. R. R. Krishnan (1967) On the Criteria for Declaring the Onset of the Southwest Monsoon over Kerala (forecasting manual, FMU Rep. No. IV–18.1). India Meteorological Department, Pune, India, 52 pp.

Annamalai, H., and J. M. Slingo (2001) Active/break cycles: Diagnosis of the intraseasonal variability of the Asian summer monsoon. *Climate Dyn.*, **18**, 85–102.

Bhat, G., S. Gadgil, S. Kumar, P. V. Hareesh, Kalsi, S. R., Madhusoodanan, P. Murty, V. S. N. Prasada Rao, C. V. K. Babu, and V. Ramesh Rao (2001) BOBMEX – The Bay of Bengal monsoon experiment. *Bull. Amer. Meteor. Soc.*, **82**, 2217–2243.

Blanford, H. F. (1884) On the connection of the Himalaya snowfall with dry winds and seasons of drought in India. *Proc. Roy. Soc. London*, **37**, 3–22.

Blanford, H. F. (1886) Rainfall of India. *Mem. India Meteorol. Dep.*, **2**, 217–448.

Charney, J. G. and J. Shukla (1981) Predictability of monsoons. In: J. Lighthill and R. P. Pearce (eds), *Monsoon Dynamics*. Cambridge University Press, Cambridge, UK, pp. 99–108.

Chatterjee, P. and B. N. Goswami (2004) Structure, genesis and scale selection of the tropical quasi-biweekly mode. *Quart. J. Roy. Meteor. Soc.*, **130**, 1171–1194.

Chen, T.-C., and J.-M. Chen (1993) The 10–20-day mode of the 1979 indian monsoon: Its relation with the time variation of monsoon rainfall. *Mon. Wea. Rev.*, **121**, 2465–2482.

Dakshinamurthy, J. and R. N. Keshavamurthy (1976) On oscillations of period around one month in the Indian summer monsoon. *Ind. J. Meteorol. Geophys.*, **27**, 201–203.

Delsole, T. and J. Shukla (2002) Linear prediction of Indian monsoon rainfall. *J. Climate*, **15**(24), 3645–3658.

Duchon, C. (1979) Lanczos filtering in one and two dimensions. *J. Appl. Meteor.*, **18**, 1016–1022.

Fennessy, M. and J. Shukla (1994) Simulation and predictability of monsoons. *Proceedings of the International Conference on Monsoon Variability and Prediction, WMO/TD619, Trieste, pp. 567–575.*

Ferranti, L., J. M. Slingo, T. N. Palmer, and B. J. Hoskins (1997) Relations between interannual and intraseasonal monsoon variability as diagnosed from AMIP integrations. *Quart. J. Roy. Meteor. Soc.*, **123**, 1323–1357.

Flatau, M., P. Flatau, and D. Rudnick (2001) The dynamics of double monsoon onsets. *J. Climate*, **14**, 4130–4146.

Flatau, M., P. Flatau, J. Schmidt, and G. Kiladis (2003) Delayed onset of the 2002 Indian monsoon. *Geophys. Res. Lett.*, **30**(14), 1768, doi:10.1029/2003GL017,434.

Fu, X., B. Wang, and T. Li (2002) Impacts of air–sea coupling on the simulation of mean Asian summer monsoon in the ECHAM4 model. *Mon. Wea. Rev.*, **130**, 2889–2904.

Fu, X., B. Wang, T. Li, and J. McCreary (2003) Coupling between northward propagating, intraseasonal oscillations and sea-surface temperature in the Indian ocean. *J. Atmos. Sci.*, **60**(15), 1755–1753.

Fukutomi, Y. and T. Yasunari (1999) 10–25 day intraseasonal variations of convection and circulation over east Asia and western north Pacific during early summer. *J. Meteor. Soc. Jap.*, **77**(3), 753–769.

Gadgil, S. (1995) Climate change and agriculture: An Indian perspective. *Curr. Sci.*, **69**, 649–659.

Gadgil, S. (2003) The Indian monsoon and its variability. *Annu. Rev. Earth Planet. Sci.*, **31**, 429–467.

Gadgil, S. and P. R. S. Rao (2000) Famine strategies for a variable climate – A challenge. *Curr. Sci.*, **78**, 1203–1215.

Gadgil, S. and S. Sajani (1998) Monsoon precipitation in the AMIP runs. *Climate Dyn.*, **14**, 659–689.

Gadgil, S. and J. Srinivasan (1990) Low frequency variation of tropical convergence zone. *Met. Atmos. Phys.*, **44**, 119–132.

Gill, A. E. (1980) Some simple solutions for heat-induced tropical circulation. *Quart. J. Roy. Meteor. Soc.*, **106**, 447–462.

Gill, A. E. (1982) *Atmosphere–Ocean Dynamics* (vol. 30 of *International Geophysics Series*). Academic Press, San Diego, 666 pp.

Goswami, B. N. (1994) Dynamical predictability of seasonal monsoon rainfall: Problems and prospects. *Proc. Ind. Nat. Acad. Sci.*, **60A**, 101–120.

Goswami, B. N. (1998) Interannual variations of Indian summer monsoon in a GCM: External conditions versus internal feedbacks. *J. Climate*, **11**, 501–522.

Goswami, B. N. and R. S. Ajaya Mohan (2001) Intraseasonal oscillations and interannual variability of the Indian summer monsoon. *J. Climate*, **14**, 1180–1198.

Goswami, B. N. and J. Shukla (1984) Quasi-periodic oscillations in a symmetric general circulation model. *J. Atmos. Sci.*, **41**, 20–37.

Goswami, B. N. and P. Xavier (2003) Potential predictability and extended range prediction of Indian summer monsoon breaks. *Geophys. Res. Lett.*, **30**(18), 1966, doi:10.1029/2003GL017,810.

Goswami, B. N., J. Shukla, E. Schneider, and Y. Sud (1984) Study of the dynamics of the intertropical convergence zone with a symmetric version of the GLAS climate model. *J. Atmos. Sci.*, **41**, 5–19.

Goswami, B. N., D. Sengupta, and G. Sureshkumar (1998) Intraseasonal oscillations and interannual variability of surface winds over the Indian monsoon region. *Proc. Ind. Acad. Sci. (Earth and Planetary Sciences)*, **107**, 45–64.

Goswami, B. N., R. S. Ajaya Mohan, P. K. Xavier, and D. Sengupta (2003) Clustering of low pressure systems during the Indian summer monsoon by intraseasonal oscillations. *Geophys. Res. Lett.*, **30**, 8, doi:10.1029/2002GL016,734.

Goswami, P. and V. Mathew (1994) A mechanism of scale selection in tropical circulation at observed intraseasonal frequencies. *J. Atmos. Sci.*, **51**, 3155–3166.

Goswami, P. and Srividya (1996) A novel neural network design for long-range prediction of rainfall pattern. *Curr. Sci.*, **70**, 447–457.

Gowarikar, V., V. Thapliyal, R. P. Sarker, G. S. Mandel, and D. R. Sikka (1989) Parametric and power regression models: New approach to long range forecasting of monsoon rain in India. *Mausam*, **40**, 125–130.

Hastenrath, S. and P. Lamb (1979) *Climatic Atlas of the Indian Ocean. Part I: Surface Climate Atmosphere Circulation*. University of Wisconcin Press, Madison.

Held, I. and A. Hou (1980) Nonlinear axially symmetric circulations in a nearly inviscid atmosphere. *J. Atmos. Sci.*, **37**, 515–533.

Hendon, H. and M. Salby (1994) The life cycle of Madden–Julian Oscillation. *J. Atmos. Sci.*, **51**, 2207–2219.

Hsu, H.-H. and C.-H. Weng (2001) Northwestward propagation of the intraseasonal oscillation in the western north Pacific during the boreal summer: Structure and mechanism. *J. Climate*, **14**, 3834–3850.

Jiang, X., T. Li, and B. Wang (2004) Structures and mechanisms of the northward propagation boreal summer intraseasonal oscillation. *J. Climate*, **17**, 1022–1039.

Jones, C., L. M. V. Carvalho, R. W. Higgins, D. E. Waliser, and J. K. E. Schemm (2004) Climatology of tropical intraseasonal convective anomalies: 1979–2002. *J. Climate*, **17**, 523–539.

Joseph, P., J. Eischeid, and R. Pyle (1994) Interannual variability of the onset of the Indian summer monsoon and its association with atmospheric features, El Niño and sea surface temperature anomalies. *J. Climate*, **7**, 81–105.

Kalnay, E., M. Kanamitsu, R. Kistler, W. Collins, D. Deaven, L. Gandin, M. Iredell, S. Saha, G. White, J. Woollen, *et al.* (1996) The NCEP/NCAR 40-year reanalysis project. *Bull. Amer. Meteor. Soc.*, **77**, 437–471.

Kang, I.-S., K. Jin, K.-M. Lau, J. Shukla, V. Krishnamurthy, S. D. Schubert, D. E. Waliser, W. Stern, V. Satyan, A. Kitoh, *et al.* (2002a) Intercomparison of GCM simulated anomalies associated with the 1997–98 El Niño. *J. Climate*, **15**, 2791–2805.

Kang, I.-S., K. Jin, B. Wang, K.-M. Lau, J. Shukla, V. Krishnamurthy, S. D. Schubert, D. E. Waliser, W. Stern, V. Satyan, *et al.* (2002b) Intercomparison of the climatological variations of Asian summer monsoon precipitation simulated by 10 GCMs. *Climate Dyn.*, **19**, 383–395.

Kemball-Cook, S. R. and B. Wang (2001) Equatorial waves and air–sea interaction in the boreal summer intraseasonal oscillation. *J. Climate*, **14**, 2923–2942.

Keshavamurthy, R. N., S. V. Kasture, and V. Krishnakumar (1986) 30–50 day oscillation of the monsoon: A new theory. *Beitr. Phys. Atmo.*, **59**, 443–454.

Kistler, R., W. Collins, S. Saha, G. White, J. Woollen, E. Kalnay, M. Chelliah, W. Ebisuzaki, M. Kanamitsu, V. Kousky, *et al.* (2001) The NCEP–NCAR 50-year reanalysis: Monthly means CD-ROM and documentation. *Bull. Amer. Meteor. Soc.*, **82**, 247–267.

Knutson, T. and K. Weickmann (1987) 30–60 day atmospheric oscillations: Composite life cycles of convection and circulation anomalies. *Mon. Wea. Rev.*, **115**, 1407–1436.

Knutson, T. R., K. M. Weickmann, and J. E. Kutzbach (1986) Global scale intraseasonal oscillations of outgoing longwave radiation and 250 mb zonal wind during northern hemispheric summer. *Mon. Wea. Rev.*, **114**, 605–623.

Krishnamurthy, V. and J. Shukla (2000) Intraseasonal and interannual variability of rainfall over India. *J. Climate*, **13**, 4366–4377.

Krishnamurthy, V. and J. Shukla (2001) Observed and model simulated interannual variability of the Indian monsoon. *Mausam*, **52**, 133–150.

Krishnamurti, T. (1985) Summer monsoon experiment – A review. *Mon. Wea. Rev.*, **113**, 1590–1626.

Krishnamurti, T., P. Jayakumar, J. Sheng, N. Surgi, and A. Kumar (1985) Divergent circulations on the 30–50 day time scale. *J. Atmos. Sci.*, **42**, 364–375.

Krishnamurti, T., D. Oosterhof, and A. Mehta (1988) Air–sea interactions on the time scale of 30–50 days. *J. Atmos. Sci.*, **45**, 1304–1322.

Krishnamurti, T. N. and P. Ardunay (1980) The 10 to 20 day westward propagating mode and "breaks" in the monsoons. *Tellus*, **32**, 15–26.

Krishnamurti, T. N. and H. N. Bhalme (1976) Oscillations of monsoon system. Part I: Observational aspects. *J. Atmos. Sci.*, **45**, 1937–1954.

Krishnamurti, T. N. and D. Subrahmanyam (1982) The 30–50 day mode at 850 mb during MONEX. *J. Atmos. Sci.,* **39**, 2088–2095.

Krishnan, R., C. Zhang, and M. Sugi (2000) Dynamics of breaks in the Indian summer monsoon. *J. Atmos. Sci.*, **57**, 1354–1372.

Lau, K. and L. Peng (1990) Origin of low-frequency (intraseasonal) oscillations in the tropical atmosphere. Part III: Monsoon dynamics. *J. Atmos. Sci.*, **47**(12), 1443–1462.

Lau, K. M. and P. H. Chen (1986) Aspects of 30–50 day oscillation during summer as infered from outgoing longwave radiation. *Mon. Wea. Rev.*, **114**, 1354–1369.

Lau, K. M., X. Li, and H. T. Wu (2002) Change in the large scale circulation, cloud structure and regional water cycle associated with evolution of the south China Sea monsoon during May–June, 1998. *J. Meteor. Soc. Jap.*, **80**, 1129–1147.

Lau, N. C. (1985) Modelling the seasonal dependence of atmospheric response to observed El Niño in 1962–1976. *Mon. Wea. Rev.*, **113**, 1970–1996.

Lawrence, D. M. and P. J. Webster (2001) Interannual variations of the intraseasonal oscillation in the south Asian summer monsoon region. *J. Climate*, **14**, 2910–2922.

Li, C. and M. Yanai (1996) The onset and interannual variability of the Asian summer monsoon in relation to land–sea thermal contrast. *J. Climate*, **9**, 358–375.

Liebmann, B. and C. A. Smith (1996) Description of a complete (interpolated) out-going longwave radiation dataset. *Bull. Amer. Meteor. Soc.*, **77**, 1275–1277.

Liebmann, B., H. H. Hendon, and J. D. Glick (1994) The relationship between tropical cyclones of the western Pacific and Indian oceans and the Madden–Julian Oscillation. *J. Meteor. Soc. Jap.*, **72**, 401–412.

Madden, R. A. (1986) Seasonal variations of the 40–50 day oscillation in the tropics. *J. Atmos. Sci.*, **43**, 3138–3158.

Madden, R. A. and P. R. Julian (1994) Detection of a 40–50 day oscillation in the zonal wind in the tropical Pacific. *Mon. Wea. Rev.*, **122**, 813–837.

Maloney, E. D. and D. L. Hartmann (2000) Modulation of hurricane activity in the gulf of Mexico by the Madden–Julian Oscillation. *Science*, **287**, 2002–2004.

Matsuno, T. (1966) Quasi-gesotrophic motions in the equatorial area. *J. Meteor. Soc. Jap.*, **44**, 25–43.

Mooley, D. A. and J. Shukla (1987) *Tracks of Low Pressure Systems which Formed Over India, Adjoining Countries, the Bay of Bengal and Arabian Sea in Summer Monsoon Season during the Period 1888–1983*. Center for Ocean Land Atmosphere Studies, Calverton, MD, USA (available from J. Shukla, COLA, USA).

Murakami, M. (1976) Analysis of summer monsoon fluctuations over India. *J. Meteor. Soc. Jap.*, **54**, 15–31.

Murakami, T., T. Nakazawa, and J. He (1984) On the 40–50 day oscillation during 1979 northern hemisphere summer. Part I: Phase propagation. *J. Meteor. Soc. Jap.*, **62**, 440–468.

Murakami, T., L. X. Chen, and A. Xie (1986) Relationship among seasonal cycles, low frequency oscillations and transient disturbances as revealed from outgoing long wave radiation. *Mon. Wea. Rev.*, **114**, 1456–1465.

Nakazawa, T. (1986) Mean features of 30–60 day variations inferred from 8 year OLR data. *J. Meteor. Soc. Jap.*, **64**, 777–786.

Nanjundiah, R. S., J. Srinivasan, and S. Gadgil (1992) Intraseasonal variation of the Indian summer monsoon. Part II: Theoretical aspects. *J. Meteor. Soc. Jap.*, **70**, 529–550.

Parthasarathy, B. and D. Mooley (1978) Some features of a long homogeneous series of Indian summer monsoon rainfall. *Mon. Wea. Rev.*, **106**, 771–781.

Parthasarathy, B., A Munot, and D. Kothawale (1988) Regression model for estimation of Indian food grain production from Indian summer rainfall. *Agric. For. Meteorol.*, **42**, 167–182.

Pearce, R. P. and U. C. Mohanty (1984) Onsets of the Asian summer monsoon, 1979–1982. *J. Atmos. Sci.*, **41**, 1620–1639.

Raghavan, K. (1973) Break-monsoon over India. *Mon. Wea. Rev.*, **101**(1), 33–43.

Rajeevan, M., D. S. Pai, S. K. Dikshit, and R. R. Kelkar (2004) IMD's new operational models for long-range forecast of southwest monsoon rainfall over India and their verification for 2003. *Curr. Sci.*, **86**(3), 422–431.

Rajendran, K., A. Kitoh, and O. Arakawa (2004) Monsoon low frequency intraseasonal oscillation and ocean–atmosphere coupling over the Indian ocean. *Geophys. Res. Lett.*, **31**, doi:10.1029/2003GL019,031.

Ramage, C. S. (1971) *Monsoon Meteorology* (vol. 15 of *International Geophysics Series*). Academic Press, San Diego, 296 pp.

Ramamurthy, K. (1969) Monsoon of India: Some aspects of 'break' in the Indian South west monsoon during July and August (forecasting manual, Part IV.18.3). India Meteorological Department, New Delhi.

Ramaswamy, C. (1962) Breaks in the Indian summer monsoon as a phenomenon of interaction between the easterly and the sub-tropical westerly jet streams. *Tellus*, **XIV**, 337–349.

Rao, Y. P. (1976) *Southwest Monsoon* (meteorological monograph). India Meteorological Department, New Delhi, 366 pp.

Reynolds, R. W. and T. M. Smith (1994) Improved global sea surface temperature analyses using optimum interpolation. *J. Climate*, **7**, 929–948.

Sahai, A. K., A. M. Grimm, V. Satyan, and G. B. Pant (2003) Long-lead prediction of Indian summer monsoon rainfall from global SST evolution. *Climate Dyn.*, **20**, 855–863.

Schneider, E. and R. Lindzen (1977) Axially symmetric steady state models of the basic state of instability and climate studies. Part I: Linearized calculations. *J. Atmos. Sci.*, **34**, 263–279.

Sengupta, D. and M. Ravichandran (2001) Oscillations of Bay of Bengal sea surface temperature during the 1998 summer monsoon. *Geophys. Res. Lett.*, **28**(10), 2033–2036.

Sengupta, D., B. N. Goswami, and R. Senan (2001) Coherent intraseasonal oscillations of ocean and atmosphere during the Asian summer monsoon. *Geophys. Res. Lett.*, **28**(21), 4127–4130.

Shukla, J. (1981) Dynamical predictability of monthly means. *J. Atmos. Sci.*, **38**, 2547–2572.

Shukla, J. (1987) Interannual variability of monsoon. In: J. S. Fein and P. L. Stephens (eds), *Monsoons*. Wiley, New York, pp. 399–464.

Shukla, J. (1998) Predictability in the midst of chaos: A scientific basis for climate forecasting. *Science*, **282**, 728–731.

Sikka, D. R. (1980) Some aspects of large-scale fluctuations of summer monsoon rainfall over India in relation to fluctuations in planetary and regional scale circulation parameters. *Proc. Ind. Acad. Sci. (Earth and Planetary Sciences)*, **89**, 179–195.

Sikka, D. R. and S. Gadgil (1980) On the maximum cloud zone and the ITCZ over Indian longitude during southwest monsoon. *Mon. Wea. Rev.*, **108**, 1840–1853.

Singh, S. V., R. H. Kriplani, and D. R. Sikka (1992) Interannual variability of the Madden–Julian Oscillations in Indian summer monsoon rainfall. *J. Climate*, **5**, 973–979.

Soman, M. and K. KrishnaKumar (1993) Space–time evolution of meteorological features associated with the onset of Indian summer monsoon. *Mon. Wea. Rev.*, **121**, 1177–1194.

Sperber, K. R. and T. N. Palmer (1996) Interannual tropical rainfall variability in general circulation model simulations associated with atmospheric model intercomparison project. *J. Climate*, **9**, 2727–2750.

Sperber K. R., J. M. Slingo, and H. Annamalai (2000) Predictability and the relationship between subseasonal and interannual variability during the Asian summer monsoons. *Quart. J. Roy. Meteor. Soc.*, **126**, 2545–2574.

Sperber, K. R., C. Brankovic, T. Palmer, M. Déqué, C. S. Frederiksen, K. Puri, R. Graham, A. Kitoh, C. Kobayashi, W. Tennant, *et al.* (2001) Dynamical seasonal predictability of the Asian summer monsoon. *Mon. Wea. Rev.*, **129**, 2226–2248.

Srinivasan, J., S. Gadgil, and P. Webster (1993) Meridional propagation of large-scale monsoon convective zones. *Met. Atmos. Phys.*, **52**, 15–35.

Straub, K. and G. Kiladis (2003) Interactions between the boreal summer intraseasonal oscillations and higher-frequency tropical wave activity. *Mon. Wea. Rev.*, **131**, 945–960.

Tomas, R. and P. Webster (1997) The role of inertial instability in determining the location and strength of near-equatorial convection. *Quart. J. Roy. Meteor. Soc.*, **123**(541), 1445–1482.

Vecchi, G. and D. E. Harrison (2002) Monsoon breaks and subseasonal sea surface temperature variability in the Bay of Bengal. *J. Climate*, **15**, 1485–1493.

Waliser, D. E. and C. Gautier (1993) A satellite-derived climatology of the ITCZ. *J. Climate*, **6**, 2162–2174.

Waliser, D. E., K. Lau, W. Stern, and C. Jones (2003a) Potential predictability of the Madden–Julian Oscillation. *Bull. Amer. Meteor. Soc.*, **84**, 33–50.

Waliser, D. E., R. Murtugudde, and L. Lucas (2004) Indo-Pacific ocean response to atmospheric intraseasonal variability. Part II: Boreal summer and the intraseasonal oscillation. *J. Geophys. Res.*, in press.

Waliser, D. E., K. Jin, I.-S. Kang, W. F. Stern, S. D. Schubert, M. L. C. Wu, K.-M. Lau, M.-I. Lee, V. Krishnamurthy, A. Kitoh, *et al.* (2003b) AGCM simulations of intraseasonal variability associated with the Asian summer monsoon. *Climate Dyn.*, doi:10.1007/s00382-003-0337-1.

Walker, G. T. (1923) Correlation in seasonal variations of weather, viii, a preliminary study of world weather. *Mem. Indian Meteorol. Dept.*, **24**, 75–131.

Walker, G. T. (1924) Correlation in seasonal variations of weather, iv, a further study of world weather. *Mem. Indian Meteorol. Dept.*, **24**, 275–332.

Wang, B. and H. Rui (1990) Synoptic climatology of transient tropical intraseasonal convection anomalies: 1975–1985. *Meteorol. Atmos. Phys.*, **44**, 43–61.

Wang, B. and X. Xie (1997) A model for the boreal summer intraseasonal oscillation. *J. Atmos. Sci.*, **54**(1), 71–86.

Wang, B., I.-S. Kang, and J.-Y. Lee (2004) Ensemble simulations of Asian–Australian monsoon variability by 11 AGCMs. *J. Climate*, **17**, 699–710.

Webster, P., E. F. Bradley, C. W. Fairall, J. S. Godfrey, P. Hacker, R. A. Houze Jr., R. Lukas, Y. Serra, J. M. Hummon, T. D. M. Lawrence, *et al.* (2002) The JASMINE pilot study. *Bull. Amer. Meteor. Soc.*, **83**(11), 1603–1630.

Webster, P. J. (1983) Mechanism of monsoon low-frequency variability: Surface hydrological effects. *J. Atmos. Sci.*, **40**, 2110–2124.

Webster, P. J., V. O. Magana, T. N. Palmer, J. Shukla, R. T. Tomas, M. Yanai, and T. Yasunari (1998) Monsoons: Processes, predictability and the prospects of prediction. *J. Geophys. Res.*, **103**(C7), 14451–14510.

Wentz, F., C. Gentemann, D. Smith, and D. Chelton (2000) Satellite measurements of sea surface temperature through clouds. *Science*, **288**, 847–850.

Wu, M., S. Schubert, I.-S. Kang, and D. Waliser (2002) Forced and free intraseasonal variability over the south Asian monsoon region simulated by 10 AGCMs. *J. Climate*, **15**(20), 2862–2880.

Xie, P. and P. A. Arkin (1996) Analyses of global monthly precipitation using gauge observations, satellite estimates and numerical predictions. *J. Climate*, **9**, 840–858.

Yanai, M. and T. Tomita (1998) Seasonal and interannual variability of atmospheric heat sources and moisture sinks as determined from NCEP–NCAR reanalysis. *J. Climate*, **11**, 463–482.

Yasunari, T. (1979) Cloudiness fluctuation associated with the northern hemisphere summer monsoon. *J. Meteor. Soc. Jap.*, **57**, 227–242.

Yasunari, T. (1980) A quasi-stationary appearance of 30–40 day period in the cloudiness fluctuation during summer monsoon over India. *J. Meteor. Soc. Jap.*, **58**, 225–229.

Yasunari, T. (1981) Structure of an Indian summer monsoon system with around 40-day period. *J. Meteor. Soc. Jap.*, **59**, 336–354.

Zheng, Y., D. E. Waliser, W. Stern, and C. Jones (2004) The role of coupled sea surface temperatures in the simulation of the tropical intraseasonal oscillation. *J. Climate*, in press.

3

East Asian monsoon

Huang-Hsiung Hsu

3.1 INTRODUCTION

The intraseasonal oscillation (ISO) is one of the major systems affecting the summer monsoon system in East Asia and the western North Pacific (EA/WNP). This has become known to the scientific community since the late 1970s and early 1980s. Studies (e.g., Murakami, 1976; Krishnamurti and Bhalme, 1976; Yasunari, 1979; Krishnamurti and Subrahmanyan, 1982) reported the prominent northward ISO propagation at both 10–20-day and 30–60-day periods in the Asian summer monsoon region. The passage of these intraseasonal fluctuations tended to be in phase with the onsets[1] (i.e., beginning of wet phases) and breaks (i.e., beginning of dry phases) of the Indian summer monsoon. It was noted that northward movement also tended to occur simultaneously in EA/WNP (e.g., Yasunari, 1979). Other intraseasonal features in EA/WNP were also documented. For example, Murakami (1980) found 20–30-day perturbations propagating westward along 10°N–20°N and northward over the South China Sea.

Studies on the EA/WNP summer intraseasonal variation (ISV) flourished after the First GARP Global Experiment (FGGE). The summer monsoon experiment provided a data set that for the first time documented the entire Asian summer monsoon in detail. The completeness of this data set spawned a great number of studies on the summer ISO not only in South Asia but also in EA/WNP. These studies (e.g., Lorenc, 1984; Murakami, 1984; Murakami *et al.*, 1984a, b; Nakazawa, 1986; Ninomiya and Muraki, 1986; Chen, 1987; Chen and Murakami, 1988; Hirasawa and Yasunari, 1990) provided us with a basic understanding of the summer EA/WNP ISV and laid the foundation for further exploration on this subject. Interestingly and incidentally, the ISV in the summer FGGE year (i.e.,

[1] The summer monsoon is characterized by rainy and dry periods that often occur intermittently. "Onset" ("active") and "break" are often used to refer to the beginning of the rainy and dry periods.

W. K. M. Lau and D. E. Waliser (eds), *Intraseasonal Variability in the Atmosphere–Ocean Climate System.*

summer of 1979) happened to be one of the most pronounced in recent history. Although the FGGE experience might not yield a general understanding of the EA/WNP ISV, it certainly paved the way for many important findings and stimulated ideas over the past 2–3 decades. One of the intriguing characteristics of the EA/WNP ISV is its close relationship with the monsoon system. As will be discussed below, the ISO has been reported to fluctuate concurrently with the monsoon onset and withdrawal in this region. This concurrence is most evident in the boreal summer. It has been obvious for many years that the ISV in this region cannot be examined and studied independently of the monsoon system. The complex land–sea contrast and topography create a complicated monsoon system in EA/WNP that evolves through several stages in the summer. The monsoon seasonal evolution, in which the ISO embeds, inevitably affects the ISO spatial distribution and temporal evolution. This in-phase relationship with the annual cycle leads to a seasonally and regionally prominent ISV.

The East Asian summer monsoon is a system involving both tropical and extra-tropical fluctuations, a unique characteristic that can not be found in other monsoon areas. Tropical–extra-tropical interaction is an inherent property of the EA/WNP ISV. In addition to these unique features, the western North Pacific is one of the major breeding regions for tropical cyclones and typhoons. These intraseasonal fluctuations, associated with the strong variation in convective activity and large-scale circulation, are found to modulate the typhoon activity in the western North Pacific. The EA/WNP ISV is under the influence of the tropical intraseasonal oscillation (TISO), which propagates eastward from the Indian Ocean to the western Pacific (e.g., Lorenc, 1984, Lau and Chan, 1986, Chen *et al.*, 1988). However, it is also likely that some of this variability is inherently regional and independent of the TISO (Wang and Rui, 1990).

This review summarizes the major characteristics of the summertime ISV in the EA/WNP monsoon system. The content includes the general characteristics of the monsoon system; the seasonality, periodicity, and regionality in the ISV; the close relationship with the summer monsoon system, the spatial structure and temporal evolution of the ISO and its modulating effect on the typhoon activity.

3.2 GENERAL CHARACTERISTICS OF THE EA/WNP MONSOON FLOW

Before the ISV discussion, it is essential to understand the general characteristics of the background monsoon flow in which the intraseasonal fluctuations exhibit prominent seasonality and regionality. The Asian monsoon, which covers South Asia, South-east Asia, tropical and extra-tropical East Asia, and the western North Pacific, is the largest monsoon system on Earth. Although this system is basically a tropical system in South and South-east Asia, it extends well into extra-tropical East Asia. In other words, the monsoon system exhibits both tropical and extra-tropical circulation characteristics. Because of the complex land–sea contrast and the high-rising topography, the Asian monsoon exhibits prominent regionality, especially in the summer. This regional characteristic can be divided into several subsystems according to their distinctive characteristics (e.g., Murakami and Matsumoto, 1994; Wang and LinHo, 2002). The EA/WNP monsoon is the easternmost subsystem that affects the weather and climate in East Asia and the western North Pacific.

The summer season definition used here is unconventional. The four seasons concept is based on the astronomical calendar. However, a natural season, often defined based on distinct weather/climate characteristics, does not always fall into these four categories. It is common practice in climate research to define the seasons based on natural season characteristics. Since the summer monsoon can begin as early as May and begins withdrawing southward in September, the long-term circulation and convection means averaged from May to September (MJJAS) is defined here to represent the summer mean state.

Figure 3.1(a) presents the summer precipitation[2] and the 850 hPa circulation.[3] Note that large precipitation in this region during the summer is generally collocated with deep convection. Large precipitation and deep convection are therefore used interchangeably here. In contrast to the precipitation in South Asia and the Indian Ocean, which tends to cluster in relatively limited areas (e.g., the Bay of Bengal), the precipitation in EA/WNP exhibits banded structures that are zonally elongated. Two such banded structures exist in the tropics and extra-tropics. The tropical band extends eastward from the South China Sea to the dateline and the weaker extra-tropical band extends eastward from Japan to the central North Pacific. Between these two precipitation bands, a region of suppressed precipitation exists.

The large precipitation in the South China Sea and the extra-tropical precipitation band tend to lie near a region of strong low-level south-westerly winds. For example, a south-westerly wind band extends from the Arabian Sea eastward all the way to the South China Sea and the western Philippine Sea. This south-westerly wind band is associated with the continental thermal low (i.e., the cyclonic circulation in the Asian continent) and the monsoon trough in the western North Pacific (i.e., this trough-like circulation and large precipitation region extends south-eastward from the Indo-China Peninsula to the tropical western North Pacific). Another south-westerly wind band extends from south-east China to Japan. These two south-westerly wind bands collocate with the large precipitation regions. The tropical precipitation band in the tropical western North Pacific occurs in a confluent zone where the westerly and easterly winds merge. The subtropical anticyclone occupies a vast area in the western North Pacific where the convection is inactive. The existence of this anticyclone is the main reason for the separation of the two precipitation bands. The southern precipitation band is associated with the monsoon trough variation in the western North Pacific, while the northern band is associated with the Meiyu (in China), Baiu (in Japan), and Changma (in Korea), which are the major precipitation and circulation systems embedded in the East Asian summer

[2] The pentad data of the CPC Merged Analysis of Precipitation (CMAP) from 1979–1992 were used here. The precipitation data on a $2.5° \times 2.5°$ grid are a combination of gauge precipitation, various satellite observations, and numerical model outputs (Xie and Arkin, 1997).

[3] The European Centre of Medium-Range Weather Forecast (ECMWF) reanalysis from 1979–1993 (Gibson *et al.*, 1997) is used to illustrate the circulation.

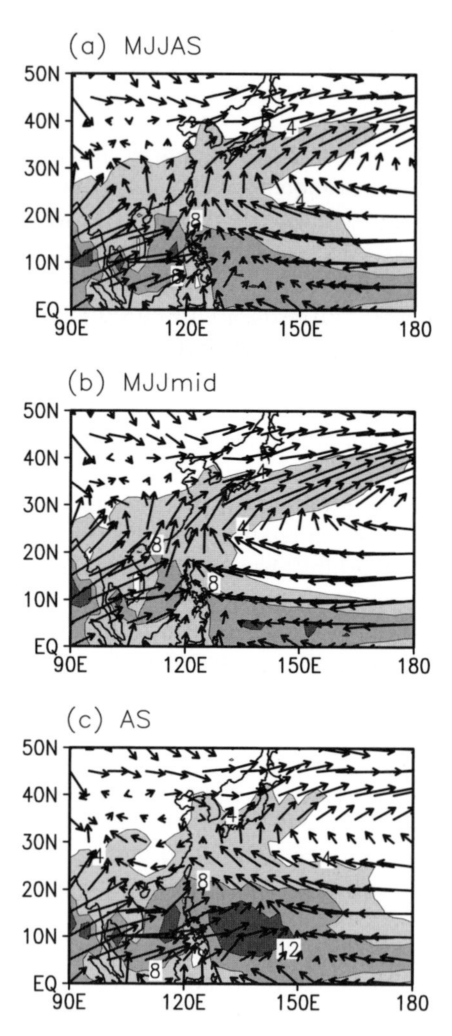

Figure 3.1. Climatological mean precipitation and 850-hPa winds during (a) May to September, (b) May to mid-July, and (c) August to September. Areas where precipitation is greater than 4 mm day^{-1} are shaded. Contour interval is 4 mm day^{-1}.

monsoon. The continental thermal low, the East Asian monsoon trough, and the subtropical Pacific anticyclone are the three major components of the EA/WNP monsoon during the northern summer. The ISV in the different regions of the EA/WNP is closely related through these circulation systems.

The EA/WNP summer monsoon, which is characterized by abrupt changes, can be divided into several stages (e.g., Matsumoto, 1992; Ueda *et al.*, 1995; Murakami and Matsumoto, 1994; Wang and Xu, 1997; Wu and Wang, 2001; Wang and LinHo, 2002). An excellent historical review on the division in natural seasons can be found in Matsumoto (1992). One of the most notable phenomena is the abrupt change in convection and circulation in late July (Ueda *et al.*, 1995). As will be discussed in the following section, this abrupt change has a significant influence on the ISV seasonality and regionality.

Figure 3.1(b, c) illustrate the contrasting circulation and convection characteristics during the first and second-half of the summer. During the first-half period (i.e., May to mid-July), the monsoon trough extends eastward only to the Philippines and the subtropical anticyclonic ridge extends westward over Taiwan and south-east China. The tropical westerly and easterly winds between the equator and 20°N merge near the Philippines where a confluent zone is located. Strong southerly and south-westerly winds are located at the western and northern flanks of the subtropical anticyclone. The south-westerly winds extending from the Indo-China Peninsula and the South China Sea reach as far north as Japan. Two notable precipitation bands are located to the south and north of the anticyclone in a strong wind region.

During the second-half period (i.e., August to September, Figure 3.1(c)), the monsoon trough penetrates as far east as 150°E while the subtropical anticyclonic ridge shifts northward to Japan. East China and Japan are under the influence of a south-easterly wind from the Pacific, in contrast to the south-westerly from the South China Sea during the first-half period. The large-scale precipitation distribution is very different from its counterpart in the first-half period. The convection in the Philippine Sea is fully developed and shifts northward to around 15°N. At the same time, the extra-tropical precipitation band appearing in the first-half of summer weakens. The precipitation characteristics shown in Figure 3.1 are consistent with the results based on an infrared equivalent blackbody temperature (e.g., Kawamura and Murakami 1995) and high cloud amount (e.g., Kang *et al.*, 1999).

3.3 PERIODICITY, SEASONALITY, AND REGIONALITY

Two frequency bands equivalent to the 30–60-day and 10–30-day periods dominate the EA/WNP ISV. These two periodicities are also characterized by strong seasonal and regional dependence. In a study on East Asian summer rainfall variability, Lau *et al.* (1988) found the coexistence of 40-day and 20-day oscillation. The 40-day oscillation occurred in the period from April to September, while the 20-day oscillation was active only from July to September. They suggested that the 40-day oscillation was associated with the TISO, while the 20-day oscillation was a local phenomenon. Tanaka (1992) found that the 20–25-day mode existed north of 13°N while the 30–60-day mode was active south of 13°N. Chen *et al.* (2000) identified the presence of two well-separated spectral peaks with periods around 30–60 days and 12–24 days in the South China Sea and the Meiyu/Baiu front (MBF) region (i.e., East China and Japan). Similar observations have been identified in many studies (e.g., Chen and Chen, 1995; Wang and Xu, 1997; Fukutomi and Yasunari, 1999; Kang *et al.*, 1999).

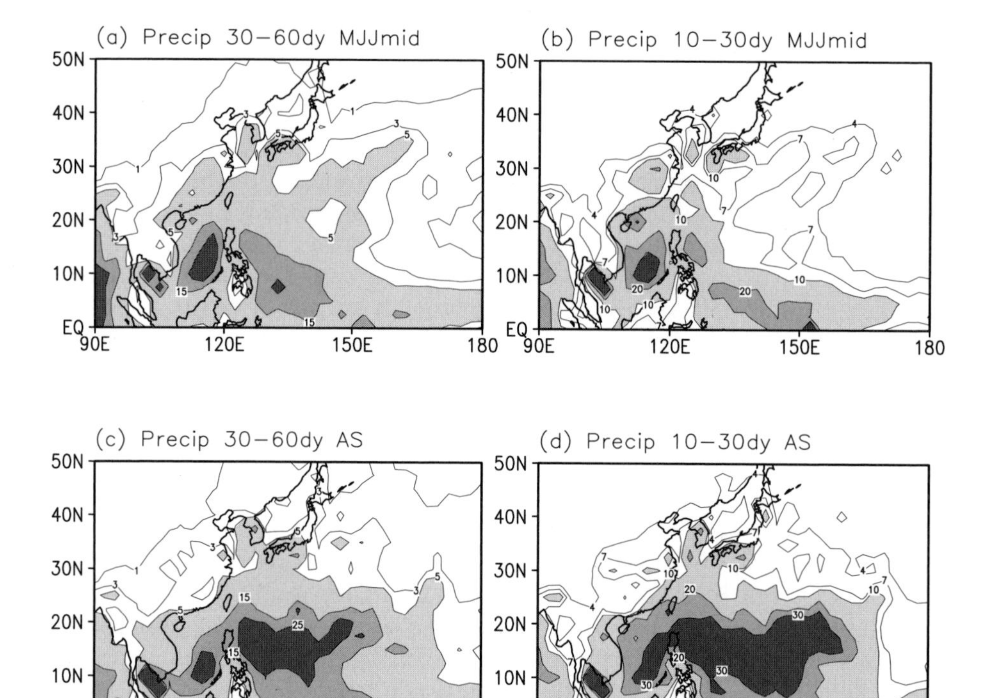

Figure 3.2. Precipitation variance for the 30–60-day (*left*) and 10–30-day (*right*) perturbations during (a, b) May to mid-July and (c, d) August to September. Contour interval is $10\,\mathrm{mm}^2\,\mathrm{day}^{-2}$ for precipitation greater than 5 (10) $\mathrm{mm}^2\,\mathrm{day}^{-2}$ for 30–60 (10–30) day perturbations. Contour lines are also drawn for smaller precipitation as indicated in the figures.

Based on the previous results, it is logical to examine the ISV in two periodicities, namely, 30–60 days and 10–30 days. A Butterworth recursive filter was used to isolate the signals in these two periodicity bands (Hamming, 1989; Kaylor, 1977). Figure 3.2 presents the 30–60-day and 10–30-day precipitation variance distributions during the first (early) and second (late) half of summer and shows that the large variability in both frequency bands is mainly restricted to south of 30°N during the entire summer. The major 30–60-day variance center in the early summer is located in the South China Sea and a secondary center is observed in the western Philippine Sea, south of 15°N (Figure 3.2(a)). Weaker variance is seen in Japan, Korea, and a band extending from Taiwan to 160°E along 25°N. These maximum variance regions are collocated with the two precipitation bands shown in Figure 3.1(b). This indicates that the precipitation fluctuates in larger amplitude in the region where the mean precipitation is large.

In the late summer (Figure 3.2(c)), the largest variability, observed in the Philippine Sea, extends eastward to 170°E. A comparison between Figure 3.2(a) and Figure 3.2(c) indicates that the major activity in the 30–60-day perturbations

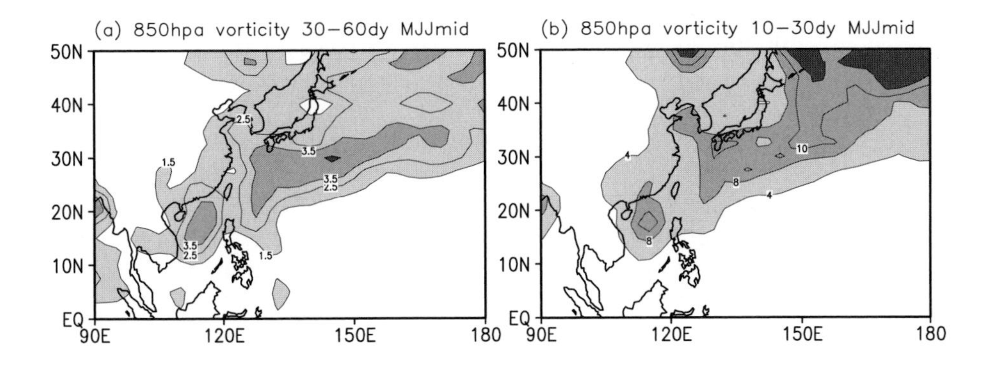

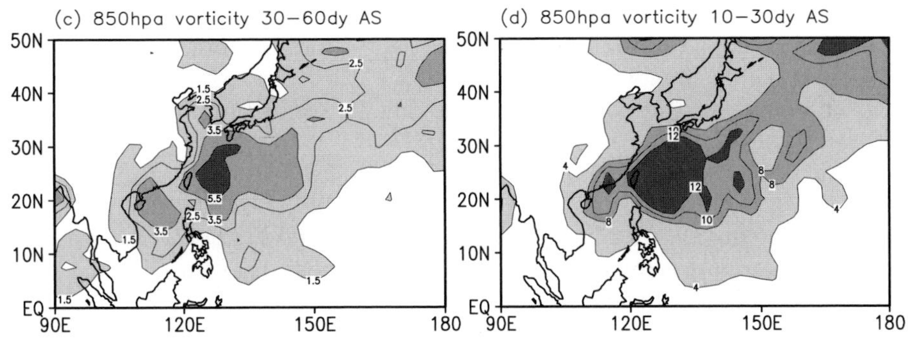

Figure 3.3. The same as in Figure 3.2 except for the 850-hPa vorticity. Contour intervals are $1 \times 10^{-10}\,s^{-2}$ and $4 \times 10^{-10}\,s^{-2}$ for the 30–60-day and 10–30-day disturbances, respectively.

shifts eastward from the Bay of Bengal and the South China Sea to the western North Pacific during the seasonal march from early to late summer. The most notable change in the western North Pacific is the northward shift of the maximum variance region from 5°N to 10°N–15°N. This change occurs concurrently with the northward shift of the tropical precipitation band and the subtropical anticyclone seen in Figure 3.1(b, c). The 10–30-day variance is about 1.5 times as large as the 30–60-day variance (Figure 3.2(b, d)). Seasonal and regional dependences similar to those seen in the 30–60-day variability are also observed in the 10–30-day variability. The contrast in the Philippine Sea between early and late summer is more prominent than that in the 30–60-day band.

The 850-hPa vorticity variance, which exhibits more pronounced extra-tropical variability, is shown in Figure 3.3 to illustrate the ISV in the low-level atmospheric circulation. During the northern summer, the 850-hPa vorticity variance in the tropics and subtropics is closely associated with the precipitation variability in both frequency bands. The major 850-hPa vorticity variance centers tend to be located to the north of the precipitation variance centers. The significant variance in the extra-tropics reflects the dominance of the extra-tropical circulation fluctuations that are less closely coupled with the deep convection.

In the early summer, a north-east–south-west oriented banded structure is observed in both the 30–60-day and 10–30-day variance in the western North Pacific, south-east of Japan (Figure 3.3(a, b)). This structure is associated with the elongated secondary maximum precipitation variance located to its south (Figure 3.2(a)). This feature reflects the strong intraseasonal activity of the quasi-stationary MBFs, which is a prominent phenomenon during the period from mid-June to mid-July in East China and Japan. Note that during this period the subtropical 850-hPa south-westerly wind prevails in the region extending from East China to the Pacific south-east of Japan (Figure 3.1(b)). This strong wind region is located near the large 850-hPa vorticity and precipitation variance regions for both 30–60-day and 10–30-day bands. Another variance maximum is located in the northern South China Sea, where the monsoon trough and the precipitation variance maximum reside (Figures 3.1(b), and Figure 3.2(a, c)). Despite the tendency for the collocation between the large 850-hPa vorticity and precipitation variance, the maximum 850-hPa vorticity variance is located mostly to the north of 20°N, while the maximum precipitation variance is located mostly to the south of 15°N. This contrast reflects the weaker coupling between the deep convection and the low-level circulation in the early summer.

In the late summer, the 30–60-day and 10–30-day vorticity variance exhibits similar spatial distributions (Figure 3.3(c, d)). The subtropical-banded structure no longer exists because of the Meiyu/Baiu withdrawal. Instead, a bullseye-shaped 850-hPa vorticity variance is found in the western North Pacific between 15°N and 35°N. This variability is associated with the large precipitation variance located to the south in a latitudinal band between 10°N and 25°N (Figure 3.2(c, d)). In contrast to the clear separation between the precipitation and 850-hPa vorticity variance maxima in the early summer, the coupling between the convection and the low-level circulation is apparently much more significant in the late summer. The phase relationship between the precipitation and vorticity variance is likely associated with the westward and north-westward propagating intraseasonal perturbations prevailing in this area (see the discussion in Sections 3.4 and 3.6). During this period, the monsoon trough and the westerly wind extend south-eastward into the Philippine Sea while the subtropical Pacific anticyclone shifts to the north with the ridge sitting around Japan. The prevailing wind in East Asia and the subtropical western North Pacific is a south-easterly wind instead of a south-westerly wind as in the early summer. The largest 850-hPa vorticity variance is embedded in the strong south-easterly wind region. These changes in ISV occur concurrently with the circulation changes between the early and late summer. It again reflects the in-phase relationship between the ISV and the monsoon seasonal evolution.

The variance distributions shown above confirm the regionality and seasonality of the EA/WNP ISV as reported in many previous studies (e.g., Lau *et al.*, 1988; Tanaka, 1992; Nakazawa, 1992; Kawamura and Murakami, 1995; Kawamura *et al.*, 1996; Wang and Xu, 1997; Wang and Wu, 1997; Fukutomi and Yasunari, 1999; Kang *et al.*, 1999). To summarize the ISV seasonal dependence, variances were computed for each calendar date (from 1 January–31 December) based on data from a certain period (or window) that centered on the corresponding calendar date. Since this is an approach similar to the way running means are calculated, the computed variance is called running variance.[4]

The longitudinal–time running variance diagrams for the precipitation averaged over 10°N–25°N in the 10–30-day and 30–60-day bands are presented in Figure 3.4(a), and Figure 3.4(b), respectively. For the 10–30-day band, the earliest activity takes place in the South China Sea (e.g., 110°E–120°E) in mid-May, followed by activity in the far western Philippine Sea (e.g., 120°E–130°E) and the west Indo-China Peninsula (e.g., 100°E) in late May, and finally the western North Pacific (e.g., 130°E–150°E) in mid-August. The most significant contrast is between the South China Sea and the western North Pacific. The 10–30-day activity in the South China Sea lasts from May to September, while its counterpart in the western North Pacific does not occur until late July and early August and lasts through November. The sudden convection flare-up in the western North Pacific is consistent with the abrupt change in the western North Pacific circulation as documented by many studies (e.g., Matsumoto, 1992; Murakami and Matsumoto, 1994; Ueda *et al.*, 1995; LinHo and Wang, 2002). This result is consistent with the findings by Hartmann *et al.* (1992). They found that the spectral peak around 20–25 days was most significant in the tropical western North Pacific and exhibited the largest amplitudes during the September–December season.

The seasonal evolution of the 30–60-day precipitation variability takes place in an order similar to its 10–30-day counterpart. However, the maximum variance in the South China Sea and the western North Pacific tends to appear earlier than the 10–30-day perturbations. The maximum variability in the western North Pacific occurs in June almost concurrently with the activity in the South China Sea. This feature is the major contrast to the 10–30-day variability, which does not become active in the western North Pacific until the late summer.

3.4 ISO PROPAGATION TENDENCY

The TISO is well known for its eastward propagation along the equator as shown in previous chapters. Similarly, the EA/WNP ISO has been known to propagate in certain directions. One of the most prominent features is the northward/northwestward propagation in the western Pacific during the boreal summer, which has been documented in many previous studies (e.g., Lau and Chan, 1986; Nitta, 1987; Chen *et al.*, 1988; Wang and Rui, 1990). These studies found that the eastward propagating TISO could trigger northward-propagating intraseasonal perturbations in the equatorial western Pacific. This time-lag relationship can also be seen clearly in Figure 4.10. However, Wang and Rui (1990) noted that many northward-propagating intraseasonal perturbations observed in the western North Pacific were spawned in

[4] Different window lengths might result in different results. Tests based on various window lengths yielded similar results except that the longer lengths yielded a smoother temporal evolution. In other words, a running variance is not particularly sensitive to the window length as long as the length is long enough compared to the corresponding intraseasonal timescales. Running variances based on 60 and 120-day window lengths are shown here for the 10–30-day and 30–60-day bands, respectively. The variance was calculated from all years and therefore contains the interannual variation of the ISV.

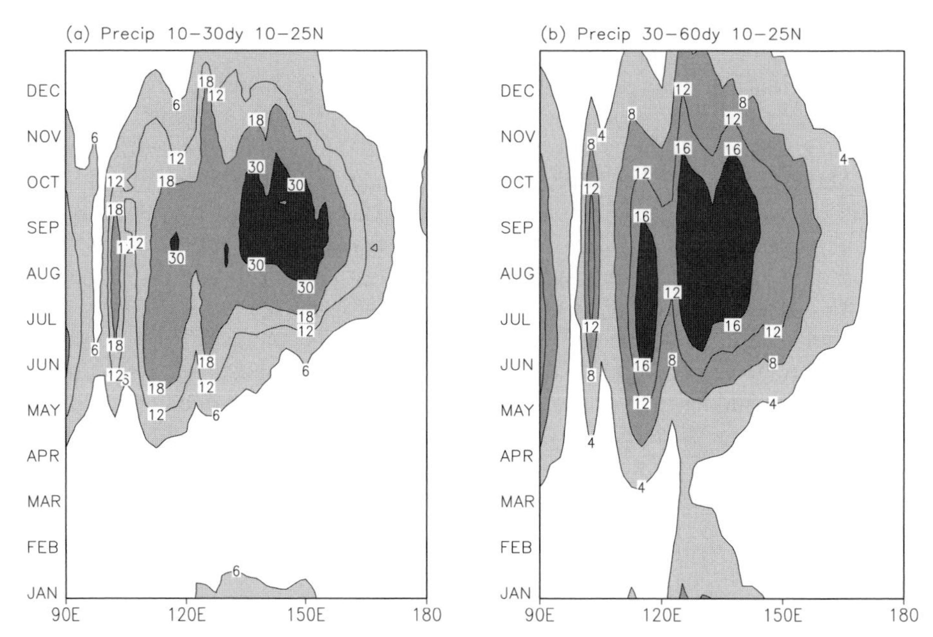

Figure 3.4. Hovmüller diagrams of running variance (see text for explanation) for the (a) 10–30-day and (b) 30–60-day precipitation perturbations averaged over 10°N–25°N. Contour intervals are $6\,mm^2\,day^{-2}$ and $4\,mm^2\,day^{-2}$ for the 10–30-day and 30–60-day perturbations, respectively.

the equatorial western Pacific and were independent of the TISO. The other prominent feature is the westward propagation from the western North Pacific to East Asia and sometimes to the Indo-China Peninsula and India (e.g., Murakami, 1980; Chen and Chen, 1995; Wang and Rui, 1990).

This propagation tendency has been explored most intensively in previous studies discussed above. Wang and Rui (1990) documented the ISO propagation paths for all seasons using a semi-subjective approach. A more objective approach taken by Jones *et al.* (2004) yielded similar results in describing the TISO eastward propagation. The other way to present the overall ISO propagation tendency is using a propagation tendency vector derived from the lagged correlation maps. This technique, used in many studies (e.g., Lau and Chan, 1986; Nitta, 1987; Hsu and Weng, 2001; LinHo and Wang, 2002; Hsu *et al.*, 2004a), provides information about the propagation tendency at every location in the most efficient way.

Previous studies reported that the 10–30-day and 30–60-day perturbations exhibit different propagation characteristics in different seasons (e.g., Wang and Xu, 1997; Fukutomi and Yasunari, 1999; LinHo and Wang, 2002). It is therefore sensible to examine the propagation tendency of both the 10–30-day and 30–60-day perturbations during the early and late summer. The propagation tendency vectors[5] for the 850-hPa vorticity anomalies shown in Figure 3.5 (see color section) were derived from the 5-day and 2-day lagged correlation maps for the 30–60-day and 10–30-day bands, respectively. In the early summer (Figure 3.5(a), see color section), the major propagating features for the 30–60-day band are the north-westward propagation in the South China Sea, the East China Sea, and the western Philippine Sea, and the northward propagation in South China. Other propagation features include the westward propagation in the eastern Philippine Sea and the southward propagation from 50°N to 20°N between 140°E and 160°E. In lagged correlation terms, the westward and north-westward propagation are the most prominent features.

In the late summer (Figure 3.5(c), see color section), the north-westward propagation, which extends well into south-east China, becomes the most dominant feature in the EA/WNP. To the east of this north-westward propagation region, a westward propagation region exists. It will be shown in Section 3.7 that the north-westward-propagating features in the Philippine Sea often originate from the westward-propagating features occurring between 150°E and the dateline. Propagation in other directions, seen in the early summer, are then no longer evident. The southward propagation (e.g., to the east of Japan) in the early summer is replaced by a northward propagation region that extends all the way to North Japan.

For the 10–30-day band during the early summer (Figure 3.5(b), see color section), westward propagation is the predominant feature in the whole domain between 5°N and 15°N. South-westward propagation is found in the region from Japan to the northern Philippine Sea, indicating an extra-tropical origin for the 10–30-day ISO. In the late summer (Figure 3.5(d), see color section), the westward propagation in the Philippine Sea seen in the early summer is replaced by north-westward propagation, while the southward propagation to the east of Japan is replaced by northward propagation. The clockwise-rotating propagation tendency vectors in the western North Pacific might be related to the frequent recurrence of the recurving tropical cyclones in the late summer (Schnadt *et al.*, 1998). The contrast between the early and late summer seen in the 30–60-day band is also evident in the 10–30-day band.

One of the major contrasts between the early and late summer is the northern limit for the northward tropical-origin ISO penetration. The northward propagation

[5] It should be noted that these tendency vectors represent only the local tendency and do not provide information about the full path along which an ISO propagates. Conversely, one can easily infer the ISO propagation in the regions where most vectors point in the same direction. This approach usually yields results similar to those obtained based on the approaches taken by Wang and Rui (1990) and Jones *et al.* (2004). To construct such a map, one must compute the lagged correlation coefficients between every grid point (i.e., base point) and all grid points in the chosen domain with the time series at the base point leading other points by 5 (2) days. One can then draw an arrow from a base point to the point where the lagged correlation coefficient is the largest. This arrow indicates the most probable path for an anomaly at the base point to propagate in the next 5 (2) days. A larger lagged correlation coefficient indicates a stronger propagation tendency. The results shown here were calculated based on the band-pass filtered data already smoothed using a T24 spectral smoothing technique (Sardeshmukh and Hoskins, 1984). The spatial smoothing yielded a more coherent propagation tendency spatial distribution.

is restricted mostly to the tropical and subtropical regions in the early summer but penetrates farther north to the extra-tropics in the late summer. Conversely, the southward propagation of the extra-tropical-origin ISO was only active in the early summer. This contrast indicates the different nature of the ISV in the early and late summer. Extra-tropical influence is still evident in the early summer, while tropical influence dominates in the late summer. All these contrasts seem to occur concurrently with the northward penetration of the East Asian summer monsoon (Tao and Chen, 1987; LinHo and Wang, 2002). It once again illustrates that the nature of the ISO is modulated by the seasonal evolution of the East Asian summer monsoon.

3.5 RELATIONSHIP WITH THE MONSOON ONSETS AND BREAKS

Prominent active and break periods, which are closely associated with the ISO, characterize the EA/WNP summer monsoon (e.g., Lau and Chan, 1986; Lau *et al.*, 1988; Tanaka, 1992; Ding, 1992; Nakazawa, 1992; Wang and Xu, 1997; Kang *et al.*, 1989, 1999; Chen *et al.*, 2000; Wu and Wang, 2001). For example, two ISO spells propagated northward ISOs in East Asia during the 1998 summer, coinciding with the Yangtze River floods in June and July (Mu and Li, 2000; Xu and Zhu, 2002; Wang *et al.*, 2003; Ding *et al.*, 2004; Hsu *et al.*, 2004b). Similar events occur from year to year.

This northward propagation was found to be often associated with the TISO. Lau and Chan (1986) identified the leading intraseasonal outgoing long-wave radiation (OLR) pattern that propagated northward in the Indian Ocean to the Indian monsoon region, north-westward from the Philippine Sea to the northern South China Sea and Taiwan, and eastward along the equator from the Indian Ocean to the maritime continent. The OLR anomaly near Borneo also propagated northward to the northern South China Sea. A similar evolution can be seen clearly in Figure 4.10. This result indicates the close relationship between the TISO and the fluctuation in the EA/WNP summer monsoon. Many studies confirmed this relationship. For example, it was suggested (e.g., Chen and Murakami, 1988; Chen *et al.*, 1988; Chen and Chen, 1993b) that the Meiyu/Baiu Convergence Zone fluctuated over an intraseasonal timescale in response to the eastward TISO propagation.

The onset of the monsoon in the South China Sea signals the beginning of the East Asian summer monsoon (Tao and Chen, 1987; Lau and Yang, 1997; Wang and LinHo, 2002). It is likely that a TISO occurring in mid-May could trigger the onset of the East Asian summer monsoon as suggested by Lau and Chan (1986). This TISO effect on the East Asian summer monsoon onset has been observed for many years. Chen and Chen (1995) studied the onset and life cycle of the South China Sea monsoon during the 1979 summer. They found that the onset (break) was triggered by the simultaneous arrival of a westward-propagating 12–24-day monsoon low (high) and a northward-propagating 30–60-day monsoon trough (ridge) in the northern section of the South China Sea. They pointed out that the northward-propagating 30–60-day monsoon perturbation was coupled with the eastward-propagating 30–60-day ISO in the tropics.

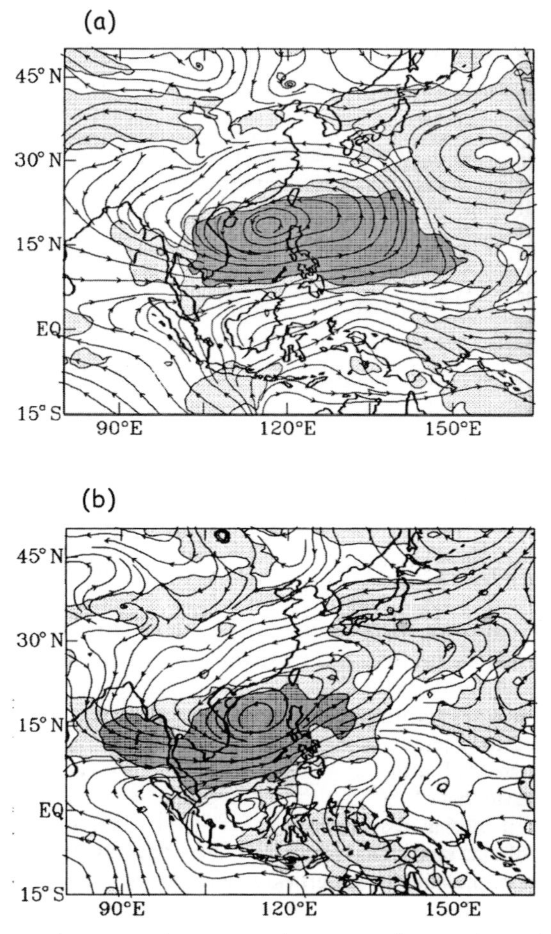

Figure 3.6. Differences between the composite streamline and equivalent black-body temperature anomaly ($\Delta T_{BB} = 270\,\text{K} - T_{BB} \geq 0$ or $= 0$ if $270\,\text{K} - T_{BB} < 0$) during the active and break phases of the westerly anomaly in the South China Sea for (a) the 30–60-day and (b) the 12–24-day mode. Values of $0 \leq \Delta T_{BB} \leq 5\,\text{K}$ and $5\,\text{K} \leq \Delta T_{BB}$ are lightly and heavily stippled, respectively. Positive ΔT_{BB} denote stronger convection.
Adapted from Chen *et al.* (2000).

Since the ISO is a large-scale feature, different parts of the monsoon flow often oscillate concurrently with one another under the influence of the ISO. One example is the interaction between the summer convection in East Asia and the South China Sea at the intraseasonal timescale. Chen *et al.* (2000) identified a coherent intraseasonal north–south oscillation existing between the MBF (representing the East Asia convection) and the Intertropical Convergence Zone (ITCZ) representing the South China Sea convection). In between the MBF and the ITCZ is a subtropical anticyclone. The convection activities at the MBF and the ITCZ tended to fluctuate out of phase. Figure 3.6(a) shows the differences between the anomalous 850-hPa streamlines and convection activity composites for the 30–60-day ISO during the strong (active) and weak (break) monsoon periods in the South China Sea. As seen, when the MBF convection was suppressed and the ITCZ convection was active, a cyclonic circulation anomaly appeared between 5°N and 30°N, signaling an enhanced monsoon trough in the region. The reversed situation, which can be inferred from Figure 3.6(a) by reversing the signs, was characterized by the enhanced MBF convection, the weakened ITCZ convection, and an anticyclonic circulation anomaly in between. This anomalous circulation corresponded to the south-westward penetration of the subtropical anticyclonic ridge or a weaker monsoon trough. Figure 3.6(b) shows a similar composite except for the 10–24-day ISO. The circulation and convection characteristics are essentially the same as in the 30–60-day counterpart, shown in Figure 3.6(a), except that the circulation centered above the South China Sea tended to tilt in the north-east–south-west direction and exhibited a larger zonal scale and a smaller meridional scale. Apparently, the oscillation in the MBF and ITCZ convection was associated with the ISO in the anticyclone at both the 30–60-day and 12–24-day timescales. It was found that the opposite phase variation between the MBF and ITCZ convection was caused by the anomalous circulation associated with the northward-propagating 30–60-day monsoon trough/ridge from the equator to 20°N and the westward-propagating 12–24-day monsoon low/high along the 15°N–20°N latitude. Similar northward-propagating 30–60-day oscillation and westward-propagating 10–30-day oscillation, clearly evident in Figures 3.2, 3.3, and 3.5, were found to be the two major intraseasonal features affecting the EA/WNP monsoon. More examples will be discussed below.

Despite the large interannual variability (IAV) of the EA/WNP summer monsoon, the climatological intraseasonal variation tends to be in-phase with the climatological seasonal evolution of the EA/WNP monsoon (i.e., long-term data averages at the same date of the year). Many studies reported this interesting characteristic (e.g., Lau. *et al.*, 1988; Kang *et al.*, 1989, 1999; Tanaka, 1992; Nakazawa 1992; Wang and Xu, 1997; Wu and Wang, 2001). Figure 3.7, adopted from Lau *et al.* (1988), presents a time–latitude diagram of the climatological 10-day mean precipitation averaged over East China (100°E–115°E). The rainy season begins in South China (25°N–30°N) around mid-May, signaling the onset of the East Asian summer monsoon. The northward propagation of the maximum precipitation region in June coincides with the onset of the Meiyu in Central China and the Baiu in Japan. The rain band continues moving to about 40°N in early July when the Meiyu suddenly ceases, indicating the Meiyu/Baiu withdrawal in Central East China and Japan. The northward propagation weakens slightly in late July and revives again in North China in early August, coinciding with the beginning of the rainy season in the region. The precipitation south of 30°N recovers around early August about 40–50 days after the first onset in June. During the period from mid-July to early September, an oscillation with a period around 20 days is the major feature observed south of 35°N, while the rainy season ends and the dry period begins in the northern region.

Many studies examined this climatological in-phase relationship. In an exam-

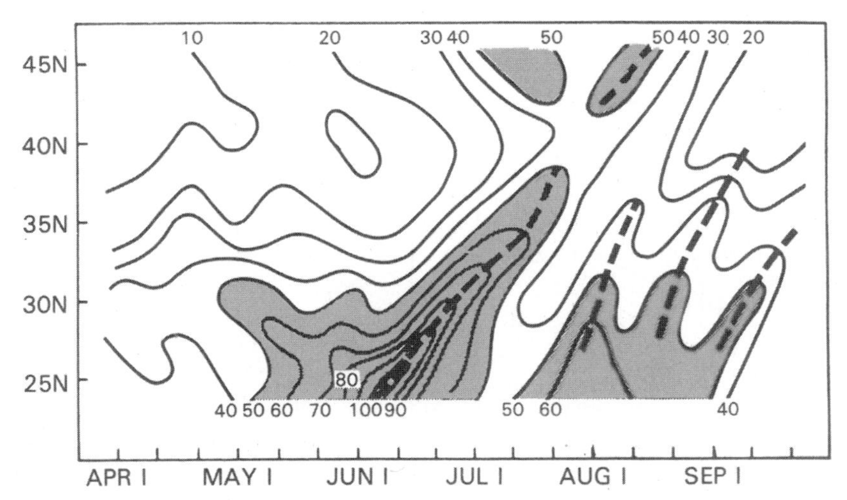

Figure 3.7. Time–longitude section of 10-day mean rainfall over East China (110°E–115°E; units are in mm). Regions of heavy rain (>50 mm) are shaded. The heavy dashed lines indicate northward propagation of the rainbands.
Adapted from Lau *et al.* (1988).

ination of the high-cloud amount fluctuations in East Asia, South-east Asia, and the western North Pacific, Tanaka (1992) found that the climatological onsets and retreats of the convection were associated with the northward-propagating 30–60-day ISO from the equator to 13°N, the westward-propagating 20–25-day ISO north of 13°N, and the seasonal evolution. After close examination, Tanaka (1992) divided the seasonal evolution of the EA/WNP monsoon into seven stages. Each stage corresponded to distinctive characteristics of the monsoon flow and convection. Wang and Xu (1997) named this climatological intraseasonal feature, embedded in the seasonal evolution of the EA/WNP monsoon, the climatological intraseasonal oscillation (CISO). To isolate CISO, Wang and Xu (1997) first computed the long-term averages of pentad-mean data (e.g., OLR) at the same pentad for every year to construct the climatological annual cycle. The first four harmonics were then removed from the time series to filter out the fluctuations with periods longer than 90 days. This procedure isolates the ISO signals with periods of 10–90 days. Kang *et al.* (1999) examined the high cloud amount data, representing deep convection, and found that the CISO explained more variance than the smoothed seasonal variation (i.e., sum of the first four harmonics) in regions of weak seasonal variation (e.g., the western Pacific between 20°N and 30°N, near Japan, and the southern part of the South China Sea).

Figure 3.8 shows some CISO features in time–latitude and time–longitude plots adopted from Wang and Xu (1997). For the OLR CISO averaged over 122.5°E–132.5°E, there were four spells of northward propagation from 10°S to 20°N during the May–October period (Figure 3.8(a)). The northward propagation was most active from May to July and became less organized in mid-July. During August

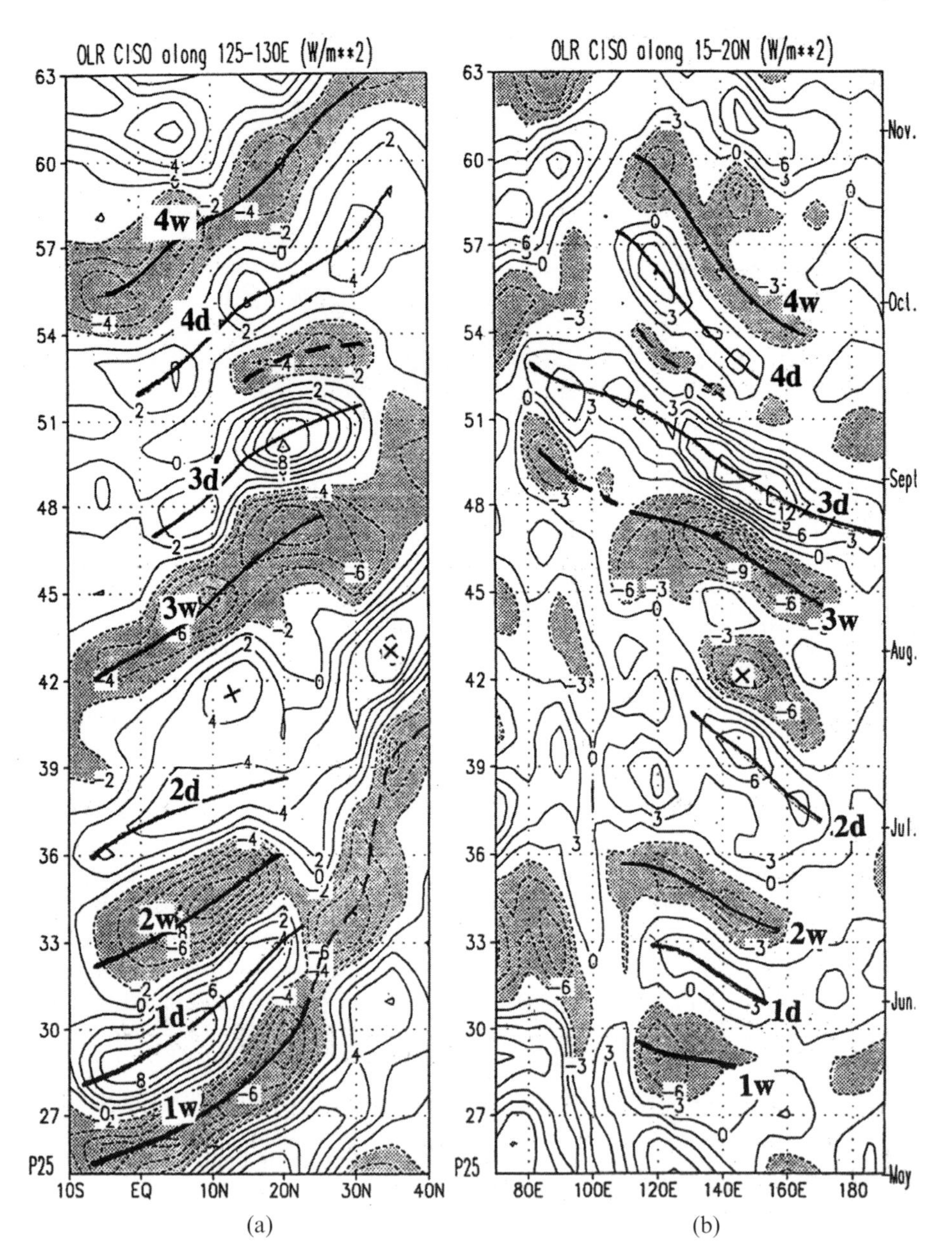

Figure 3.8. Hovmüller diagrams of the OLR CISO averaged over (a) 122.5°N–132.5°N and (b) 12.5°N–22.5°N. Four major propagation episodes are indicated by heavy lines and labels. The labeled numbers indicate the sequence of events and the "d" and "w" denote the dry and wet phases, respectively. The numbers labeled on the left-hand side of each panel denote the pentad number (e.g., P25 denotes the 25th pentad (1–5 May) of the year). Note that one year is divided into 72 pentads.
Adapted from Wang and Xu (1997).

and September, the northward propagation occurred mostly in the subtropics. These propagation events tended to occur for every 30–40 days. For the OLR CISO, averaged over 12.5°N–22.5°N, four spells of westward propagation associated with the northward propagation were observed (Figure 3.8(b)). In contrast to the northward propagation, the westward propagation was more prominent in the late summer than in the early summer. The OLR CISO in August and September propagated westward all the way to the Bay of Bengal, while the propagation in the early summer occurred mostly to the east of 120°E. The contrasting propagation characteristics between the early and late summer are associated with the abrupt change occurring in late July.

The seven stages and four cycles classified by Tanaka (1992) and Wang and Xu (1997), respectively, are similar although the data type and length are different. The corresponding monsoon characteristics in the four cycles (marked by "w" and "d" in Figure 3.8) defined by Wang and Xu (1997) are described as follows. The peak wet phase of the first CISO cycle occurring between 16–20 May reflects the onset of the South China Sea–Philippine summer monsoon. The following dry phase between 26 May–4 June corresponds to the pre-monsoon dry period in the Indian summer monsoon, the western North Pacific summer monsoon, and the Meiyu/Baiu region. The peak wet and dry phases of cycle II, occurring between 15–19 June and 10–14 July respectively, coincide with the simultaneous monsoon onsets and breaks in the above three regions. The extremely wet phase of cycle III between 14–18 August marks the peak of the western North Pacific monsoon, while the dry phase of cycle III is characterized by the prominent westward propagation shown in Figure 3.7, which coincides with the second break in the western North Pacific monsoon and the Indian summer monsoon. The wet phase of cycle IV in mid-October is associated with the last active monsoon in the western North Pacific and terminates the Indian summer monsoon.

While the ISO and the monsoon seasonal evolution tend to synchronize, the two do not always evolve in phase. The tempo between the two could vary from year to year. This means that one needs to understand how much ISV is explained by the CISO. The CISO variance distribution for the precipitation is shown in Figure 3.9(a). It exhibits maximum variance in the South China Sea, the western North Pacific around 20°N, and near Japan. It is interesting to note that the CISO variance tends to be larger over the ocean than the land. Kang *et al.* (1999) showed a similar spatial distribution using the high-cloud amount. Both results are derived from data representing convection, which is generally larger over the ocean in this area, therefore emphasize the variation over the ocean.

For comparison, the intraseasonal variation (20–100-day ISO) was extracted from the CMAP data for every summer. The ratio between the ISO variance and the total variance is shown in Figure 3.9(b). The total variance was computed relative to the long-term mean without any filtering. It therefore includes the inter-annual, seasonal, and intraseasonal variability. The ISO explains more than 50% of the total variance in the South China Sea and the western North Pacific around 20°N, indicating again the importance of the ISO in the EA/WNP monsoon region. While the ISO contributes a large amount of variance, it would be interesting to

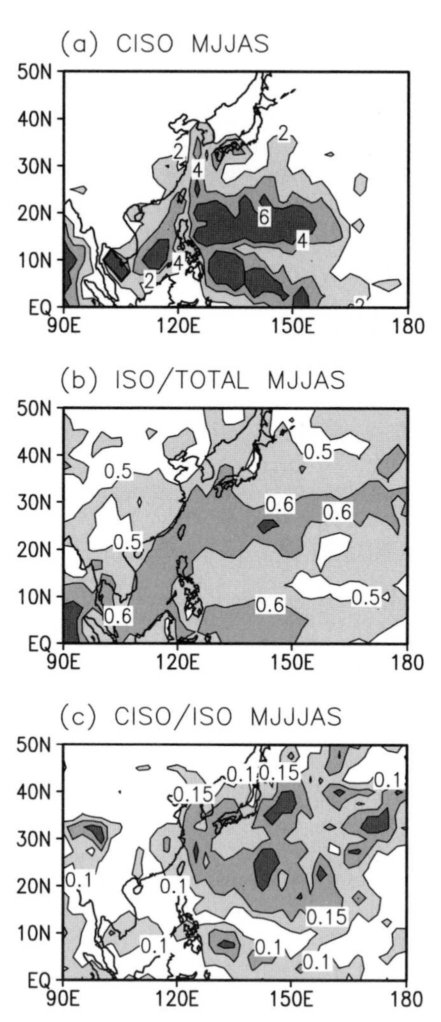

Figure 3.9. (a) Precipitation CISO variance during the May–September period, (b) ratio of precipitation ISO variance to total variance, and (c) ratio of precipitation CISO variance to ISO variance. Contour intervals are $2\,\mathrm{mm^2\,day^{-2}}$ for (a), and 0.1 and 0.05 for (b) and (c), respectively.

know how much of that portion is explained by the CISO. Figure 3.9(c), which presents the percentage of the ISO variance explained by the CISO, reveals that the CISO explains less than 20% of the ISO variance in most areas, which is equivalent to less than 10% of the total variance. The largest percentage is seen near Japan where the CISO is most active. This result indicates that the IAV of the ISO is much larger than the CISO variability. Interestingly, only a few studies on the IAV have been published so far. One of the recent studies is Teng and Wang (2003). They showed that the westward and north-westward-propagating ISO in the western North Pacific are enhanced in July–October during the developing El Niño. This enhancement is due to the increased easterly vertical shears over the tropical western Pacific, which favors the north-westward emanation of the Rossby waves from the equatorial western–central Pacific (Wang and Xie, 1996; Xie and Wang, 1996). The IAV of the ISO and its cause, which are not yet well understood, are two of the top issues for exploration in the future EA/WNP ISV studies.

The cause and effect between the CISO and the background smoothed annual cycle is an interesting and yet unsolved problem. To understand this, we have to consider the multi-scale nature of the ISO and the interaction between phenomena with different time and spatial scales. The processes involved in these multi-scale interactions are poorly understood. The following is a scenario based on intuitive thinking. It has been shown that the TISO tends to flare in favorable background conditions (e.g., high sea surface temperature (SST), abundant moisture content). During the seasonal evolution of the EA/WNP summer monsoon, the large-scale circulation and moisture distribution evolves slowly and creates a breeding ground for the ISO at different regions in different stages of this seasonal evolution. The ISO is therefore more likely to spawn in these regions at the right time of the season. The in-phase relationship between the ISO and the seasonal evolution of the EA/WNP monsoon would then occur naturally. The right conditions needed to breed the ISO and how the ISO feeds back into the background monsoon flow are other interesting topics for future study.

3.6 THE 10–30-DAY AND 30–60-DAY BOREAL SUMMER ISO

3.6.1 The 30–60-day northward–north-westward propagating pattern

The north-westward propagation in the Philippine Sea is clearly documented in the propagation tendency vectors shown in Figure 3.5 (see color section). This propagation, which occurs concurrently with the northward propagation in South Asia, as seen in Figure 4.10, often appears as one component of a large-scale see-saw pattern dominating the summertime ISV in the Asian summer monsoon region (Lau and Chan, 1986; Chen and Murakami, 1988; Zhu and Wang, 1993). The evolution of the 850-hPa vorticity and OLR anomalies associated with the north-westward propagating 30–60-day disturbances, documented by Hsu and Weng (2001), is presented in Figure 3.10 (see color section). Before the appearance of a positive vorticity anomaly in the Philippine Sea at day −5 (Figure 3.10(c)), two negative OLR anomalies (one from the west along the equator, the other one from the east in the subtropics), which are located to the east of positive vorticity anomalies, merge in the Philippine Sea (Figure 3.10(b)). This merging results in the large OLR anomaly in the Philippine Sea and enhancement of the 850-hPa vorticity anomaly, which has been propagating westward in the subtropics and is a part of the wave-like structure extending north-eastward from the South China Sea to the central North Pacific (Figure 3.10(c)). The coupled convection–circulation system then propagates north-westward toward Taiwan and south-east China and dissipates when it approaches the land area (Figure 3.10(d–e)).

Kawamura *et al.* (1996) documented the northward propagation in the 110°E–160°E longitudinal band and presented results similar to those shown in Figure 3.10 (see color section). Kawamura *et al.* (1996) summarized the circulation and convection characteristics during the period when the 30–60-day deep convection is active in the South China Sea and the Philippine Sea, equivalent to Figure 3.10(c). At this stage, both the south-westerly wind from the Indian Ocean and the south-easterly wind from the Pacific were enhanced. This indicates the enhancement of the east–west vertical circulations across the Indian Ocean and the western Pacific, corresponding to the enhanced convection in the South China Sea and the Philippine Sea. The north–south vertical circulations between 110°E and 160°E, with rising motion between 10°N and 20°N, and sinking motion to the north and south, are also enhanced.

Many mechanisms have been proposed to explain the eastward-propagating TISO. In comparison, the mechanisms responsible for the northward–north-westward propagation are poorly understood. Nitta (1987) documented a similar feature in the western North Pacific and suggested that the propagation was probably due to the advection by the south-easterly prevailing in the Philippine Sea during the summer. However, Hsu and Weng (2001) found that the pattern propagated at a speed much faster than the background wind speed. Wang and Xie (1997) simulated the north-westward propagation in a shallow water model. They suggested that the propagating disturbance was the equatorial Rossby wave breaking away from the Kelvin–Rossby wave packet, which propagated from the Indian Ocean into the Pacific and dissipated near the equatorial Central Pacific. However, an equatorial Rossby wave would not propagate north-westward unless there is a strong potential vorticity (PV) gradient in the north-east–south-west direction. Since such a PV gradient was not observed, other mechanisms have to be considered to explain the north-westward propagation of the ISO.

Hsu and Weng (2001) suggested that the frictional convergence associated with the Rossby-wave-like circulation might result in north-westward propagation of the system. During the evolution, surface friction results in frictional convergence near the center of the Rossby-wave-like cyclonic circulation, which is located to the north-west of the deep convection (e.g., day 0 in Figure 3.10(c), see color section). The anomalous south-westerly in the south-western quarter of the 850-hPa vorticity anomaly extracts surface latent heat flux from the Indian Ocean and the South China Sea and transports moisture into the center of the anomalous circulation. The anomalous moisture convergence north-west of the deep convection not only fuels the anomalous convection and circulation but also helps create a less stable condition in the lower troposphere, a precondition for further north-westward propagation. This interpretation is consistent with Kawamura *et al.* (1996) in which they concluded that the moisture convergence occurring north of the deep convection was responsible for the northward propagation.

Atmosphere–ocean interaction has been proposed as an important mechanism for the eastward-propagating TISO (e.g., Flatau *et al.*, 1997). Hsu and Weng (2001) explored the relationship between the atmospheric circulation, SST, and surface heat fluxes. They found that, although positive SST anomalies were found in the region located to the north-west of the anomalous convection, an analysis of the surface heat flux anomalies indicated less heat fluxes from the ocean surface to the atmosphere in this positive SST anomaly region. Instead, the evaporation in the oceans located to the south-west of the cyclonic circulation, where the anomalous south-westerly winds prevailed (e.g., day −5 to day 10 in Figure 3.10, see color section), was the major source of moisture, which was transported to the center of the cyclonic circulation by the south-westerly wind anomalies. This result suggests that the ocean north-west of the anomalous convection does not play an active role in destabilizing the lower troposphere by heating the lower atmosphere. Hsu and Weng (2001) concluded that the atmosphere–ocean interaction might help maintain the north-westward-propagating ISO, but the ocean played a passive role by supplying moisture in response to the atmospheric forcing.

The evolution shown in Figure 3.10 (see color section) exhibits certain characteristics that are not observed in the circulation and convection associated with the TISO (e.g., Figure 4.10). One of the most notable features is the Rossby-wave-like perturbation extending from the South China Sea to the extra-tropical EA/WNP. This suggests that the 30–60-day ISO in the EA/WNP is not entirely a response to the TISO as suggested in previous studies. Instead, it is also affected by perturbations originated in EA/WNP. The presence of the wave-like structure indicates that the tropical–extra-tropical interaction could be an important process affecting the 30–60-day ISO in the EA/WNP. Nitta (1987) noted a similar wave-like pattern, which exhibits strong variability in both intraseasonal and interannual timescales. The wave-like pattern exhibits a phase reversal between the 850-hPa and 200-hPa circulation in the South China Sea and the Philippine Sea, where the tropical heating perturbations were located, and essentially a barotropic vertical structure in the extra-tropics (Kawamura *et al.*, 1996). Because of its similarity to the theoretical results (e.g., Hoskins and Karoly, 1981), this wave pattern was interpreted as the Rossby wave dispersion forced by the northward-propagating convection in the Philippine Sea (e.g., Kawamura *et al.*, 1996; Hsu and Weng, 2001). While the Rossby-wave-like packet develops continuously downstream into the extra-tropical North Pacific, the individual cyclonic (anticyclonic) anomaly along with the active (suppressed) convection to its south-east propagates westward like a Rossby wave. When these individual circulation and convection anomalies reach the South China Sea, they trigger another wave-like packet with a reversed polarity, which again emanates into the extra-tropical North Pacific. Another round of ISO with opposite signs is then ready to repeat its predecessor's path.

Kawamura and Murakami (1995) studied the interaction between the mean summer monsoon flow and the 45-day perturbations similar to those shown in Figure 3.10 (see color section). They found that the 45-day waves in the western North Pacific amplified barotropically by weakening the sheared mean zonal and meridional flow, while the corresponding extra-tropical 45-day perturbations were maintained by the moist baroclinic instability poleward of the Pacific anticyclone. Their study indicates that the wave pattern described above contributes to the development of the extra-tropical ISO. Through this tropical–extra-tropical interaction, the 30–60-day ISO becomes one of the most prominent features that strongly interact with the EA/WNP summer monsoons.

3.6.2 The 10–30-day westward-propagating pattern

As shown in Figure 3.2(b, d), the westward propagation between 5°N and 20°N is one of the most prominent characteristics associated with the 10–30-day ISO (e.g., Nakazawa, 1986; Tanaka, 1992; Chen and Chen, 1993a, 1995; Wang and Xu, 1997; Chen and Weng, 1999; Fukutomi and Yasunari, 1999). These perturbations, which are most evident in the early summer (Figure 3.2(b)), often propagate from the tropical western North Pacific, through the South China Sea and the Indo-China Peninsula, and to the Bay of Bengal (Chen *et al.*, 2000). The 12–30-day westward-propagating ISO documented by Chen *et al.* (2000) exhibits a spatial structure similar to the one shown in Figure 3.6(b). This type of propagation exhibits the characteristics resembling those of an equatorial Rossby wave and is often accompanied by convection fluctuations at both its northern and southern sides, which tend to be out of phase with each other. When viewed in variables representing convection, two westward propagation paths, one near 15°N and the other near the equator, were observed (e.g., figure A2 in Chen *et al.*, 2000). Since the vorticity anomaly is located between the northern and southern convection anomalies, the westward propagation of the vorticity anomaly predominantly between 5°N and 15°N (Figures 3.2(b, d)) is consistent with the propagation paths of the convection anomalies.

As shown in Figure 3.2, the 10–30-day ISO exhibits a maximum variance in the South China Sea. Fukutomi and Yasunari (1999, 2002) used the 10–25-day filtered OLR averaged over 10°N–20°N and 110°E–120°E as an index for composites, and examined the spatial and temporal evolution of the corresponding 10–25-day intraseasonal perturbations during the June–August season. The result turns out to be another type of westward-propagating 10–30-day ISO, which prevails in higher latitudes around 30°N. In contrast to the westward-propagating ISO in the lower latitudes, which is essentially tropical in nature, this westward-propagating ISO exhibited both tropical and extra-tropical characteristics.

Fukutomi and Yasunari (1999, 2002) divided a complete cycle of the corresponding 10–30-day ISO evolution into 8 categories. Figure 3.11(a) shows the OLR, 850-hPa streamfunction, and wind anomalies at category 3, which correspond to the phase when the convection in the South China Sea is most active as indicated by the dark shading in the figure. The negative OLR anomaly (representing anomalous convection) in the South China Sea is accompanied by a cyclonic circulation located to the north-west. Downstream is a wave-like pattern extending eastward along the 20°N–40°N latitudinal band. The corresponding 200-hPa circulation is shown in Figure 3.11(b). The 200-hPa circulation anomalies tend to be of the same signs as their 850-hPa counterparts, indicating an equivalent barotopic vertical structure in the extra-tropics. However, the 850-hPa and 200-hPa circulation anomalies near East China and the South China Sea tend to be out-of-phase, indicating a first baroclinic mode vertical structure. During the period from

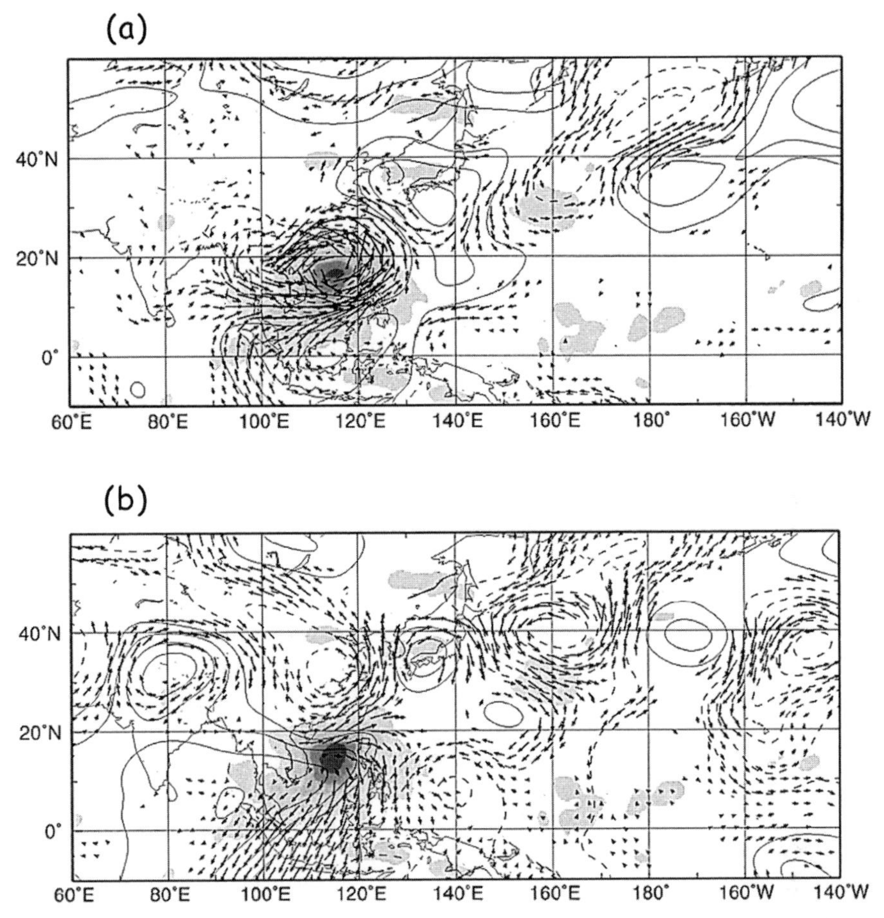

Figure 3.11. Spatial distribution of composite OLR, vector wind, and (a) 850-hPa and (b) 200-hPa streamfunction anomalies in the 10–25-day band when the convection is strongest in the South China Sea. Contour interval is $3.0 \times 10^5\,\mathrm{m^2\,s^{-1}}$. OLR anomalies less (greater) than -5 (5) $\mathrm{W\,m^{-2}}$ are darkly (lightly) shaded. Only locally significant wind vectors are shown. Adapted from Fukutomi and Yasunari (2002).

category 3 to category 6, the wave pattern moves westward, while the anomalous low-level anticyclonic circulation, originally located in the 20°N–40°N and 120°E–140°E region, moves south-westward to the South China Sea where the OLR anomaly becomes positive (not shown). This anticyclonic anomaly continues moving westward near 20°N to the Bay of Bengal at category 8 (not shown).

It is interesting to note that the anomaly circulation associated with the wave pattern switches from an equivalent barotropic vertical structure to a first baroclinic mode vertical structure when it reaches the South China Sea. Fukutomi and Yasunari (1999) suggested that the south-westward movement of the anomalous anticyclonic (cyclonic) circulation into the South China Sea initiated a convection inactive (active) state in the South China Sea, which in turn triggered the downstream Rossby-wave-like pattern. The existence of this wave-like pattern modulated the Pacific anticyclone, the monsoon trough, and the convection. They suggested that the mutual interaction between the tropics and the extra-tropics on the 10–25-day timescale played an important role in the variability of monsoon convection and circulation in EA/WNP.

Fukutomi and Yasunari (2002) found that the wave-like pattern and the tropical–extra-tropical interaction were most pronounced during the June–July period and less pronounced in August. This result is consistent with the south-westward propagation in the region between Japan and the northern Philippine Sea, which is observed in early summer but not in late summer (Figure 3.2(b, d)). The wave-like pattern occurred in the background south-westerly wind, which served as a wave guide for the downstream development of the Rossby wave. They also suggested that the south-westward propagation of the circulation anomaly into the South China Sea was similar to the retrograding Rossby wave along a westerly duct.

3.7 RELATIONSHIP WITH TROPICAL CYCLONE FORMATION

The EA/WNP monsoon trough extends south-eastward into the tropical western North Pacific where the SST is higher than 26°C in the boreal summer and has abundant moisture in the lower troposphere. In addition, the region occupied by the monsoon trough is a region of cyclonic relative vorticity in the lower troposphere and anticyclonic relative vorticity in the upper troposphere. The vertical shear of horizontal winds is also smaller than the surrounding regions. These characteristics are among the favorable environmental factors for tropical cyclone formation (Gray, 1968, 1998; Elsberry, 2004). Many studies confirmed that most typhoons occurred in an active monsoon trough environment (e.g., McBride, 1995; Harr and Elsberry, 1995; Elsberry, 2004). Other studies have found that the eastern end of the monsoon trough, where the tropical westerlies and easterlies meet to result in a confluent zone in the lower troposphere, is also a region favorable for tropical cyclone formation (e.g., Harr and Elsberry, 1995; Briegel and Frank, 1997). Tropical disturbances with a period of 8–9 days, originating in this confluent zone, were found to propagate north-westward in the western North Pacific (Lau and Lau, 1990; Chang *et al.*, 1996; Kuo *et al.*, 2001). These results all indicate that the monsoon trough in the western North Pacific is a breeding ground for tropical cyclones. Its fluctuation in both structure and amplitude can significantly affect the tropical cyclone formation and track. Elsberry (2004) gives an informative review on this subject.

Since the ISV in EA/WNP is closely associated with the monsoon trough fluctuation (e.g., the contraction and expansion of the trough in the east–west direction and/or the meridional shifts of the trough), it is likely that the ISO modulates the typhoon activity in the western North Pacific. Gray (1978) noted that tropical cyclogenesis tended to cluster in an active period of 1–2 weeks and was separated by a 2–3-week inactive period. Similar clustering phenomenon was also observed in the western North Pacific. In a study of the intraseasonal variations of the tropical

OLR during the FGGE year, Nakazawa (1986) found that the generation and growth of tropical cyclones tended to occur during the convection–active phase of the ISO at both the 15–25-day and 30–60-day timescales in both northern and southern hemisphere summers. Heta (1990) studied the relationship between the tropical wind and the typhoon formation during the 1980 summer (July–October) in the tropical western Pacific. In that particular year, the confluent region in the western North Pacific moved westward and eastward with a timescale of 10–30 days and most of the tropical cyclones and storms appeared when the westerly wind region expanded eastward (i.e., a strengthened monsoon trough). This finding was consistent with the above discussion on the effect of the confluent zone on the tropical cyclogenesis. When the confluent zone shifts zonally with an intraseasonal timescale, the region of tropical cyclogenesis might shift in a similar manner.

Hartmann *et al.* (1992) found that the 20–25-day OLR anomalies in the western North Pacific were most active during the September–December period. The 20–25-day interval appeared to be the most preferred recurrence frequency for the tropical cyclone in the western North Pacific, although other recurrence frequencies also existed. It was concluded that the 20–25-day oscillation played a modest role, although not a dominant role, in setting the pace of typhoon development in this region and season. Schnadt *et al.* (1998) also identified the relationship but found that the 15–25-day spectral peak did not exist in some years.

While the findings discussed above might be affected by the EA/WNP ISO that was not necessarily associated with the TISO, other studies have documented the modulating effect of the TISO on the typhoon activity in the western North Pacific. For example, Liebmann *et al.* (1994) found that tropical cyclones in the northern Indian Ocean and western Pacific tended to spawn in the wet phase of the 35–95-day TISO, which propagated eastward along the equator. The development of tropical cyclones was associated with the low-level vorticity and divergence anomalies located to the north-west of the anomalous TISO convection. The intraseasonal perturbation strongly modulated the large-scale tropical convection fluctuations to create an environment favoring the development of tropical cyclones. However, they noted that increased tropical cyclone activity also occurred in active monsoon troughs, which were not associated with the TISO.

The findings discussed above are consistent with the observations that the 3–10-day tropical disturbances activity increased during the westerly phase of the TISO (Nakazawa, 1986; Yamazaki and Murakami, 1989; Sui and Lau, 1992; Maloney and Hartmann, 2001; Straub and Kiladis, 2003; Maloney and Dickinson, 2003). Figure 3.12(a–c) presents the 2.5–12-day 850-hPa vorticity variance during the entire June–August period, the TISO westerly events, and the TISO easterly events, respectively. Among the three periods, the variance was the largest during the westerly events and smallest during the easterly events. The maximum variance during the westerly events was located in a north-west–south-east elongated region from South China to the south-eastern Philippine Sea. The monsoon trough in the Philippine Sea was deepened during the westerly phases while almost missing during the easterly phases (not shown). The energetic analysis done by Maloney and Dickinson (2003) indicated an enhanced barotropic energy conversion and an enhanced conversion

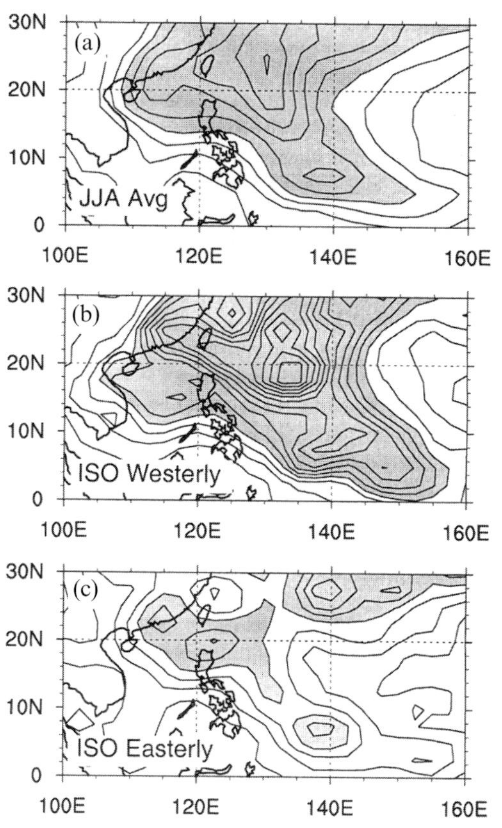

Figure 3.12. The 850-hPa perturbation vorticity variance in the 2.5–12-day band averaged during (a) June–August, (b) ISO westerly events, and (c) ISO easterly events. Contour interval is $2 \times 10^{-11}\,s^{-2}$. Values greater than $10 \times 10^{-11}\,s^{-2}$ are shaded.
Adapted from Maloney and Dickinson (2003).

from perturbation available potential energy to perturbation kinetic energy during the westerly phases. This result suggested favorable conditions for the growth of tropical disturbances during the TISO westerly events. The deepened monsoon trough during the westerly phases apparently acts as a breeding ground for tropical synoptic disturbances, which might develop into tropical cyclones.

3.8 FINAL REMARKS

This review summarized the interesting characteristics of the EA/WNP ISV during the boreal summer. We can identify more phenomena than we can explain. This simply reflects the fact that many outstanding issues remain to be solved. Some of these issues are discussed below.

3.8.1 Close association with the EA/WNP monsoon

The close relationship between the monsoon onsets/breaks and the ISO is remarkable. The good similarity in timing between the ISO and the abrupt changes in the monsoon circulation does not necessarily yield the cause and effect. The passage of an ISO certainly affects the convective activity and precipitation in various regions. However, whether it leads to the abrupt changes in circulation remains to be seen.

The ISO behaves differently during different periods of the EA/WNP summer monsoon. We need to understand how the background monsoon flow affects the ISO characteristics. The ISO propagation tendency exhibits strong geographical dependence in the northern summer: northward in the South China Sea, north-westward in the Philippine Sea, and westward in the subtropical western North Pacific. Although there are already some explanations proposed for these characteristics (e.g., mechanisms for propagation), little has been said about the reason for the geographical dependence. Background flow property, land–sea contrast, ocean–atmosphere interaction, scale interaction, etc., are among the possible mechanisms.

3.8.2 CISO vs. IAV

The CISO helps us understand the in-phase relationship between the ISO and the annual cycle. This in-phase relationship is often referred to as phase lock. However, the term "phase lock" does not necessarily warrant us a better understanding of the interaction between the ISO and the annual cycle. Is there an interaction between the ISO and the smoothed background monsoon evolution? Or, could it be simply that the slowly evolved monsoon flow sets up an environment favorable for the ISO development? Moreover, the CISO explains less than 20% of the total ISV and is prominent only in certain regions. The IAV of the ISO is much larger than the CISO variability, but little is known about the cause for this large IAV. It is probably affected by the interannual fluctuation of the background flow, which in turn is induced by the ocean–atmosphere interaction such as the El Niño Southern Oscillation (e.g., Teng and Wang, 2003), the atmosphere–land interaction in the Eurasian continent (e.g., the Tibetan Plateau heating effect, snow cover), the internal dynamics in the atmosphere, etc.

3.8.3 Multi-periodicities and multi-scale interaction

Another interesting phenomenon is the existence of two spectral peaks around 30–60 days and 10–30 days in the summer. Are these two the intrinsic modes in the monsoon background flow? What are the reasons for their prominent seasonality and regionality? The ISO variability is also closely related to the tropical storm activity in the western North Pacific. All these facts indicate that the ISO is likely one component of a multi-scale system, which involves the interannual, annual, intraseasonal, and synoptic scale variability in the EA/WNP. The multi-scale interaction between features of these timescales appears to be one of the most important issues that must be addressed in future EA/WNP ISO studies.

3.8.4 Others

One area that has been hardly touched for the EA/WNP ISO is the cloud–radiative effect. Several studies (e.g., Slingo and Madden, 1991; Hu and Randall, 1994, 1995; Mehta and Smith, 1997) reported on the possible effects of the cloud–radiative process on the TISO. Wang *et al.* (2004) recently identified the dramatic weakening of the cloud–radiative cooling in East Asia after the East Asian summer monsoon onset in June. The cloud–radiative process could also have a strong effect on the ISO. The complexity of this problem and the lack of reliable data have hindered this research for a long time. A combination of the radiation model and the observational data seems to be a promising approach to shed light on this issue and adds another dimension toward a better understanding of the EA/WNP ISV.

3.9 REFERENCES

Briegel, L. M. and W. M. Frank (1997) Large-scale influences on tropical cyclogenesis in the western North Pacific. *Mon. Wea. Rev.*, **125**, 1397–1413.

Chang, C.-P., J. M. Chen, P. A. Harr, and L. E. Carr (1996) Northwestward-propagating wave patterns over the tropical western North Pacific during summer. *Mon. Wea. Rev.*, **124**, 2245–2266.

Chen, T.-C. (1987) 30–50 day oscillation of 200-mb temperature and 850-mb height during the 1979 northern summer. *Mon. Wea. Rev.*, **115**, 1589–1605.

Chen, T.-C. and M. Murakami (1988) The 30–50-day variation of convective activity over the western Pacific Ocean with emphasis on the northwestern region. *Mon. Wea. Rev.*, **116**, 892–906.

Chen, T.-C., M.-C. Yen, and M. Murakami (1988) The water vapor transport associated with the 30–50-day oscillation over the Asian monsoon regions during 1979 summer. *Mon. Wea. Rev.*, **116**, 1983–2002.

Chen, T.-C. and J.-M. Chen (1993a) The 10–20-day mode of the 1979 Indian monsoon: Its relation with the time variation of monsoon rainfall. *Mon. Wea. Rev.*, **121**, 2465–2482.

Chen, T.-C. and J.-M. Chen (1993b) The intraseasonal oscillation of the lower-tropospheric circulation over the western Pacific during the 1979 northern summer. *J. Meteor. Soc. Jap.*, **71**, 205–220.

Chen, T.-C. and J.-M. Chen (1995) An observational study of the South China Sea monsoon during the 1979 summer: Onset and life cycle. *Mon. Wea. Rev.*, **123**, 2295–2318.

Chen, T.-C. and S.-P. Weng (1999) Interannual and intraseasonal variations in monsoon depressions and their westward-propagating predecessors. *Mon. Wea. Rev.*, **127**, 1005–1020.

Chen, T.-C., M.-C. Yen, and S.-P. Weng (2000) Interaction between the summer monsoon in East Asia and South China Sea: Intraseasonal monsoon modes. *J. Atmos. Sci.*, **57**, 1373–1392.

Ding, Y. (1992) Summer monsoon rainfalls in China. *J. Meteor. Soc. Jap.*, **70**, 373–396.

Ding, Y., C. Li, and Y. Liu (2004) Overview of the South China Sea Monsoon Experiment (SCSMEX). *Adv. Atmos. Sci.*, in press.

Elsberry, R. L. (2004) Monsoon-related tropical cyclones in East Asia. In: C.-P. Chang (ed), *East Asian Monsoon* (World Scientific Series of East Asia, Vol 2). World Scientific, Singapore, 560pp.

Flatau, M., P. Flatau, P. Phoebus, and P. Niller (1997) The feedback between equatorial convection and local radiative and evaporation processes: The implications for intraseasonal oscillation. *J. Atmos. Sci.*, **54**, 2373–2386.

Fukutomi, Y. and T. Yasunari (1999) 10–25-day intraseasonal variations of convection and circulation over East Asia and western North Pacific during early summer. *J. Meteor. Soc. Jap.*, **77**, 753–769.

Fukutomi, Y. and T. Yasunari (2002) Tropical–extratropical interaction associated with the 10–25-day oscillation over the western Pacific during the northern summer. *J. Meteor. Soc. Jap.*, **80**, 311–331.

Gibson, J. K., P. Kallberg, S. Uppala, A. Hernandez, A. Nomura, and E. Serrano (1997) ECMWF re-analysis project report series. 1: ERA description. European Centre for Medium-Range Weather Forecasts, Shinfield, Reading, U.K., 72 pp.

Gray, W. M. (1968) Global view of the origin of tropical disturbances and storms. *Mon. Wea. Rev.*, **96**, 669–700.

Gray, W. M. (1978) Hurricanes: Their formation, structure and likely role in the tropical circulation. In: D. B. Show (ed), *Meteorology over the Tropical Oceans. Roy. Meteor. Soc.*, 155–218.

Gray, W. M. (1998) The formation of tropical cyclones. *Meteor. Atmos. Phys.*, **67**, 37–69.

Hamming, R. W. (1989) Digital filters. Prentice-Hall International, Inc., New Jersey, 284 pp.

Harr, P. A. and R. L. Elsberry (1995) Large-scale circulation variability over the tropical western North Pacific. Part I: Spatial patterns and tropical cyclone characteristics. *Mon. Wea. Rev.*, **123**, 1225–1246.

Hartmann, D. L., M. L. Michelsen, and S. A. Klein (1992) Seasonal variations of tropical intraseasonal oscillations: A 20–25-day oscillation in the western Pacific. *J. Atmos. Sci.*, **49**, 1277–1289.

Heta, Y. (1990) An analysis of tropical wind fields in relation to typhoon formation over the western Pacific. *J. Meteor. Soc. Jap.*, **68**, 65–76.

Hirasawa, N. and T. Yasunari (1990) Variation in the atmospheric circulation over Asia and the western Pacific associated with the 40-day oscillation of the Indian summer monsoon. *J. Meteor. Soc. Jap.*, **68**, 129–143.

Hoskins, B. J. and D. J. Karoly (1981) The steady linear response of a spherical atmosphere to thermal and orographic forcing. *J. Atmos. Sci.*, **38**, 1179–1196.

Hsu, H.-H. and C.-H. Weng (2001) Northwestward propagation of the intraseasonal oscillation in the Western North Pacific during the boreal summer: Structure and mechanism. *J. Climate*, **14**, 3834–3850.

Hsu, H.-H., C.-H. Weng, and C. H. Wu (2004a) Contrasting characteristics between the Northward and Eastward Propagation of the Intraseasonal Oscillation during the Boreal Summer. *J. Climate*, **17**, 727–743.

Hsu, H.-H., Y.-C. Yu, W.-S. Kau, W.-R. Hsu, and W.-Y. Sun (2004b) Simulation of the 1998 East Asian summer monsoon using the Purdue regional model. *J. Meteor. Soc. Jap.*, in press.

Hsu, H.-H. and M.-Y. Lee (2004c) Topographic effects on the eastward propagation and initiation of the Madden-Julian Oscillation. *J. Climate*, in press.

Hu, Q. and D. A. Randall (1994) Low-frequency oscillations in radiative–convective systems. *J. Atmos. Sci.*, **51**, 1089–1099.

Hu, Q. and D. A. Randall (1995) Low-frequency oscillations in radiative–convective systems. Part II: An idealized model. *J. Atmos. Sci.*, **52**, 478–490.

Jones, C., L. M. V. Carvalho, R. W. Higgins, D. E. Waliser, and J.-K. E. Schemm (2004) Climatology of tropical intraseasonal connective anomalies: 1979–2002. *J. Climate*, **17**, 523–539.

Kang, I. S., S.-I. An, C.-H. Joung, S.-C. Yoon, and S.-M. Lee (1989) 30–60 day oscillation appearing in climatological variation of outgoing longwave radiation around East Asia during summer. *J. Korean Meteor. Soc.*, **25**, 149–160.

Kang, I. S., C.-H. Ho, and Y.-K. Lim (1999) Principal modes of climatological seasonal and intraseasonal variations of the Asian summer monsoon. *Mon. Wea. Rev.*, **127**, 322–340.

Kawamura, B. and T. Murakami (1995) Interaction between the mean summer monsoon flow and 45-day transient perturbations. *J. Meteor. Soc. Jap.*, **73**, 1087–1114.

Kawanura, B., T. Murakami, and B. Wang (1996) Tropical and mid-latitude 45-day perturbations over the western Pacific during the northern summer. *J. Meteor. Soc. Jap.*, **74**, 867–890.

Kaylor, R. E. (1977) Filtering and decimation of digital time series (Tech. Note BN 850). Institute of Physical Science Technology, University of Maryland, 42 pp.

Krishnamurti, T. N. and H. N. Bhalme (1976) Oscillations of a monsoon system. Part I: Observational aspects. *J. Atmos. Sci.*, **33**, 1937–1954.

Krishnamurti, T. N. and D. Subrahmanyam (1982) The 30–50 day mode at 850 mb during MONEX. *J. Atmos. Sci.*, **39**, 2088–2095.

Kuo, H.-C., J.-H. Chen, R. T. Williams, and C.-P. Chang (2001) Rossby waves in zonally opposing mean flow: Behavior in northwest Pacific summer monsoon. *J. Atmos. Sci.*, **58**, 1035–1050.

Lau, K.-H. and N.-C. Lau (1990) Observed structure and propagation characteristics of tropical summertime synoptic scale disturbances. *Mon. Wea. Rev.*, **118**, 1888–1913.

Lau, K.-M. and P. H. Chan (1986) Aspects of the 40–50 day oscillation during the northern summer as inferred from outgoing longwave radiation. *Mon. Wea. Rev.*, **114**, 1354–1367.

Lau, K.-M., G. J. Yang, and S. H. Shen (1988) Seasonal and intraseasonal climatology of summer monsoon rainfall over East Asia. *Mon. Wea. Rev.*, **116**, 18–37.

Lau, K.-M. and S. Yang (1997) Climatology and interannual variability of the southeast Asian summer monsoon. *Adv. Atmos. Sci.*, **14**, 141–162.

Liebmann, B., H. H. Hendon, and J. D. Glick (1994) The relationship between tropical cyclones of the western Pacific and Indian Ocean and the Madden–Julian Oscillation. *J. Meteor. Soc. Jap.*, **72**, 401–412.

LinHo and B. Wang (2002) The time–space structure of the Asian-Pacific summer monsoon: A fast annual cycle view. *J. Climate*, **15**, 2001–2018.

Lorenc, A. C. (1984) The evolution of planetary scale 200-mb divergences during the FGGE year. *Quart. J. Roy. Meteor. Soc.*, **110**, 427–441.

Maloney, E. D. and D. L. Hartmann (2001) The Madden–Julian Oscillation, barotropic dynamics, and North Pacific cyclone formation. Part I: Observations. *J. Atmos. Sci.*, **58**, 2545–2558.

Maloney, E. D. and M. J. Dickinson (2003) The intraseasonal oscillation and the energetics of summertime tropical western North Pacific synoptic-scale disturbances. *J. Atmos. Sci.*, **60**, 2153–2168.

Matsumoto, J. (1992) The seasonal changes in Asian and Australian monsoon regions. *J. Meteor. Soc. Jap.*, **70**, 257–273.

McBride, J. L. (1995) *Tropical Cyclone Formation* (*Chapter 3*: *Global Perspectives on Tropical Cyclones* (Tech. Doc. WMO/TD No. 693). World Meteorological Organization, Geneva, Switzerland, pp. 63–105.

Mehta, A. V. and E. A. Smith (1997) Variability of radiative cooling during the Asian summer monsoon and its influence on intraseasonal waves. *J. Atmos. Sci.*, **54**, 941–966.

Mu, M. and C. Li (2000) On the outbreak of South China Sea summer monsoon in 1998 and activities of atmospheric intraseasonal oscillation. *Journal of Climate and Environmental Research*, **5**, 375–387 [in Chinese].

Murakami, M. (1976) Analysis of summer monsoon fluctuations over India. *J. Meteor. Soc. Jap.*, **54**, 15–32.

Murakami, M. (1984) Analysis of the deep convective activity over the western Pacific and South-east Asia. Part II: Seasonal and intraseasonal variations during northern summer. *J. Meteor. Soc. Jap.*, **62**, 88–108.

Murakami, T. (1980) Empirical orthogonal function analysis of satellite-observed outgoing longwave radiation during summer. *Mon. Wea. Rev.*, **108**, 205–222.

Murakami, T., T. Nakazawa, and J. He (1984a) On the 40–50 day oscillations during the 1979 Northern Hemisphere summer. Part I: Phase propagation. *J. Meteor. Soc. Jap.*, **62**, 440–468.

Murakami, T., T. Nakazawa, and J. He (1984b) On the 40–50 day oscillations during the 1979 Northern Hemisphere summer. Part II: Heat and moisture. *J. Meteor. Soc. Jap.*, **62**, 469–484.

Murakami, T. and J. Matsumoto (1994) Summer monsoon over the Asian continent and western North Pacific. *J. Meteor. Soc. Jap.*, **72**, 719–745.

Nakazawa, T. (1986) Intraseasonal variations of OLR in the tropics during the FGGE year. *J. Meteor. Soc. Jap.*, **64**, 17–34.

Nakazawa, T. (1992) Seasonal phase lock of intraseasonal oscillation during the Asian summer monsoon. *J. Meteor. Soc. Jap.*, **70**, 597–611.

Ninomiya, K. and H. Muraki (1986) Large-scale circulation over East Asia during Baiu period of 1979. *J. Meteor. Soc. Jap.*, **64**, 409–429.

Nitta, T. (1987) Convective activities in the tropical western Pacific and their impact on the Northern Hemisphere summer circulation. *J. Meteor. Soc. Jap.*, **65**, 373–390.

Sardeshmukh, P. D. and B. J. Hoskins (1984) Spatial smoothing on the sphere. *Mon. Wea. Rev.*, **112**, 2524–2529.

Schnadt, C., A. Fink, D. G. Vincent, J. M. Schrage, and P. Speth (1998) Tropical cyclones, 6–25 day oscillations, and tropical–extratropical interaction over the North-western Pacific. *Meteor. Atmos. Phys.*, **68**, 151–169.

Slingo, J. M. and R. A. Madden (1991) Characteristics of the tropical intraseasonal oscillation in the NCAR community climate model. *Quart. J. Roy. Soc.*, **117**, 1129–1169.

Straub, K. H. and G. N. Kiladis (2003) Interactions between the boreal summer intraseasonal oscillation and high frequency tropical wave activity. *Mon. Wea. Rev.*, **131**, 781–796.

Sui, K.-H. and K.-M. Lau (1992) Multiscale phenomenon in the tropical atmosphere over the western Pacific. *Mon. Wea. Rev.*, **120**, 407–430.

Tanaka, M. (1992) Intraseasonal oscillation and onset and retreat dates of the summer monsoon over east, southeast Asia and the western Pacific region using GMS high cloud amount data. *J. Meteor. Soc. Jap.*, **70**, 613–629.

Tao, S. and L. Chen (1987) A review of recent research on the East Asian summer monsoon in China. In: C.-P. Chang and T. N. Krishnamurti (eds), *Monsoon Meteorology*. Oxford University Press, New York, 60–92.

Teng, H. and B. Wang (2003) Interannual variations of the boreal summer intraseasonal oscillation in the Asian–Pacific region. *J. Climate*, **16**, 3572–3584.

Ueda, H., T. Yasunari, and R. Kawamura (1995) Abrupt seasonal changes of large-scale convective activity over the western Pacific in the northern summer. *J. Meteor. Soc. Jap.*, **73**, 795–809.

Wang, B. and H. Rui (1990) Synoptic climatology of transient tropical intraseasonal convection anomalies: 1975–1985. *Meteor. Atmos. Phys.*, **44**, 43–61.

Wang, B. and X. Xie (1996) Low-frequency equatorial waves in vertically sheared zonal flow. Part I: Stable waves. *J. Atmos. Sci.*, **53**, 449–467.

Wang, B. and X. Xie (1997) A model for the boreal summer intraseasonal oscillation. *J. Atmos. Sci.*, **54**, 72–86.

Wang, B. and X. Xu (1997) Northern hemisphere summer monsoon singularities and climatological intraseasonal oscillation. *J. Climate*, **10**, 1071–1085.

Wang, B. and R. Wu (1997) Peculiar temporal structure of the South China Sea summer monsoon. *Adv. Atmos. Sci.*, **14**, 177–194.

Wang, B. and LinHo (2002) Rainy season of the Asian–Pacific summer monsoon. *J. Climate*, **15**, 386–398.

Wang, W.-C., W. Gong, W.-S. Kau, C.-T. Chen, H.-H. Hsu, and C.-H. Tu (2004) Characteristics of cloud radiative forcing over East China. *J. Climate*, in press.

Wang, Y., O. L. Sen, and B. Wang (2003) A highly resolved regional climate model (IPRC-RegCM) and its simulation of the 1998 severe precipitation event over China. Part I: Model description and verification of simulation. *J. Clim.*, **16**, 1721–1738.

Wu, R. and B. Wang (2001) Multi-stage onset of the summer monsoon over the western North Pacific. *Clim. Dyn.*, **17**, 277–289.

Xie, P. and P. A. Arkin (1997) Global precipitation: A 17-year monthly analysis based on gauge observations, satellite estimates and numerical model outputs. *Bull. Amer. Meteor. Soc.*, **78**, 2539–2558.

Xie, X. and B. Wang (1996) Low-frequency equatorial waves in vertically sheared flow. Part II: Unstable waves. *J. Atmos. Sci.*, **53**, 3589–3605.

Xu, G. and Q. Zhu (2002) Feature analysis of summer monsoon LFO over SCS in 1998. *Journal of Tropical Meteorology*, **18**, 309–316.

Yamazaki, N. and M. Murakami (1989) Intraseasonal amplitude modulation of the short-term tropical disturbances over the western Pacific. *J. Meteor. Soc. Jap.*, **67**, 791–807.

Yasunari, T. (1979) Cloudiness fluctuations associated with the northern hemisphere summer monsoon. *J. Meteor. Soc. Jap.*, **57**, 227–242.

Zhu, B. and B. Wang (1993) The 30–60 day convection see-saw between the tropical Indian and western Pacific Oceans. *J. Atmos. Sci.*, **50**, 184–199.

4

Pan-America

Kingtse C. Mo and Julia Nogues-Paegle

4.1 INTRODUCTION

Rains have strong socio-economic impact for the 850 million inhabitants of the American continents. Both continents depend on rainfall to sustain agriculture, hydroelectric power, and to maintain their waterways. Rainfall over Pan-America has large interannual variability (IAV) and intraseasonal variability (ISV). In the interannual band, El Niño Southern Oscillation (ENSO) has strong impact on total seasonal rainfall (Ropelewski and Halpert, 1987, 1989) over the region, while the occurrence of extreme rainfall episodes is more likely modulated by intraseasonal oscillations (ISOs). Persistence of atmospheric patterns during episodes of strong intraseasonal events raises expectations of converting this information into predictability enhancement beyond the current limitation of about one week for weather forecasts. This would be of great value to optimize crop management, particularly in South America, where regional economies are largely based on agriculture and livestock.

The impact of intraseasonal oscillations on Pan-America within a global perspective was firstly discussed by Weickmann (1983) and Weickmann *et al.* (1985), who identified large-scale patterns of variability in tropical outgoing long-wave radiation (OLR). The atmospheric responses to the tropical ISO (TISO) are stronger in winter and are in expansions and contractions of subtropical and extra-tropical jets. They found that the tropical–extra-tropical linkages in the intraseasonal and the interannual bands are remarkably similar. The responses of circulation anomalies to the Madden–Julian Oscillation (MJO) are similar to the responses associated with tropical sea surface temperature (SST) anomalies (SSTA) in the Pacific Ocean over interannual timescales (Horel and Wallace, 1981). Kousky (1985a) also found similarities between the interannual (ENSO) and intraseasonal signatures in relationships between north-east Brazil rains and circulation patterns over the Atlantic and the subtropical jet. The jet extends from the Caribbean

W. K. M. Lau and D. E. Waliser (eds), *Intraseasonal Variability in the Atmosphere–Ocean Climate System.*

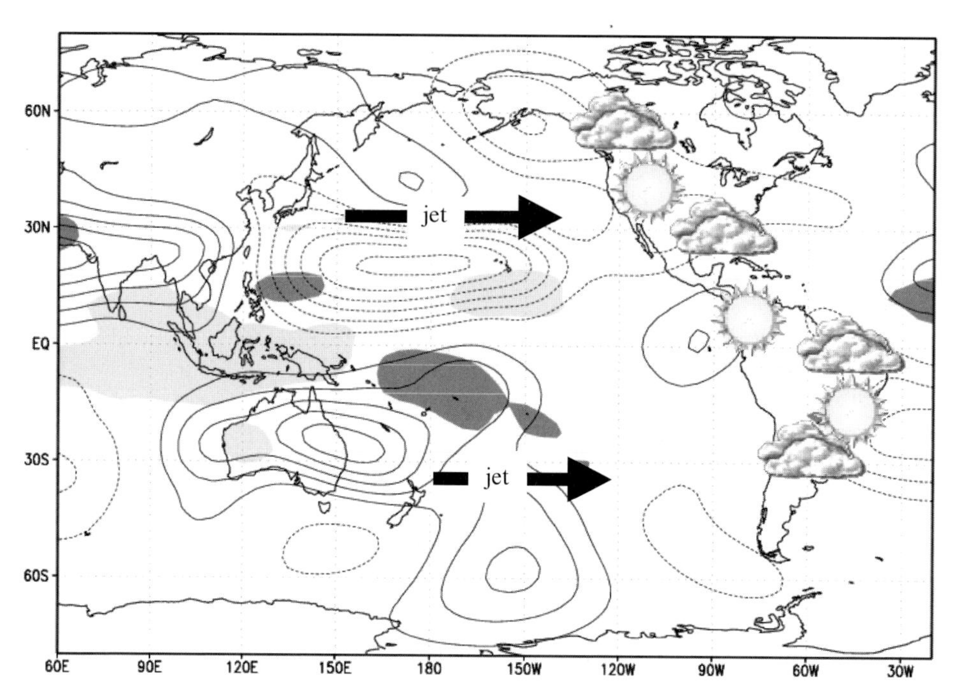

Figure 4.1. Schematic drawing of the atmospheric responses and downstream impact of the TISO.

towards the east–north-east across the Atlantic into North Africa, and it is also a component of the global patterns of ISV (Weickmann 1983; Weickmann *et al.*, 1985). Other intense weather events, such as the Atlantic hurricanes, appear to preferentially form and develop within the intraseasonal regimes with convection patterns that resemble those during an ENSO cold phase (Mo, 2000b).

Carvahlo *et al.* (2004) found that the MJO modulates the persistence (enhanced convection lasting more than 4 days) of the South Atlantic Convergence Zone (SACZ), which is defined as an elongated cloud band that originates in the Amazon basin, extending into the subtropical Atlantic Ocean. In addition to the MJO, the persistence of the oceanic part of the SACZ is also modulated by warm ENSO events. There is a 25–30% increase of rainfall over eastern tropical Brazil for the MJO phase with enhanced convection over the Central Pacific (Figure 4.1). These extreme rainfall events can provoke states of emergency in south-eastern Brazil due to an increase in floods and mud slides. The implications of these studies are that there may be internal patterns of atmospheric variability associated with diabatic heating over the intraseasonal timescale that play a role similar to that of SSTA in the interannual scale. Jones *et al.* (2004) find no statistical significant differences in the frequency of the MJO during warm and cold ENSO phases. That does not rule out the possibility that the warm phase of ENSO sets up conditions over the equatorial Pacific, which are conducive to the excitation of ISOs or vice versa (Chapter 9).

The convection associated with the TISO often exhibits a dipole pattern (Figure 4.1). The atmospheric responses to convection are the expansion and contraction of subtropical and extra-tropical jets mentioned above. They are components of wave trains with alternating positive and negative upper air height anomalies that extend from the tropical Pacific into mid-latitudes as indicated in Figure 4.1. These wave trains are referred to as the Pacific–North American (Wallace and Gutzler, 1981) and the Pacific–South American modes (e.g., Mo and Nogues-Paegle, 2001). These wave trains affect the weather over North and South America through modulations of the Pacific storm tracks, a modulation that is strongest during boreal winter (Figure 4.1). During this season, they regulate precipitation on the west coast of North America. Precipitation over the tropical continent of South America is also regulated by a rich spectrum of frequencies contained within the intraseasonal band (Figure 4.1).

In the intraseasonal band, the wave trains previously mentioned can be excited by two different modes. One mode is the well-known MJO, an eastward propagating mode with convective amplitude that dampens as the convection moves over the cold water of the eastern Pacific Ocean (Chapter 1). This mode has characteristics typical of Kelvin and tropical Rossby modes (Nogues-Paegle *et al.*, 1989). In addition to the MJO, there are sub-monthly modes with smaller spatial scales and periods of 20–28 days. They were originally observed in the Indian Ocean and the Pacific sector (Krishnamurti and Bhalme, 1976; Fukutomi and Yasunari, 1999). These modes do not contribute to a large percentage of variance in the Tropics, but are known to modulate the Indian and the Asian monsoons. Over the Pan-American region, they influence rainfall over both South and North America. For example, strong flooding in California is often associated with the sub-monthly mode. Figure 4.2(a) shows the 5-day running mean precipitation averaged over 9 stations during the 1996/1997 winter (Mo, 1999). Five coastal stations (Brookings, Los Angeles, Pendleton, San Diego, and San Francisco) and four inland stations (Blue Canyon, Fresno, Stockton, and Thermal) were used. There were four wet episodes roughly 20 days apart with breaks in between. These alternating wet and dry events occur often in California during winter. Over South America, the sub-monthly oscillations are also found over the SACZ. Liebmann *et al.* (1999) examined the power spectra of the 90-day high-pass filtered OLR anomalies (OLRA) at several grid points. They found peaks near 48, 27, 16, 10, and 8 days for a point in the SACZ (solid thick line in Figure 4.2(b)). In contrast, there is no 27-day peak for the point in the Amazon basin (dashed thin line in Figure 4.2(b)). This is consistent with other studies that find strong diurnal and seasonal cycles in Amazon convection, with weaker signals in the intraseasonal band (e.g., Kiladis and Weickmann, 1992a, b). The following sections focus on the influence of TISO over the Americas, based on empirical orthogonal function (EOF) analysis of OLRA and related atmospheric circulation anomalies.

4.2 VARIATIONS IN THE INTRASEASONAL BAND

Seasonal variations of convection associated with ISOs are depicted in Figure 4.3. This figure shows standard deviations of OLRA in the low pass greater than 10 days,

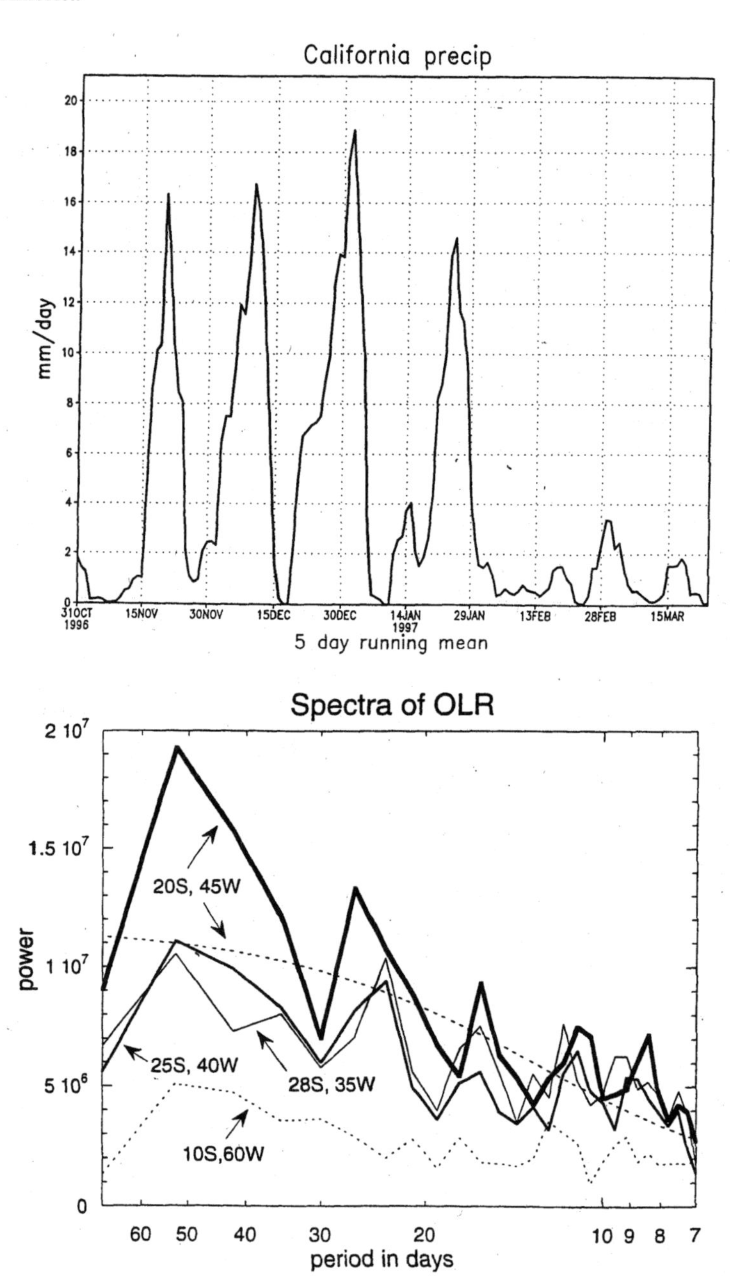

Figure 4.2. (a) 5-day running mean of California precipitation averaged over 9 stations (Brookings, Los Angeles, Pendleton, San Diego, San Francisco, Blue Canyon, Fresno, Stockton, and Thermal) during the 1996/1997 winter (the units used are mm day^{-1}) and (b) average power spectra of OLRA for points at locations marked. Also shown is the red noise spectrum for the uppermost curve, computed from the lag-one autocorrelation.
(a) from Mo (1999), (b) from Liebmann *et al.* (1999).

10–90-day and 10–30-day bands for summer and winter. The largest standard deviations of the low-pass filtered OLRA follow the annual evolution of convection and are concentrated mostly in the summer hemisphere (Figure 4.3). For DJFM, large standard deviations are located over the Indian Ocean and the western and central Pacific predominantly south of the equator (Figure 4.3(a)). As the season progresses, largest values show a northward displacement of 20 to 30 degrees of latitude over the same longitude from winter (DJFM) to summer (JJAS) (Figure 4.3(d)). The largest standard deviations in the 10–90-day band (Figure 4.3(b, e)) are located in the Indian Ocean and the western and central Pacific, where the contribution to the low-pass filtered OLRA variance can be as large as 50–75%. Such contributions are smaller at equatorial latitudes east of the dateline, where convection is largely due to IAVs related to ENSO. In North America, largest values are located along the west coast of North America. Variations of the Intertropical Convergence Zone (ITCZ) contribute to the large standard deviations in the eastern Pacific just above the equator. For JJAS, large values are also located over Mexico and the southern U.S.A. Over South America, large contributions from the 10–90-day band are evident in the SACZ and subtropical plains, where the contributions to the low-pass filtered OLRA variance can be as large as 50–75%. Figure 4.3(c, f) show the contribution from the sub-monthly timescales. They exhibit a similar global imprint as that of the 10–90-day band and contribute less than 50% to the low-pass variance.

For DJFM, convection related to the TISO modulates the Australian monsoon over the western Pacific and Australia (Chapter 5). Large standard deviations also extend into the South Pacific Convergence Zone (SPCZ) and the SACZ. For JJAS, the TISO modulation of the Indian monsoon is evident in standard deviations higher than $30\,\mathrm{W\,m^{-2}}$ that extend from the Indian Ocean to the Indian continent (see Chapters 2 and 3). The band that extends from the South China Sea to Japan indicates the TISO modulation of the Asian monsoon and East Asian Mei-yu. Over North America, large values over Mexico and the south-western U.S.A. during JJAS represent the intraseasonal modulation of the North American monsoon. Over South America, large standard deviations during boreal winter are also found over the SACZ and the subtropical plains, where both the 30–60-day band, related to the MJO, and the sub-monthly oscillation modulate the South American monsoon.

4.3 ISV IN DECEMBER–MARCH

The ISV is described next through EOF analysis of OLRA. This methology has been extensively used to isolate dominant patterns of large-scale variability. The technique is not as useful when the interest is on local phenomena (e.g.,Carvalho *et al.*, 2004).

4.3.1 EOF modes

An EOF analysis was performed on the OLRA from 40°S to 50°N for DJFM. The OLRA were filtered to retain variability on a 10–90-day band and the resolution was

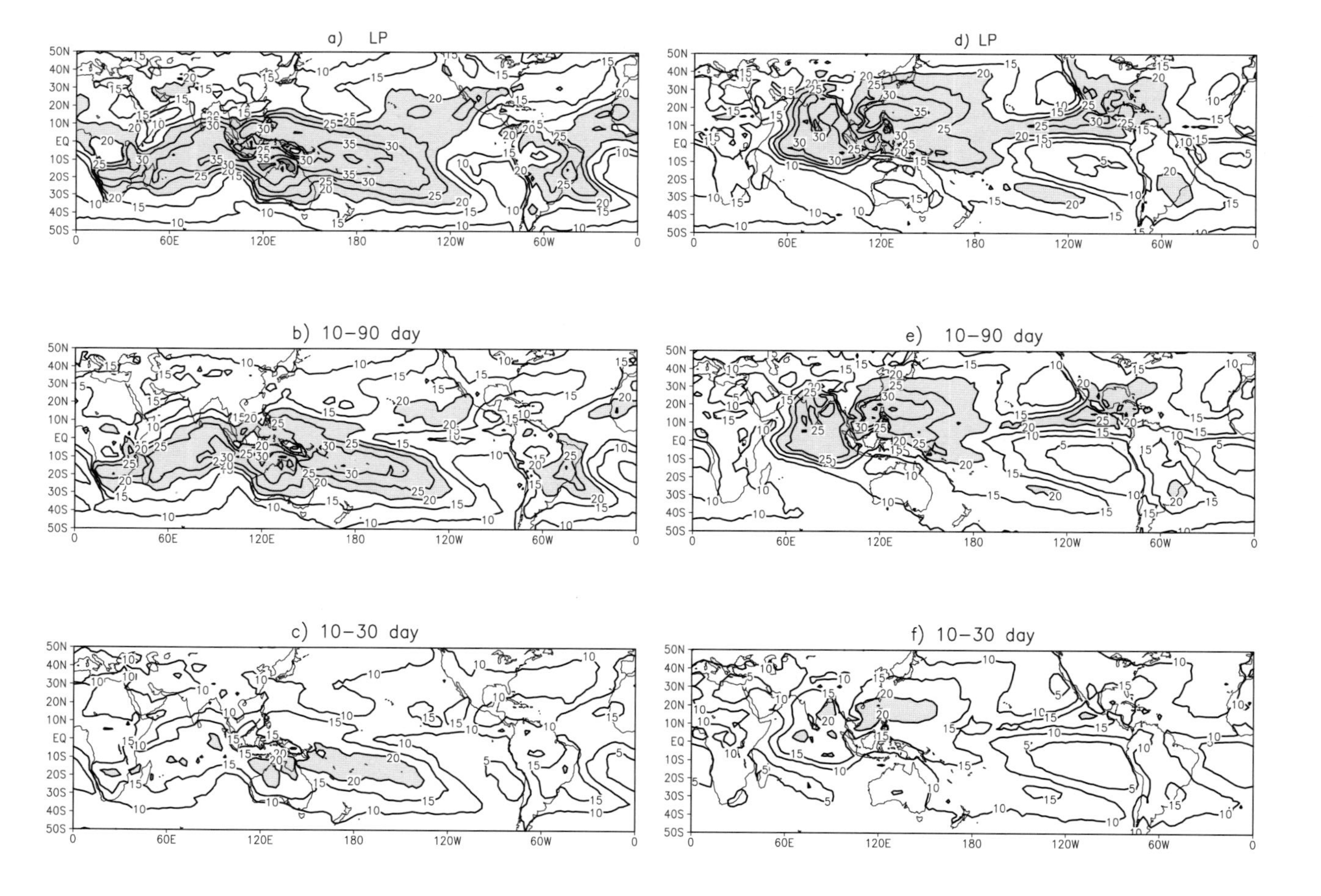

Figure 4.3. (a) Standard deviation for the low-pass filtered (>10 days) OLRA obtained from daily averages from NOAA satellites (Liebmann and Smith, 1996) for the period between 1 January, 1979–31 December, 2001, for DJFM (boreal winter). Anomalies are computed as departures from the seasonal cycle defined as the grand mean plus the annual and semi-annual cycles (contour interval is $5\,\mathrm{W\,m^{-2}}$). Values greater than $20\,\mathrm{W\,m^{-2}}$ are shaded. (b) Same as (a), but for the 10–90-day filtered OLRA. (c) Same as (a), but for the 10–30-day filtered OLRA. (d)–(f) Same as (a)–(c), but for JJAS (boreal summer). Anomalies are filtered using the minimum bias window developed by Papoulis (1973).

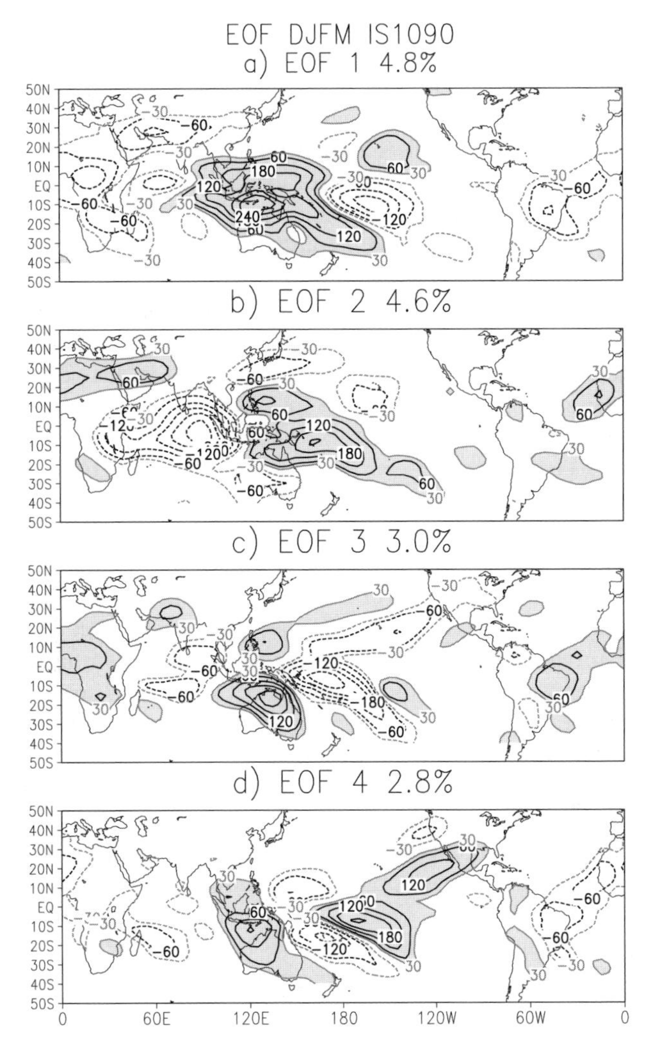

Figure 4.4. (a) EOF 1, (b) EOF 2, (c) EOF 3, and (d) EOF 4 for 10–90-day filtered OLRA over the domain from 40°S to 50°N for DJFM. Contour interval is 60 non-dimensional units. Zero contours are omitted. Contours −30 and 30 non-dimensional units are added. Positive values are shaded.

reduced to 5° prior to EOF analysis. Both the spectral analysis and the singular spectrum analysis (Vautard and Ghil, 1989) identify the leading modes (EOF 1 and EOF 2) with a period of 40–48 days. They are in quadrature in time as well as space and explain nearly the same percentage of variance (about 4.7%). Together, they represent the MJO (Figure 4.4(a, b)). In the Indian–Pacific sector, they are similar to the patterns isolated by the pioneering work of Lau and Chan (1985) and the extended EOF (EEOF) (Chapter 1). EOF 1 (Figure 4.4(a)) shows suppressed convection over the western Pacific accompanied by enhanced convection in the central Pacific and the Indian Ocean. Over the Pan-American region, enhanced convection extends from north-eastern Brazil through the tropical Atlantic to the west coast of Guinea. EOF 2 (Figure 4.4(b)) shows a longitudinal dipole with two centers located at 90°E and 165°E respectively. The positive loadings are also located in the SPCZ and the SACZ. Rainfall patterns do not map into OLRAs over the African desert north of 10°N, and therefore OLRA should not be interpreted as precipitation anomalies in this region (Waliser *et al.*, 1993).

While the first two EOFs project strongly onto a wave number 1 structure in longitude with centers over the Indian Ocean and Pacific sector (Figure 4.4(a, b)), the next two EOFs exhibit a more complex structure with at least two positive and two negative centers (Figures 4.4(c, d)). EOFs 3 and 4 are also orthogonal to each other and explain nearly all the 2.9% of the total variance. EOF 5 explains only 1.6% of the total variance. They are well separated from EOF 2 and EOF 5 by the North criterion (North *et al.*, 1982). The second pair of EOFs represents oscillations with timescales of 22–28 days. They are similar to the leading EOFs in the 10–30-day band (Mo, 1999). Even though they do not explain a large percentage of the total variance, the sub-monthly modes are stronger than the MJO for certain years. During these periods, they have large influence on rainfall over Pan-America. As discussed in the introduction of this chapter, strong sub-monthly oscillations are often responsible for winter floods in California (Mo, 1999). Carvalho *et al.* (2002) and Liebmann *et al.* (2004) related the occurrence of extreme wet events over tropical south-eastern South America with contributions from the TISO.

Both EOFs 3 and 4 (Figures 4.4 (c, d)) show high loadings extending from the centers at the equator near the dateline to the west coast of North and Central America, indicating their large influence on rainfall over the Pan-American region. EOF 3 and EOF 4 both show a dipole with positive loadings over California and negative loadings to the south. Over South America, they exhibit a three-cell structure, with large values over north-eastern Brazil, flanked by an arch-shaped pattern with loadings of opposite sign that extends from northern South America into the South American subtropics.

4.3.2 The MJO

The evolution of OLRA and atmospheric circulation anomalies associated with the MJO were examined with a compositing approach obtained as follow: the 10–90-day filtered OLRA were projected onto EOF modes 1 and 2 to obtain a time series of principal components (PCs). For each PC, the standard deviation was computed. A positive (negative) day was selected when the PC for that day was above 1.2 (below −1.2) standard deviations. This date is also defined as the onset day. Composites of rainfall anomalies, the 10–90-day filtered 200-hPa eddy streamfunction with zonal means removed, and OLRA were formed from 20 days before to 20 days after the onset day. There are more than 300 days in each composite. To assure that OLRA composites represent rainfall, they were compared with composites of the pentad rainfall anomalies from CPC Merged Analysis of Precipitation (CMAP) (Xie and Arkin, 1997; Xie *et al.*, 2003). From the above daily PC time series, the 5-day means were computed. The same composite procedures were used to obtain the pentad rainfall anomalies. Overall, OLRA and rainfall composites are similar except in areas over West Africa, north of 10°N, where OLRA do not represent rainfall (Waliser *et al.*, 1993). The statistical significance of each map was tested using a Student's t-test. The degrees of freedom are determined by assuming 6 days as the de-correlation time. Composites for positive and negative events are similar with a sign reversal, therefore, composite differences between positive and negative events are presented to amplify the signal. The OLRA composite evolution is shown in Figure 4.5 based on PC-1. Areas with anomalies in the corresponding rainfall composite and daily OLRA composites with statistical significance at the 5% level are shaded.

Figure 4.5 shows an eastward propagating pulse and a stationary component that is most evident over South America between the equator and 20°S. A dipole with enhanced convection (negative OLRA) to the east propagates from the western Pacific to the central Pacific in 15 days (Figures 4.5(a–d) and 4.6(a)). South America acts as a bridge linking the convection from the central Pacific to the tropical Atlantic. As negative OLRA shift towards the central Pacific east of the dateline, a link is established between the center over eastern Brazil at 20°S and the western African coast of Guinea (Figure 4.5(d)). The compensatory branch of suppressed convection is found over the tropical Atlantic (Figure 4.5(f–g)). This is indicative of meridional displacements of the Atlantic ITCZ. As the negative OLRA proceed eastward further into the central Pacific east of the dateline, negative anomalies extend south-eastward from West Africa to South Africa and connect to anomalies in the Indian Ocean (Figure 4.5(f)). With enhancement of convection in the Indian Ocean, positive OLRA are found in the western Pacific and another cycle starts (Figure 4.5(g, h)). At the same time, the Atlantic convective branch between South America and Africa weakens. The OLRA composite at day −10 resembles EOF 2 (Figure 4.4(b)), showing that PC-1 has evolved into PC-2 in 10 days, in about one-quarter of the total period.

The circulation anomalies and rainfall over the Pan-American region depend on the location of tropical convection. The strongest large-scale upper-level response to the MJO in the Pan-American region is at day 5 based on the PC-1 composite (Figure 4.6(c)), when enhanced convection is located east of the dateline centered at 140°W–150°W with suppressed convection located over the western Pacific (Figure 4.5(f)). The response represented by the 200-hPa streamfunction difference shows that anomalies are symmetric in the tropics and exhibit a four-cell pattern with high-pressure centers (i.e., anticyclonic flow) flanking strengthened convective activity in the Pacific. There is also a wave train extending from the convective region in the tropics to both North and South America (Weickmann *et al.*, 1985). Over North America, there are negative anomalies near the west coast of the U.S.A., positive anomalies over Canada and negative anomalies over the east coast of North America. This closely represents a Pacific–North American (PNA) pattern.

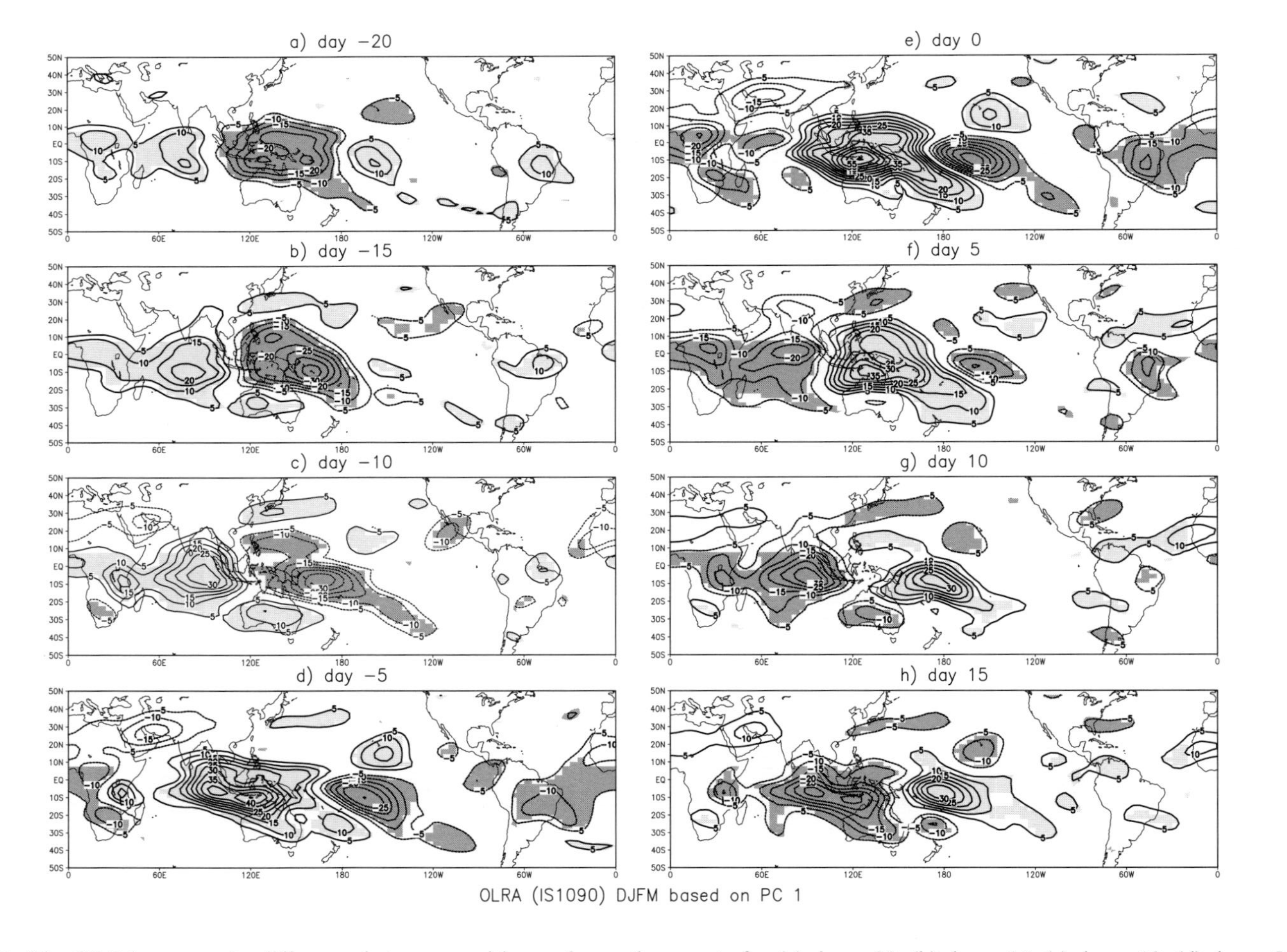

Figure 4.5. The OLRA composite difference between positive and negative events for (a) day −20, (b) day −15, (c) day −10, (d) day −5, (e) day 0, (f) day 5, (g) day 10, and (h) day 15 based on PC 1 for DJFM. The OLRA are 10–90-day filtered (contour interval is $5\,\mathrm{W\,m^{-2}}$; zero contours are omitted). Areas where positive (negative) anomalies are statistically significant at the 5% level are shaded light (dark).

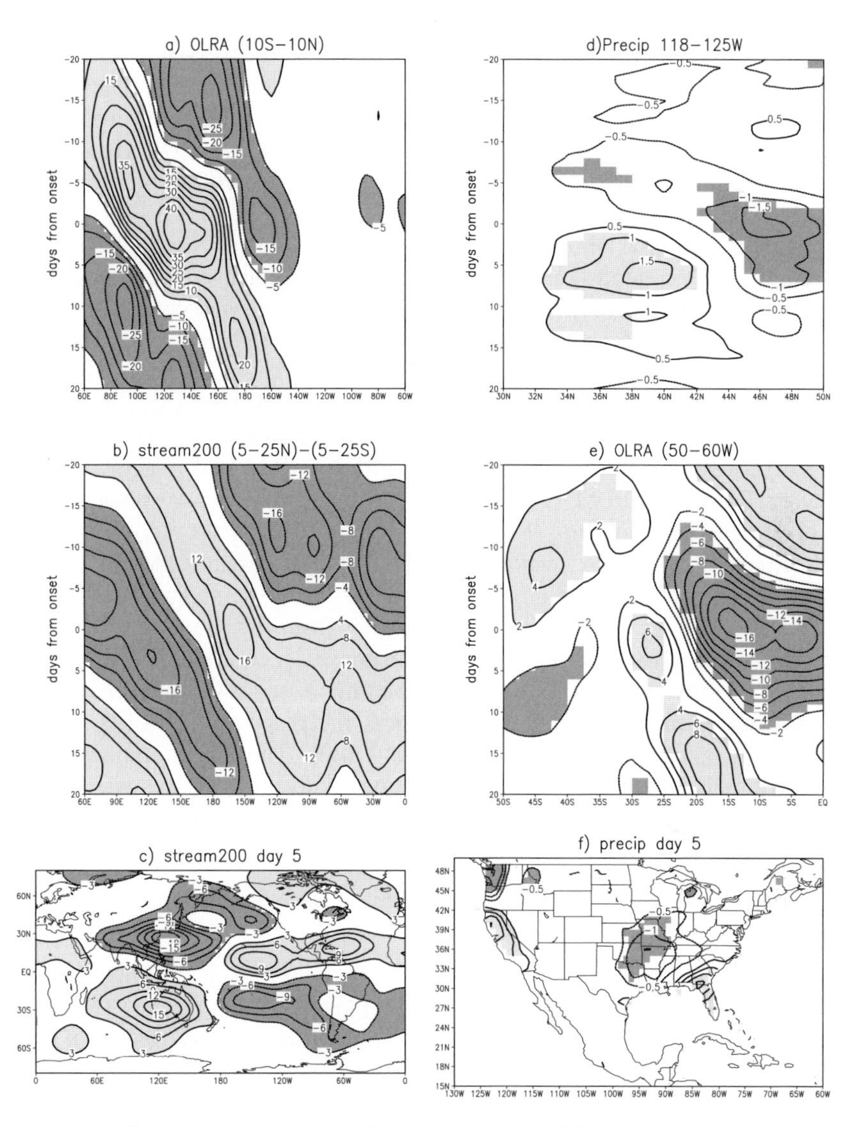

Figure 4.6. (a) Time–longitude plot of OLRA composite difference averaged from 10°S to 10°N between positive and negative events from 20 days before to 20 days after the onset of PC 1 (contour interval is $5\,\mathrm{W\,m^{-2}}$). Areas where positive (negative) values are statistically significant at the 5% level are shaded light (dark). (b) Same as (a), but for a composite of 200-hPa streamfunction difference between mean from 5°N–25°N and 5°S–20°S. (c) Composite of 200-hPa streamfunction between positive and negative events for day 5 based on PC 1 for DJFM (contour interval is $3 \times 10^6\,\mathrm{m^2\,s^{-1}}$; zero contours are omitted). Areas where positive (negative) anomalies are statistically significant at the 5% level are shaded light (dark). (d) Same as (a), but the time–latitude plot of precipitation difference averaged from 118°W–125°W over land (contour interval is $0.5\,\mathrm{mm\,day^{-1}}$). (e) Same as (d), but for OLRA averaged from 50°W–60°W. (f) Same as (c), but for precipitation difference (contour interval is $0.5\,\mathrm{mm\,day^{-1}}$). The 200-hPa streamfunction was obtained from the NCEP–NCAR reanalysis (Kalnay *et al.*, 1996). The daily observed precipitation over the U.S.A. and Mexico was derived from the unified gridded daily data (Higgins *et al.*, 2000).

Liebmann and Hartmann (1984) examined tropical–extra-tropical connections and found a similar wave train pattern as that of day 5 (Figure 4.6(c)). The location of anomalies in the North Pacific is also the region where blocking and persistent anomaly events are most likely to form (Higgins and Mo, 1997). It is likely that the tropical forcing related to the MJO sets up favorable conditions for persistent weather anomalies.

Consistent with the above discussion, the response of North American west coast rainfall also depends on the location of the tropical convection anomalies. At day 5, the rainfall response exhibits a dipole pattern with centers located over California and the Pacific north-west (Figure 4.6(f)), which resembles the response during warm ENSO events. In addition to the dipole, there are negative anomalies centered over the Southern Plains. The rainfall response is partly due to a westward shift of the storm track to California, which contributes to dry conditions in the Pacific north-west and wet conditions over California (Mo and Higgins, 1998(a, b)).

The evolution of the MJO cycle and the downstream responses are shown in terms of Hovmöller diagrams (Figure 4.6). When tropical OLRA move eastward (Figure 4.6(a)), the dipole response in the 200-hPa streamfunction propagates eastward in concert (Figure 4.6(b)) (Weickmann, 1983; Knutson and Weickmann, 1987). In mid-latitudes, the wave train propagates from the convective region downstream to North America. The rainfall response is consistent with circulation anomalies. It shows a dipole pattern depicting a see-saw between the Pacific north-west and California. In South America, the responding OLRA propagates from the subtropics to the equator (Figure 4.6(e)). The largest influence is over north-east Brazil when suppressed convection is located in the western Pacific (Figure 4.5(f)). The OLRA pattern also shows a dipole with centers over north-east Brazil and the SACZ (Figure 4.6(e)).

4.3.3 The sub-monthly oscillation

In addition to the influence of the MJO, the sub-monthly oscillation, depicted by EOFs 3 and 4 also influences rainfall over the Pan-American region. The evolution of OLRA and streamfunction anomalies was determined by projecting OLRA into EOFs 3 and 4 to obtain time series of PCs. These composites were computed following the methodology described in the previous section for the MJO. Selected time sequences of OLRA, 200-hPa streamfunction, and precipitation anomaly composites are depicted in Figures 4.7 and 4.8 to illustrate its temporal behavior. The typical eastward propagation of the MJO is not apparent for submonthly timescales. Instead, westward propagation of OLRAs is dominant along 10°–20°N (Figure 4.7(a, d) shading) from the central Pacific into the Indian Ocean. This may be better represented by the Hovmöller diagrams of OLRA based on both PC-3 and PC-4 (Figure 8(a, b)). In addition to propagation, there are standing components with centers located at 150°E and 150°W for PC-4 (Figure 4.8(a)) and 120°E and 150°W for PC-3 (Figure 4.8(b)). The timescale of the oscillation is about 20–22 days. When convection associated with the submonthly mode moves westward, the circulation anomaly composites (Figure 4.7(a–d), contours) show a northward

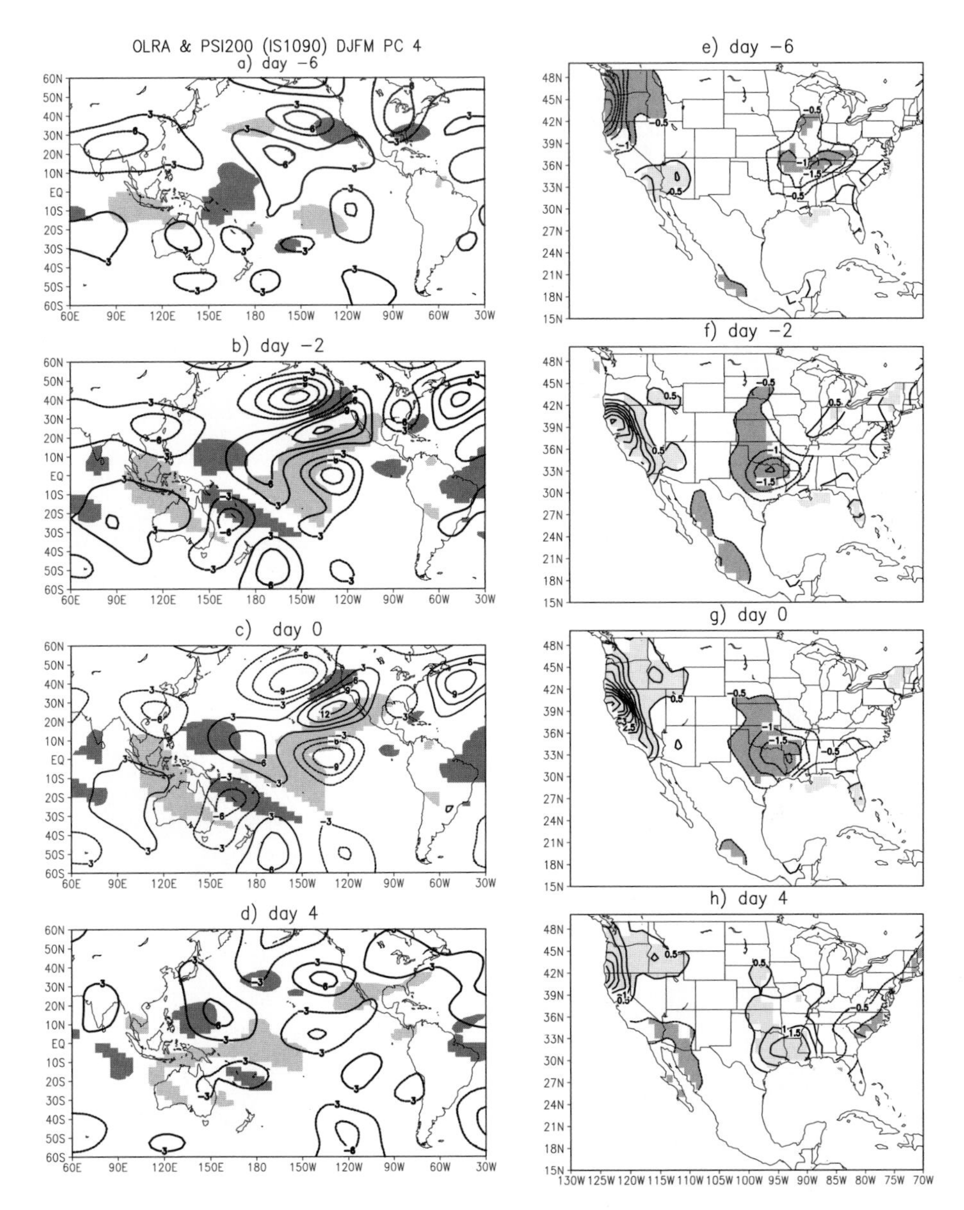

Figure 4.7. The OLRA (shaded) and 200-hPa streamfunction (contoured) composite difference between positive and negative events for (a) day −6, (b) day −2, (c) day 0, and (d) day 4 based on PC-4 for DJFM. Areas where the OLRA is greater (less) than 6 (−6) $W\,m^{-2}$ and anomalies are statistically significant at the 5% level are shaded light (dark). The contour interval for the 200-hPa streamfunction composite is $3 \times 10^6\,m^2\,s^{-1}$. Zero contours are omitted. (e–h) Same as (a–d), but for precipitation (contour interval is $0.5\,mm\,day^{-1}$). Areas where positive (negative) anomalies are statistically significant at the 5% level are shaded light (dark).

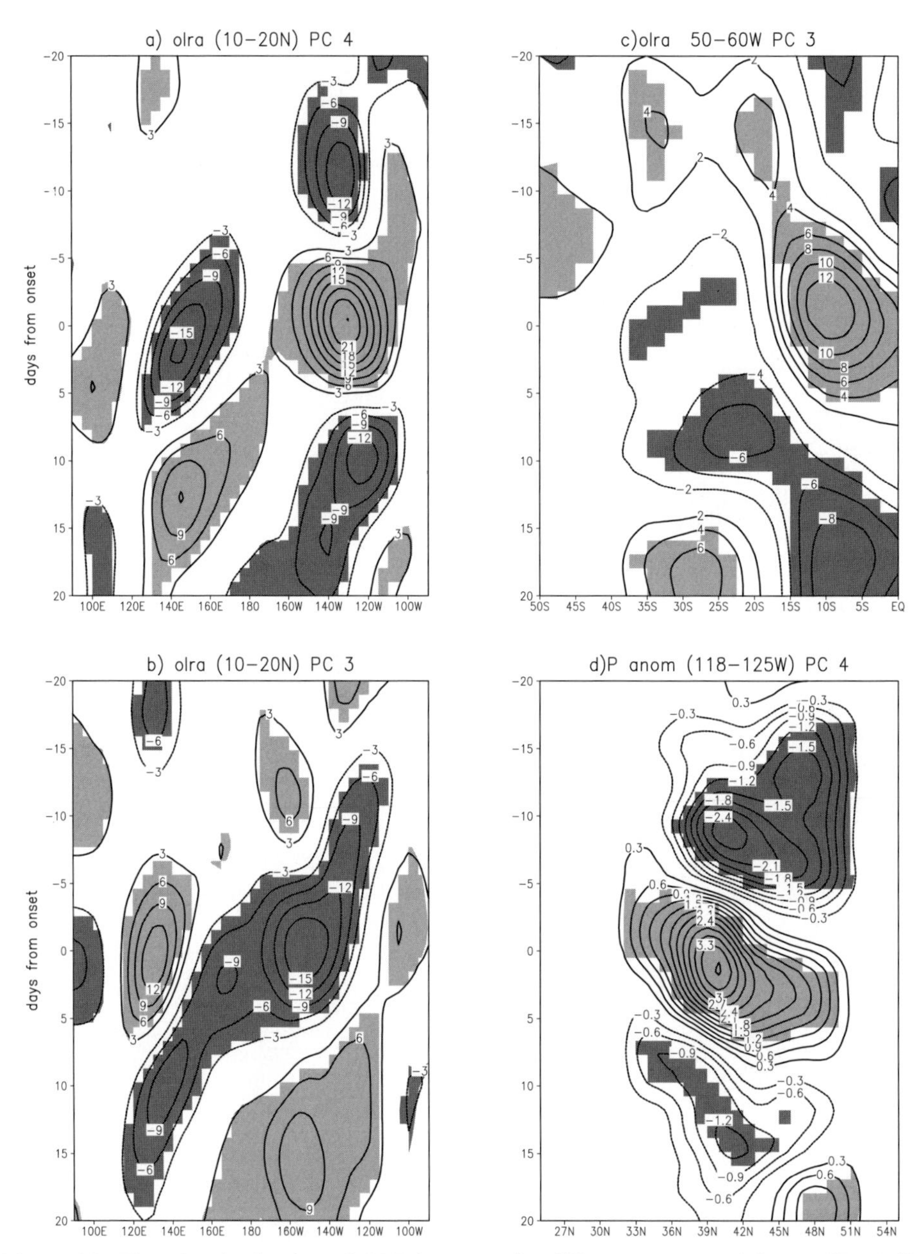

Figure 4.8. Time–longitude plot of OLRA composite difference averaged from 10°N–20°N between positive and negative events from 20 days before to 20 days after onset of PC-4 for DJFM (contour interval is $3\,\mathrm{W\,m^{-2}}$). Areas where positive (negative) values are statistically significant at the 5% level are shaded light (dark). (b) Same as (a), but based on PC-3. (c) Same as (a), but for time–latitude plot of OLRA composite difference averaged from 50°W–60°W based on PC-3 (contour interval $2\,\mathrm{W\,m^{-2}}$). (d) Same as (c), but for precipitation difference averaged from 118°W–125°W over land based on PC-4 (contour interval $3\,\mathrm{mm\,day^{-1}}$).

displacement into mid-latitudes of anomalies, extending from the subtropical eastern Pacific through Mexico, California, the Pacific north-west to the Gulf of Alaska (Figure 4.7). The north-westward traveling of circulation anomalies were firstly reported by Banstator (1987) when he analyzed circulation patterns in 1979. The composites also indicate a modulation in timescales of 20–28 days of the so called "pineapple express", characterized by moisture plumes that feed wet conditions over western North America.

The evolution of precipitation over the west coast of North America shows a three-cell pattern with negative anomalies over the Pacific north-west and positive anomalies over southern California at day −6. The rainfall pattern shifts northward along the west coast in response to the circulation changes. For example, the positive anomalies originated from southern California move to northern California at day −2 and reach the Pacific north-west at day 4. The time–latitude plot of rainfall anomalies averaged over the west coast demonstrates the northward movement of rainfall (Figure 4.8(d)). Over the U.S.A., composites (Figure 4.7(e–h)) also show a phase reversal of rainfall anomalies between southern California and the southern plains.

In contrast, OLRAs over tropical South America originate in the subtropics and propagate into the deep tropics (Figure 4.8(c)). This is related to the sub-monthly modulation of the wave train response to the MJO. This timescale has been identified by several studies (e.g., Nogues-Paegle and Mo, 1997; Liebmann *et al.*, 1999; Nogues-Paegle *et al.*, 2000; Mo and Nogues-Paegle, 2001). It is also linked to Rossby trains over the Pacific Ocean with low-level cold air moving northward channeled by the Andes and triggering enhanced convection along the SACZ. Liebmann *et al.* (1999) pointed out that convection over the south-western Amazon basin on sub-monthly timescales appears to propagate from the south, while convection over the south-eastern Amazon is accompanied by disturbances moving from the Atlantic. This is consistent with the early results of Kousky (1985b) that relate rainfall anomalies over tropical Brazil to cold fronts moving from the south with an enhanced Atlantic subtropical high, and enhanced easterlies over the continent that persist for periods commensurate with those of the ISO.

The synoptic picture described above is complemented by previous studies related to other tropical convective bands. Kodama (1992, 1993) discussed common characteristics of the SPCZ, SACZ and the Baiu frontal zone over South Asia. These convergence zones originate from equatorial convection extending poleward and eastward. Moisture has a dominant monsoonal origin in the tropical portion of these bands and it is advected poleward by subtropical highs. In the case of South America, the poleward moisture flux is modified by the steep orographic relief of the Andes, which deflect the prevailing trade winds southward, transporting large amounts of water vapor into subtropical South America. These characteristics of the time-mean South American climate have been shown to also typify the sub-monthly oscillation (Nogues-Paegle and Mo, 1997). The SACZ and the fertile plains of South America located towards the south of the SACZ constitute a dipole of convection, such that the low-level moisture-laden flow from the tropics fuels convection over the plains prior to cold air moving northward and triggering enhancements of the SACZ. The central South American low-level jet (LLJ) constitutes an integral part of the South American convective dipole, and it is strongly modulated by the ISOs (Nogues-Paegle and Mo, 1997).

4.4 INTRASEASONAL VARIABILITY IN JUNE–SEPTEMBER

4.4.1 EOF modes

An EOF analysis was performed on the OLRA for June–September (JJAS). The procedures are the same as those used to obtain EOFs for DJFM. The first two EOFs are nearly in quadrature with each other and the singular spectrum analysis indicates that the leading temporal mode has a period of 40–48 days. Together, they represent the MJO (Figure 4.9(a–c)). EOF 3 also has a spectrum peak with a period of 40–48 days, which suggests an additional modulation of the MJO. The complexity in OLRA patterns introduced by convection associated with the Asian monsoon requires three EOFs during boreal summer to adequately represent the MJO during this season, unlike the case during boreal winter when most of the convection is found over oceanic areas. EOF 1 (Figure 4.9(a)) shows positive loadings north of the equator extending from the western Pacific to the central Pacific with negative loadings in the Indian Ocean. Over the Pan-American region, the largest negative loadings are found over Central America. The largest loading in the vicinity of Central America indicates modulations of the North American monsoon by the much stronger intraseasonal anomalies of the eastern hemisphere. Weak negative loadings extend across South America into the Atlantic. EOF 2 (Figure 4.9(b)) has a four-cell structure in the eastern hemisphere with two dipoles opposite in phase straddling the equator. EOF 3 (Figure 4.9(c)) is similar to EOF 1 over the Americas, but it exhibits sign reversals from that of EOF 1 over the eastern hemisphere. EOF 1 over the Indian–Pacific sector resembles EOF 2 for the 20–60 day filtered OLRA (Lau and Chan, 1986). Such agreement is lacking in subsequent EOFs, possibly due to differences in the analysis domain and width of the time filter.

EOF 4 (Figure 4.9(d)) explains about 2.6% of the total variance. This mode is found in the 10–30-day band and the dominant temporal mode has a period of 20–24 days. This mode, together with the MJO, establishes the linkages between convection over the South China Sea, associated with the Asian monsoon, and summer precipitation over the Pan-American region in the intraseasonal band. The dominant feature is a four-cell pattern with two dipoles, opposite in sign, located over the western and central Pacific. There are positive loadings over the South China Sea (5°N–15°N, 100°E–130°E) and negative loadings to the east with a wave of alternating positive and negative anomalies towards the north-east. A three-cell OLRA pattern along the west coast of North America is noted. In South America, there are positive loadings over the SACZ.

Composites of the 10–90-day filtered OLRA and 200-hPa streamfunction and precipitation anomalies were formed for JJAS using the same compositing procedures as for DJFM. The composites for positive and negative events are similar with

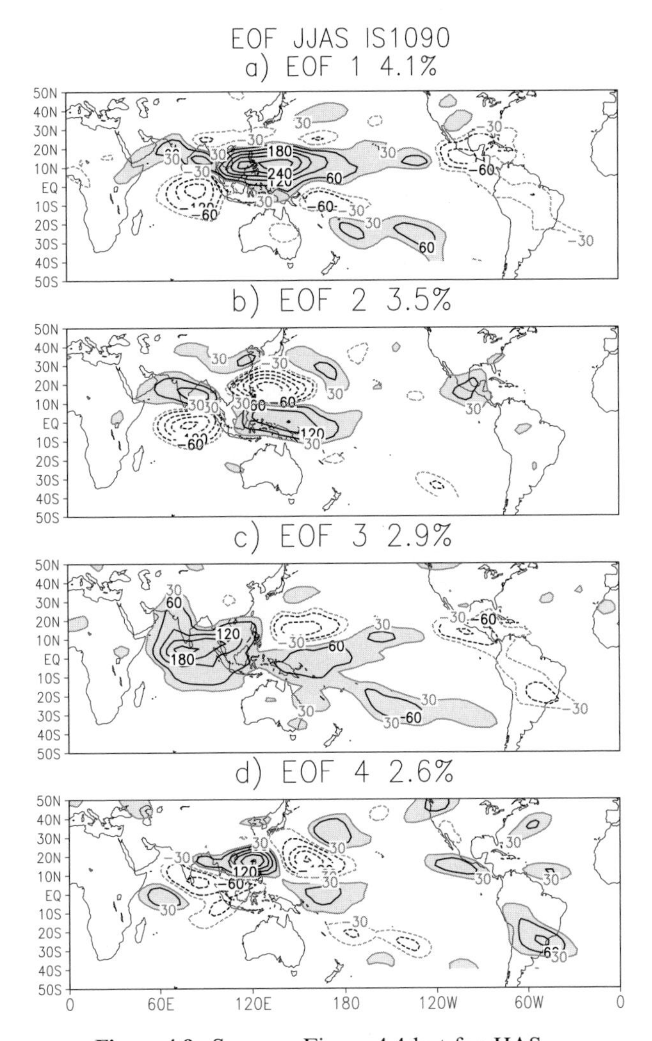

Figure 4.9. Same as Figure 4.4 but for JJAS.

a sign reversal so the composite differences between positive and negative events are given. Areas where features are statistically significant at the 5% level are shaded.

4.4.2 MJO

The OLRA composites (Figure 4.10) show the evolution of EOF 1. In addition to eastward propagation there is also a northward shift of OLRA in the Indian Ocean. The positive OLRA centered in the Indian Ocean at day −20 (Figure 4.10(a)) shift from 5°S to 10°N at day −5 (Figure 4.10(d)) and extend into the western Pacific. From day −5 onwards, a negative center is established in the equatorial Indian

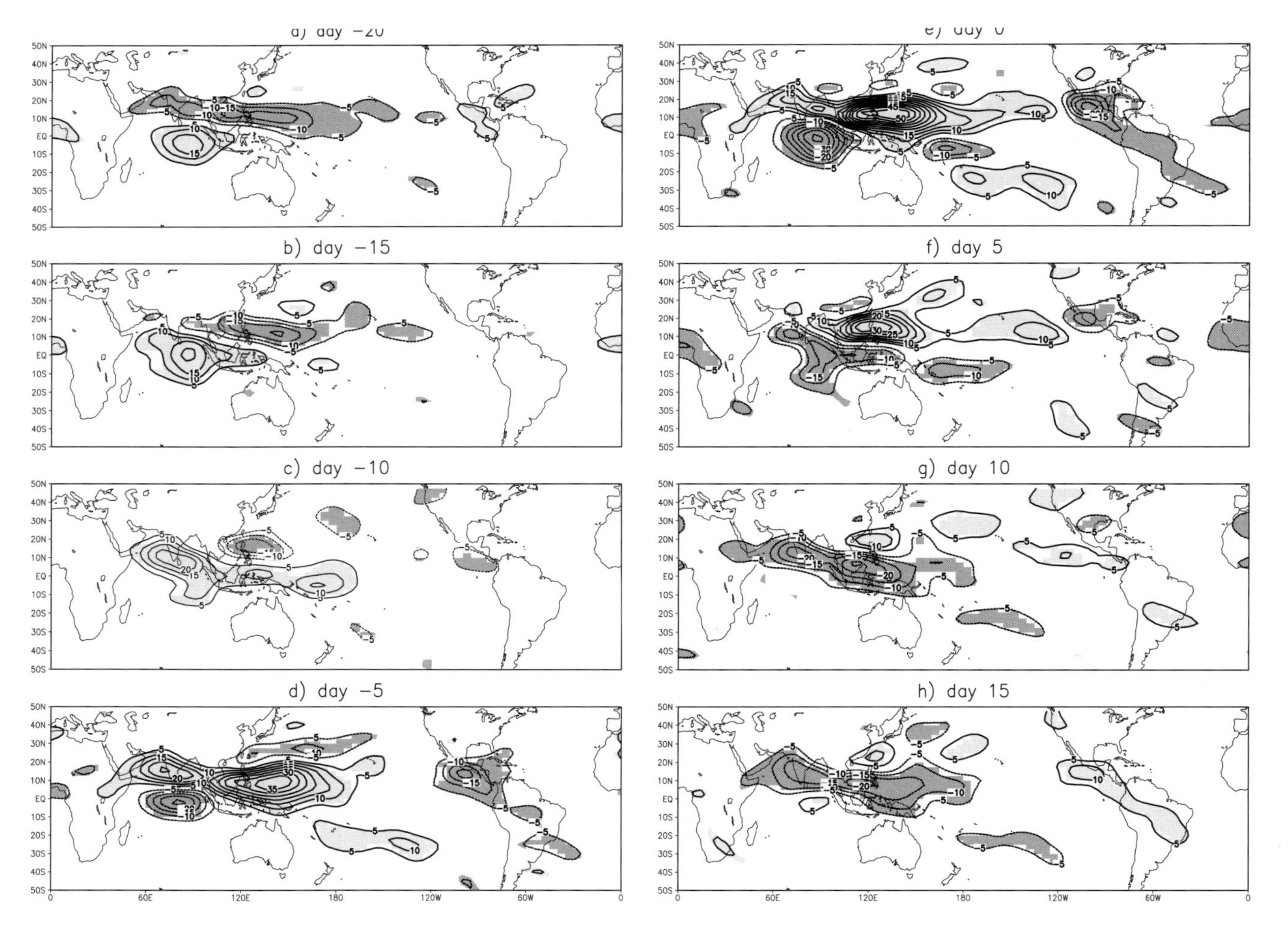

Figure 4.10. Same as Figure 4.5 but for JJAS, based on PC-1.

Ocean with positive OLRA extending from the western Pacific to the eastern Pacific just north of the equator. The connection with the Americas is evident only after positive OLRAs are established east of the dateline (day −5–day 15). This takes the form of a band of convection that extends from Mexico into South America (day 0, Figure 4.10(e)), which evolves into an opposite phase by day 15 (Figure 4.10(h)). The evolution over the eastern hemisphere is consistent with well-known features associated with the Asian monsoon (Gadgil, 1983; Krishnarmurti *et al.*, 1985; Lau and Chan, 1986, see chapters 2 and 3 for a review). The composite at day 0 is an amplified version of the composite at day −20 with a phase reversal, indicating a period near 40–48 days. The composite at day −10 is similar to EOF 3 (Figure 4.9(c)) west of the dateline in agreement with the notion that all three leading EOFs represent different phases of the MJO.

Figure 4.11(b) shows rainfall anomaly patterns for days 0–2 based on PC-1, when the strongest anomalies are found over North America, consistent with the OLRA composites shown in Figure 4.10(e). The MJO modulation of the North American monsoon is characterized by anomalies that extend from the core of the monsoon over southern Mexico (Higgins and Shi, 2001) across the Gulf of Mexico. Rainfall anomalies with an opposite sign are found centered over the Great Plains and New Mexico. Anomalies over the south-eastern U.S.A., except for areas near the Gulf of Mexico), are not significant at the 5% level. The circulation response to the MJO is shown in Figure 4.11(a) in terms of the 200-hPa streamfunction difference between positive and negative PC 1 events. It is characterized by a global wave number 1 with maximum values located in subtropical latitudes. The pattern indicates high (low) pressure in the western (eastern) hemisphere at this time, with an enhancement of equatorial easterlies and subtropical westerlies over the Americas and the eastern Pacific. These results are consistent with those of Nogues-Paegle and Mo (1987) for the 1979 FGGE year. That study linked planetary divergence and rotational circulation and concluded that the seasonal northward displacement of OLRA from South America into Central America was triggered by the passage of the ISOs over the Americas during 1979. Results presented here suggest the validity of this conclusion for a longer time period.

The strongest TISO influence on South America can be represented by two dominant wave train patterns known as the Pacific–South American modes (PSA; Mo and Nogues-Paegle, 2001) given here (Figure 4.12) as the leading two EOFs of the 200-hPa streamfunction eddy anomalies over the Southern hemisphere (Mo and Higgins, 1998c). The wave trains extend from the tropics into mid-latitudes and bend northward into South America. The two wave trains are in quadrature of each other and are present in all seasons. The PSA 1 pattern shows a region of negative streamfunction anomalies centered about 120°W and 60°S, a region of frequent blocking in the southern hemisphere (e.g., Kiladis and Mo, 1998), predominantly in austral winter. Composites of 10–90-day filtered OLRA and 200-hPa eddy streamfunction based on PCs associated with PSA patterns were obtained with the same technique as composites based on OLRA PCs.

There is a good correspondence between evolution of the PSA modes and tropical convection related to the TISO (Mo and Higgins, 1998c) with major

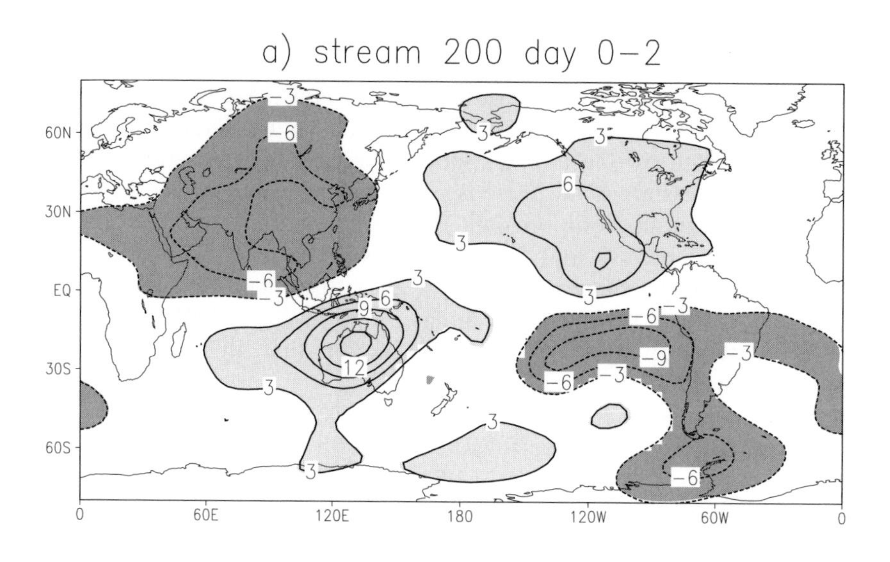

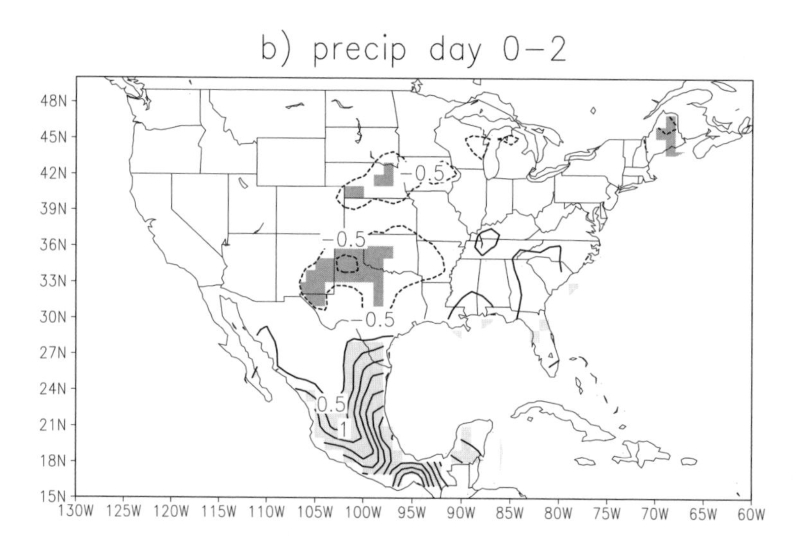

Figure 4.11. The 200-hPa streamfunction composite difference between positive and negative events averaged from day 0 to 2 for JJAS based on PC 1 (contour interval is $3 \times 10^6\,\mathrm{m^2\,s^{-1}}$, zero contours are omitted). Areas where positive (negative) anomalies are statistically significant at the 5% level are shaded light (dark). (b) Same as (a), but for precipitation (contour interval is $0.5\,\mathrm{mm\,day^{-1}}$).

contributions from the MJO. The connection to the TISO is demonstrated by the composites of the 10–90-day filtered OLRA based on the PSA PCs (Figure 4.12(c, d)). For JJAS, the PSA 1 is excited when the OLRA EOF 1 is strong (Figure 4.9(a)) with enhanced convection centered on the western Pacific and

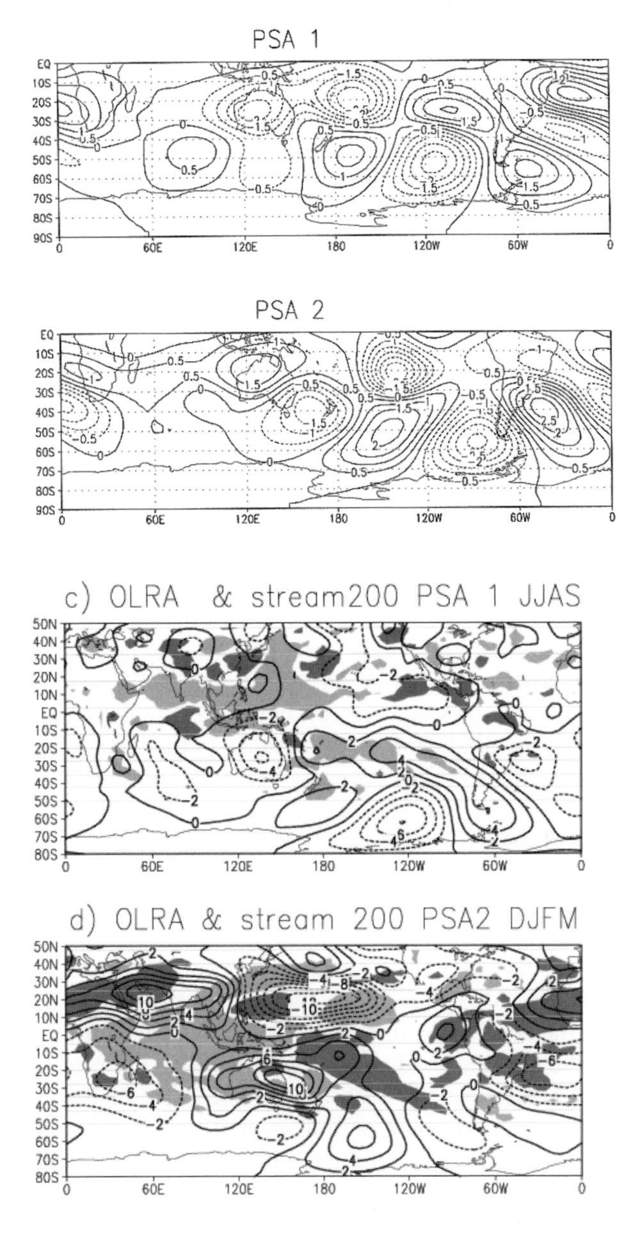

Figure 4.12. (a) PSA 1 and (b) PSA 2 from the first two EOFs for the 200-hPa streamfunction with zonal means removed. EOFs are normalized to 1 and time 100. Contour interval is 0.5 non-dimensional units. (c) Composites of 10–90-day filtered OLRAs (shading) and 200-hPa streamfunction eddy difference (contour) between positive and negative PSA 1 events for JJAS. Areas where OLRAs greater (less) than $3\,\mathrm{W\,m^{-2}}$ are shaded dark (light). Contour interval for the 200-hPa streamfunction composite is $2 \times 10^6\,\mathrm{m^2\,s^{-1}}$. (d) Same as (c) but for PSA 2 events for DJFM.

(a, b) From Mo and Higgins (1998c).

suppressed convection in the Indian Ocean. This pattern is associated with suppressed convection over north-eastern Brazil. For DJFM, the PSA 2 is linked to enhanced convection (light shading) in the western Pacific and suppressed convection in the central Pacific. In South America, PSA 2 is associated with suppressed rainfall over the SACZ and enhanced convection over the subtropical plains. This dipole pattern is the prominent convection pattern during South American winter (Nogues-Paegle and Mo, 1997).

4.4.3 Sub-monthly oscillation

The sub-monthly modes with periods around 10–28 days play an important role in modulating the Indian monsoon (Krishnamurthi and Andanuy, 1980) and the Asian monsoon (Krishnamurthi and Bhalme, 1976; Krishnamurthi *et al.*, 1985; Chen and Chen, 1993; Wu *et al.*, 1999). These are reviewed in Chapters 2, 3, and 5. Recently, a sub-monthly mode with fluctuations in the 10–25-day range has also been reported by Fukutomi and Yasunari (1999). When enhanced convection is located over the South China Sea, a wave train extends from the convective region to the North Pacific. Fukutomi and Yasunari (1999) suggested the interactions between the tropics and the subtropics play important roles on the Baiu front development. Linkages between the Asian monsoon and OLRA over the Americas are described with composites based on PC 4 at day 0 to 2 (Figure 4.13).

The OLRA composite (Figure 4.13(a)) indicates suppressed convection over the South China Sea flanked by enhanced convection in the Indian Ocean to the south-west and over the Pacific to the north-east. From the convective area, a wave train extends to the U.S.A. with positive anomalies over the Pacific north-west, negative anomalies over the central U.S.A. and positive anomalies over the east coast (Figure 4.13(b)). This is consistent with rainfall anomalies and OLRAs. Over the west coast of North America, there is suppressed convection over southern Mexico (Figure 4.13(c)), enhanced convection over northern Mexico and Arizona, and negative anomalies over the Pacific north-west. The linkages between rainfall over the South China Sea and North America also exist in the interannual band (Lau and Weng, 2000), but the wave train there has a longer wave length in comparison with the wave train in the intraseasonal band (Figure 4.13(b)). During JJAS, the influence of the MJO on rainfall over the south-west is not large. The dominant mode that influences rainfall active and break periods over the south-west is a mode of 20–28 days (Mo, 2000a), which is near the range of PC to 4. Over the southern hemisphere, there is also a wave train extending from the Indian Ocean downstream to South America. The corresponding OLRAs show positive anomalies over central South America and negative anomalies to the south.

4.5 THE INTRASEASONAL MODULATION ON HURRICANES

Previous sections have emphasized the atmospheric responses to the TISO and downstream impact on rainfall over the Pan-American region. There are also

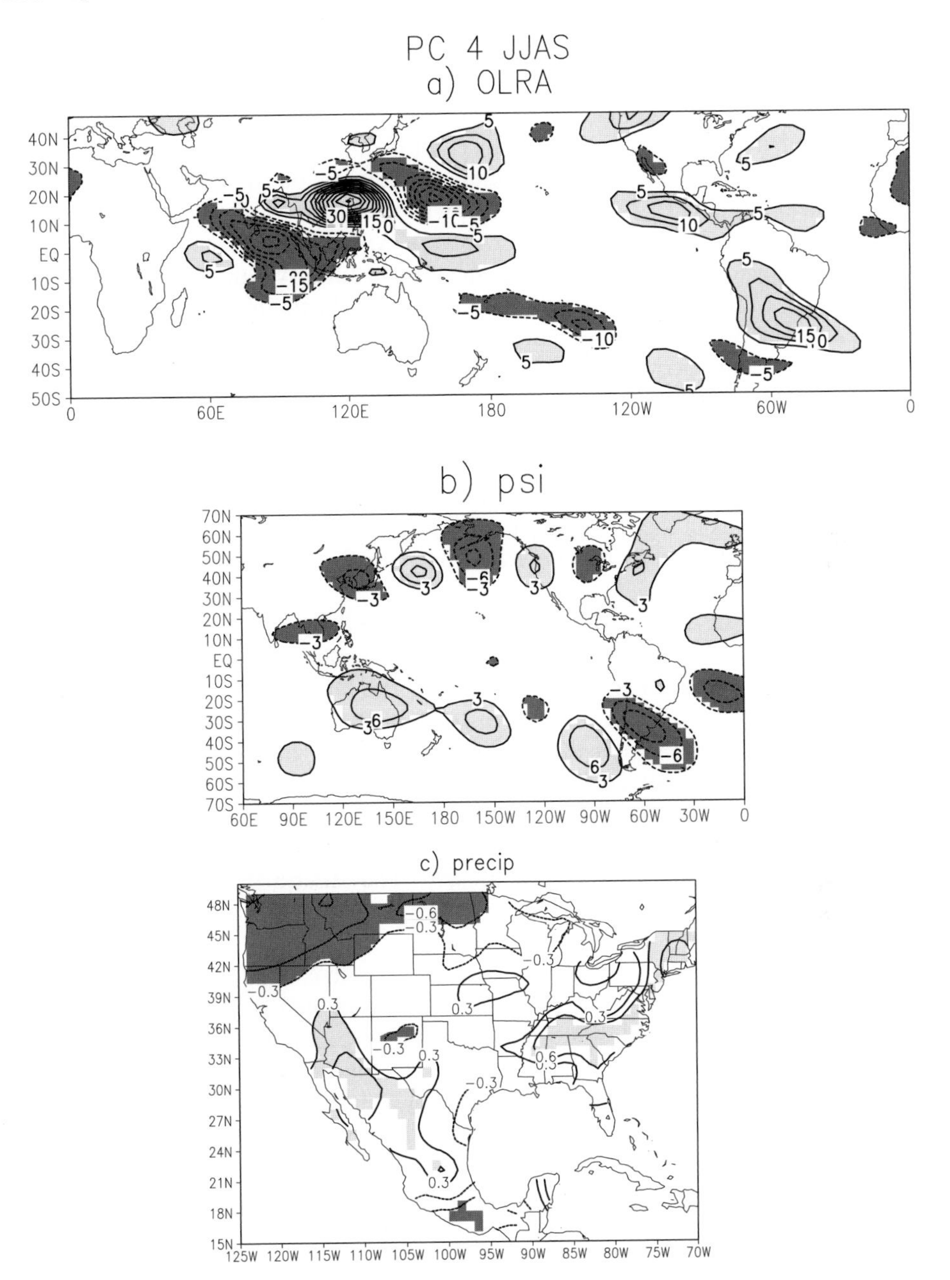

Figure 4.13. (a, b) The OLRA composite difference between positive and negative events averaged from day 0 to 2 for JJAS based on PC-4 (contour interval is $5\,\mathrm{W\,m^{-2}}$, zero contours are omitted). Areas where positive (negative) anomalies are statistically significant at the 5% level are shaded light (dark). (b) Same as (a) but for the 200-hPa streamfunction (contour interval is $3 \times 10^6\,\mathrm{m^2\,s^{-1}}$). (c) Same as (a) but for precipitation (contour interval is $0.3\,\mathrm{mm\,day^{-1}}$).

impacts on severe weather events such as in the occurrence of hurricanes. The following discussion documents this for North America since hurricanes are not observed over South America. In the Atlantic, hurricanes tend to occur during August–October when SSTAs are warm (Landsea, 1993; Gray, 1984). The hurricanes are most likely to develop in the area extending from the west coast of Africa to the tropical Atlantic (5°N–20°N), which was identified by Goldberg and Shapiro (1996) as the main development region for the Atlantic hurricanes. In the eastern Pacific, the tropical storm/hurricane season starts in June over the region close to the west coast of Mexico (10°N–20°N) (Maloney and Hartmann, 2000a). The major factor that controls the development of hurricanes is the vertical wind shear, with low values favoring hurricane formation (Goldberg and Shapiro, 1996).

In the interannual band, the occurrence of hurricanes or tropical storms in the Atlantic are modulated by ENSO (Gray, 1984), SSTAs in the Atlantic (Shapiro and Goldberg, 1998), and decadal SSTA trends in the Pacific. Enhanced convection related to warm SSTAs in the central Pacific caused by warm ENSO or by the decadal warm trends in the 1990s generates high wind shear in the main development region of the Atlantic sector, and below normal hurricane occurrence. The situation reverses for cold SSTAs.

While the MJO does not influence the total number of tropical storms and hurricanes in the Atlantic or eastern Pacific, it influences the periods when storms are most likely to occur. The positive phase of the OLRA EOF 1 (Figure 4.9(a)), with suppressed convection in the central tropical Pacific and enhanced convection over the Indian Ocean, favors more tropical storms in the Atlantic (Mo, 2000b) similar to the influence of cold ENSO events. The atmospheric response to this convection pattern is suppressed vertical wind shear in the Atlantic. Maloney and Hartmann (2000a, b) examined the impact of the MJO on tropical storms in the eastern North Pacific and the Gulf of Mexico. They found that the occurrence of hurricanes and tropical storms are regulated by the MJO related convection over the eastern Pacific.

The MJO modulation of tropical storm occurrence is not limited to the eastern Pacific and the Atlantic. Higgins and Shi (2001) examined the points of origin of tropical cyclones that developed into hurricanes/tropical storms in the western Pacific, the eastern Pacific, and the Atlantic (Figure 4.14). They composited numbers of tropical storms originated at any given location according to the phases of the MJO as indicated by the 200-hPa velocity potential from July–September. Strong tropical storms (open circles) are more likely to develop in regions where the MJO favors enhanced convection. As the MJO moves eastward, the favored region of storm development also moves eastward from the western Pacific to the eastern Pacific and into the Atlantic basin.

4.6 SUMMARY

Convection associated with the TISO is an important source to the circulation and precipitation variability in the vicinity of the region of maximum variability in

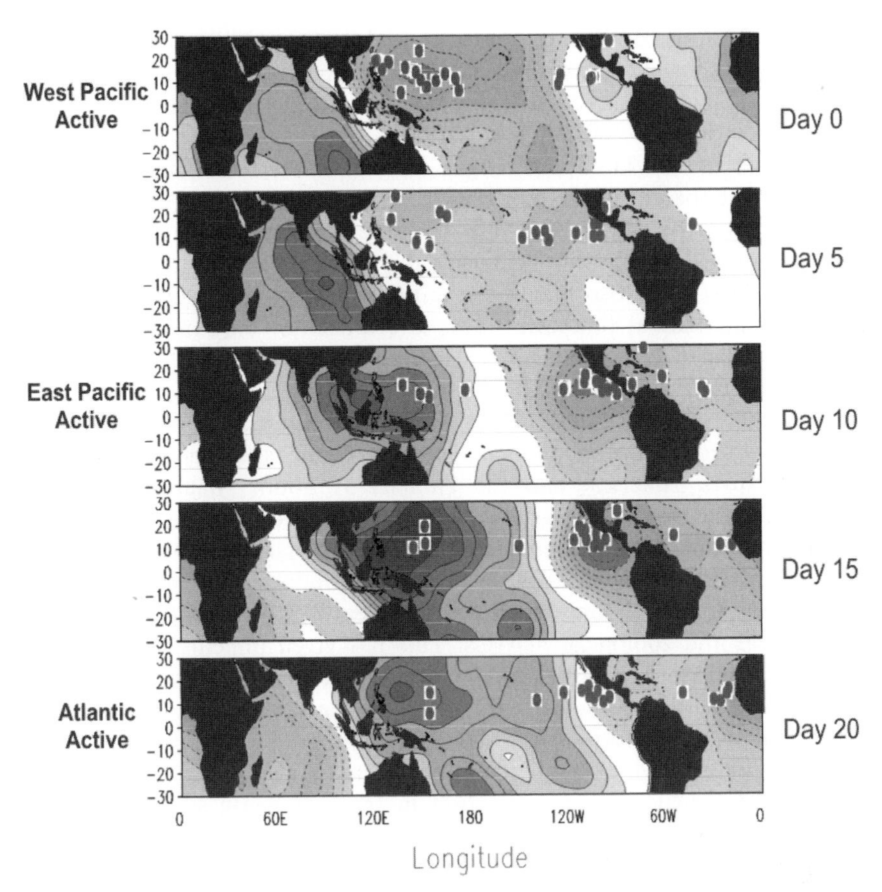

Figure 4.14. Composite evolution of the 200-hPa velocity potential anomalies together with points of origin of tropical cyclones that developed into hurricanes/typhoons (open circles) for the 35-day period from day -15 to day $+15$. Composites are based on 21 events. Hurricane track data for JAS 1979–1997 were used (contour interval is $0.5 \times 10^6\,\mathrm{m^2\,s^{-1}}$, negative contours are dashed, and zero contours are omitted).
From Higgins and Shi (2001).

intraseasonal timescales (see Chapters 2, 3, and 5). It also has an effect on the Pan-American region. This chapter shows linkages to the Pan-American region that are established through wave trains to both North and South America (Mo and Nogues-Paegle, 2001). The geography of these two continents is quite different. The North America land mass lies mostly in mid-latitudes, while South America is a tropical continent. Though both continents have extensive meridionally oriented mountain ranges, the Rockies gently slope towards the east, while the Andes abruptly decrease from a high-level plateau with averaged heights of 4,000 m to sea level in a distance of a few hundred kilometers. In spite of these differences there are remarkable similarities in circulation and rainfall response over Pan-America and its vicinity.

The PNA and PSA patterns act as guides to modulate weather anomalies over these continents in synchronization with variations in tropical convection. Over both the North and South Pacific, there is a center of marked response to the TISO in regions of frequent blocking. The OLRA over the two continents exhibit dipole structures and the impact of the TISO is stronger on the winter hemisphere. Over South America, this is apparent in the modulation of the SACZ and the establishment of the rainfall dipole between the SACZ and the subtropical plains. Over North America, the rainfall dipole with centers at the Pacific north-west and California is regulated by the TISO. Also of interest is the similarity of the response at intraseasonal and interannual timescales. This is evident not only in the modulations of dipole patterns of convection, but also in the generation and development of hurricanes.

Identification of sources of variability in the two continents is of great social and economic significance. Both continents possess fertile subtropical plains (Great Plains over North America, and the Pampas of South America) with strong weather dependence on their agriculture and hydroelectric industries. There are also some important waterways in these plains: the Mississippi and La Plata basin and their tributaries provide passageways to move goods downstream. The Amazon basin of South America is one of the world's treasures of natural resources, with intact botanical and zoological marvels. This has given impetus to the efforts to better understand and measure the processes responsible for large weather anomalies that result in extended periods of floods and draughts. The expectation is that this will result in improved predictions with the consequent positive impact on regional economies. Nogues-Paegle *et al.* (1998) find that errors that develop early in the forecast during suppressed convection in the SACZ are smaller than errors developed when convection is enhanced. Since the SACZ events last between 5–10 days, they are influenced by both the MJO and the sub-monthly mode. Jones and Schemm (2000) extended their results to examine 50-day forecasts for five years. They confirmed that forecast skill depends on the TISO regimes. The model has relatively high skill during strong SACZ dominated by the MJO. However, the skill is low if the SACZ is dominated by the sub-monthly mode.

There are several national and international initiatives aimed at improved long-term predictability. International efforts have been organized under the Climate Variability and Predictability/Variability of the American Monsoon Systems (CLIVAR/VAMOS) panel to address a number of issues related to improved prediction of the summer rains. Within the U.S.A. community, the USCLIVAR/PANAM (Pan-American) panel has parallel objectives with a focus on North America. Two important field experiments, NAME (North American Monsoon Experiment) and SALLJEX (South American Low-Level Jet Experiment) have collected unique data sets that will help understand, among other goals, moisture fluxes that fuel massive convective systems over the continents. These experiments will also contribute to answer questions relevant to the Global Water Cycle, and thus of relevance to Global Energy and Water Cycle Experiment (GEWEX), to quantify the variability of the water cycle and to assess the degree to which it is predictable. SALLJEX was a three-month experiment (15 November, 2002–15 February, 2003) and included a marked ISO that appears to have had a positive impact on the two-week prediction by global models over South America. Unique data sets collected during these experiments offer a scientific challenge to translate observational improvements into consistent predictability gains by numerical model improvements. Such gains will be realized when models are capable of correctly simulating diurnal cycles, the formation and development of tropical convection through better parameterizations, and the ISV observed in nature.

4.7 REFERENCES

Branstator, G. (1987) A striking example of the atmosphere leading traveling pattern. *J. Atmos. Sci.*, **44**, 2310–2323.

Carvahlo, L. M. V., C. Jones, and B. Liebmann (2002) Extreme precipitation events in southeastern South America and large-scale convective patterns in the South Atlantic Convergence Zone. *J. Climate*, **15**, 2377–2394.

Carvahlo, L. M. V., C. Jones, and B. Liebmann (2004) The South Atlantic Convergence Zone: Intensity, form, persistence and relationships with intraseasonal to interannual activity and extreme rainfall. *J. Climate*, **17**, 88–108.

Chen T. C. and J. M. Chen (1993) The 10–20-day mode of the 1979 Indian monsoon: Its relation with the time variation of monsoon rainfall. *Mon. Wea. Rev.*, **121**, 2465–2482.

Fukutomi, Y. and T. Yasunari (1999) 10–25-day intraseasonal variations of convection and circulation over east Asia and western North Pacific during early summer. *J. Met. Soc. Jap.*, **77**, 753–769.

Gadgil, S. (1983) Recent advances in monsoon research with particular reference to the Indian monsoon. *Aust. Met. Mag.*, **36**, 193–204.

Goldberg, S. B. and L. J. Shapiro (1996) Physical mechanisms for the association of El Niño and West African rainfall with Atlantic major hurricane activity. *J. Climate*, **9**, 1169–1187.

Gray, W. M. (1984) Atlantic seasonal hurricane frequency. Part I: El Niño and 30-mb quasi biennial oscillation influences. *Mon. Wea. Rev.*, **112**, 1649–1668.

Higgins, R. W. and W. Shi (2001) Intercomparison of the principal modes of intraseasonal and interannual variability of the North American monsoon system. *J. Climate*, **14**, 403–417.

Higgins, R. W. and K. C. Mo (1997) Persistent North Pacific circulation anomalies and the tropical intraseasonal oscillation. *J. Climate*, **10**, 223–244.

Higgins, R. W., W. Shi, E. Yarosh, and R. Joyce (2000) *Improved United States Precipitation Quality Control System and Analysis* (NCEP/Climate Prediction Center ATLAS No. 7). NCEP /NWS/ NOAA, available from NCEP, 47 pp.

Horel, J. D. and M. J. Wallace (1981) Planetary scale atmospheric phenomena associated with the Southern Oscillation. *Mon. Wea. Rev.*, **109**, 813–929.

Jones, C. and J.-K. E. Schemm (2000) The influence of intraseasonal variations on medium to extended range weather forecasts over South America. *Mon. Wea. Rev.*, **128**, 486–494.

Jones, C., L. M. V. Carvalho, R. W. Higgins, D. E. Waliser, and J.-K. E. Schemm (2004) Climatology of tropical intraseasonal convective anomalies: 1979–2002. *J. Climate*, **17**, 523–539.

Kalnay E., M. Kanamitsa, R. Kistler, W. Collins, D. Deaven, W. Gandin, M. Iredell, S. Jaha, G. White, J. Wollen, Y. Zhu, *et al.* (1996) The NMC/NCAR CDAS/Reanalysis Project. *Bull. Amer. Meteor. Soc.*, **77**, 437–471.

Kiladis, G. N. and K. M. Weickmann (1992a) Circulation anomalies associated with tropical convection during northern winter. *Mon. Wea. Rev.*, **120**, 1900–1923.

Kiladis, G. N. and K. M. Weickmann (1992b) Extratropical forcing of tropical Pacific convection during northern winter. *Mon. Wea. Rev.*, **120**, 1924–1938.

Kiladis, G. N. and K. C. Mo (1998) Interannual and intraseasonal variability in the Southern Hemisphere. *Meteorology of the Southern Hemisphere.* In: D. Karoly and D. G. Vincent (eds), AMS, Boston, pp. 307–335.

Knutson, T. R. and K. W. Weickmann (1987) 30–60-day atmospheric oscillations: Composite life cycle of convection and circulation anomalies. *Mon. Wea. Rev.*, **115**, 1407–1436.

Kodama, Y.-M. (1992) Large-scale common features of subtropical precipitation zones (the Baiu frontal zone, the SPCZ and the SACZ). Part I: Characteristics of subtropical frontal zones. *J. Meteor. Soc. Jap.*, **70**, 813–835.

Kodama, Y.-M. (1993) Large-scale common features of subtropical precipitation zones (the Baiu frontal zone, the SPCZ and the SACZ). Part II: Conditions of the circulations for generating the STCZs. *J. Meteor. Soc. Jap.*, **71**, 581–610.

Kousky, V. (1985a) Atmospheric circulation changes associated with rainfall anomalies over tropical Brazil. *Mon. Wea. Rev.*, **113**, 1951–1957.

Kousky, V. (1985b) Frontal influences on north-east Brazil. *Mon. Wea. Rev.*, **107**, 1140–1153.

Krishnamurti, T. N. and H. Bhalme (1976) Oscillations of a monsoon system. Part 1: observational aspects. *J. Atmos. Sci.*, **33**, 1937–1954.

Krishnamurti, T. N. and P. Andanuy (1980) The 10–20-day westward propagating mode and breaks in the monsoon. *Tellus*, **32**, 15–26.

Krishnamurti, T. N., P. K. Jayaakaumar, J. Sheng, N. Surgi, and A. Kumar (1985) Divergent circulation on the 30–50-day timescale. *J. Atmos. Sci.*, **42**, 364–375.

Landsea, C. W. (1993) A climatology of intense (major) Atlantic hurricanes. *Mon. Wea. Rev.*, **121**, 1703–1713.

Lau, K. M. and P. H. Chan (1985) Aspects of the 40–50-day oscillations during the northern winter as inferred from outgoing long-wave radiation. *Mon. Wea. Rev.*, **113**, 1889–1909.

Lau, K. M. and P. H. Chan (1986) Aspects of the 40–50-day oscillation during the northern summer as inferred from outgoing long-wave radiation. *Mon. Wea. Rev.*, **114**, 1345–1367.

Lau, K. M. and H. Weng (2000) Recurrent teleconnection patterns linking summer time precipitation variability over East Asia and North America. *J. Meteor. Soc. Jap.*, **80**, 1309–1324.

Liebmann, B. and D. L. Hartmann (1984) An observational study of tropical–mid-latitude interaction on intraseasonal timescales during winter. *J. Atmos. Sci.*, **41**, 3333–3350.

Liebmann, B. and C. A. Smith (1996) Description of a complete (interpolated) outgoing longwave radiation dataset. *Bull. Amer. Meteor. Soc.*, **77**, 1275–1277.

Liebmann, B., G. N. Kiladis, J. A. Marengo, T. Ambrizzi, and J. D. Glick (1999) Submonthly convective variability over South America and the South Atlantic convergence zone. *J. Climate*, **12**, 1899–1891.

Liebmann, B., G. N. Kiladis, C. S. Vera, A. C. Saulo, and L. M. V. Carvalho (2004) Subseasonal variations of rainfall in South America in the vicinity of the low level jet east of the Andes and comparison to those in the South Atlantic Convergence Zone. *J. Climate*, **17**, 3829–3842.

Maloney, E. D. and D. L. Hartmann (2000a) Modulation of eastern North Pacific hurricanes by the Madden–Julian Oscillations. *J. Climate*, **13**, 1451–1460.

Maloney, E. D. and D. L. Hartmann (2000b) Modulation of hurricane activity in the Gulf of Mexico by the Madden–Julian Oscillations. *Science*, **287**, 2002–2004.

Mo, K. C. (1999) Alternating wet and dry episodes over California and intraseasonal oscillations. *Mon. Wea. Rev.*, **127**, 2759–2776.

Mo, K. C. (2000a) Intraseasonal modulation of summer precipitation over North America. *Mon. Wea. Rev.*, **128**, 1490–1505.

Mo, K. C. (2000b) The association between intraseasonal oscillations and tropical storms in the Atlantic basin. *Mon. Wea. Rev.*, **128**, 4097–4107.

Mo, K. C. and R. W. Higgins (1998a) Tropical convection and precipitation regimes in the western United States. *J. Climate*, **11**, 2404–2423.

Mo, K. C. and R. W. Higgins (1998b) Tropical influences on California precipitation. *J. Climate*, **11**, 412–430.

Mo, K. C. and R. W. Higgins (1998c) The Pacific South American modes and tropical convection during the Southern Hemisphere winter. *Mon. Wea. Rev.*, **126**, 1581–1596.

Mo, K. C. and J. Nogues-Paegle (2001) The Pacific South American modes and their downstream effects. *Int. J. of Climatology.*, **21**, 1211–1229.

North, G. R., T. L. Bell, R. F. Cahalan, and F. J. Moeng (1982) Sampling errors in the estimation of empirical orthogonal functions. *Mon. Wea. Rev.*, **110**, 699–702.

Nogues-Paegle, J. and K. C. Mo (1987) Spring-to-summer transitions of global circulations during May–July 1979. *Mon. Wea. Rev.*, **115**, 2088–2102.

Nogues-Paegle, J. and K. C. Mo (1997) Alternating wet and dry conditions over South America during summer. *Mon. Wea. Rev.*, **125**, 279–291.

Nogues-Paegle, J., A. Byerle, and K. C. Mo (2000) Intraseasonal modulation of South American summer precipitation. *Mon. Wea. Rev.*, **128**, 837–850.

Nogues-Paegle, J. and K. C. Mo (1998) Predictability of the NCEP–NCAR reanalysis model during austral summer. *Mon. Wea. Rev.*, **126**, 3135–3152.

Nogues-Paegle, J., B.-C. Lee, and V. Kousky (1989) Observed modal characteristics of the intraseasonal oscillation. *J. Climate*, **2**, 496–507.

Papoulis, A. (1973) Minimum bias windows for high resolution spectral estimates. *IEEE Trans Infor. Theory*, **19**, 9–12.

Roplewski, C. F. and H. S. Halpert (1987) Global and regional precipitation patterns associated with the El Niño/Southern Oscillation. *Mon. Wea. Rev.*, **115**, 1606–1626.

Roplewski, C. F. and H. S. Halpert (1989) Precipitation patterns associated with the high index phase of the Southern Oscillation. *J. of Climate*, **2**, 268–284.

Shapiro, L. J. and S. B. Goldberg (1998) Atlantic sea surface temperatures and tropical cyclone formation. *J. Climate*, **11**, 578–590.

Vautard, R. and M. Ghil (1989) Singular spectrum analysis in non linear dynamics with applications to paleoclimatic time series. *Physica D.*, **35**, 392–424.

Wallace, J. M. and D. S. Gutzler (1981) Teleconnections in the geopotential height field during the Northern Hemisphere center. *Mon. Wea. Rev.*, **109**, 784–812.

Waliser, D. E., N. E. Graham, and C. Gautier (1993) Comparison of the high reflective cloud and outgoing longwave radiation data sets for use in estimating tropical deep convection. *J. Climate*, **6**, 331–353.

Weickmann, K. M. (1983) Intraseasonal circulation and outgoing longwave radiation modes during Northern Hemisphere winter. *Mon. Wea. Rev.*, **111**, 1838–1858.

Weickmann, K. M., G. R. Lussky, and J. Kutzbach (1985) Intraseasonal circulation and outgoing longwave radiation modes during Northern Hemisphere winter. *Mon. Wea. Rev.*, **111**, 1838–1858.

Wu, M. L., S. Schubert, and N. E. Huang (1999) The development of the South Asian summer monsoon and intraseasonal oscillation. *J. of Climate*, **12**, 2054–2075.

Xie, P. and P. A. Arkin (1997) Global precipitation: A 17-year monthly analysis based on gauge observations, satellite estimates, and numerical model outputs. *Bull. Amer. Met. Soc.*, **78**, 2539–2558.

Xie, P. P., J. E. Janowiak, P. A. Arkin, R. Alder, A. Gruber, R. Ferraro, G. J. Huffmann, and S. Curtis (2003) GPCP pentad precipitation analyses: An experimental data set based on gauge observations and satellite estimates. *J. Climate*, **16**, 2197–2214.

5

Australian–Indonesian monsoon

M. C. Wheeler and J. L. McBride

5.1 INTRODUCTION

Like its northern hemisphere counterparts (e.g., Asian monsoons of Chapters 2 and 3), the region of northern Australia and nearby longitudes, and the area immediately to its north (primarily within Indonesia), experience a marked seasonal cycle in winds and precipitation characteristic of a monsoon (e.g., Troup, 1961; McBride, 1987; Suppiah, 1992). At lower tropospheric levels, the mean winds shift from being easterly in austral winter, with correspondingly small rain totals, to westerly in summer, with much enhanced cumulonimbus convection and rainfall (e.g., Figure 5.1). This monsoonal character of the region has long been recognized. Indeed, for both northern Australia and Indonesia, reference to this nature dates back at least as far as the early 19th century.[1]

Given these defining monsoon characteristics, there is understandably a large climatic influence on the lifestyles and practices of the people of the area. In particular, the monsoon seasonal cycle has a governing influence on agriculture, and in times past has had a large influence on navigation and trade. Such influences have undoubtedly played an important role for the highly populated islands of Indonesia (e.g., Java and Bali), making research on year-to-year variability of the monsoon clearly important. Indeed, there has been a long history of studies of interannual variability (IAV), with multiple examples from both the colonial period (e.g.,

[1] For northern Australia, the climatologists of the early 20th century used the word monsoon to describe the climate of the region (e.g., Hunt *et al.*, 1913). During the same period the term monsoon (or in Dutch, "moesson") was used by government meteorologists of the Netherlands Indies (modern-day Indonesia) to describe the climate of the southern hemisphere parts of Indonesia. For example, Braak (1919) described the region of the "Malay Archipelago" as "the most typical monsoon region of the world". The term was sufficiently entrenched that it appears in even earlier writings about the region such as in the journals of the explorers Matthew Flinders (1814) and Alfred Russel Wallace (1891). Indeed, the Indonesian language provides no distinction between the words for "season" and "monsoon", for which they use the word "musim".

W. K. M. Lau and D. E. Waliser (eds), *Intraseasonal Variability in the Atmosphere–Ocean Climate System*.

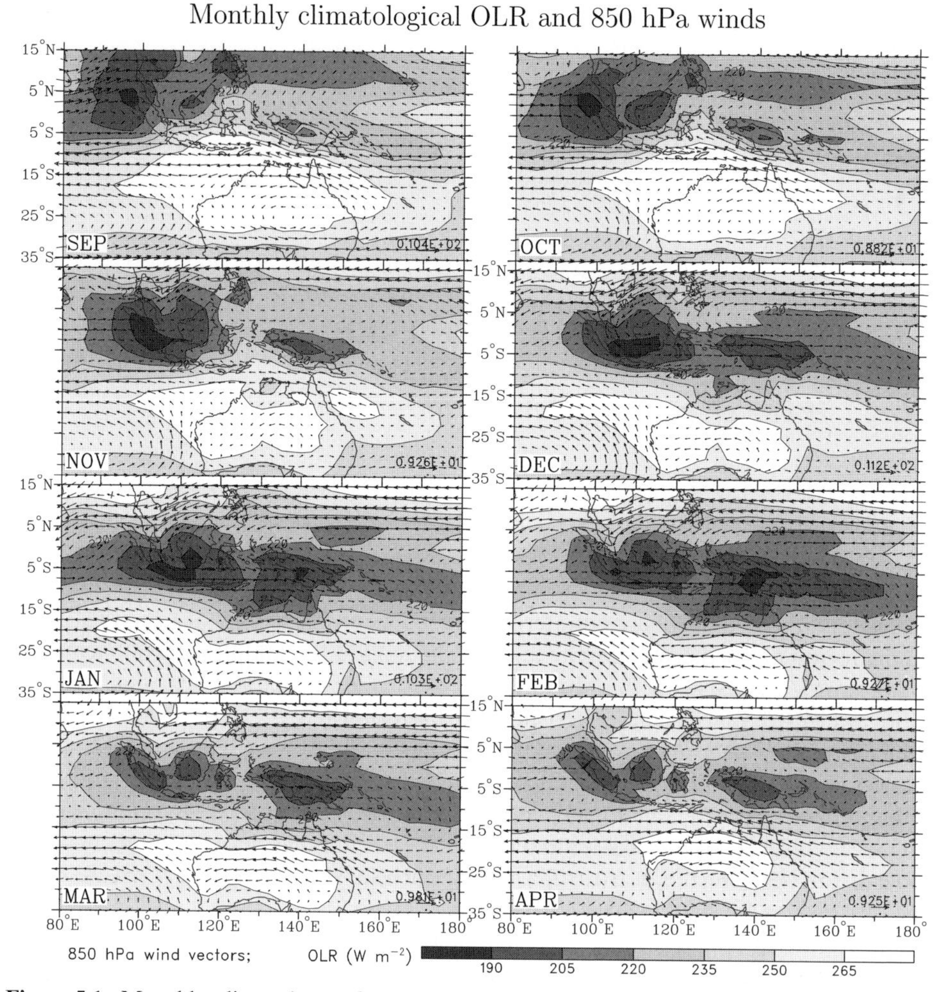

Figure 5.1. Monthly climatology of NOAA satellite-observed outgoing long-wave radiation (OLR) and NCEP/NCAR Reanalysis 850-hPa level winds, based on data from 1979–2002. All wind vectors are scaled the same, and the maximum vector for each panel is as displayed in the bottom-right corner. Low values of OLR are indicative of cold cloud tops as produced by precipitating cumulonimbus convection.

Berlage, 1927; de Boer, 1947) and the modern research era (e.g., Nicholls, 1981; Hastenrath, 1987; Naylor *et al.*, 2001 and references therein). Yet, as is the case for the other monsoon areas of the world (Chapters 2–4), intraseasonal variability (ISV) is also prominent. Research on ISV within the region, however, has had a comparatively shorter history, and for Indonesian meteorological parameters in particular, is relatively absent.

The beginning of published research on ISV in the region appears, to our knowledge, to be the seminal paper by Troup (1961). Troup, concentrating on the

Australian component of the monsoon system, used the term "bursts" to describe spells of excessively wet or low-level westerly conditions occurring for periods shorter than the overall summer season. "Onset" was then naturally defined as being the beginning of the first westerly burst in each wet season. Such bursts, by definition, are the manifestation of sub-seasonal variability within the monsoon, for which a very large variance component falls within the range we classify here as ISV. Since Troup's paper, many authors have referred to active (burst) and break events in the context of the Australian monsoon (e.g., Murakami and Sumi, 1982; McBride, 1983; Holland, 1986; Gunn *et al.*, 1989; Drosdowsky, 1996; McBride and Frank, 1999), and the current operational methodologies to define onset and active vs. break periods follow the framework Troup developed.

Yet there have been other influences upon the development of the region's research. Due partly to the utility of Darwin in northern Australia as a base for meteorological field experiments, the Australian monsoon has received focussed study in the last two decades (e.g., see review chapters by McBride, 1987, 1998; Suppiah, 1992; Manton and McBride, 1992). In parallel with this development, internationally there has been a recognition of the importance of variability on intraseasonal timescales in tropical weather and climate, and in particular on the role of the Madden–Julian Oscillation (MJO) as a large-scale control (e.g., Madden and Julian, 1971, 1994; Weickmann *et al.*, 1985; Lau and Chan, 1985; Knutson and Weickmann, 1987; Wang and Rui, 1990; Hendon and Salby, 1994; see also Chapter 1). Despite this, with the notable exception of the work by Hendon and collaborators (Hendon *et al.*, 1989; Hendon and Liebmann, 1990a, b; Wheeler and Hendon, 2004), the research literature on bursts and breaks in the Australian monsoon has seemed to largely ignore or downplay the influence of the MJO. Indeed, a seemingly great diversity of views about the importance of the MJO can be found. This is particularly evident in the case of monsoon onset, for which Hendon and Liebmann (1990a) implied that nearly all onsets result from the passage of a convectively-active MJO phase, while others have implied that the MJO has little or no impact (e.g., Davidson *et al.*, 1983; Drosdowsky, 1996).

Even more recently there has been an increased recognition of other "modes" of ISV within the monsoons, besides just the MJO. In particular, ever-increasing attention is being paid to the pure shallow-water-like equatorial waves, as have now been well-observed to couple to convection, and produce prominent perturbations in the near-equatorial monsoons (e.g., Takayabu, 1994; Numaguti, 1995; Wheeler and Kiladis, 1999; Wheeler *et al.*, 2000; Wheeler and Weickmann, 2001; Straub and Kiladis, 2002). Comparatively less, however, is known about such modes, especially within the Australian–Indonesian region.

Within this context then, this chapter provides a timely review and synthesis on the topic of ISV within this monsoon region. Included in the contents is a general description of the climatological seasonal cycle of the region, as it is this that forms the necessary background state about which ISV appears (Section 5.2). Although exceptions can be found, ISV is usually defined as covering all timescales of variability beyond the synoptic limit ($\sim$10 days) to less than a season ($\sim$90 days). We adopt this definition here. The earliest work concerning variability within this broad range in the Australian–Indonesian region is discussed in Section 5.3. Frequency-only power and coherence spectra of the standard monsoon variables are examined in Section 5.4. While the MJO is the only prominent spectral peak, the fact that much variance does exist in the intraseasonal band, and much of it is highly coherent, is important.

The local manifestation in the monsoon of the spectrally-broad ISV, irrespective of its source, are the bursts and breaks, the general meteorology of which is discussed in Section 5.5. Section 5.6 is then devoted to the character and influence specifically of the MJO in the region. The discussion of the previous sections is then contrasted and compared to the evolution of two individual monsoon years, 1983/1984 and 1987/1988 (Section 5.7). Importantly, the fact that all years show at least some degree of burst or break activity (and hence have ISV) while not necessarily showing strong MJO variability, is highlighted. Section 5.8 then addresses the contrasting views that have been expressed on the importance of the role of the MJO, especially with regards to monsoon onset. As such, it is our attempt to reconcile the above-mentioned differences of opinion. We then turn our attention to the other sources and modes of ISV in the region (Section 5.9), the modulation of extreme events in the region by ISV (Section 5.10), and the region's intraseasonal extratropical–tropical interaction (Section 5.11).

5.2 SEASONAL CYCLE OF BACKGROUND FLOW

In this chapter we focus on the tropical region south of the equator and within the longitude bounds of about 100°E–170°E. Included in the region are the islands of New Guinea, Timor, Flores, Celebes, Bali, Java, and much of Sumatera, along with northern Australia. As has been presented in several papers (e.g., Troup, 1961; Meehl, 1987; Drosdowsky, 1996), the mean seasonal cycle of the region is characterized by a reversal in lower tropospheric winds and a marked change in rainfall. An appreciation of such seasonal changes can be made with reference to the climatological monthly mean fields of Figure 5.1, showing the 850-hPa level winds, as representative of winds of the lower troposphere, and the satellite-observed OLR field, as representative of the cold cloud tops of rain-producing deep-convective systems. Twenty-four years of data have been used.

In September, the prevailing mean 850-hPa winds across the defined region are south-easterly trade winds emanating from the subtropical ridge lying along about 25°S. The strength of these trade winds is about 5 to 9 $\mathrm{m\,s^{-1}}$. At this time of year, the strongest convective activity (as represented by OLR of less than 220 $\mathrm{W\,m^{-2}}$) is restricted to being mostly north of the equator, especially in the north-west of the plotted domain. This convective activity is associated with the northern hemisphere summer monsoon still being active; and the line of convection (low OLR) along the northern boundary of the domain denotes the location of the Intertropical Convergence Zone (ITCZ). As the seasonal cycle progresses into the months of October and November, there is an overall decrease in the low-level easterlies together with a shift of the strongest convective activity southward and eastward. This is first marked by a build up of convection over the large islands of Sumatra, Borneo, and New Guinea during October to November, coinciding with monsoon onset on those islands by some definitions (e.g., Tanaka, 1994). By December, the convective activity has transformed into a continuous line across the region centred near 5°S, and represents the ITCZ having shifted south of the equator.

As the ITCZ becomes established south of the equator in December, westerly winds appear between the equator and 10°S, while the trade winds retreat southwards over the Australian continent. At this time a small portion of the strong convective activity (OLR $< 220\,\mathrm{W\,m^{-2}}$) has reached the Australian continent at Darwin (12.5°S, 130.9°E), consistent also with the late December mean monsoon onset date there (Holland, 1986; Hendon and Liebmann, 1990a; Drosdowsky, 1996). By January to February, the peak of the southern hemisphere summer season tropical convective activity is reached. This peak occurs earlier in the west (e.g., Java) than the east (e.g., New Guinea). The defined region of interest (i.e., essentially the Australian–Indonesian tropical latitudes) is now mostly occupied by westerlies, with magnitudes up to $9\,\mathrm{m\,s^{-1}}$. At this stage the trade wind easterlies have also strengthened across the southern part of the plotted domain. Consequently, along about 10°S–15°S there is a well-defined monsoon shear line marked by the line of strong cyclonic ($-\partial u/\partial y$) shear separating lower latitude westerlies from higher latitude easterlies (McBride and Keenan, 1982; McBride, 1995).

For much of the domain of interest, the overall character of the sequence from October through to February is thus a replacement of dry easterlies with convective westerlies. This seasonal character, like for other monsoons, is mostly thought to occur due to the existence of the land–sea thermal contrast, resulting in this case from the location of the off-equatorial Australian continent (e.g., Webster *et al.*, 1998). As modeled by Yano and McBride (1998), however, there is also a forcing due to the seasonal excursion of the warmest sea surface temperatures (SSTs) south of the equator during the southern summer months. At the location of Darwin, the actual changes in convection that occur correspond to a climatological mean rainfall peak of around $12\,\mathrm{mm\,day^{-1}}$ in February, to less than $1\,\mathrm{mm\,day^{-1}}$ from June to September (Drosdowsky, 1996). Further seasonal changes that are known to occur are an increase in upper-level easterlies around Darwin during the summer months, and a southward movement of the southern hemisphere subtropical jet (Troup, 1961). As will be discussed next, however, the appearance of any particular year can be quite different to this slowly evolving mean seasonal cycle.

5.3 BROADBAND INTRASEASONAL BEHAVIOR: BURSTS AND BREAKS

Among the first to appreciate and document elements of intraseasonal behavior of the Australian–Indonesian monsoon region was Troup (1961). In examining time series of daily mean winds and rainfall in the Darwin area, he demonstrated that each wet season consisted of a number of spells of heavy rains, or "bursts", each lasting the order of a few days to a week or more. Similarly, the seasonal changes in winds in each year were the result of more frequently occurring, and longer lasting wind spells, and not necessarily from the establishment of a steady regime. The heavy rain bursts tended to coincide with low-level westerly wind bursts, and in the 4 years he examined, there were between 1 and 6 bursts per summer season (viz., during December–March), the in-between dry spells being called "breaks". The yearly monsoon onset he defined corresponded to the first such wet or westerly burst, and when viewed relative to these monsoon onset dates, the transition to active monsoon conditions took only a matter of days.

A very similar appreciation of the intraseasonal behavior of the region can be obtained from the multiple panels of Figure 5.2. The upper curve in each panel shows the satellite OLR field averaged over the indicated area encompassing a large section of the monsoon region. The lower curve shows rainfall averaged over a much smaller area in northern Australia. The only time smoothing that has been applied is that of a 3-day running mean.[2] Superimposed upon the OLR curve is the climatological seasonal cycle (dashed curve) computed from the long-term mean and three harmonics, with shading to denote anomalies. Given that downward excursions of the OLR curve represent convectively active conditions, a reasonably close correspondence between the large-scale convective conditions, and the smaller scale rainfall is apparent. Also apparent are the characteristic monsoon bursts, many appearing in both the OLR and rainfall. For example, during the 1984/1985 monsoon season there are two notable bursts in the OLR, both showing similarly-timed peaks in the rainfall. In 1987/1988, there are three notable bursts, also with matching peaks in the rainfall. Much of the variance of these bursts can be identified as being that of ISV (i.e., having a timescale of about 10 to 90 days).

Given our identification of the bursts in Figure 5.2 as ISV, one important aspect of the region's ISV is revealed, and this is that its amplitude appears just as large or larger than that of the seasonal cycle. One way that this can be appreciated is through consideration of the absolute minimum OLR value reached in each year. By the seasonal cycle alone, this value is only 200 $\mathrm{W\,m^{-2}}$, but due to the presence of the ISV, it actually reaches minimum values below 180 $\mathrm{W\,m^{-2}}$ and usually below 160 $\mathrm{W\,m^{-2}}$, and this can occur at any time between December and the end of March. Conversely, in at least half of the years, the mid-summer monsoon breaks are strong enough to fully negate the effects of the seasonal cycle (in the OLR), causing conditions equivalent to the dry season. The break of late February/early March 1988 is a good example.

Within the framework of the "burst" definition by Troup (1961), there also came a natural definition of the "onset" of the monsoon as being the first such burst of each wet season. Following this methodology, a number of authors have developed objective definitions of onset for the Australian portion of the monsoon system, as also presented in Figure 5.2. Hendon and Liebmann (1990a) based their definition on the simultaneous satisfaction of two criteria: that the low-pass filtered (1-2-3-2-1)

[2] A 3-day running mean effectively serves as a low-pass filter with a half-power point near a period of 6 days, and passing 90% power at a period of 9 days.

daily time series of 850-hPa wind at Darwin become positive (westerly), and that the average rainfall rate for Australian stations equatorward of 15°S exceed 7.5 mm day^{-1}. Thus their definition was based on the concept of the first occurrence of "wet westerlies". Hung and Yanai (2004) had a similar definition based on 850-hPa zonal wind over the region 2°S–15°S and 115°E–150°E, and simultaneous satisfaction of a criterion that OLR over the same box be less than a threshold value. Drosdowsky's (1996) definition, on the other hand, is based purely on the zonal winds at Darwin, requiring the mean lower tropospheric (surface to 500 hPa) wind to be westerly, with an additional requirement being that the upper tropospheric (300–100 hPa) wind be easterly. Obviously, each definition is subtly different, but for the sake of the current discussion, we have simply taken all their dates of onset and indicated them on Figure 5.2 with arrows. The main point portrayed is that for each of the years shown, as we have also confirmed is generally the case for all years, all the defined onset dates coincide with a large-scale OLR-measured intraseasonal burst. Further, as the intraseasonal bursts have no apparent phase-locking to the seasonal cycle, they introduce an important interannual variation in the defined monsoon onsets. Indeed, in the 6 years shown, the onsets vary anywhere from early December to mid-January.

Yet a further consequence of the occurrence of strong ISV within the monsoon, as can also be inferred from Figure 5.2, is that the changes occurring at the time of each year's monsoon onset (as defined above), are invariably much more rapid than that implied by the mean seasonal cycle alone. A natural extension of defining a monsoon onset is thus to view the circulation changes with respect to that onset date, and not just with respect to the seasonal cycle. Each of the studies of Hendon and Liebmann (1990a), Drosdowsky (1996), and Hung and Yanai (2004), employed such a view. An example, as adapted from Drosdowsky (1996), is seen in the time–height cross section of zonal winds at Darwin in Figure 5.3. In the first panel is shown the 35-year mean composite of the entire seasonal cycle, while the second shows a 30-day time slice of the same winds when composited with respect to the onset dates. The second panel is thus an average view of the first intraseasonal burst of the wet season. In the seasonal cycle plot, the wet season low-level westerlies do not exceed a value of 2 m s^{-1} at heights greater than 800 hPa. In the plot depicting the onset burst, however, westerlies of a magnitude of 2 m s^{-1} reach as high as 400 hPa. Thus, not only is the transition to westerlies much more rapid when viewed with respect to the first intraseasonal burst, but their vertical extent into the troposphere is greater, and considerably unlike that seen even at the peak of the seasonal cycle. It is this deep westerly view of the active monsoon that seems the most applicable to that observed at any instant during a burst. Indeed, the collective experience gained from forecasters in Darwin (Drosdowsky having been one), and that gained from a number of field experiments (e.g., Gunn *et al.*, 1989 figure 12), substantiates the view of an active monsoon as being characterized by deep westerlies near Darwin.

From the combination of observations discussed above, we are left with a somewhat different perspective on the region's monsoon system than that obtained from the previous section. Like that of the other monsoon regions (e.g., Chapters 2–4), instead of being wholly a consequence of the seasonal cycle, it has

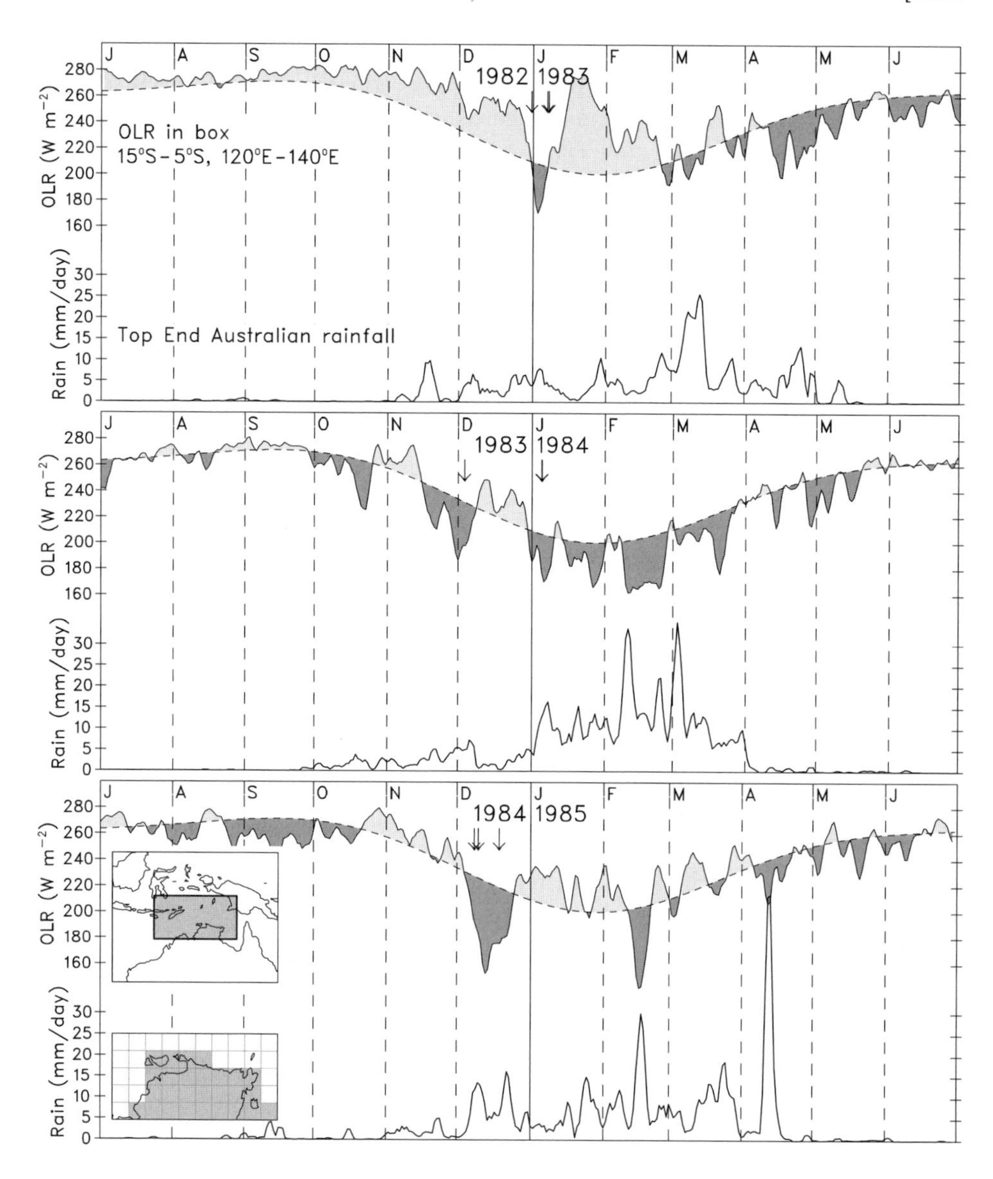

Figure 5.2. 3-day running mean time series of NOAA satellite-observed OLR, averaged for the box 15°S to 5°S and 120°E to 140°E, and Australian "Top End" rainfall, averaged for all available Australian Northern Territory stations north of 15°S. Dashed line (for the OLR) shows the climatological seasonal cycle created by taking the mean and first three harmonics of the 1979–2001 climatology. Dark/light shading (for the OLR) are indicative of times of anomalously active/inactive convection. Also shown, by arrows above the OLR curve, are the monsoon onset dates as defined by Hendon and Liebmann (1990a), Drosdowsky (1996), and Hung and Yanai (2004). Key maps for the areas used for the OLR and rainfall are shown on the bottom left.

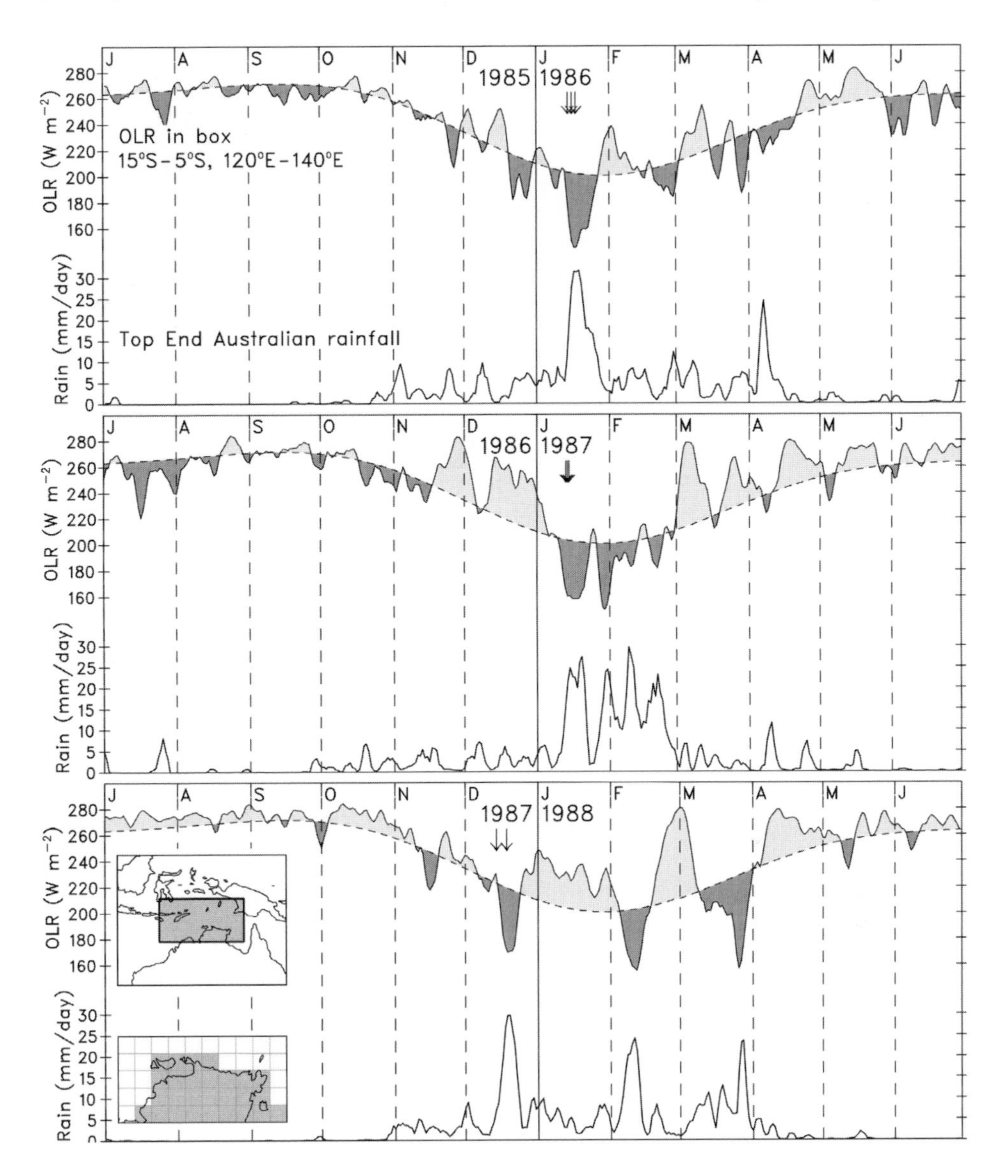

been indicated to have the broad range of ISV as an intrinsic component. Indeed, many aspects of the monsoon system that we take for granted (e.g., the bursts of deep westerlies), would appear not to occur without the presence of ISV.

5.4 BROADBAND INTRASEASONAL BEHAVIOR: SPECTRAL ANALYSIS

Returning to Figure 5.2, it appears that the region's intraseasonal events should be characterized by a fairly wide range of timescales. Quantification of this character

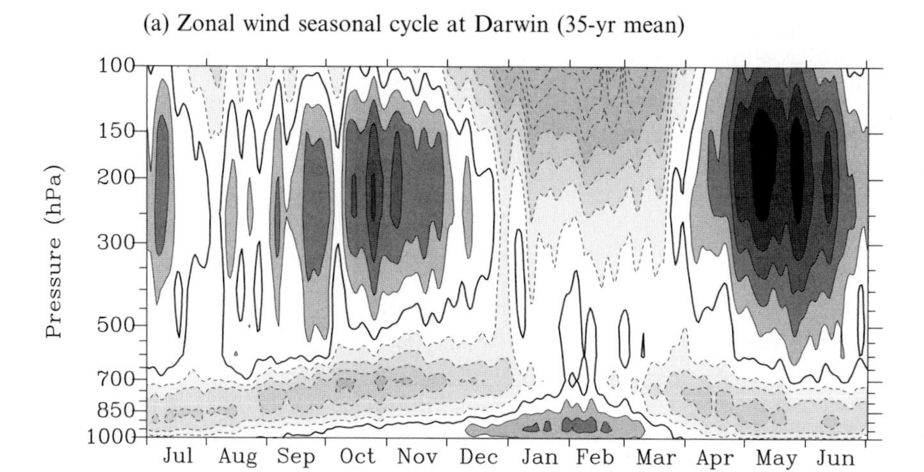

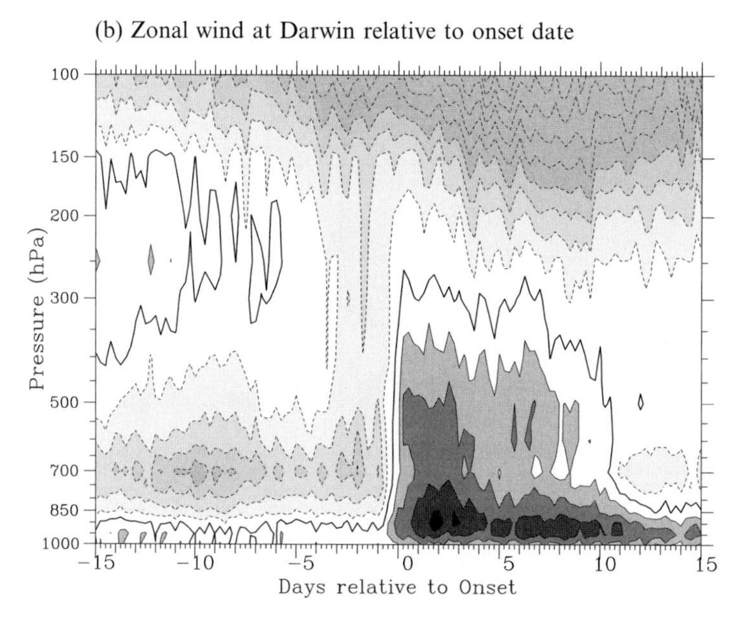

Figure 5.3. Time–height sections of station zonal wind at Darwin when composited for (a) 35-year mean seasonal cycle (July 1957–June 1992), and (b) relative to the 35 different onset dates of Drosdowsky (1996). Contour interval is $2\,\mathrm{m\,s^{-1}}$, the zero contour is a heavy solid line, and negative (easterly) contours are dashed.
Adapted from Drosdowsky (1996).

has been obtained through spectral analysis techniques. Among the most enlightening use of such techniques has been that of Lau and Chan (1988), Hendon and Liebmann (1990a), and Drosdowsky (1996). Confirming the impression given above of the occurrence of strong ISV, Lau and Chan showed that the Australian–Indonesian monsoon region is close to being at the global maximum of intraseasonal OLR variance during austral summer. By comparison, the region is neither at the global maximum for variance in interannual nor 1–5-day bands. They further showed in spectra that the region's ISV is particularly enhanced in the 30–60-day band, as indicated by a rather broad spectral peak in the OLR. Hendon and Liebmann (1990a) showed a similar spectral peak in the Darwin 850-hPa zonal wind, although they did not find any significant peaks in the spectrum of north Australian rainfall.

Confirmation of some of these results is provided in power spectra computed for the OLR and rainfall data of Figure 5.2, as displayed in Figure 5.4. Here we have computed spectra in the same manner as Hendon and Liebmann (1990a). Starting from anomaly data (smoothed seasonal cycle removed), spectra were calculated for segments of length 212 days starting on 1 October each year. As such, the spectra represent the sub-seasonal variability acting during the months of October–April in all available years. Twenty-seven years/segments are used for the OLR data (Figure 5.4(a)), and forty-nine for the rainfall (Figure 5.4(b)). Each segment was padded with zeroes to 256 days, giving a bandwidth of 1/256 cpd. The resulting spectral power from each year was then averaged, providing multiple degrees of freedom (d.o.f.) for each spectral estimate (up to 54 d.o.f. for the OLR spectrum, and 98 d.o.f. for the rainfall spectrum). The displayed "AR1 noise" reference spectra are computed by performing the same procedure on a very long (20,000-year) time series generated from a first-order autoregressive model with the same lag-1 autocorrelation as the segmented data. The 99% confidence curves are based on this noise spectrum and the chi-squared distribution. The spectrum of the OLR closely follows the AR1 noise curve at the higher intraseasonal frequencies (0.04 to 0.12 cpd), has a broad spectral peak in the 30–80-day range (which exceeds the computed 99% signficance around 45 days), and then has a sharp reduction in power for longer periods. The spectrum of rainfall, on the other hand, shows no statistically significant spectral peaks.

Why a broad 30–80-day spectral peak exists for the multi-year OLR data set, but not for the rainfall, is difficult to understand. It appears to not just be a question of scale; the spectrum of a smaller area-average of OLR still shows a statistically-significant spectral peak around 40 days (not shown). It is perhaps related to the greater degree of noise inherent in the rainfall field (e.g., see Figure 5.2). Despite their different-shaped spectra, the OLR and rainfall still show a close correspondence when viewing their individual time series, as evidenced in Figure 5.2, and this correspondence is even stronger when the same areas are taken for both the OLR and rainfall.

Quantification of the relationship between two time series, as a function of frequency, can be obtained by the application of cross-spectra. Hendon and Liebmann (1990b), for example, were able to show that although rainfall shows no obvious preference for a particular frequency, very coherent rainfall fluctuations accompany the pronounced 30–80-day fluctuations that exist in other monsoon variables. Two examples of cross-spectra are presented in Figure 5.5. They have been computed using total fields (i.e., the seasonal cycle was retained), and using all days of the available multi-year time series. The first of the presented cross-spectra is between the large-scale OLR and large-scale 850-hPa zonal wind

(a) Spectrum of OLR in 15°S–5°S, 120°E–140°E

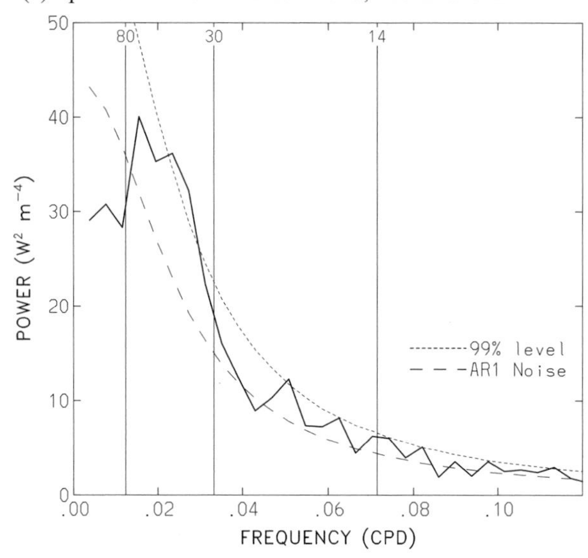

(b) Spectrum of "Top End" rainfall, 1950–1999

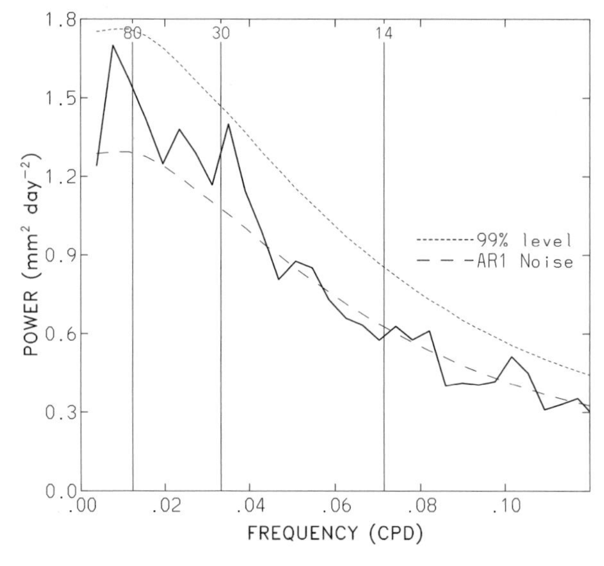

Figure 5.4. (a) Power spectrum of daily OLR anomalies averaged for the box 15°S to 5°S and 120°E to 140°E, using October–April data for all available seasons from 1974–2003. Bandwidth is 1/256 cpd, and the reference frequencies corresponding to periods of 80, 30, and 14 days are marked with a vertical line. "AR1 noise" refers to the similarly-computed spectrum of a time series generated from a first-order autoregressive model, and "99% level" is the significance level. (b) As in (a), except using area-weighted station rainfall data from all available stations in the "Top End" region (approximately 15°S to 10°S and 128°E to 138°E, as in Figure 5.2) of northern Australia for the years 1950–1999.

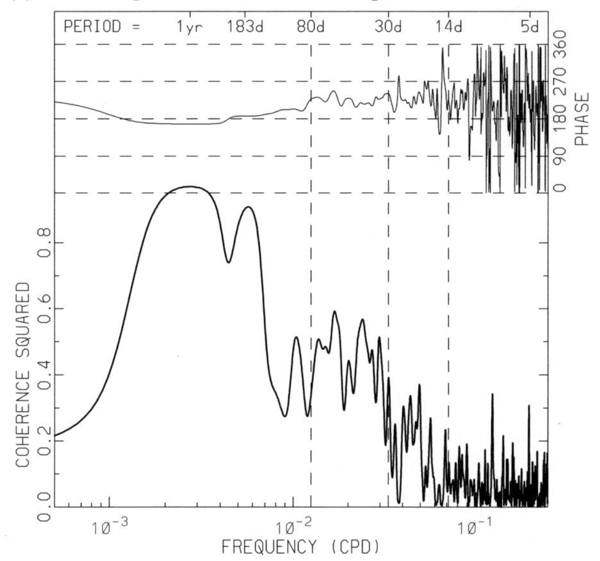

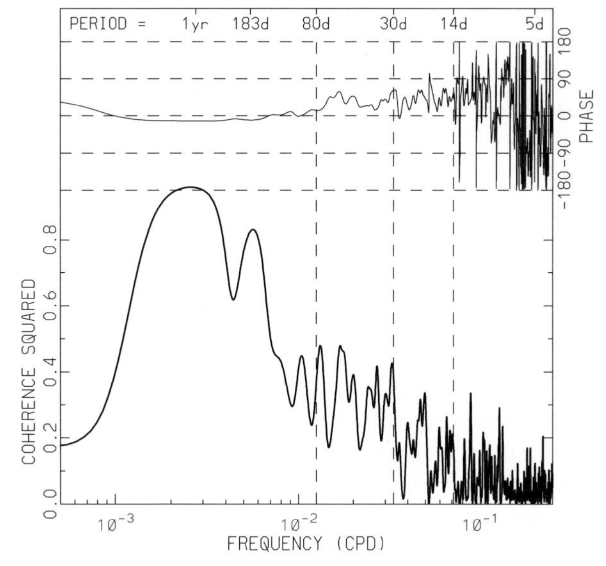

Figure 5.5. (a) Coherence-squared and phase between multi-year time series (using all days) of OLR and 850-hPa zonal wind both averaged for the box 15°S–5°S and 120°E–140°E. Multiple passes of a 1-2-1 filter were applied to the co-spectra and quadrature-spectra before computing the phase and coherence resulting in an effective bandwidth of 2.5×10^{-3} cpd. A 90° phase relationship means that OLR is leading the wind by a quarter cycle. (b) As in (a), except between "Top End" averaged rainfall and 850-hPa zonal wind averaged over the box 15°S–10°S and 130°E–135°E.

(Figure 5.5(a)). Most striking is the high coherence between the two fields for frequencies around the annual cycle and its harmonics. The phase relationship at these annual harmonics is such that positive zonal wind anomalies tend to occur in conjunction with negative OLR anomalies (i.e., enhanced convection). Relatively high coherence also exists in the 30–80-day range, especially when compared to the coherence for frequencies just outside it. The phase within the 30–80-day range is such that enhanced convection (negative OLR) slightly leads (by about one-eighth of a cycle) the westerly zonal winds, the same relationship that generally occurs across the whole ISV range. The results for Australian "Top End" rainfall and zonal wind (Figure 5.5(b)) are essentially the same; coherence steadily increases with increasing period from 5 days to 1 year, but with notable bumps around the annual harmonics and within the 30–80-day range. In the whole ISV range, rainfall slightly leads the westerly wind.

On the whole then, multi-year frequency spectra of fields like OLR and rainfall in the Australian–Indonesian monsoon region are probably most of note for their general lack of statistically-significant spectral peaks; the only robust peak being the broad 30–80-day peak in the OLR. Nonetheless, the spectra of Figure 5.4 still indicate the existence of a great deal of ISV in these fields, occurring over a rather broad range of frequencies. Much of this variance, however, appears to be a consequence of the continuum of scales from "weather" to "climate", being well-characterized by the lag-1 autoregressive model of red noise. The spectra also show the subjectivity that must be involved in any definition of ISV itself, as there is no spectral gap between the intraseasonal and higher frequency scales. The increased wind–rain coherence seen at ISV scales, however, is at least one distinguishing aspect of ISV, and the broad 30–80-day spectral peak appears particularly coherent. This spectral peak is with little doubt associated with the global-scale MJO (Madden and Julian, 1971, 1994; Chapter 1), arguably the most important defined "mode" of ISV for the region. We return to its discussion in Section 5.6.

5.5 METEOROLOGY OF THE BURSTS AND BREAKS

The first views of the structural details of the monsoon bursts and breaks in the vicinity of Australia, irrespective of their source, were presented in papers of the 1980s. Using infrared imagery from the Geostationary Meteorological Satellite (GMS) during the summer of 1978/1979, McBride (1983) found the bursts to be associated with large-scale envelopes of enhanced convective activity, each spanning approximately 35° longitude and 15° latitude, moving either eastward or westward.[3] This finding motivated a number of others. Davidson (1984) and Keenan and Brody (1988) both studied the composite structure of the active (burst) and break periods, subjectively defined in terms of regions of convectively-active or suppressed con-

[3] McBride (1983), as was common with other authors at the time, used the term "synoptic scale" to describe the envelope of convective activity, of order 35° longitude across, that he observed. In this chapter, as is common with more recent treatments (e.g., Hendon and Liebmann, 1990b), we reserve the use of the term "synoptic" to refer to the timescale that is shorter than that of ISV.

ditions as determined from Hovmöller (time–longitude) representations of satellite imagery. Davidson's study emphasized the differences in structure of the tropical wind flow, demonstrating an enhanced southern hemisphere Hadley circulation during the active periods and a movement of the ITCZ closer to the equator during breaks. Keenan and Brody concluded that a major factor governing the tropical convection was the location of upper level (200 hPa) subtropical troughs and ridges associated with the circumpolar westerly flow at upper levels. In particular, the regions of tropical convection in their composite were associated with the eastern side of an upper trough. Whether these results will stand the test of time is difficult to gauge, however, as neither study has been followed up with more recent data. Importantly, all the authors emphasized that the large-scale envelopes of tropical convective and suppressed regions underwent slow eastward and westward movements, with speeds of between 4 and $10\,\mathrm{m\,s^{-1}}$ (3 and 8° $\mathrm{day^{-1}}$). This primarily zonal movement of the convective envelopes is notably unlike the poleward movement as seen in the northern hemisphere monsoon at India's longitudes (e.g., Yasunari, 1979; Sikka and Gadgil, 1980; Chapter 2).

Since the time of the above-mentioned studies, at least some of the *eastward movement* of the large-scale envelopes of convection has become identified as being a result of the MJO, as will be described in the following section. The predominance of the MJO in current mainstream thought, however, due largely to the work of Hendon and Liebmann (1990a,b), is such that the *westward moving* envelopes have since been little discussed. Yet, looking back at the time–longitude strips for the 1978/1979 season in McBride's (1983) paper (also shown in Davidson and Hendon, 1989), and the strips for 1983/1984 shown in Keenan and Brody (1988), the westward movements definitely exist. Most likely some of the westward cloud features are associated with, and have the dynamics of, internal equatorially-trapped Rossby waves, as will be discussed in Section 5.9.

Further details on the broadband active and break periods have come from McBride and Frank (1999). They studied the thermodynamic structure of the active vs. break periods using data from a radiosonde array surrounding the Gulf of Carpentaria in tropical Australia during the Australian Monsoon Experiment (AMEX). They found major changes in the lapse rate of virtual temperature, and that the differences between the large-scale active and break regions far exceeded any higher frequency differences between convective and non-convective soundings within the active envelope. Thus it was interpreted that mid-tropospheric temperatures are primarily adjusted by dynamical processes acting over large scales, rather than by *in situ* processes acting on the scale of individual convective cells. This is illustrated schematically in Figure 5.6. As shown, based on their observations, the monsoon active regions are characterized by a warm upper troposphere and a cold lower troposphere, with the reverse in the suppressed or break regions. That is, *active* monsoon conditions are actually associated with *increased* mid-tropospheric static stability.

Hendon and Liebmann (1990b) studied the intraseasonal structure of the Australian monsoon specifically in the 30–60-day band. They defined active events as having a peak in both wind and rain in their band-pass filtered data. Once events

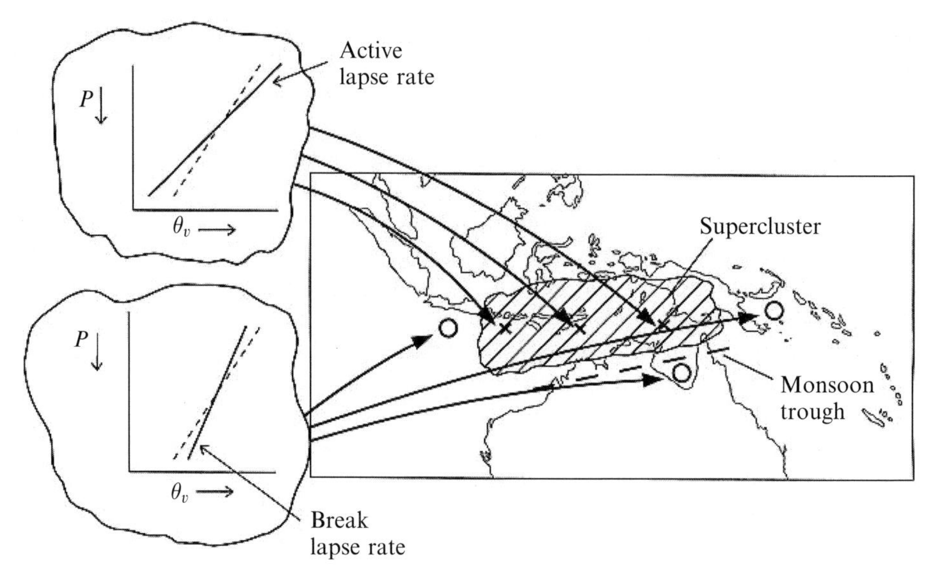

Figure 5.6. A schematic overview of the Australian–Indonesian monsoon region showing the envelope of active convection, or "supercluster" (although we now prefer not to use this overly bandied term), with the break regions being outside the envelope. The nature of the variations in the lapse rate of virtual potential temperature are indicated to the left of the map, with X marks on the map showing active regions and O marks showing break regions.
Reproduced from McBride and Frank (1999).

were defined, the composite structure from 20 days prior to 20 days after the westerly maximum was obtained through the averaging of unfiltered grid-point analysis data. Their composite structure in both wind and rainfall was very similar to the eastward moving fields typically associated with the MJO, as will be described in the following section. Further, their thermodynamic structure was similar to that described by McBride and Frank (1999), with a warm anomaly of the order of 0.5 K in the upper troposphere and a cold anomaly of similar size in the lower troposphere (see figure 5(d), Hendon and Liebmann (1990b)).

Following the methodology of Troup (1961), Drosdowsky (1996) produced a bar chart showing periods of deep westerly flow, and also of area-averaged rain events, for the Darwin area over all monsoon seasons from 1957/1958 though to 1991/1992. Despite a "less than obvious" relationship between the two, both he and Hendon and Liebmann (1990a) did calculate a statistically significant coherence between wind and rain over a wide range of intraseasonal timescales, especially beyond about 20 days (see also Figure 5.5). As will be discussed in the following section, the zonal wind vs. rain phase relationship is now well established as a feature of the MJO, yet these coherence calculations show that this is in fact a common property of much of the intraseasonal band.

Since the mid-1980s, Darwin has been the base for a series of both international and Australian meteorological field experiments. It has also been the site of numerous radar deployments. This has led to a number of research papers on the fundamental properties of tropical convection (e.g., Frank and McBride, 1989; Webster and Houze, 1991; Mapes and Houze, 1992; Keenan and Carbone, 1992; Rutledge *et al.*, 1992; Williams *et al.*, 1992; McBride and Frank, 1999; May and Rajopadhyaya, 1999; Keenan *et al.*, 2000). Usually in these papers, the convective and non-convective periods are interpreted as being active vs. break monsoon conditions. Unfortunately, however, for one of the major field experiments (AMEX in 1986/1987) the "break" period was only of the order of several days (see figure 15 of Gunn *et al.*, 1989). Thus, it is not clear to what extent some of these studies are relevant to the intraseasonal (viz., greater than ~ 10 day) timescale. Various authors did state that the properties of convection in the break (dry-easterlies) period during AMEX were the same as in the long period of easterlies prior to the monsoon onset (e.g., Danielsen, 1993; Selkirk, 1993; Gunn *et al.*, 1989).

One consistent and useful theme to emerge from the various convection studies utilizing data from the Darwin site is that the convection has very different properties depending on whether the background flow is deep westerly monsoonal as distinct from deep easterly (break) trade flow. The convective cells present during the pre-monsoon and break periods have higher vertical development, higher reflectivities above the melting level, a lack of large stratiform decks, more intense updrafts, and higher electrical activity (e.g., Keenan and Carbone, 1992; Williams *et al.*, 1992). In comparison, the convection present during the monsoon westerly bursts is often associated with squall-like structures within large mesoscale stratiform decks, warm rain coalescence processes, and weak updrafts (Mapes and Houze, 1992; Keenan and Rutledge, 1993). Thus, while active monsoon conditions result in more rain, the individual convective cells involved are generally less intense. To some extent these differences can be attributed to a continental (for the break) as compared with a maritime (for the monsoon burst) source of the airstream in which the convection is embedded. However, such differences are also consistent with the general large-scale static stability changes found by McBride and Frank (1999), thus it is likely that the observed active vs. break differences in convective type may also apply to oceanic conditions away from the continent.

5.6 CHARACTERISTICS AND INFLUENCE OF THE MJO

As revealed in the OLR spectrum in Figure 5.4(a), there is at least one feature of the region's broad spectrum of ISV that has been shown to be robust and associated with a particular phenomenon; and that is that of the MJO. Although its statistically significant spectral peak appears at around 40–50 days, hence its traditional name (Madden and Julian, 1971, 1972), it is broader band than this, being shown by a number of studies to contribute variance in the range of about 30–80 days (e.g., Weickmann, 1983; Lau and Chan, 1985; Salby and Hendon, 1994). Importantly however, not all variance that is within these frequency bands need be associated with the MJO. As described in Chapter 1, one of the essential characteristics of the

MJO is its large-scale coherent eastward propagation. Calculations show that the large-scale eastward propagating component of tropical variability (e.g., eastward planetary wave numbers 1–6) accounts for only about one-half the variance in the 30–80-day band, in fields of OLR and zonal wind across the equatorial Indian and western Pacific Oceans (Hendon *et al.*, 1999). Outside the near-equatorial band, the portion of atmospheric 30–80-day variance that is attributable to the MJO is likely to be even less. In the Australian–Indonesian monsoon region, even without the restriction to large-scale eastward propagation, the coherence calculations of Figure 5.5(a) suggest that only one-half the 30–80-day variance may be attributed to a coherent mode. When considering the influence and characteristics of the MJO in any region then, such attribution must be kept firmly in mind.

In this context, the results of the earliest studies that pertain to the influence of the MJO on the Australian–Indonesian monsoon region must be viewed with some caution. As was common practice at the time, McBride (1987) used a rather narrow band-pass filter (half power response at 37 and 54.7 days) on the local wind at Darwin, and interpreted the resulting time series as being the signal of the MJO. Although this interpretation is not entirely correct, McBride's work did serve to demonstrate the out-of-phase character between upper (100 hPa) and lower (850 hPa) level zonal winds for this frequency band, and the general correspondence between peaks and troughs of the filtered winds and the active and break periods of the monsoon.

Another early piece of work that is often quoted in the context of the MJO's influence on the Australian monsoon is that of Holland (1986). Holland defined active bursts of the monsoon by way of filtering the 850-hPa zonal wind at Darwin, and found that the mean period between active bursts was 40 days, with a standard deviation of 10 days. Holland interpreted this recurrence interval as evidence of an influence by the MJO. As discussed by Drosdowsky (1996), however, Holland's cubic spline filter was effectively low-pass, with a cut-off at approximately 30 days. Consequently, Holland's result of a 40-day mean recurrence interval could have equally as well been made with an input of red noise, and hence did not necessarily imply an MJO influence.

Beyond the earliest studies pertaining to the MJO, the work of Hendon and Liebmann (1990b) has provided a much more detailed picture. Using the techniques of cross-spectra and compositing relative to a 30–60-day-filtered time series around Darwin, they were effectively able to show the dominance of large-scale eastward-propagation in this frequency range. Again, however, some caution needs to be given with regards to the attribution of the structures presented to the MJO. Indeed, as Hendon and Liebmann used a 30–60-day filtered time series of winds and rain near Darwin as the "predictor" about which to composite, their composites may contain contributions from more aspects of variability than just the MJO. One likely symptom of this is that the anomaly of enhanced convection within their composite was seen to be traceable from the southern Indian Ocean at a latitude around 10°S (their figure 10), rather than the usual equatorial Indian Ocean, as seen in composites made with more global indicators of the MJO (e.g., Knutson and Weickman, 1987; Wang and Rui, 1990; Jones *et al.*, 2004; Wheeler and Hendon,

2004; Figure 5.8). Nevertheless, presumably due to the relative incoherence of non-MJO fluctuations on this timescale, other aspects of Hendon and Liebmann's results provide a picture of the influence of the MJO that has remained accurate to this day. Quantifying this picture, with a composite comprising 91 events from 30 wet seasons (1957–1987), they found the amplitude of the oscillation at Darwin to be about $5\,m\,s^{-1}$ in zonal wind, $0.75\,m\,s^{-1}$ in meridional wind, 5 mm rainfall per day, and 10% in relative humidity. The deep baroclinic structure in zonal wind had its node at about 300 hPa, not unlike that of the westerly burst of Figure 5.3(b). The amplitude of the associated OLR anomaly near Darwin was found to be about $30\,W\,m^{-2}$, placing the broadband ISV of Figure 5.2 in perspective.

Compared to these earlier pieces of work, more recent studies on the influence of the MJO have tended to use more global MJO indicators. One common approach for identifying the MJO is to employ filtering in both frequency and wave number, which allows one to discriminate for only the lowest zonal wave numbers, and those propagating to the east (e.g., Wheeler and Kiladis, 1999; Hendon *et al.*, 1999). Another common approach is to employ empirical orthogonal function (EOF) analysis of tropical-wide fields such as OLR and zonal wind (e.g., Hendon *et al.*, 1999; Hall *et al.*, 2001; Waliser *et al.*, 2003; Wheeler and Hendon, 2004). Provided the seasonal cycle and IAV are prior removed, structures akin to our current view of the MJO are well described by the leading EOF pair, and projection of the global data onto those EOFs extracts the large-scale, predominantly eastward-propagating, signal of the MJO. While such approaches appear superior to using a locally-defined index for the MJO, they are not without their own caveats, and there will always be the concern that an index may contain contributions from non-MJO variability and/or be missing some of that variability. Notwithstanding the caveats, here we present results, as adapted mostly from Wheeler and Hendon (2004), of a regional analysis of the MJO's influence when employing the tropical-wide EOF approach for identifying the MJO. This identification will be used throughout the rest of this chapter, and will serve to provide a useful comparison with earlier work, and for making new calculations.

Wheeler and Hendon's (2004) MJO-defining EOFs are as displayed in Figure 5.7(a). They are the leading EOF pair of the combined analysis of near-equatorially-averaged fields of OLR, 850-hPa zonal wind, and 200-hPa zonal wind using daily data for all seasons. As can be seen, the EOFs show the predominantly zonal wave number 1 and 2 structure of the MJO, and when taken as a pair, can describe its propagation. Indeed, as shown in Figure 5.7(b), projection of the daily observed data onto the multiple-variable EOFs yields principal component (PC) time series that vary mostly on the 30–80-day timescale of the MJO only, and these PCs are approximately in quadrature, implying predominantly eastward propagation. Taken together, the two PC time series are used as the MJO index, and as this particular index was developed for real-time use, they have been called the Real-time Multivariate MJO series 1 (RMM1) and 2 (RMM2). By the analysis of Wheeler and Hendon (2004), RMM1 and RMM2 are able to represent the essential characteristics of what most scientists consider as the "MJO". In particular, the state of the MJO can be measured in the phase space defined by RMM1 and RMM2

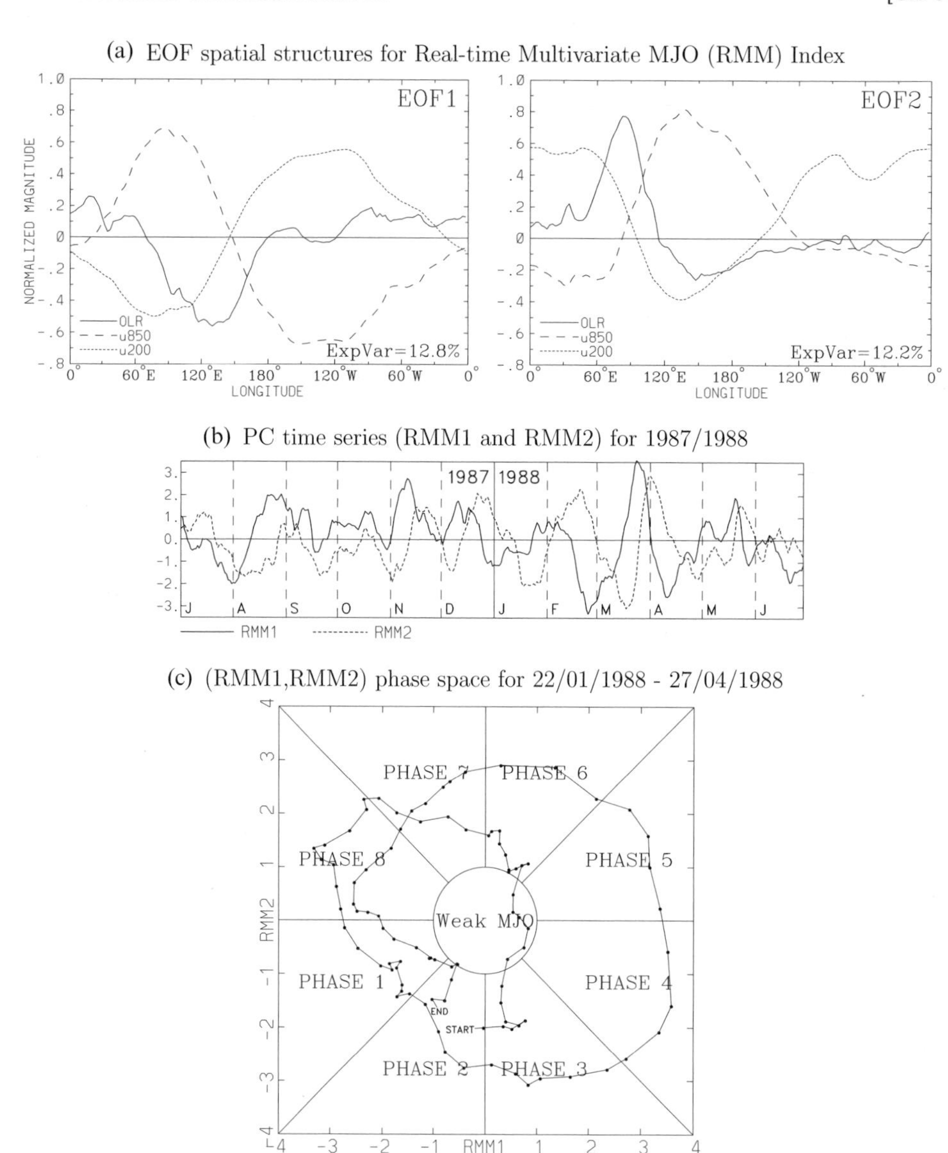

Figure 5.7. (a) Structure of EOFs designed to isolate the signal of the MJO. 15°S–15°N-averaged fields are used. See text for further details. (b) Example series of RMM1 (PC-1) and RMM2 (PC-2). (c) (RMM1,RMM2) phase space for 22 January, 1988–27 April, 1988. Adapted from Wheeler and Hendon (2004).

(Figure 5.7(c)): eight different phases are used for when the MJO is considered relatively strong, and a "weak MJO" phase for when the (RMM1,RMM2) vector has an amplitude of less than 1.0. When individual sequences of days of strong MJO activity are viewed in the phase space, their paths trace large anticlockwise circles.

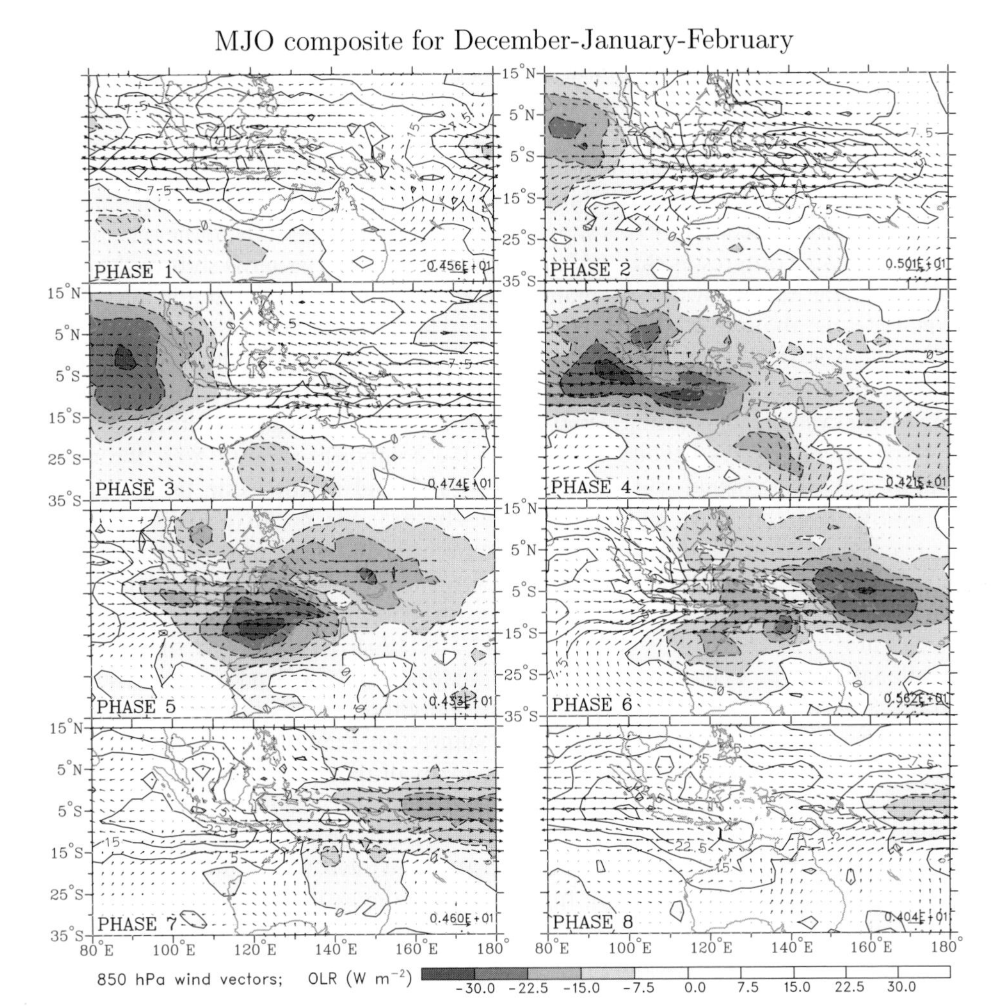

Figure 5.8. Composited OLR and 850-hPa wind anomalies for eight phases of the MJO during December/January/February (DJF). OLR contour interval is $7.5\,\mathrm{W\,m^{-2}}$, with negative contours dashed and negative values shaded. Black vectors indicate wind anomalies that are statistically significant at the 99% level, based on their local standard deviation and the Student's t-test. Grey vectors do not pass the 99% level test. The magnitude of the largest vector is shown on the bottom-right of each panel.

On average, each phase lasts for about 6 days, but there can be considerable variability in this number from event to event.

Given the description of the MJO by the (RMM1,RMM2) phase space, composites may be formed by averaging the observed anomaly fields occurring for the days that fall within each of the defined phases (Figure 5.7(c)). Wheeler and Hendon (2004) produced such composites on a global scale for both the southern summer and northern summer seasons. Here, in Figure 5.8, we present such a composite for the December/January/February season focussed on the region of interest, showing the fields of OLR and the 850-hPa wind. About 200 days of data were averaged for each of the presented eight MJO phases, while 961 days were rejected for being at a time of weak MJO amplitude (representing 37% of the time). The composite of Phase 1 shows suppressed convection over the tropical part of the whole domain, with positive OLR anomalies reaching values greater than 22.5 W m^{-2}. Anomalous tropical easterlies accompany this suppressed convection, especially in the west of the domain. As time progresses through each of the phases, the patterns shift to the east. In Phase 2 the easterly wind anomalies are at their greatest around 130°E, with a magnitude of about 5 m s^{-1}. By Phase 3, a large negative OLR anomaly has appeared over the eastern equatorial Indian Ocean. At the same time, there is evidence of tropical–extra-tropical interaction, with a negative OLR anomaly associated with cyclonically-turning winds over extra-tropical western Australia.

As time progresses beyond Phase 3 in the composite (Figure 5.8), the main tropical convective signal shifts further eastward, and at the same time expands southward over the north of Australia. In Phase 5, the negative OLR anomaly has reached its most southward extent, being centered at 12°S, with a value of less than −30 W m^{-2}. Maximum westerly anomalies are placed coincident, and somewhat to the west, of the enhanced convection, with maximum values of about 5 m s^{-1}. In the vicinity of Darwin (12.5°S, 130.9°E) there is a signal in the meridional wind as well, with northerly anomalies (Phase 4) leading the westerlies (Phases 5, 6, and 7). Interestingly, the largest amplitude signals in the OLR tend to be over the sea. The island of Papua New Guinea, for example, is clearly visible as having reduced amplitude OLR anomalies in many of the phases. Also of note is the lesser degree to which the convection moves poleward when compared to what occurs over the Indian monsoon region in the opposite season (Chapter 2). On the whole, many of these features of this composite are equivalent to those of the study of Hendon and Liebmann (1990b). Importantly, the magnitudes are favorably high when compared to the broadband ISV bursts and breaks of Figures 5.2 and 5.3(b), thus suggesting the relative importance of the MJO for producing some of that variability.

The relationship of the MJO to the traditional broadband ISV bursts and breaks deserves more attention. One question that seems rather important to ask then, and which has been little addressed in the literature to date, is the strength of this relationship. We explore this question in Figure 5.9, which shows the same OLR time series as Figure 5.2, except solid bars have been added to indicate times when the MJO is diagnosed to be in either Phases 4, 5, or 6 (i.e., times when the composite MJO has negative OLR anomalies in the region of interest). Thus, if the MJO/burst relationship is strong, the solid bars should coincide with the negative excursions of the OLR time series below the seasonal cycle. Sometimes the relationship seems strong (e.g., during 1987/1988), and at other times it seems weak (e.g., 1982/1983 and 1983/1984). Quantification of the relationship is given by the multiple correlation coefficient squared, R^2, between the 3-day running-mean OLR anomaly and the RMM1 and RMM2 values, calculated for the November–April months only. The value of $R^2 = 0.58$ in 1987/1988 indicates that 58% of the variations of the 3-day

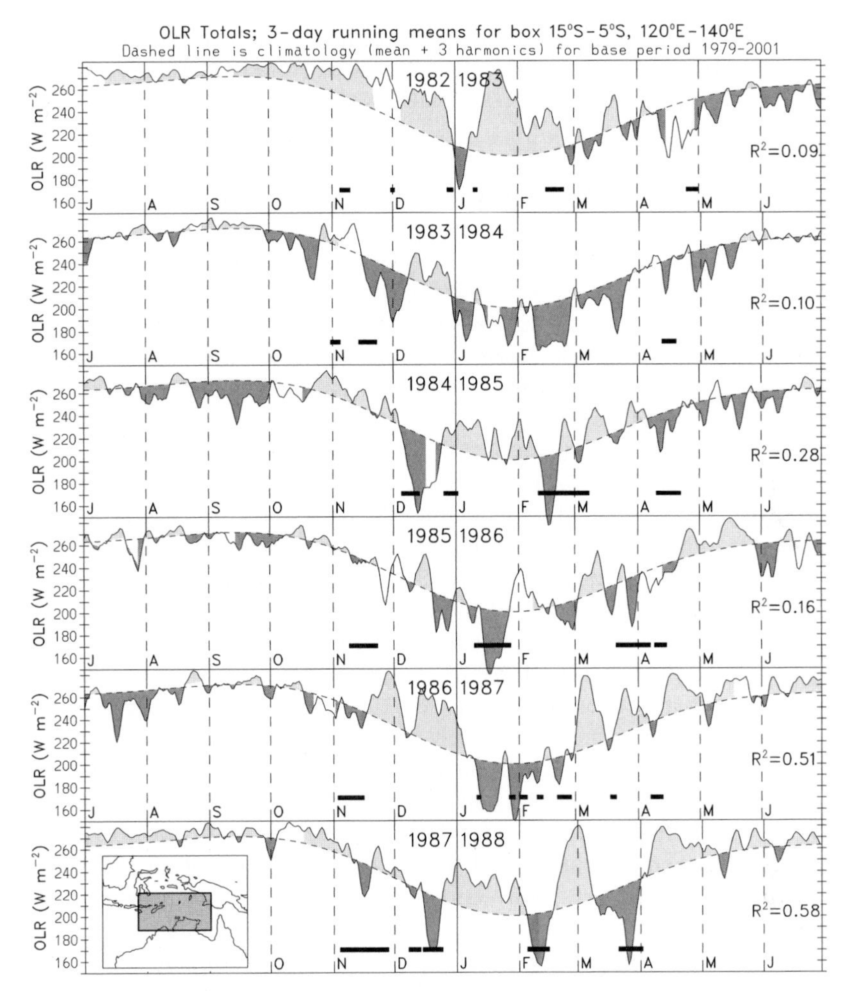

Figure 5.9. As in Figure 5.2, except showing OLR series only, and showing a solid bar when the phase of the MJO, as defined by the (RMM1,RMM2) phase space, was within either Phase 4, 5, or 6, for the months of November–April only. Also given is the square of the multiple correlation coefficient between the OLR anomaly time series and the RMM1 and RMM2 values calculated for the November–April months.

running mean OLR can be linearly accounted for by the variations in this EOF-based global measure of the MJO.[4] In 1982/1983 and 1983/1984, the amount is less than 10%. Thus it is indicated that there is a great deal of interannual modulation of the MJO/ISV burst relationship. Some of the interannual modulation is likely related

[4] In some sense the calculated R^2 are an overestimate of the variability accounted for by the MJO, and in another it is an underestimate. It is an overestimate in the sense that some of the fluctuations of the RMM indices, especially at higher frequencies, are presumably not associated with the MJO. Underestimation comes from the fact that just two spatial structures (EOF1 and EOF2) presumably cannot capture all the real world MJO variability.

to the great deal of interannual modulation in the strength of the MJO itself (e.g., Hendon *et al.*, 1999). In some years the MJO is essentially absent (e.g., 1982/1983 and 1983/1984), and in those years the relationship almost certainly has to be weak.

5.7 1983/1984 AND 1987/1988 CASE STUDIES

Many of the subtleties of the MJO, and its relation to ISV bursts and breaks and the monsoon seasonal cycle, are best appreciated by the way of case studies. Here we provide a fuller view of two cases, the summers of 1983/1984, and 1987/1988. They were chosen because the monsoon ISV of 1983/1984 shows little relation to the MJO, as we define it by the pair of RMM EOFs, while that of 1987/1988 shows a relatively strong relationship (Figure 5.9).

As has been mentioned, much of the ISV of the region tends to be characterized by either eastward or westward movement of large-scale envelopes of convection (e.g., McBride, 1983). The MJO is characterized as showing large-scale eastward movement (Chapter 1). Time–longitude plots, as presented for the two case years in Figures 5.10 and 5.11, are thus a useful way to view such movements. Here we show the 15°S to equator average of the OLR and 850-hPa vector wind field. Most striking about the two figures is the presence of three large-scale eastward-moving envelopes of convection in 1987/1988, oscillating with a period around 50 days, and the absence of such a clear oscillation in 1983/1984. The three dramatic intraseasonal events of the 1987/1988 summer are with no doubt produced by the MJO. Indeed, the 1987/1988 summer has often been used in previous studies to provide a clear example of the MJO (e.g., Hendon and Liebmann, 1994; Matthews, 2000). Further, many of the features of the composite MJO, as presented in Figure 5.8, are present for the individual events in Figure 5.11. For example, for each of the three dominant events, the regions of strongest convection (lowest OLR) tend to be coincident with, or somewhat leading, the regions of strongest westerly winds, and on the whole, the coupled convection–wind pattern tends to move across the domain to the east. Even still, there are some exceptions to this rule. For example, in the first 2 weeks of January, there is a coupled convection–wind signal in the eastern part of the plotted domain (between 130°E and 160°W) that is moving to the west. As previously computed (Figure 5.9), the squared multiple correlation coefficient, R^2, tells us that there is still a sizeable portion ($\sim$42%) of the sub-seasonal 3-day running-mean variance in the 120°E–140°E box that cannot be accounted for by the MJO in the 1987/1988 season. This westward moving feature would comprise one small component of that unaccounted variance.

Compared to the 1987/1988 season, the 1983/1984 season (Figure 5.10) provides an interesting contrast. Rather than the nearly 50-day periodic bursts and breaks, the 1983/1984 season shows no clear periodicity. Instead, there appears to be a greater amount of high-frequency variability, with around a 1–2-week timescale, imbedded within a low-frequency envelope of convection lasting for much of the monsoon season. An early wet-westerly burst of the monsoon occurs at the longitudes of Australia and Indonesia (near 130°E) at the end of November, followed by a

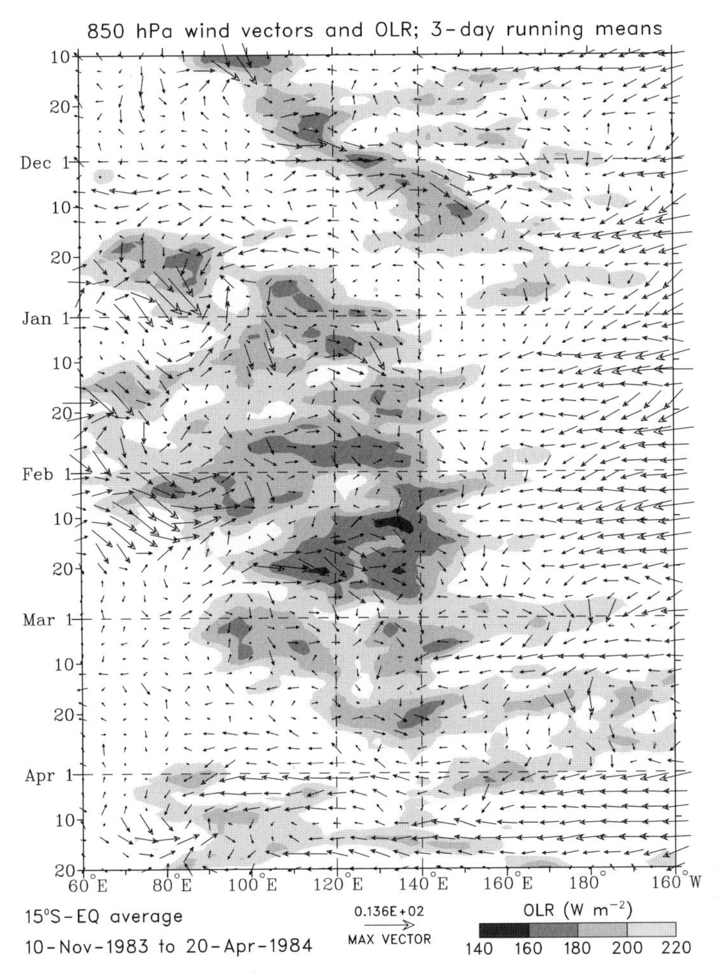

Figure 5.10. Time-longitude plot of 3-day running mean total 850-hPa wind and OLR, averaged from 15°S to the equator, for the monsoon season of 1983/1984.

break in mid-December. Monsoon westerlies then return at the end of December, and, except for a brief time in February, don't change back to easterly until late March. Yet there is still a lot of variability, of an intraseasonal timescale, occurring during this time. For example, the OLR returns to values greater than 220 W m^{-2} on four occasions (at 125°E) during the three months of sustained westerlies. Interestingly, much of this latter variability appears to be associated with westward propagation originating in the Pacific. Some of this westward propagating variability can be identified with equatorially-trapped Rossby waves, as we will return to in our discussion in Section 5.9. Except in the early monsoon burst in November/December, large-scale eastward propagation, like the MJO, appears to be playing little role during the season.

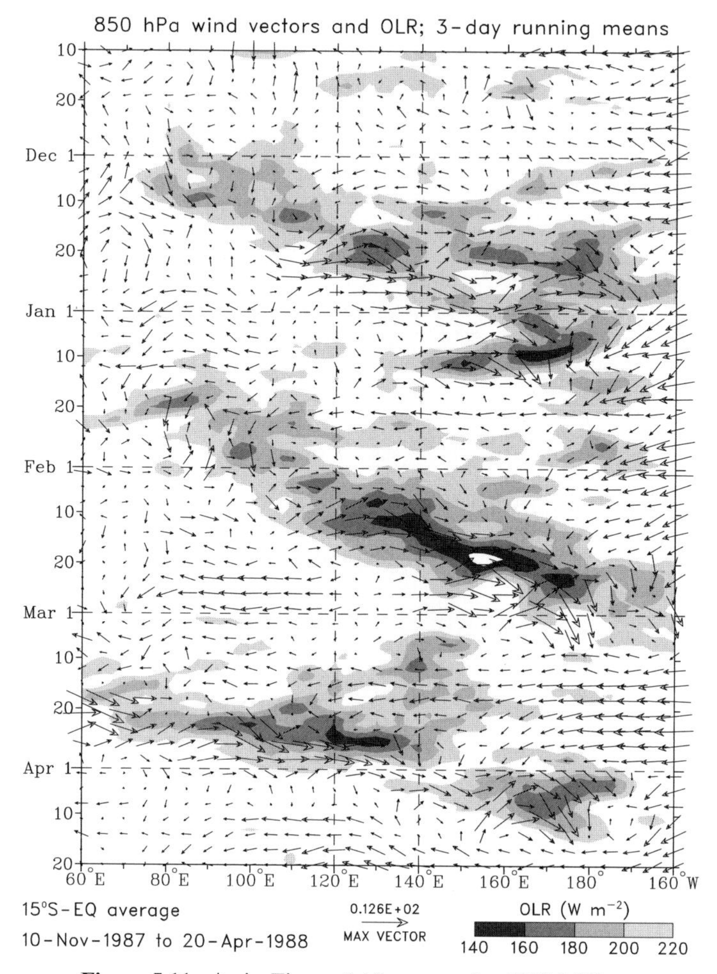

Figure 5.11. As in Figure 5.10, except for 1987/1988.

A further view of the two periods can be had by looking at time series of rainfall, as shown in Figure 5.12 for the locations of Bali (approximately 115°E–116°E, 8°S–9°S) and the northern Australian Top End region (as defined in Figure 5.2). The Bali series is an average of 10 stations on the island,[5] while the Australian series comprises information from about 100 stations which were averaged in an area-weighted fashion. As has already been mentioned, station rainfall data tends to be more noisy than the satellite observed OLR data, yet many of the same features observed in the OLR can be seen. For example, at both locations the 1983/1984 season has an early burst of rainfall in late November/early December, then a more

[5] The Bali stations used were Abiansemal, Bangli, Bebandem, Bukti, Candikuning, Gerokgak, Gianyar, Rambutsiwi, Ngurah Rai, and Kubu.

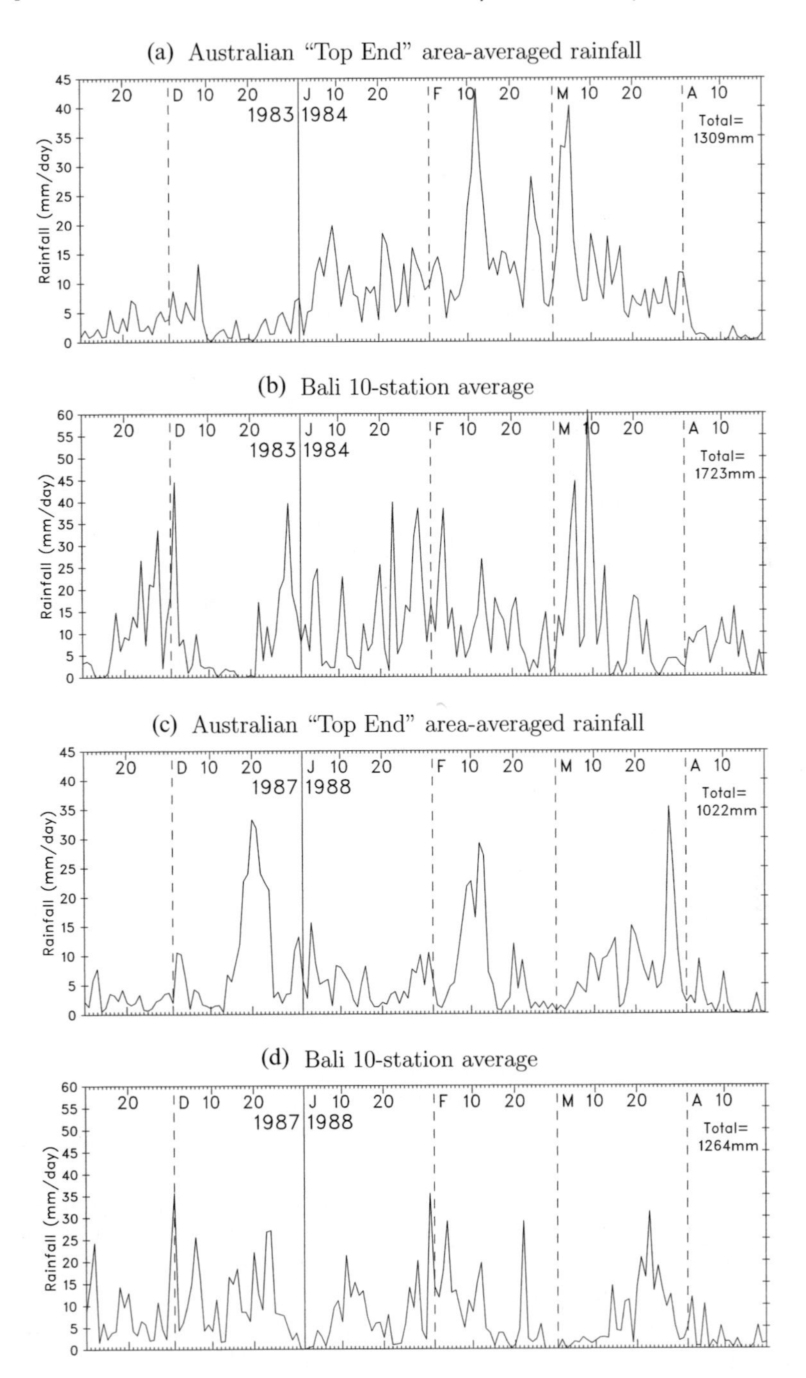

Figure 5.12. Daily precipitation averaged for the "Top End" region (northern Australia, as in Figure 5.2) and for the island of Bali (approximately 115°E–116°E, 8°S–9°S) for the 1983/1984 and 1987/1988 wet seasons. The total rainfall amount in each season is given.

continuous period of rainfall from late December through to late March. In the 1987/1988 season, on the other hand, the Australian series is seen to be comprised of three major bursts spaced about 50 days apart, characteristic of the MJO. Those three bursts are also somewhat apparent in the 1987/1988 Bali series as well, albeit occuring a little earlier, and obscured by a much greater component of noise. Thus, knowing that 1987/1988 was a time of strong MJO activity (e.g., Slingo *et al.*, 1999; Wheeler and Hendon, 2004), and that activity had a strong correlation with large-scale variations in the monsoon (e.g., Figure 5.9), this example shows how strong MJOs can still be very difficult to observe in single-station, or small area-averaged, rainfall data.

One further point that can be made with regard to Figure 5.12, is the fact that in both locations the total season rainfall is greater in 1983/1984 than 1987/1988. This is consistent with the relationship found by Hendon *et al.* (1999), who found that enhanced MJO activity tends to occur in conjunction with positive season-mean OLR anomalies over northern Australia. Thus there is the suggestion that it is the strong (i.e., highly suppressed) break periods produced by the MJO that distinguish those monsoon years with strong MJO activity from others. In comparison, Hendon *et al.*'s (1999) calculations show that the relationship of the season-mean (November–March) OLR with the El Niño Southern Oscillation (ENSO) is comparitively weaker than the relationship with season-mean MJO activity. Further north over Indonesia, however, the relationship of season-mean OLR with ENSO is much stronger, such that during El Niño there are strong positive anomalies over the Indonesian islands. Some evidence for this can be seen in Figure 5.9, when noting that a strong El Niño occurred in 1982/1983, and a weaker event in 1986/1987.

5.8 MJO INFLUENCE ON MONSOON ONSET: RECONCILING THE RESEARCH LITERATURE DIFFERENCES

As mentioned in the introduction, the research literature pertaining to the region's monsoon onset gives the impression of widely differing roles for the MJO. On the one hand, Hendon and Liebmann (1990a) suggested that the monsoon onset in each year is "strongly influenced by the 40–50-day oscillation", and that in 27 out of 30 years the onset fell within 4 days of the passage of the oscillation. Composites of atmospheric fields constructed relative to their onset dates showed that onset coincided with the arrival of an eastward propagating convective anomaly originating in the Indian Ocean. On the other hand, the study by Davidson *et al.* (1983) examined the evolution of the large-scale flow at monsoon onset in six years, and although they suggested an important role for a number of different triggers, no relationship was suggested with the propagation of MJO-like perturbations from the Indian Ocean. Part of the reason of the omission of the MJO in the Davidson *et al.* study was that, although the original Madden and Julian (1971, 1972) studies long-preceded their work, the MJO was still relatively unknown among researchers at the time (Chapter 1). Yet later studies on the monsoon onset have also seemingly down-

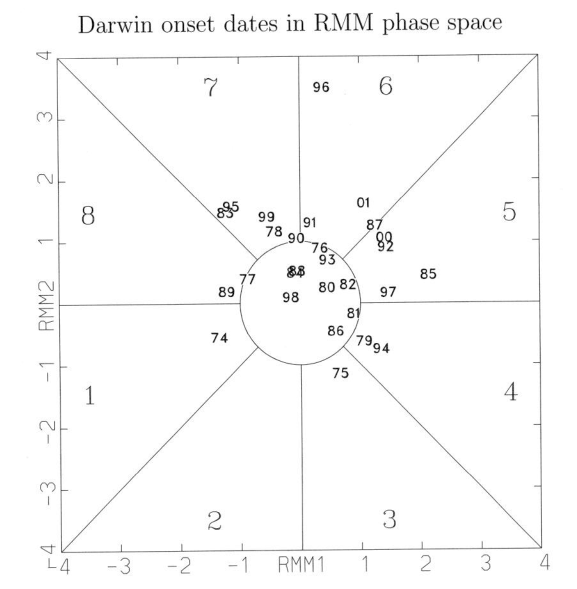

Figure 5.13. (RMM1,RMM2) phase space points (as marked by the dual numbers) for the days on which the monsoon was defined to onset, based on the daily deep-layer mean zonal wind, at Darwin, Australia. The (dual) number plotted refers to the monsoon year, being that of the nearest December. The monsoon dates are defined and taken (with updates) from Drosdowsky (1996), covering all wet seasons from 1974/1975 to 2001/2002.
Reproduced from Wheeler and Hendon (2004).

played the influence of the MJO. Drosdowsky (1996), for example, in relation to the similarly-defined "active periods" stated, "in contrast to a number of recent studies that have highlighted the so-called 40–50-day oscillation in the Australian summer monsoon, no dominant timescales are found in the length of the active periods or in the recurrence time between active phases." Of course, this statement doesn't specifically refer to the monsoon onset, but given that monsoon onset is universally defined as the beginning of the first active period each wet season, it can be difficult to reconcile this statement given the earlier work of Hendon and Liebmann (1990a).

A recent re-evaluation of the onset/MJO relationship has been made by Wheeler and Hendon (2004), based on the information as presented here in Figure 5.13. Using the objective, independent, wind-only, monsoon onset definition of Drosdowsky (1996), the figure shows the relationship of the onset dates to the globally-defined MJO, as determined in the (RMM1,RMM2) phase space (introduced here in Section 5.6). A definite relationship with Drosdowsky's onset dates appears. Considering only the dates that lie outside the central unit circle (i.e., those occurring when the MJO is non-weak), more than 80% (15) of the dates occur in Phases 4–7 (when MJO low-level westerlies are in the vicinity of northern Australia, Figure 5.8) and less than 20% (3) of the onset dates occur in the other phases (when northern Australia is under the influence of MJO easterlies). Further, given all 28 years presented in the figure, in more than 60% (>17) of years the MJO appears to be having a positive impact.

Yet the spread of onsets from Phases 4–7 covers a time window of half the period of the MJO (i.e., about 20 to 30 days), a signficantly greater spread than the ±4 days found by Hendon and Liebmann (1990a). Thus it appears that while the globally-defined MJO is limiting monsoon onset to be within its active half-cycle, as lasting for a few weeks, the actual day of onset (as defined at the point location of Darwin) is often being set by other, presumably shorter timescale, phenomena.[6] This view is in essence the same as that of Hung and Yanai (2004) and the much earlier study of Hendon *et al.* (1989), both of whom found the large-scale, low-frequency, influence of the MJO to be only one of a number of factors that determined the monsoon onset.

How then did Hendon and Liebmann (1990a) get a much closer correspondence of the MJO and Australian monsoon onset? The reason is because of their less stringent definition of monsoon onset, and especially that for the MJO. For monsoon onset they used a 1-2-3-2-1 low-pass filter of Darwin winds and northern-Australian rainfall, and for their "40–50-day oscillation" they used a 30–60-day filtered series of the same winds and rainfall (i.e., a very local MJO definition). Obviously, dates defined from such similar series are likely to have a closer correspondence than those coming from unfiltered winds (as in the case of Drosdowsky's (1996) onset dates), and a series using information from planetary-scale fields of winds and OLR (as in the case of the RMM MJO index). Thus while the MJO was a notable oversight as a potential onset mechanism in Davidson *et al.*'s (1983) work, its dominance, we feel, was overstated by Hendon and Liebmann (1990a).

5.9 OTHER MODES AND SOURCES OF ISV

As we have already mentioned, much of the current mainstream thinking on modes of ISV in the Australian–Indonesian monsoon appears concentrated on the MJO (e.g., as conveyed in the recent review chapter by Webster *et al.*, 1998). No doubt, this is partly a reflection of the fact that the MJO is responsible for the most robust intraseasonal peak in frequency spectra of the standard atmospheric variables (e.g., Figure 5.4(a)). When looking at time–longitude plots of atmospheric fields at the appropriate latitudes for the region (e.g., Figures 5.10 and 5.11), much of the large-scale coherent eastward propagation can be identified as MJO events, and at least one MJO cycle can perhaps be identified in the region in most years. Yet, as has also been mentioned (Sections 5.5 and 5.7), cases of large-scale coherent westward propagation have also been observed to influence the region (see also McBride, 1983;

[6] We also examined the phase portrait of monsoon onsets (i.e., as in Figure 5.13) using the onset dates of both Hendon and Liebmann (1990a) and Hung and Yanai (2004). Although there were fewer points to examine, the general appearance of a large spread of the onset days across a number of MJO phases was still apparent.

Davidson *et al.*, 1983; Keenan and Brody, 1988; Hendon *et al.*, 1989). Although it can be quite varied, the phase speed of the westward convective envelopes is often about the same, except in the opposite direction, as that of the eastward MJO, being of the order of $5\,\mathrm{m\,s^{-1}}$ ($\sim 4^\circ\,\mathrm{day}^{-1}$). Usually the westward envelopes are maximized off the equator, and can be accompanied by large variations in the rotational wind as well. Such characteristics are suggestive of an influence on the monsoon by internal equatorial Rossby (ER) waves, the "mode" that is perhaps the next most important for this monsoon region.

Since the mid-1990s, evidence for the lowest order ($n = 1$) ER waves has become relatively prolific in studies of the tropical Pacific (e.g., Numaguti, 1995; Kiladis and Wheeler, 1995; Pires *et al.*, 1997; Wheeler and Kiladis, 1999). The evidence presented is of westward propagating disturbances with structures much like the theoretical shallow-water ER waves of Matsuno (1966). With the two exceptions of McBride (1983) and Hendon *et al.* (1989), however, they have been little mentioned in literature pertaining specifically to the Australian–Indonesian monsoon. While the waves do not readily appear in frequency spectra of individual time series of monsoon variables (e.g., OLR, as in Figure 5.4(a)), wave number–frequency spectra, using data from many longitudes, has had success at showing their existence above the red-noise background (e.g., Wheeler and Kiladis, 1999). The suggested broad frequency band of the waves is 1/7–1/40 cpd, which corresponds well with what we define as intraseasonal. Filtering for their specific zonal wave numbers and frequencies has revealed that they have their maximum variance in OLR near 150°E, and account for a good portion of the variance across the whole Australian–Indonesian monsoon region (figure 3(c) in Wheeler *et al.*, 2000). Concentrating on this signal, Wheeler *et al.* (2000) showed a typical structure of the "convectively-coupled" ER waves near their longitude of maximum variance, as reproduced here in Figure 5.14. It was constructed by computing lagged linear regressions of predictand variables against a reference time series of wave number–frequency filtered OLR. The waves' structure of symmetric circulation cells on either side of the equator is clearly depicted, as is a clear indication of an influence on winds and convection within the region of interest.

A case of $n = 1$ ER waves in the region is presented in Figure 5.15, as taken from a small portion of the 1983/1984 season that was previously shown and discussed in relevance to Figure 5.10. In particular, we focus on late February to early April, 1984, during which a number of westward-propagating features were previously identified. Given the $n = 1$ ER wave's symmetry about the equator, averages over the latitude band of 15°S–15°N are shown. In the total OLR data of Figure 5.15(a), there is the clear indication of flare-ups of convection occurring in the 120°E–140°E domain with a period between 10 and 20 days. Applying the wave number–frequency filtering of Wheeler and Kiladis (1999), as shown by the contours, indicates a close correspondence between those flare-ups and the ER wave's canonical wave number–frequency spectral peak (from which the filtering was defined). Yet further consistency with the ER wave is given in Figure 5.15(b), showing a clear signal of the waves in the intraseasonally-filtered (10–90 days) antisymmetric meridional wind [$(V_{\mathrm{NORTH}} - V_{\mathrm{SOUTH}})/2$]. Obviously, much of the broadband intraseasonal variance

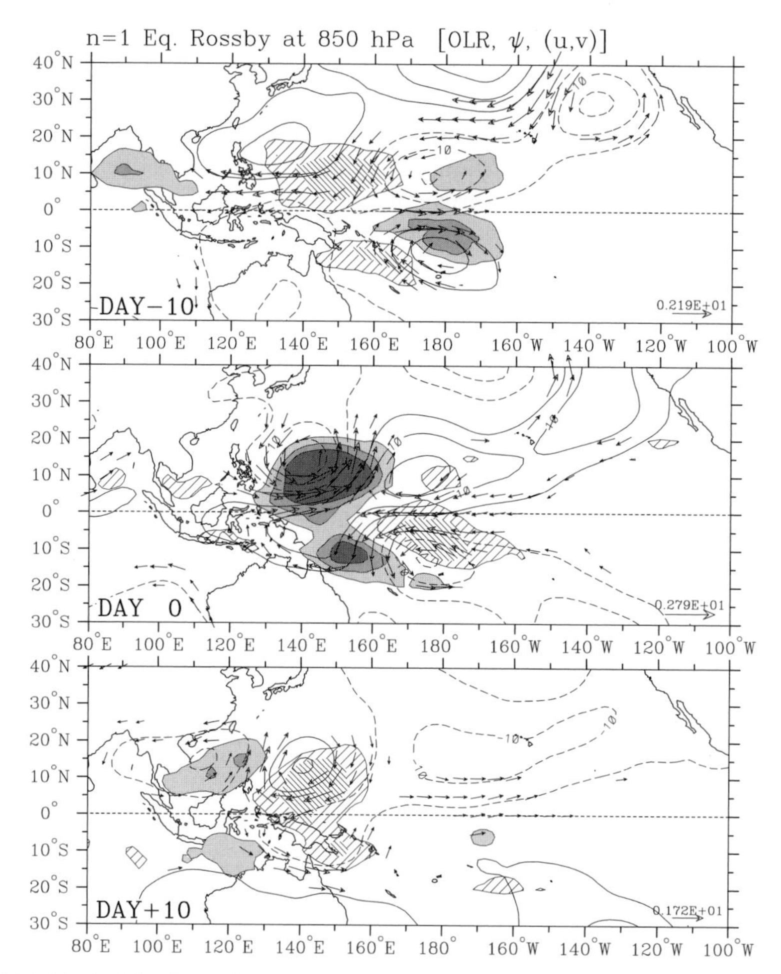

Figure 5.14. Typical horizontal structure of a convectively coupled $n = 1$ ER wave over a sequence spanning 21 days, as computed using lagged regression based on a two standard deviation anomaly in the ER wave filtered OLR series at 10°S, 150°E. Shading/cross-hatching show the negative/positive OLR anomalies at the levels of −15, −10, −5, 5, and $10\,\mathrm{W\,m^{-2}}$. Contours are streamfunction at the 850-hPa level, having an interval of $5 \times 10^5\,\mathrm{m^2\,s^{-1}}$, with negative contours dashed and the zero contour omitted. Vectors are the 850-Pa wind anomalies, plotted only where the local correlation of either wind component is statistically significant at the 99% level.
Reproduced from Wheeler *et al.* (2000).

can be accounted for by the ER waves during this particular period, similar cases of which can be found in many other years as well.

Other defined modes of large-scale tropical ISV also exist. Like the $n = 1$ ER wave, many of the shallow-water equatorially-trapped wave modes occur as convectively-coupled signals (Wheeler and Kiladis, 1999). Of the others, it is only

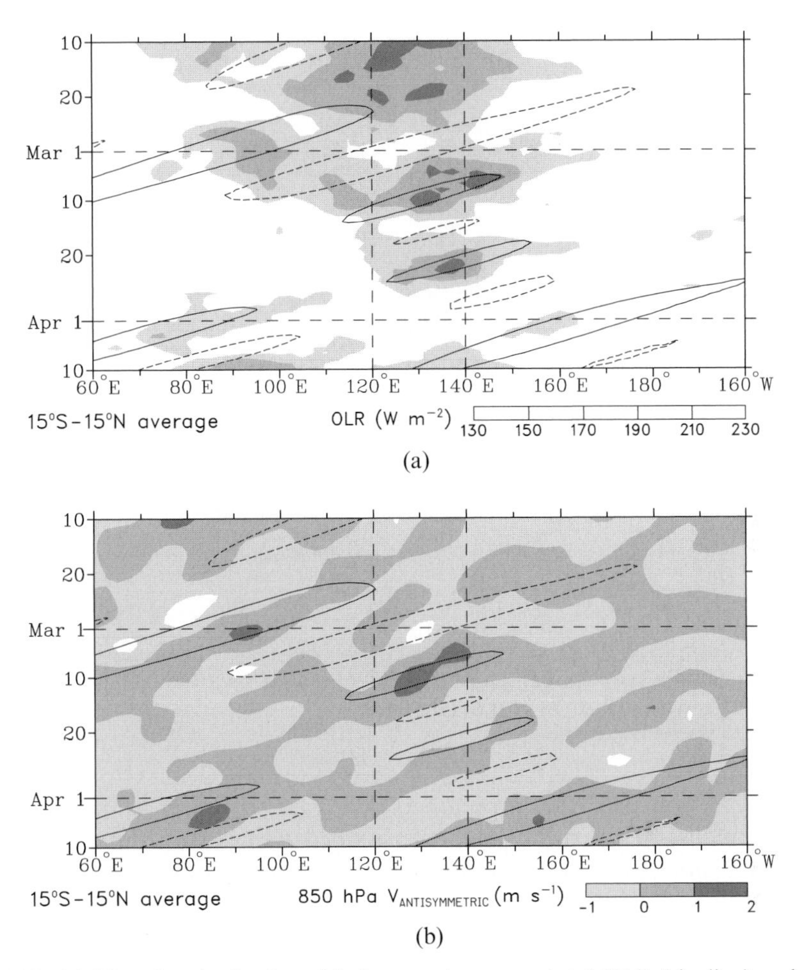

Figure 5.15. (a) Time–longitude plot of 3-day running mean total OLR (shading) and $n = 1$ ER wave filtered OLR (contours) averaged from 15°S to 15°N for a 2-month period in 1984. Contour interval for filtered anomalies is $10\,\mathrm{W\,m^{-2}}$, with the positive contour dashed, and the zero contour ommitted. (b) As in (a), except showing shading for the antisymmetric component of the 10–90-day band-pass filtered 850-hPa meridional wind [$(V_{\mathrm{NORTH}} - V_{\mathrm{SOUTH}})/2$].

the convectively-coupled Kelvin wave that is expected to have a direct influence in the intraseasonal frequency band, the others having frequencies higher than the defined intraseasonal limit (i.e., higher than 1/10 cpd). Using the same regression procedure as that applied for the ER wave above, Figure 5.16 shows the typical structure of the convectively-coupled Kelvin wave, as computed at its location of maximum variance (0°, 90°E) by Wheeler *et al.* (2000). It can be seen to be eastward moving, faster than the MJO, with its convective signal more confined near the equator, and primarily has a signal in the zonal component of the wind only.

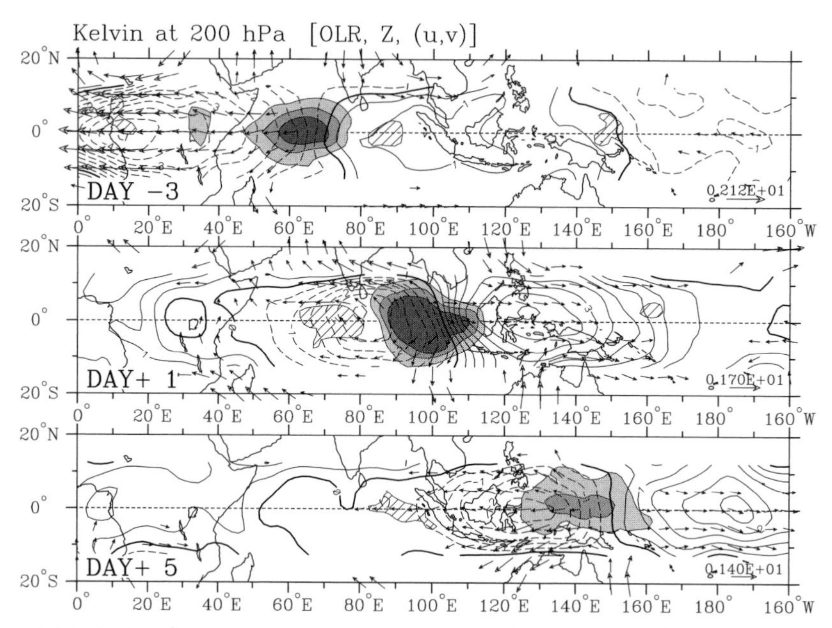

Figure 5.16. As in Figure 5.14, except for the convectively-coupled Kelvin wave, and showing contours of 200-hPa geopotential height and 200-hPa wind vector anomalies, and using the Kelvin wave filtered OLR series at 0°S, 90°E.
Reproduced from Wheeler *et al.* (2000).

Obviously, the Kelvin wave is expected to have a greater influence in the Indonesian part of the monsoon system. A typical phase speed of this convectively-coupled wave is around $15\,\mathrm{m\,s^{-1}}$, and its structure is a close match to disturbances previously described within the observational papers of Nakazawa (1986, 1988), Hayashi and Nakazawa (1989), Takayabu and Murakami (1991), and Dunkerton and Crum (1995). Terminology for the wave has not always been consistent, however, with some of these earlier studies using the much-bandied term "supercluster", when referring to them in the context of rapid eastward components embedded within the MJO. They need not always be associated with the MJO, however, as cases have shown.

An example case period for the convectively-coupled Kelvin waves is shown in Figure 5.17. The first panel is a time series of area-averaged OLR over the period from late 1997 to mid-1998. Crosses are drawn for days when the Kelvin wave-filtered OLR (as in Wheeler and Kiladis, 1999) reaches a value less than $-15\,\mathrm{W\,m^{-2}}$. Despite the fact that the OLR is generally higher than climatology, owing to the occurrence of the strong 1997/1998 El Niño, the Kelvin wave episodes correspond quite well with the total area-averaged convective variations during this whole period. Concentrating on the three episodes during January and February, the latter two panels (Figure 5.17(b, c)) give a clear indication of the fast eastward propagation of the waves, not only in the convection (Figure 5.17(b)), but in the intraseasonally-filtered zonal wind field as well (Figure 5.17(c)). As calculated

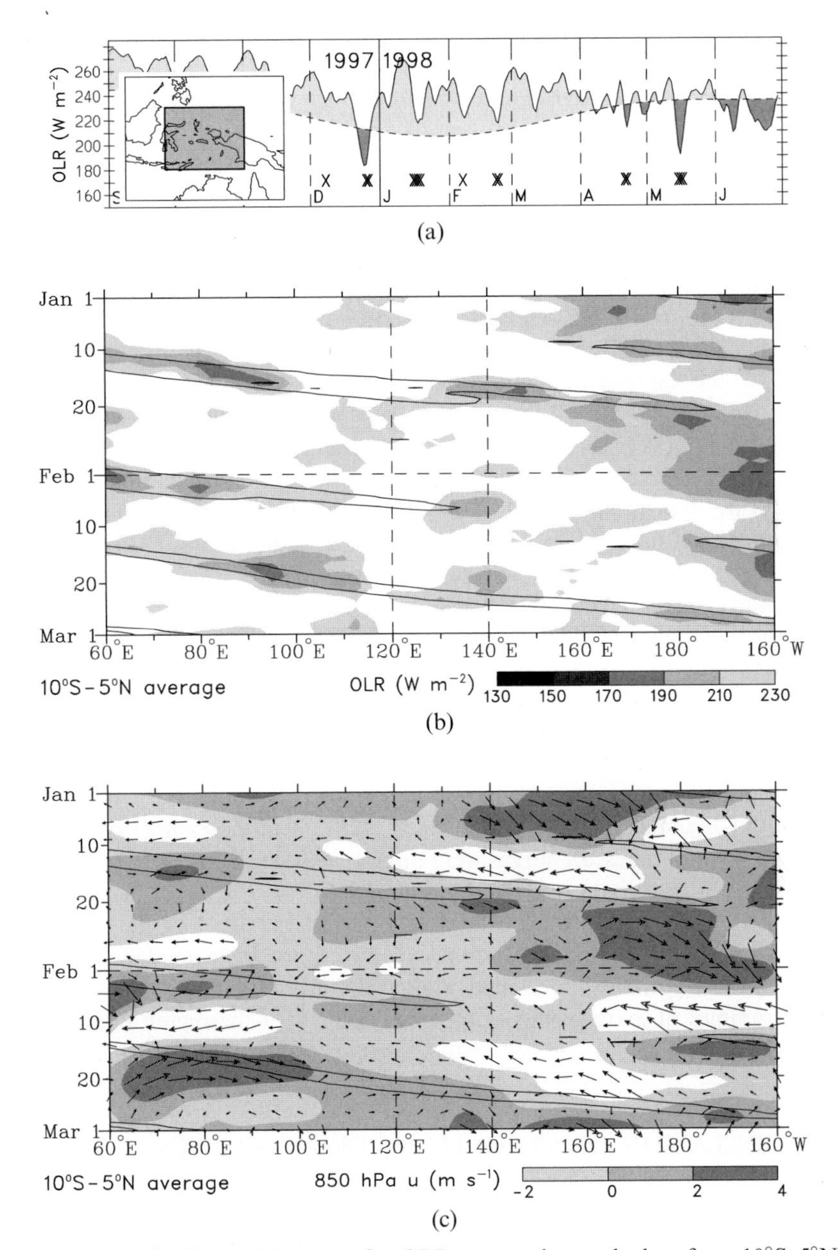

Figure 5.17. (a) As in Figure 5.2, except for OLR averaged over the box from 10°S–5°N and 120°E–140°E, and the monsoon period in 1997/1998, only. Also shown are crosses marking the days on which the Kelvin wave filtered component of the area-averaged OLR reaches a value of less than $-15\,\mathrm{W\,m^{-2}}$. (b) As in Figure 5.15(a), except for a 2-month period in 1998, an average from 10°S–5°N, and showing contours (negative only, contour interval of $15\,\mathrm{W\,m^{-2}}$) of the Kelvin wave filtered OLR. (c) As in (b), except showing shading for the 10–90-day band-pass filtered 850-hPa zonal wind, and vectors for the 10–90-day filtered 850-hPa total wind.

by Wheeler *et al.* (2000), February–July is typically the time that such Kelvin waves are most active.

Besides the contribution of variance from the above-mentioned "modes" of ISV, there may also be a significant source of ISV originating from tropical cyclones (TCs). Though these systems are primarily considered "synoptic" in both time and space ($\sim$5 days, $\sim$1,000 km), their large amplitude perturbations can significantly project onto lower frequencies, and the largest scales, as well. The same can be said for any other large-amplitude, non-periodic "weather"-like disturbance. Thus for example, Gunn *et al.* (1989) and Webster and Houze (1991) noted the simultaneous existence of a number of cyclones and monsoon depressions during AMEX and the Equatorial Mesoscale Experiment (EMEX), and while present, they dominated the structure of the large-scale flow. Moreover, in examining the large-scale flow averaged over the 1978/1979 season, McBride (1987) noted that the location of the mean upper level velocity potential maximum (or center of the large-scale divergent outflow) was within 6° longitude of the genesis location of five TCs. Thus McBride hypothesized that a major contribution to the three month-mean velocity potential map came from these cyclones, amounting to a projection of these synoptic-scale events on to the large-scale, low-frequency, structure. It is possible, however, that such observations are the result of ISV modulating TCs, rather than the reverse, a topic discussed in the next section.

Another potential source of ISV within the Australian–Indonesia monsoon region is the cross-equatorial influence of cold surges in the South China Sea. Occurring during the northern hemisphere winter, these surges are characterized by periods (of the order of several days) of strong northerly winds, anomalously low temperatures, and an increase in surface pressure, from the East Asian continent to the South China Sea (e.g., Lau and Chang, 1987). Compo *et al.* (1999) found sub-monthly (6–30 day) timescale surges were directly related to convective activity south of Indonesia. Sumathipala and Murakami (1988), on the other hand, found no contribution of lower frequency, 30–60-day, northerly surges of East Asian origin to convection in the Australian–Indonesian monsoon. Instead, they found a contribution from north-easterly flows originating in the subtropical north Pacific. Further examples of extra-tropical influences on ISV of the region are discussed in Section 5.11.

5.10 MODULATION OF SYNOPTIC EXTREME EVENTS

Besides the relationship to monsoon onset, bursts, and breaks, the modulation of synoptic extreme events by tropical ISV, and in particular by the MJO, is an aspect of this variability that has great potential for application given its direct relevance to society. The application arises due to the inherent predictability implied by the relatively long timescale of the various "modes" of ISV, like the MJO, a topic that is proving to be a fruitful avenue of research itself (e.g., Waliser *et al.*, 1999; Lo and Hendon, 2000; Wheeler and Weickman, 2001; see also Chapter 12).

Early results pertaining to the modulation of synoptic extremes were presented in the papers of Hendon and Liebmann (1990b) and Liebmann *et al.* (1994). In the former study they computed composites of the "synoptic" (2.5–8-day) kinetic energy and rainfall variance relative to their index of the "MJO" near Darwin (as based on their local 30–60-day filtered fields). Enhanced transient kinetic energy accompanied the wet phase of the oscillation, along with enhanced rainfall variance. The enhanced kinetic energy occurred through the depth of the troposphere but with maxima around 850 hPa and 100 hPa, consistent with the baroclinic structure that is common to many tropical systems. Of course, the existence of the enhanced synoptic rainfall variance was somewhat expected because of the nature of tropical rainfall (i.e., when there is no rainfall there is no variance, and when it is raining it is highly variable). But the enhanced synoptic-scale kinetic energy was not necessarily expected, and provided the impetus for their later study. Indeed, in Liebmann *et al.* (1994) they looked specifically at the modulation of the occurrence of TCs by the MJO, and showed a roughly 2 : 1 modulation between wet and dry MJO phases for TCs of the Indian and western Pacific Oceans. The hypothesis for such a modulation is that the large-scale MJO anomalies alter the "climatologically favourable factors" for TC development (e.g., low-level cyclonic vorticity, low vertical wind shear; Gray, 1979) on a timescale that is slow enough that they act in the same way as those climatological base-state factors.

Given their obvious impact, the modulation of the occurrence of TCs by the MJO has since gained a great deal of attention (e.g., Maloney and Hartmann, 2000a, b; Hall *et al.*, 2001; see also Chapters 2–4). The study of Hall *et al.* (2001) focused on the Australian–Indonesian region. By defining more categories of the MJO than Liebmann *et al.* (1994) (four instead of two), and looking at TCs forming over smaller regions ($\sim 40^\circ$ longitude wide), they were able to demonstrate a greater modulation than that found by the former study, being up to 4 : 1 to the north-west of Australia.

In a similar way to that employed by Hall *et al.* (2001), Figure 5.18 demonstrates the modulation of TC genesis across the whole southern hemisphere, except using a definition of MJO categories from the RMM-defined phases as used throughout the rest of this chapter (e.g., compare with Figure 5.7(c)). During the "weak MJO" category, the locations of TC genesis are evenly spread across the ocean basins, and when taking into account the number of days in the category, the chance of TC genesis is not significantly altered from normal. For the four MJO categories, on the other hand, the shift of the TC genesis locations from the west to east, in accordance with the phase of the MJO, is clear. To the west of Australia (west to about 90°E), TC genesis is significantly enhanced in Phases 4 and 5, and reduced in Phases 8 and 1. To the east of Australia (east to about the dateline), TC genesis is significantly enhanced in Phases 6 and 7. When comparing to the composite MJO structure of Figure 5.8, the TC genesis locations tend to cluster slightly poleward and westward of the main large-scale equatorial convective anomaly. It is there that the low-level cyclonic vorticity anomalies of the MJO tend to be greatest, as were found by Hall *et al.* (2001) to be an excellent diagnostic for TC modulation.

Interestingly, the modulation of TC numbers by low-frequency (relative to the TC) variability was found by Liebmann *et al.* (1994) to be an effect that is not

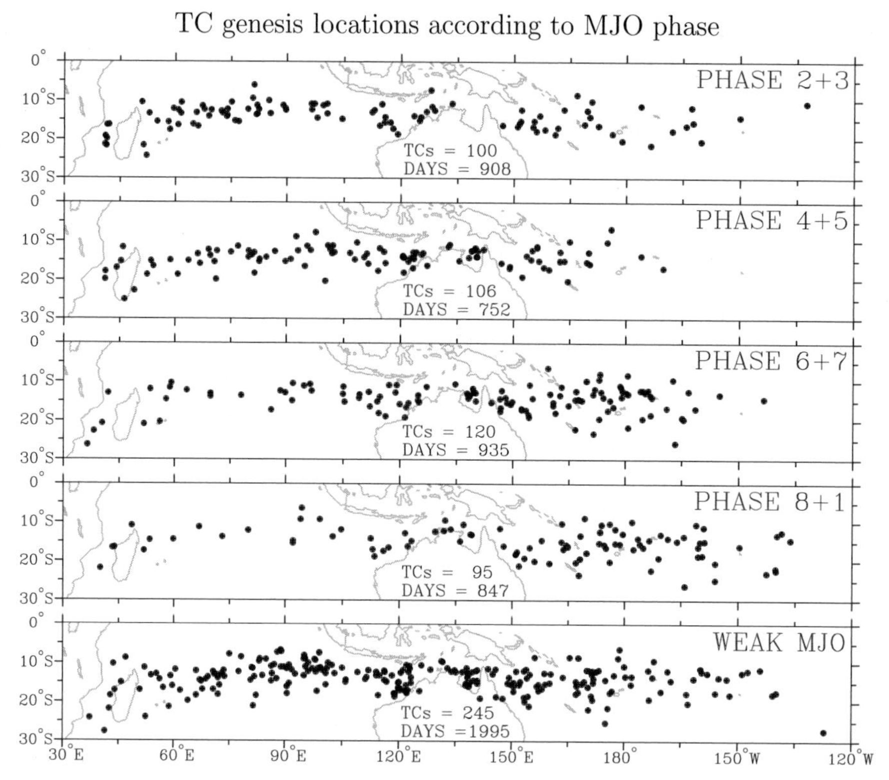

Figure 5.18. TC genesis (defined as occuring when a developing system reaches an estimated central pressure of 995 hPa) locations plotted according to the phase of the MJO as described by the daily (RMM1,RMM2) value (as in Figure 5.7(c)). November–April data used only, for all years from 1969–1999. "Weak MJO" refers to days when the amplitude of the (RMM1,RMM2) vector is less than 1.0. The number of TCs counted within each phase is given, as are the number of days for which that MJO phase occurred.

restricted to the MJO band. A higher frequency band produced the same degree of modulation. Hence it was suggested that any form of low-frequency (relative to the TC), large-scale variability, that alters the dynamical factors favorable for cyclogenesis, can also modulate TC activity. Indeed, investigation of the modulation of TCs by the other defined modes of ISV in the region is proving to be a fruitful avenue of current research. Of those modes, the convectively-coupled $n = 1$ ER wave has been found to provide the most significant modulation, in accordance with its relatively large amplitude perturbations of the low-level vorticity (as yet unpublished work[7]). No doubt, more research will be done on this in the future.

Yet further evidence of the modulation of the region's extreme synoptic events by the MJO has been provided by Wheeler and Hendon (2004). They looked at the contemporaneous relationship between the occurrence of the highest quintile of

[7] For discussion of real-time cases by the authors and others, see `http://www.tstorms.org`

weekly rainfall across Australia and the MJO as measured by the (RMM1,RMM2) phase, specifically for the December/January/February (DJF) season. Using a 1° gridded data set of weekly rainfall totals, the threshold value for the highest quintile weekly rainfall for the season ranges from only 10 mm in the central Australian desert up to greater than 130 mm along parts of the northern and eastern coasts. The normal probability of a weekly rainfall total in DJF exceeding this value is 20%. The conditional probability of the weekly rainfall exceeding this threshold, stratified by the phase of the MJO, is displayed in Figure 5.19. As can be seen, the probability of such an extreme event varies greatly with the phase of the MJO. For example, in the Top End region around Darwin, the probability varies from less than 12% in Phases 1 and 2 to greater than 36% in Phases 5 and 6. This represents more than a tripling of the likelihood of extreme rainfall from the extreme dry to wet phase of the MJO. In some locations (e.g., Arnhem Land and the northern tip of Cape York) the relative change of probabilities is as great as 6 times. No doubt, some of these swings in the occurrence of extreme weekly rainfall would arise due to the landfall of the already discussed TCs.

Besides the probability swings diagnosed across northern Australia, Figure 5.19 shows evidence of the same interesting extra-tropical rainfall patterns as discussed in Section 5.6 in reference to Figure 5.8. Increased rainfall probability in Phase 3 begins in the latitude band south of about 25°S, especially in the west, and progresses to the north and east by Phase 4. It is not until Phase 5 that the direct tropical signal of the MJO engulfs the north-western part of the continent (north of about 20°S); the tropical convection subsequently shifts further east by Phase 6. Tropical–extra-tropical interaction processes appear to produce such increased rainfall ahead of the main tropical signal, as will be discussed in the next section.

5.11 EXTRA-TROPICAL–TROPICAL INTERACTION

A number of the papers already discussed have presented evidence that the extra-tropics have an effect on convection in the Australian–Indonesian monsoon. In their study of the monsoon onset during the winter Monsoon Experiment (MONEX), Davidson *et al.* (1983) presented evidence that the trigger mechanism for onset lay in the evolution of highs and lows over the oceans to the south and west of Australia. In particular, they hypothesized that "prior to onset the seasonal build-up of planetary-scale and land–sea temperature gradients has reached a critical stage such that the troposphere is in a state of readiness for the monsoon. Before the onset can take place, however, it must wait for the southern hemisphere mid-latitude traveling highs and lows to be in such a configuration that trade wind easterlies are prevalent across the Australian continent." While not invoking the exact same mechanism, twenty years later Hung and Yanai (2004) also listed the "intrusion of mid-latitude troughs" as one of their four major factors contributing to the onset of the Australian monsoon. Other evidence for a role played by extra-tropical systems was discussed in Keenan and Brody (1988). They showed evidence

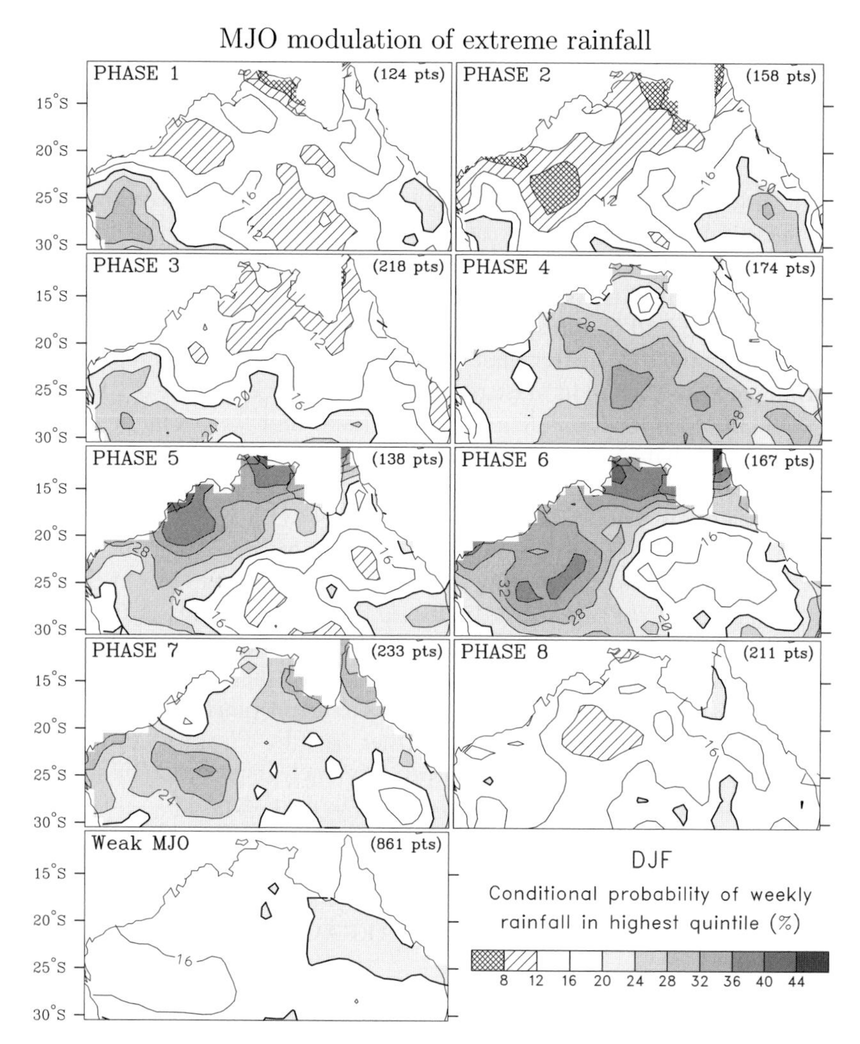

Figure 5.19. Probability of weekly rainfall in highest quintile conditioned upon the phase of the MJO when described by the daily (RMM1,RMM2) value. The rainfall weeks are centered on the day upon which the (RMM1,RMM2) phase is determined. The probabilities are computed for overlapping weeks in DJF for all available data in the 1974–1999 period. "Weak MJO" refers to days when the amplitude of the (RMM1,RMM2) vector is less than 1. Reproduced from Wheeler and Hendon (2004).

for a modulation of the tropical convection by upper level troughs in the higher latitude westerly flow.

Several of the case studies for the year of the AMEX/EMEX experiments also provide evidence for an influence on the monsoon from the extra-tropics. McBride and Frank (1999) found that the AMEX break period coincided with horizontal advection of dry air from higher latitudes at the levels of the middle to lower troposphere. Danielsen (1993) proposed a mechanism whereby the passage of mid-latitude cold fronts south of the Australian continent spread cold air northward across the continent which in turn interacted with the continental-scale sea breeze lying across the northern part of the continent. Such a description bares much similarity to that for the South China Sea cold surges mentioned in Section 5.9 (see also Love, 1985). Danielsen showed that changes in lower tropospheric stability and low-level convergence, factors that contribute to the triggering of convection, were synchronized with the passages of these higher latitude cold fronts.

None of the above-mentioned observational papers on the extra-tropics influencing the tropics discriminate between the "synoptic" and "intraseasonal" timescales, however. Another negative note is that even though they are consistent in the view that tropical convection responds to higher latitude systems, there appears little consistency from paper to paper on the actual mechanisms involved. Besides the gravity-current-like mechanism involved in the propagation of low-level cold surges to the equator (e.g., Love, 1985; Compo *et al.*, 1999; Section 5.9), one other obvious theoretically-derived mechanism is the meridional propagation of upper-level extra-tropical Rossby wave energy into the tropics. Although such Rossby wave propagation should primarily be limited to regions of upper-level westerlies (e.g., Webster and Holton, 1982), eastward-moving extra-tropical forcings have been shown to be capable of generating an equatorial response in the presence of upper-level easterlies, as is the case in the Australian–Indonesian region (Hoskins and Yang, 2000). This mechanism was invoked by Straub and Kiladis (2003), when explaining the observed connection between eastward extra-tropical Rossby wave activity in the southern hemisphere subtropical jet, and the initiation of convectively-coupled Kelvin waves over Indonesia. Another theoretical mechanism is found in the work of Frederiksen and Frederiksen (1997). Based on a normal-mode analysis employing the basic state of January, 1979, they demonstrated a link between intraseasonal activity in the Australian monsoon and baroclinic instability of the higher latitude flow. Little has been done, however, to employ this latter theoretical mechanism to understanding the observations of ISV in the Australian–Indonesian monsoon.

We turn now to the influence in the opposite direction, that is, the influence of the tropics on the extra-tropics. Here we concentrate our discussion on the near-field response, specifically near the longitudes of Australia. The interaction most commonly noted in the early monsoon-specific papers is the southward shift in the location of the southern hemisphere subtropical jet at the time of monsoon onset. This was first shown by Troup (1961) and later by Murakami and Sumi (1982), McBride (1983), Davidson *et al.* (1983), and Hendon and Liebmann (1990a). Again, however, these results only apply to ISV to the extent that monsoon onset is occurring on the intraseasonal timescale.

Since those early papers, more specific work has been done on the extra-tropical response to tropical convection, specifically on the intraseasonal timescale. Among the most pertinent work is that of Kiladis and Weickmann (1992), Berbery and Nogues-Paegle (1993), and Tyrrell *et al.* (1996). As has been well documented in the recent review by Kiladis and Mo (1998), the strong relationship between tropical convection and the Australian subtropical jet, as described for the monsoon onset above, holds for the intraseasonal timescale of the MJO (viz., 30 to 70 days) as well. During the phase of the MJO when convection is at the longitudes of northern Australia (Phase 5 in Figure 5.8), the response at the 200-hPa level (in summer) is that of an anomalous anticyclone centered at the latitude of about 30°S, and stretching in the zonal direction from the central Indian Ocean to Australia (not shown). That is, an upper-level anticyclone located poleward and westward of the tropical heating, not too dissimilar to the response expected from the theoretical equilibrium solution of Gill (1980), but perhaps better described by the wavy basic-state barotropic model of Sardesmukh and Hoskins (1988). When compared to the seasonal mean flow in summer, this represents an expansion of the upper-level equatorial easterlies, and a strengthening of the jet south of 30°S. At higher intraseasonal frequencies than the MJO (viz., less than 30 days), the result appears similar, albeit more localized near the region of convection (Kiladis and Weickmann, 1992, 1997).

One other example of tropical–extra-tropical interaction has been suggested in this chapter, and that was in relation to the extra-tropical rainfall signal in Australia placed ahead of the main tropical convective signal of the MJO, as seen in Figures 5.8 and 5.19 (Phases 3 and 4). While this particular signal in Australia has had little study, elsewhere in the world there appear similar features, with plausible explanations (e.g., Matthews *et al.*, 1996; Matthews and Kiladis, 1999). Such explanations invoke the generation of upper-level subtropical Rossby wave trains, and the favoring of precipitation ahead (to the east) of the Rossby wave's upper-level trough. As with all work on extra-tropical–tropical interaction processes, however, the configuration of the basic state is critical.

5.12 FINAL REMARKS

Overall, this review has demonstrated that there has been much progress during recent decades in the documentation and understanding of ISV within the Australian–Indonesian monsoon. The following key points can be made:

1. In comparison to other frequency bands, the seasonal cycle, and with other regions of the globe, the ISV of the Australian–Indonesian monsoon has particularly large amplitude.
2. The presence of ISV appears necessary for many of the defining characteristics of the monsoon, for example, the monsoon's sudden onset, its breaks, and deep westerly bursts.
3. Frequency-only power spectra of most monsoon parameters (excluding rainfall) show a robust signature of the MJO, but are otherwise predominantly "red", with no spectral gap between intraseasonal and higher frequencies with which to distinguish them. There is, however, high coherence between monsoon parameters (e.g., zonal wind vs. rain) at intraseasonal scales, particularly beyond 20 days, not seen at higher frequencies.

4. The coherence and phase of the zonal wind vs. rain also demonstrate the similarity in structure between the seasonal cycle and ISV. That is, both frequency bands share the useful, and traditional, concept of a two-state system: viz., wet-westerlies vs. dry-easterlies. This disappears at higher frequencies.
5. Convection studies have revealed that while the large-scale monsoon bursts involve more rain, the individual convective cells within the bursts are generally less intense than their counterparts in the breaks.
6. The MJO is the strongest mode of ISV within the Australian–Indonesian region, with a discernible influence on many aspects of the monsoon, including onset. Given its sometimes loose definition in older literature, however, we feel there has often been a tendency to overstate that influence.
7. Other well-defined modes of ISV also exist in the region. Of particular note are the convectively-coupled Kelvin and $n = 1$ ER waves, both of which, during cases, also have a discernible influence in many of the total fields.
8. Besides that accounted for by the above-mentioned modes of tropical ISV, there is still much variance in the intraseasonal band. Given its close correspondence to the red spectrum of a lag-1 autoregressive model, some of this variance appears best described as simply a consequence of the continuum of scales from "weather" to "climate". An isolated tropical cyclone, for example, given its large amplitude, will project significantly onto the intraseasonal frequencies.
9. The intrusion of disturbances and energy from the extra-tropics is yet another source of ISV in the Australian–Indonesian monsoon. Likewise, there is an influence of the region's tropical ISV on the extra-tropics.

This review also highlights a number of areas where there is a need for further research. The imbalance in research activity between the Australian vs. the Indonesian component of the monsoon system, especially in terms of using station-observed variables, is an area of concern. This is partly driven by the difficulty of obtaining station data, but can be addressed. The authors have a high level of interaction with Indonesian meteorologists on a "weather service to weather service" basis. From these interactions we know that the Indonesian hydrometeorological service (Badan Meteorogi dan Geofisika) makes seasonal forecasts of monsoon onset across the country. The operational onset definitions within Indonesia are based on cumulative rainfall totals over consecutive 10-day periods. These definitions are of major use for agriculture and are coherent across the country, as well as having strong relationships with large-scale low-frequency parameters such as the Southern Oscillation Index; and indeed these relationships are the basis of their seasonal forecasts. However, from some preliminary analysis, these rain-based definitions appear unrelated to the phase of the MJO. Yet, as we have shown in this chapter, 1987/1988 is at least one monsoon season in which the MJO appears to have influenced Bali's rainfall. Clearly, the role of ISV within Indonesia needs to be further explored.

Another research area in need of further consideration is that on the detailed nature of the monsoon burst vs. break events. As reviewed in this chapter, a large body of work on this topic has made a number of fundamental findings on the different modes of convection, on vertical stratification of the atmosphere, and on the electrical properties of the convection. However, the specific timescale involved for the bursts and breaks has hardly been a consideration in this literature, and thus it is extremely difficult to interpret in the context of the differences between intra-seasonal vs. higher frequency variations in the structure of the large-scale monsoon.

A third area for additional research is the further documentation and understanding of modes of ISV other than the MJO, particularly the shallow-water-like equatorial waves, as they appear in both the winds and convection. Lastly, as discussed here, ISV of the Australian–Indonesian monsoon can be partly understood in terms of its strong structural similarity to the seasonal cycle, in terms of the spectral properties of red noise, in terms of influences of large-scale modes such as the MJO, and in terms of influences from the extra-tropics. The level of understanding, however, is still primitive, and can be improved through further research. In particular, the relationships between time and space scales of all the sub-seasonal fluctuations need to be explored, as do the nature and cause of the geographical, seasonal and interannual variations in the level of ISV.

5.13 ACKNOWLEDGMENTS

Thanks to Wasyl Drosdowsky and Harry Hendon for providing data and/or various insights. Anne Leroy contributed to the production of Figure 5.18. Reviews of the original version of this chapter, as provided by George Kiladis, Duane Waliser, and Bill Lau, helped improve it.

5.14 REFERENCES

Berbery, E. H. and J. Nogues-Paegle (1993) Intraseasonal interactions between the tropics and extra-tropics in the Southern Hemisphere. *J. Atmos. Sci.*, **50**, 1950–1965.

Berlage, H. P. (1927) East-monsoon forecasting in Java. *Verhandelingen* (No. 20). Koninklijk Magnetisch en Meteorologisch Observatorium te Batavia, 42 pp.

Braak, C. (1919) Atmospheric variations of short and long duration in the Malay archipelago and neighbouring regions, and the possibility to forecast them. *Verhandelingen* (No. 5). Koninklijk Magnetisch en Meteorologisch Observatorium te Batavia, 57 pp.

Compo, G. P., G. N. Kiladis, and P. J. Webster (1999) The horizontal and vertical structure of east Asian winter monsoon pressure surges. *Quart. J. Roy. Meteor. Soc.*, **125**, 29–54.

Danielsen, E. F. (1993) In situ evidence of rapid, vertical irreversible transport of lower tropospheric air into the lower tropical stratosphere by convective cloud turrets and by large-scale upwelling in tropical cyclones. *J. Geophys. Res.*, **98**, 8665–8681.

Davidson, N. E. (1984) Short term fluctuations in the Australian monsoon during winter MONEX. *Mon. Wea. Rev.*, **112**, 1697–1708.

Davidson, N. E., J. L. McBride, and B. J. McAvaney (1983) The onset of the Australian monsoon during Winter MONEX: Synoptic aspects. *Mon. Wea. Rev.*, **111**, 496–516.

Davidson, N. E. and H. H. Hendon (1989) Downstream development in the Southern Hemisphere monsoon during FGGE/WMONEX. *Mon. Wea. Rev.*, **117**, 1458–1470.

de Boer, H. J. (1947) On forecasting the beginning and end of the dry monsoon in Java and Madura. *Verhandelingen* (No. 32). Koninklijk Magnetisch en Meteorologisch Observatorium te Batavia, 20 pp.

Drosdowsky, W. (1996) Variability of the Australian summer monsoon at Darwin: 1957–1992. *J. Climate*, **9**, 85–96.

Dunkerton, T. J. and F. X. Crum (1995) Eastward propagating ~2- to 15-day equatorial convection and its relation to the tropical intraseasonal oscillation. *J. Geophys. Res.*, **100**, 25781–25790.

Flinders, M. (1814) *A Voyage to Terra Australia*. G & W. Nicol, London. Edited reproduction available as *Terra Australis*. T. Flannery (ed.) (2000). Text Publishing, Melbourne.

Frank, W. M. and J. L. McBride (1989) The vertical distribution of heating in AMEX and GATE cloud clusters. *J. Atmos. Sci.*, **46**, 3464–3478.

Frederiksen, J. S. and C. S. Frederiksen (1997) Mechanisms of the formation of intraseasonal oscillations and Australian monsoon disturbances: The roles of latent heat, barotropic, and baroclinic instability. *Contributions to Atmospheric Physics*, **70**, 39–56.

Gill, A. E. (1980) Some simple solution for heat induced tropical circulations. *Quart. J. Roy. Meteor. Soc.*, **106**, 447–462.

Gray, W. M. (1979) Hurricanes: Their formation, structure and likely role in the tropical circulation. In: D. B. Shaw (ed.), *Meteorology Over the Tropical Oceans*. Royal Meteorological Society, U.K., pp. 155–218.

Gunn, B. W., J. L. McBride, G. J. Holland, T. D. Keenan, N. E. Davidson, and H. H. Hendon (1989) The Australian summer monsoon circulation during AMEX Phase II. *Mon. Wea. Rev.*, **117**, 2554–2574.

Hall, J. D., A. J. Matthews, and D. J. Karoly (2001) The modulation of tropical cyclone activity in the Australian region by the Madden–Julian Oscillation. *Mon. Wea. Rev.*, **129**, 2970–2982.

Hastenrath, S. (1987) Predictability of Java monsoon rainfall anomalies: A case study *J. Climate Appl. Meteor.*, **26**, 133–141.

Hayashi, Y. and T. Nakazawa (1989) Evidence of the existence and eastward motion of superclusters at the equator. *Mon. Wea. Rev.*, **117**, 236–243.

Hendon, H. H., N. E. Davidson, and B. Gunn (1989) Australian summer monsoon onset during AMEX 1987. *Mon. Wea. Rev.*, **117**, 370–390.

Hendon, H. H. and B. Liebmann (1990a) A composite study of onset of the Australian summer monsoon. *J. Atmos. Sci.*, **47**, 2227–2240.

Hendon, H. H. and B. Liebmann (1990b) The intraseasonal (30–50 day) oscillation of the Australian summer monsoon. *J. Atmos. Sci.*, **47**, 2909–2923.

Hendon, H. H. and B. Liebmann (1994) Organization of convection within the Madden–Julian Oscillation. *J. Geophys. Res.*, **99**, 8073–8083.

Hendon, H. H. and M. L. Salby (1994) The life cycle of the Madden–Julian Oscillation. *J. Atmos. Sci.*, **51**, 2225–2237.

Hendon, H. H., C. Zhang, and J. D. Glick (1999) Interannual variation of the Madden–Julian Oscillation during Austral Summer. *J. Climate*, **12**, 2538–2550.

Holland, G. J. (1986) Interannual variability of the Australian summer monsoon at Darwin: 1952–82. *Mon. Wea. Rev.*, **114**, 594–604.

Hoskins, B. J. and G.-Y. Yang (2000) The equatorial response to higher-latitude forcing. *J. Atmos. Sci.*, **57**, 1197–1213.

Hung, C.-W. and M. Yanai (2004) Factors contributing to the onset of the Australian summer monsoon. *Quart. J. Roy. Meteor. Soc.*, **130**, 739–758.

Hunt, H. A., G. Taylor, and E. T. Quayle (1913) *The Climate and Weather of Australia*. Commonwealth Bureau of Meteorology, Melbourne, Australia, 93 pp.

Jones, C., L. M. V. Carvalho, R. W. Higgins, D. E. Waliser, and J.-K. E. Schemm (2004) Climatology of tropical intraseasonal convective anomalies. *J. Climate*, **17**, 523–539.

Keenan, T. D. and L. R. Brody (1988) Synoptic-scale modulation of convection during the Australian summer monsoon. *Mon. Wea. Rev.*, **116**, 71–85.

Keenan, T. D. and R. E. Carbone (1992) A preliminary morphology of precipitation systems in tropical northern Australia. *Quart. J. Roy. Meteor. Soc.*, **118**, 283–326.

Keenan, T. D. and S. A. Rutledge (1993) Mesoscale characteristics of monsoonal convection and associated stratiform precipitation. *Mon. Wea. Rev.*, **121**, 352–374.

Keenan, T., S. Rutledge, R. Carbone, J. Wilson, T. Takahashi, P. May, N. Tapper, M. Platt, J. Hacker, S. Sekelsky *et al.* (2000) The Maritime Continent Thunderstorm Experiment (MCTEX): Overview and some results. *Bull. Amer. Meteor. Soc.*, **81**, 2433–2455.

Kiladis, G. N. and K. C. Mo (1998) Interannual and intraseasonal variability in the Southern Hemisphere. In: D. Karoly and D. Vincent (eds), *Meteorology of the Southern Hemisphere*. American Meteorological Society, Boston, U.S.A., pp. 307–336.

Kiladis, G. N. and K. M. Weickmann (1992) Circulation anomalies associated with tropical convection during northern winter. *Mon. Wea. Rev.*, **120**, 1900–1923.

Kiladis, G. N. and K. M. Weickmann (1997) Horizontal structure and seasonality of large-scale circulations associated with submonthly tropical convection. *Mon. Wea. Rev.*, **125**, 1997–2013.

Kiladis, G. N. and M. Wheeler (1995) Horizontal and vertical structure of observed tropospheric equatorial Rossby waves. *J. Geophys. Res.*, **100**, 22981–22997.

Knutson, T. R. and K. M. Weickmann (1987) 30–60 day atmospheric oscillations: Composite life cycles of convection and circulation anomalies. *Mon. Wea. Rev.*, **115**, 1407–1436.

Lau, K.-M. and P. H. Chan (1985) Aspects of the 40–50 day oscillation during the northern winter as inferred from outgoing longwave radiation. *Mon. Wea. Rev.*, **113**, 1889–1909.

Lau, K.-M. and P. H. Chan (1988) Intraseasonal and interannual variations of tropical convection: A possible link between the 40–50 day oscillation and ENSO? *J. Atmos. Sci.*, **45**, 506–521.

Lau, K. M. and C.-P. Chang (1987) Planetary scale aspects of the Winter monsoon and atmospheric teleconnections. In: C. P. Chang and T. N. Krishnamurti (eds), *Monsoon Meteorology*. Oxford University Press, New York, pp. 161–202.

Liebmann, B., H. H. Hendon, and J. D. Glick (1994) The relationship between tropical cyclones of the western Pacific and Indian Oceans and the Madden–Julian Oscillation. *J. Meteor. Soc. Jap.*, **72**, 401–412.

Lo, F. and H. H. Hendon (2000) Empirical extended-range prediction of the Madden–Julian Oscillation. *Mon. Wea. Rev.*, **128**, 2528–2543.

Love, G. (1985) Cross-equatorial influence of winter hemisphere subtropical cold surges. *Mon. Wea. Rev.*, **113**, 1487–1498.

Madden, R. A. and P. R. Julian (1971) Detection of a 40–50 day oscillation in the zonal wind in the tropical Pacific. *J. Atmos. Sci.*, **28**, 702–708.

Madden, R. A. and P. R. Julian (1972) Description of global-scale circulation cells in the tropics with a 40–50-day period. *J. Atmos. Sci.*, **29**, 1109–1123.

Madden, R. A. and P. R. Julian (1994) Observations of the 40–50-day tropical oscillation – A review. *Mon. Wea. Rev.*, **122**, 814–837.

Maloney, E. D. and D. L. Hartmann (2000a) Modulation of hurricane activity in the Gulf of Mexico by the Madden–Julian Oscillation. *Science*, **287**, 2002–2004.

Maloney, E. D. and D. L. Hartmann (2000b) Modulation of eastern North Pacific hurricanes by the Madden–Julian Oscillation. *J. Climate*, **13**, 1451–1460.

Manton, M. J. and J. L. McBride (1992) Recent research on the Australian monsoon. *J. Meteor. Soc. Jap.*, **70**, 275–284.

Mapes, B. and R. A. Houze, Jr. (1992) An integrated view of the 1987 Australian monsoon and its mesoscale convective systems. Part I: Horizontal structure. *Quart. J. Roy. Meteor. Soc.*, **118**, 927–963.

Matsuno, T. (1966) Quasi-geostrophic motions in the equatorial area. *J. Meteor. Soc. Jap.*, **44**, 25–43.

Matthews, A. J. (2000) Propagation mechanisms for the Madden–Julian Oscillation. *Quart. J. Roy. Meteor. Soc.*, **126**, 2637–2652.

Matthews, A. J., B. J. Hoskins, J. M. Slingo, and M. Blackburn (1996) Development of convection along the SPCZ within a Madden–Julian Oscillation. *Quart. J. Roy. Meteor. Soc.*, **122**, 669–688.

Matthews, A. J. and G. N. Kiladis (1999) The tropical–extratropical interaction between high-frequency transients and the Madden–Julian Oscillation. *Mon. Wea. Rev.*, **127**, 661–677.

May, P. and D. K. Rajopadhyaya (1999) Vertical velocity characteristics of deep convection over Darwin, Australia. *Mon. Wea. Rev.*, **127**, 1056–1071.

McBride, J. L. (1983) Satellite observations of the southern hemisphere monsoon during Winter MONEX. *Tellus*, **35A**, 189–197.

McBride, J. L. (1987) The Australian summer monsoon. In: C. P. Chang and T. N. Krishnamurti (eds), *Monsoon Meteorology*. Oxford University Press, New York, 203–231.

McBride, J. L. (1995) Tropical cyclone formation. *Global Perspectives on Tropical Cyclones*, WMO Tech. Doc. WMO/TD 693, World Meteorological Organization, 63–105.

McBride, J. L. (1998) Indonesia, Papua New Guinea, and tropical Australia: The Southern Hemisphere monsoon. In: D. Karoly and D. Vincent (eds), *Meteorology of the Southern Hemisphere*. American Meteorological Society, Boston, U.S.A., pp. 89–99.

McBride, J. L. and W. M. Frank (1999) Relationships between stability and monsoon convection. *J. Atmos. Sci.*, **56**, 24–36.

McBride, J. L. and T. D. Keenan (1982) Climatology of tropical cyclone genesis in the Australian region. *J. Climatol.*, **2**, 13–33.

Meehl, G. A. (1987) The annual cycle and interannual variability in the tropical Pacific and Indian Ocean regions. *Mon. Wea. Rev.*, **115**, 27–50.

Murakami, T. and A. Sumi (1982) Southern Hemisphere monsoon circulation during the 1978–79 WMONEX. Part II: Onset, active and break monsoons. *J. Meteorol. Soc. Jap.*, **60**, 649–671.

Nakazawa, T. (1986) Mean features of 30–60 day variations as inferred from 8-year OLR data. *J. Meteor. Soc. Jap.*, **64**, 777–786.

Nakazawa, T. (1988) Tropical super clusters within intraseasonal variations over the western Pacific. *J. Meteor. Soc. Jap.*, **66**, 823–839.

Naylor, R. L., W. P. Falcon, D. Rochberg, and N. Wada (2001) Using El Niño/Southern Oscillation climate data to predict rice production in Indonesia. *Clim. Change*, **50**, 255–265.

Nicholls, N. (1981) Air–sea interaction and the possibility of long-range weather prediction in the Indonesian Archipelago. *Mon. Wea. Rev.*, **109**, 2435–2443.

Numaguti, A. (1995) Characteristics of 4- to 20-day-period disurbances observed in the equatorial Pacific during the TOGA COARE IOP. *J. Meteor. Soc. Jap.*, **73**, 353–377.

Pires, P., J.-L. Redelsperger, and J.-P. Lafore (1997) Equatorial atmospheric waves and their association to convection. *Mon. Wea. Rev.*, **125**, 1167–1184.

Rutledge, S. A., E. R. Williams, and T. D. Keenan (1992) The Down Under Doppler and Electricity Experiment (DUNDEE): Overview and preliminary results. *B. Amer. Meteor. Soc.*, **73**, 3–16.

Salby, M. L. and H. H. Hendon (1994) Intraseasonal behavior of clouds, temperature and motion in the Tropics. *J. Atmos. Sci.*, **51**, 2207–2224.

Sardesmukh, P. D. and B. J. Hoskins (1988) The generation of global rotational flow by steady idealized tropical divergence. *J. Atmos. Sci.*, **45**, 1228–1251.

Selkirk, H. B. (1993) The tropopause cold trap in the Australian monsoon during STEP/AMEX 1987. *J. Geophys. Res.*, **98**, 8591–8610.

Sikka, D. R. and S. Gadgil (1980) On the maximum cloud zone and the ITCZ over Indian longitudes during the southwest monsoon. *Mon. Wea. Rev.*, **108**, 1840–1853.

Slingo, J. M., D. P. Rowell, K. R. Sperber, and F. Nortley (1999) On the predictability of the interannual behaviour of the Madden–Julian Oscillation and its relationship with El Niño. *Quart. J. Roy. Meteor. Soc.*, **125**, 583–609.

Straub, K. H. and G. N. Kiladis (2002) Observations of a convectively coupled Kelvin wave in the eastern Pacific ITCZ. *J. Atmos. Sci.*, **59**, 30–53.

Straub, K. H. and G. N. Kiladis (2003) Extratropical forcing of convectively coupled Kelvin waves during Austral winter. *J. Atmos. Sci.*, **60**, 526–543.

Sumathipala, W. L. and T. Murakami (1988) Intraseasonal fluctuations in low-level meridional winds over the south China Sea and the western Pacific and monsoonal convection over Indonesia and northern Australia. *Tellus A*, **40**, 205–219.

Suppiah, R. (1992) The Australian summer monsoon: A review. *Progress in Physical Geography*, **16**, 283–318.

Takayabu, Y. N. (1994) Large-scale cloud disturbances associated with equatorial waves. Part I: Spectral features of the cloud disturbances. *J. Meteor. Soc. Jap.*, **72**, 433–448.

Takayabu, Y. N. and M. Murakami (1991) The structure of super cloud clusters observed on 1–20 June 1986 and their relationship to easterly waves. *J. Meteor. Soc. Jap.*, **69**, 105–125.

Tanaka, M. (1994) The onset and retreat dates of the Austral summer monsoon over Indonesia, Australia and New Guinea. *J. Meteor. Soc. Jap.*, **72**, 255–267.

Troup, A. J. (1961) Variations in upper tropospheric flow associated with the onset of the Australian summer monsoon. *Indian J. Meteor. Geophys.*, **12**, 217–230.

Tyrrell, G. C., D. J. Karoly, and J. L. McBride (1996) Links between tropical convection and variations of the extratropical circulation during TOGA COARE. *J. Atmos. Sci.*, **53**, 2735–2748.

Wallace, A. R. (1891) *The Malay Archipelago; The Land of the Orang-utan and the Bird of Paradise; A Narrative of Travel With Studies of Man and Nature* (Tenth Edition). Macmillan & Co., London. Reprinted as *The Malay Archipelago*, A. R. Wallace (ed.) (2000). Periplus (HK) Limited, Singapore.

Waliser, D. E., C. Jones, J. K. E. Schemm, and N. E. Graham (1999) A statistical extended-range tropical forecast model based on the slow evolution of the Madden–Julian Oscillation. *J. Climate*, **12**, 1918–1939.

Waliser, D. E., K. M. Lau, W. Stern, and C. Jones (2003) Potential predictability of the Madden–Julian Oscillation. *Bull. Amer. Meteor. Soc.*, **84**, 33–50.

Wang, B. and H. Rui (1990) Synoptic climatology of transient tropical intraseasonal convection anomalies: 1975–1985. *Met. Atmos. Phys.*, **44**, 43–61.

Webster, P. J. and J. R. Holton (1982) Cross-equatorial response to middle-latitude forcing in a zonally-varying basic state. *J. Atmos. Sci.*, **39**, 722–733.

Webster, P. J. and R. A. Houze, Jr. (1991) The Equatorial Mesoscale Experiment (EMEX): An overview. *Bull. Amer. Soc.*, **72**, 1481–1505.

Webster, P. J., V. O. Magana, T. N. Palmer, J. Shukla, R. A. Tomas, M. Yanai, and T. Yasunari (1998) Monsoons: Processes, predictability, and the prospects for prediction. *J. Geophys. Res.*, **103**, 14451–14510.

Weickmann, K. M. (1983) Intraseasonal circulation and outgoing longwave radiation modes during Northern Hemisphere winter. *Mon. Wea. Rev.*, **111**, 1838–1858.

Weickmann, K. M., G. R. Lussky, and J. E. Kutzbach (1985) Intraseasonal (30–60 day) fluctuations of outgoing longwave radiation and 250-mb streamfunction during Northern Winter. *Mon. Wea. Rev.*, **113**, 941–961.

Wheeler, M. C. and H. H. Hendon (2004) An all-season real-time multivariate MJO index: Development of an index for monitoring and prediction. *Mon. Wea. Rev.*, **132**, 1917–1932.

Wheeler, M. and G. N. Kiladis (1999) Convectively coupled equatorial waves: Analysis of clouds and temperature in the wavenumber–frequency domain. *J. Atmos. Sci.*, **56**, 374–399.

Wheeler, M., G. N. Kiladis, and P. J. Webster (2000) Large-scale dynamical fields associated with convectively coupled equatorial waves. *J. Atmos. Sci.*, **57**, 613–640.

Wheeler, M. and K. M. Weickmann (2001) Real-time monitoring and prediction of modes of coherent synoptic to intraseasonal tropical variability. *Mon. Wea. Rev.*, **129**, 2677–2694.

Williams, E. R., S. A. Rutledge, S. G. Geotis, N. Renno, E. Rasmussen, and T. Rickenbach (1992) A radar and electrical study of tropical "hot towers". *J. Atmos. Sci.*, **49**, 1386–1395.

Yano, J.-I. and J. L. McBride (1998) An aqua-planet monsoon. *J. Atmos. Sci.*, **55**, 1373–1399.

Yasunari, T. (1979) Cloudiness fluctuations associated with the Northern Hemisphere summer monsoon. *J. Meteor. Soc. Jap.*, **57**, 227–242.

6

The oceans

William S. Kessler

6.1 INTRODUCTION

There is a very wide variety of intraseasonal variability (ISV) in the oceans, due to many different processes beyond forcing by tropical intraseasonal winds and heat fluxes. The main focus of this chapter, however, is on the upper ocean response to the tropical atmospheric ISV that is discussed in the other chapters of this book and is most germane in this context. The prominent oceanic ISV signatures generated by other mechanisms (largely intrinsic to the ocean), and those found in other regions are briefly reviewed in Section 6.7.

Episodic wind events on intraseasonal timescales affect the ocean through three main mechanisms: increased evaporation, the generation of equatorial jets and waves that produce advective changes remotely, and enhanced mixing and entrainment. As Webster and Lukas (1992) note, these responses are proportional to the windspeed u, u^2, and u^3 respectively, and therefore depend very differently on the background wind and the structure of its variance. Much of the forcing by tropical intraseasonal oscillations (TISO) occurs over the warm pools of the Indian and west Pacific Oceans where the thermocline is usually deeper than the mixed layer. Thus, the near-surface density structure is relatively unconstrained by large-scale ocean dynamics and can easily be modulated by the winds and the heat and moisture fluxes due to the ISV, providing the opportunity for air–sea feedbacks, non-linear effects, and the retention of an oceanic memory of previous forcing. The dynamic response depends on the thickness of the accelerating layer, which is a function both of the background stratification and of local precipitation and mixing. Thus a principal focus of this chapter (Sections 6.2 and 6.3) is the factors controlling the upper ocean stratification under rapidly changing windspeed and precipitation sufficient for salinity variation to determine the mixed layer depth. The correlation of ISV of solar short-wave forcing with the wind fluctuations can also lead to significant

W. K. M. Lau and D. E. Waliser (eds), *Intraseasonal Variability in the Atmosphere–Ocean Climate System.*

effects on mixed layer temperature structure, with a variety of consequences (Section 6.5 and Chapter 7).

Because most of the work on oceanic ISV has been done in the Pacific, while the Indian Ocean is relatively poorly sampled, the processes of the oceanic response are described in the Pacific context, and factors specific to the Indian Ocean are discussed in Section 6.6.

Much of the interest in tropical ISV in recent years has concerned its possible interaction with the El Niño Southern Oscillation (ENSO) cycle, which has been a controversial element of the debate over the nature of ENSO. While coupled models without realistic ISV have been successful in reproducing aspects of ENSO statistics, it remains in question whether ENSO is a disturbance to a stable background state, in which case an initiating perturbation would be required, or is a self-sustained mode on an unstable background. After satellite sampling established the occurrence of strong Madden–Julian Oscillation (MJO) events penetrating into the western Pacific during the onset stages of El Niño, several mechanisms have been proposed by which rectification of intraseasonal forcing in the ocean could interact constructively with the ENSO cycle; these are discussed in Section 6.5. (see also Chapters 9 and 12 for additional discussion of ISV–ENSO interactions).

6.2 HEAT FLUXES

Intraseasonal ocean–atmosphere heat fluxes are discussed in several other parts of this book, especially Chapters 7, 10, and 11; this section will focus on changes in the structure of the ocean in response to those fluxes, especially within the west Pacific warm pool that has been extensively studied (see Godfrey *et al.*, 1998, for a review of the Tropical Ocean Global Atmosphere (TOGA) Coupled Ocean–Atmosphere Response Experiment (COARE) program). The west Pacific warm pool differs from most other open-ocean regions because its heavy precipitation and generally weak winds mean that the seasonal background mixed layer depth is often controlled by salinity stratification. Net precipitation minus evaporation over the warm pool is 1–2 m yr^{-1} (Anderson *et al.*, 1996), leading to low surface salinity that plays a major role in determining the vertical stability of the warm pool. As a result, the thick warm layer above the thermocline can be split by a halocline, and its lower part thereby uncoupled from surface forcing. Much of the precipitation occurs during convection associated with TISO events, which also produce strong shortwave and windspeed variability as convective systems pass across the region (see Figure 1.2 and Chapter 7).

6.2.1 Salinity and the barrier layer

The isothermal layer beneath the halocline has become known as the "barrier layer" (Lukas and Lindstrom, 1991; Sprintall and Tomczak, 1992) since it inhibits the communication between the surface and the thermocline that dominates sea surface temperature (SST) change in the central and eastern equatorial Pacific (e.g., Kessler and McPhaden, 1995b; Zhang, 2001). A thin mixed layer tends to

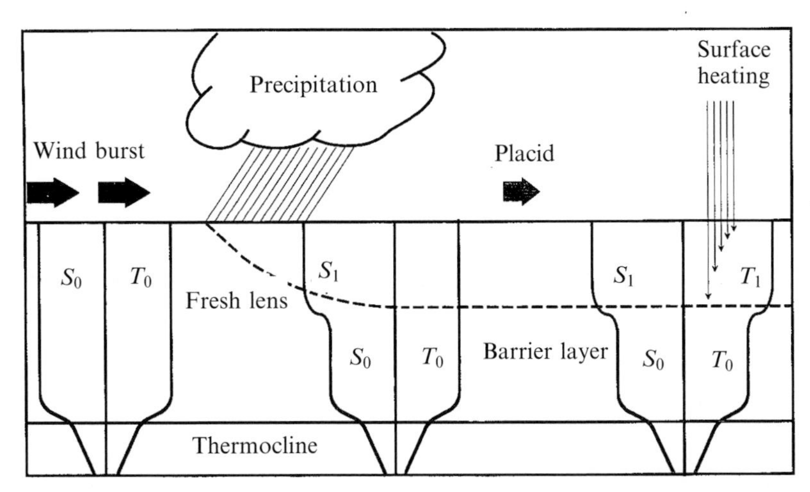

Figure 6.1. Schematic diagram showing the Lukas–Lindstrom "barrier layer" theory. During a strong wind burst, the surface mixed layer extends down to the top of the thermocline. Following the wind burst, the additional buoyancy from precipitation and strong surface heating acts to form a relatively warm and fresh thin surface layer. Below this thin layer is a strong halocline, which effectively decouples the surface forcing from the deeper waters. Further heating is trapped to vertical mixing above the barrier formed by the halocline.
After Anderson *et al.* (1996).

trap surface fluxes of heat and momentum within it, enhancing both SST variability in response to heat fluxes and the acceleration of surface currents in response to winds. Figure 6.1 shows a schematic of barrier layer formation and erosion (Anderson *et al.*, 1996). Under strong winds, the upper layer can become well-mixed down to the deep thermocline of the west Pacific (typically 150 m; e.g., Figure 6.2). Following heavy precipitation, as long as the winds aren't strong enough to mix it away, a fresh lens can cap the top of the isothermal layer (Figure 6.1, middle). If this stratification is maintained, subsequent surface heating under clear skies will mostly occur within this relatively thin layer (≤ 50 m), because the short-wave attenuation with depth is exponential (Figure 6.1, right) and not be mixed over the entire above-thermocline layer as would generally occur in other regions. (However, if the surface layer is very thin, some radiation will penetrate into the barrier layer (see next section).) Because the halocline inhibits mixing, and because additional heating enhances its stability, the barrier layer tends to persist until winds (often in the form of intraseasonal westerlies) are strong enough to mix it away. Model experiments based on the situation during TOGA COARE suggest that the observed stratification produces close to the maximum possible SST: increased precipitation or weaker winds would result in a shallower mixed layer that would lose heat through its base by penetrative radiation, whereas decreased precipitation or stronger winds would lead to fewer barrier layer occurrences and thus more entrainment cooling (Anderson *et al.*, 1996). Recent work has suggested other mechanisms that can produce or intensify barrier layers on intraseasonal timescales. For example, since west Pacific surface salinity is lower than that further east, if a

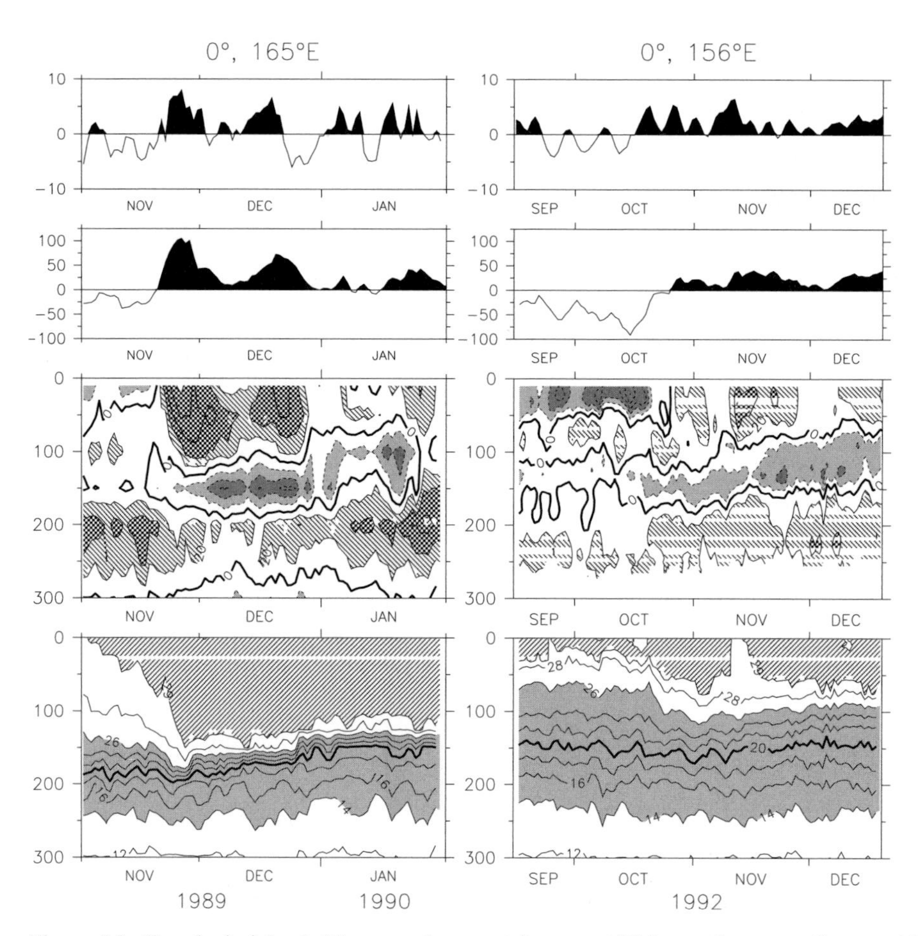

Figure 6.2. Zonal wind (*top*), 10-m zonal current (*upper middle*), zonal current (*lower middle*), and temperature (*bottom*) at 0°, 165°E during 1989–1990 (*left panels*) and at 0°, 156°E during 1992 (*right panels*). In the upper panels westerly winds and eastward currents are shaded. In the lower middle panels hatching indicates eastward currents, gray-shading westward currents, with a contour interval of $20\,\mathrm{cm\,s^{-1}}$. In the bottom panels, hatching indicates temperatures greater than 29°C, and gray-shading indicates the thermocline with temperatures between 14°C and 26°C.

rain-produced halocline and westerly winds lead to a surface-intensified eastward jet (see Section 6.3), then the resulting shear will tend to tilt the zonal salinity gradient by causing the fresh lens to run over the saltier eastern layer (Roemmich *et al.*, 1994; Cronin and McPhaden, 2002). Similarly, Ekman convergence in response to westerly winds can bring fresher northern hemisphere surface water to the equator (Cronin and McPhaden, 2002). However it is caused, the existence of a barrier layer enhances the local ocean response to both heat and momentum fluxes by concentrating it in a thin surface layer.

6.2.2 A 1-D heat balance?

Numerous studies have shown that, although advection can on occasion be important in determining near-surface temperature change, a 1-D balance dominated by surface fluxes is the principal influence determining warm pool SST variability (McPhaden and Hayes, 1991; Webster and Lukas, 1992; McPhaden *et al.*, 1992; Sprintall and McPhaden, 1994; Anderson *et al.*, 1996; Cronin and McPhaden, 1997; Shinoda and Hendon, 1998, 2001; Zhang and McPhaden, 2000). The principal surface flux terms on intraseasonal timescales are latent heat flux, which varies mostly due to windspeed since the SST is always high (Cronin and McPhaden, 1997), and short-wave radiation, varying mostly due to the thick cloudiness of convective systems; both of these have strong signatures as TISO events pass across the region (Shinoda *et al.*, 1998). Since highest wind speeds occur during westerly wind bursts (Weller and Anderson, 1996; Zhang and McPhaden, 2000), which are themselves associated with the convective phase of TISOs (Zhang, 1996; Shinoda and Hendon, 2002), the short-wave and latent heat flux terms are approximately in phase on intraseasonal timescales (Figure 7.1), and a convective event produces strong cooling (McPhaden *et al.*, 1988, 1992; Ralph *et al.*, 1997; Zhang and McPhaden, 2000). Figure 6.3 shows episodes of cooling under the intraseasonal

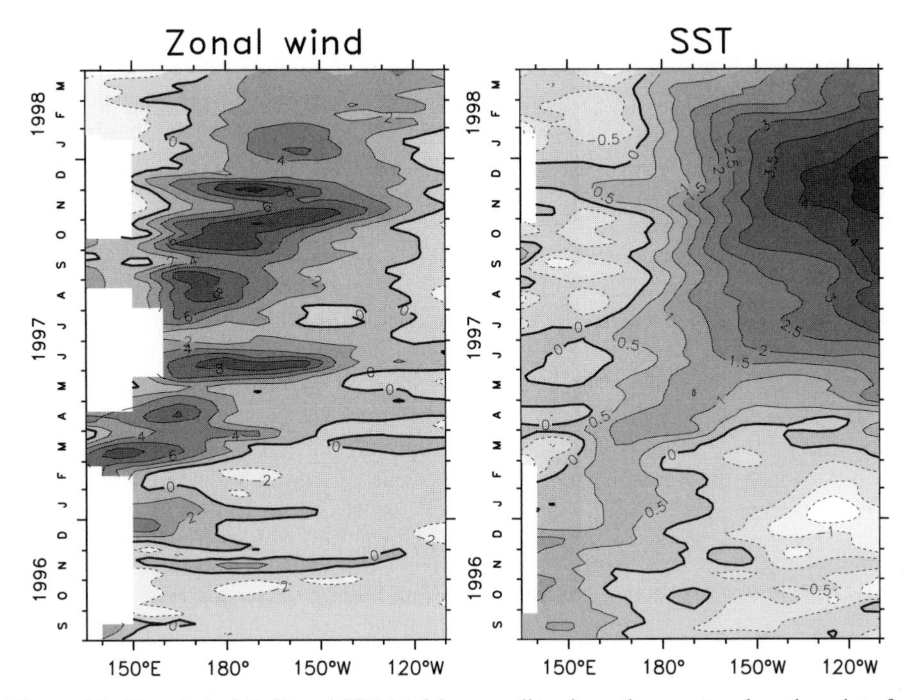

Figure 6.3. Zonal wind (*left*) and SST (*right*) anomalies along the equator, based on data from Tropical Atmosphere Ocean (TAO) moorings. Dark shading and solid contours indicate westerly wind and high SST anomalies. Contour intervals and $2\,\mathrm{m\,s^{-1}}$ and 0.5°C.

westerly wind bursts during the growth of the 1997/1998 El Niño. The implications of the net cooling of the far western Pacific as a result of TISO events will be discussed in Section 6.5. In addition to attempts to directly estimate the heat balance terms, several types of overview evidence indicate the dominance of surface flux forcing in the upper layer intraseasonal heat balance. First, the meridional scale of cooling under strong westerly winds has been observed to have the relatively broad scale of the wind, rather than that of the ocean dynamical response, which is more closely trapped to the equator (Ralph *et al.*, 1997; Shinoda and Hendon, 2001). Second, temperature anomalies subsequent to surface fluxes associated with the MJO are observed to propagate downwards, and are not in phase with deeper temperature variability (Zhang, 1997).

Entrainment from below might also contribute to SST change in a 1-D balance, and this has been considered by several investigators, although it cannot be measured directly and is often inferred from the residual of other terms (e.g., McPhaden and Hayes, 1991; Cronin and McPhaden, 1997). Entrainment could be fostered by dynamical processes like Ekman-divergence-caused upwelling bringing cooler water within the reach of wind mixing, as occurs in the eastern Pacific, or due to wind mixing itself against shallow stratification (e.g., mixing away a halocline and exposing a cooler barrier layer as would occur in Figure 6.1). The thickness of the warm layer and its frequent stabilization by salinity make entrainment relatively ineffective at cooling the SST in the west Pacific warm pool, most of the time (Meyers *et al.*, 1986; McPhaden and Hayes, 1991; Eldin *et al.*, 1994). Exceptions have been noted, however. Cronin and McPhaden (1997) used a steady-state turbulence model to show that entrainment was a cooling tendency during a period of shallow pycnocline early in COARE, though it was apparently not the main reason for changes in pycnocline depth. Sprintall and McPhaden (1994) found that during La Niña conditions in 1988/1989, with stronger than normal trades and weak rainfall at 0°, 165°E, there was no barrier layer. In this situation, SST changes were significantly influenced by upwelling (downwelling) in response to easterly (westerly) wind anomalies, much as occurs in the eastern and central Pacific. Although entrainment is generally a cooling term, salinity stratification can result in entrainment warming. Under low wind and clear-sky conditions, a very shallow halocline can lead to heating of the barrier layer by penetrative radiation (which remains stable because of the low surface salinity). With the turn to the cloudy–windy phase, surface flux cooling reduces the vertical stability while wind mixing strengthens; the result can be that entrainment produces heating of the surface (Anderson *et al.*, 1996; Shinoda and Hendon, 1998; Schiller and Godfrey, 2003).

6.2.3 The role of advection

The importance of intraseasonal heat advection in the warm pool has been controversial. On one hand, as noted above, many investigators have concluded that a 1-D balance represents the dominant physics; these arguments appear reasonable since mean SST gradients in the warm pool are small. However, several examples have demonstrated that advection can be a significant contributor to the heat balance in certain cases, involving different processes under both easterly and westerly winds. Despite the uniformity of mean SST in the warm pool, remnants of anomalous SST patches due to preceding conditions can leave significant, if transient, gradients for currents to work on, and as discussed in Section 6.3, equatorial currents can spin up rapidly in response to intraseasonal wind reversals. Two examples from the COARE experiment suggest the range of possibilities. During the early part of COARE in October 1992, cooler SSTs lay at, and west of, the 156°E mooring, presumably the residual of a westerly wind event in September. Moderate easterly winds (Figure 6.2, top right) spun up a strong westward surface current (Figure 6.2, right middle) that produced advective warming of about 1°C during the first three weeks of October (Cronin and McPhaden, 1997). Two months later, the strongest intraseasonal westerly event during COARE occurred in late December 1992, with stresses as high as $0.4\,\mathrm{N\,m^{-2}}$ (Weller and Anderson, 1996). Currents spun up by the December winds were well observed by surface drifters, which are drogued to move with the current at a 15-m depth and had been seeded extensively around the region (Ralph *et al.*, 1997). During the event, drifters within at least 2°S to 2°N converged onto the equator and accelerated eastward. The surface jet extended at least to 180°, and was about 300 km wide (Figures 7.10 and 7.11). Surface layer cooling under the strong winds was substantial, with a temperature drop of as much as 1°C along the drifter trajectories. Ralph *et al.* (1997) found that this heat loss resulted in a positive SST gradient with SSTs warmer by at least 1°C near 180° than at 145°E, and consequently eastward advection was a cooling term in the eastern part of the warm pool. They noted that the strong latent heat and shortwave cooling under the cloudy-westerly phase of TISO events means that eastward surface currents tend to be correlated with positive SST gradients ($\overline{u'T'_x} > 0$). This correlation suggests that long-term mean zonal advection is not negligible despite the small mean SST gradients, and that the time-average effect of TISO winds and clouds is cooling at the east edge of the warm pool (Ralph *et al.,* 1997). Both these examples contradict the conventional idea that mean SST advection along the equator is due to the westward mean South Equatorial Current (SEC) working on the mean negative SST gradient, and indicate the potential for ISV to produce low-frequency changes in SST (see Section 6.5).

6.3 VERTICAL STRUCTURE UNDER WESTERLY WINDS

The complex and rapidly-varying vertical structure of west Pacific zonal currents has been noticed from the earliest cruises in the region (Hisard *et al.*, 1970). During a cruise along 170°E in March 1967, trade winds prevailed and the equatorial currents had a two-layer structure, with a westward SEC above 60 m depth, and an eastward Equatorial Undercurrent (EUC) below. This is typical of the trade wind regime of the central Pacific; in the mean it has been shown to be the result of a directly wind-forced frictional current in the surface layer, with a baroclinic pressure gradient due to thermocline tilt producing an eastward current below (McPhaden and Taft, 1988). In the following month, a westerly wind burst had occurred, and Hisard *et al.*

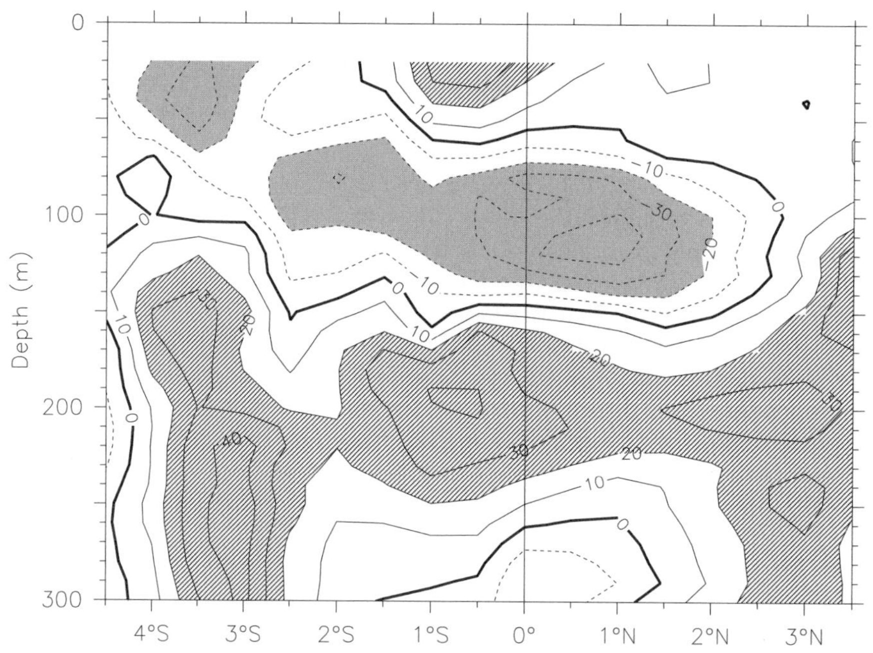

Figure 6.4. Zonal current example illustrating the subsurface westward jet (SSWJ) sandwiched between a frictional surface eastward current and the eastward EUC at 200 m. Hatching and solid contours indicate eastward current, gray-shading and dashed contours indicate westward current. The contour interval is $10\,\text{cm}\,\text{s}^{-1}$. The measurements were made by shipboard acoustic doppler current profiler (ADCP) along 156°E during 8–11 December, 1992, from the *R/V Le Noroit*. Only data obtained while the ship was on-station was used to construct the section.

(1970), observed a three-layer structure, with eastward flow above 60 m depth, westward flow from 60 m to 175 m, and the eastward EUC below that (similar to Figure 6.4); these have become known as "reversing jets" and the sandwiched westward current has been called the subsurface westward jet (SSWJ).

Since the Hisard *et al.* (1970) study, there have been numerous reports of such reversing jets in the western equatorial Pacific (McPhaden *et al.*, 1988, 1990, 1992; Delcroix *et al.*, 1993; Cronin *et al.*, 2000). In general, they are not found in the central or eastern Pacific, despite decades-long moored velocity time series at several locations; however, Weisberg and Wang (1997) showed one example of a brief reversing jet at 170°W during the eastwardmost penetration of westerly winds in January 1992, at the height of the El Niño of that year. It seems likely that a deep thermocline and a thick, weakly stratified surface layer is necessary for the complex structure of velocity to exist, thus restricting the reversing jets to the warm pool, except occasionally during El Niños when these conditions spread eastward.

It was soon recognized that the surface eastward current associated with the reversing jets was an example of a Yoshida Jet (Yoshida, 1959). A Yoshida Jet has the simple, accelerating balance:

$$u_t - fv = \tau^x / h \tag{6.1a}$$

$$fu = -p_y \tag{6.1b}$$

$$p_t + hv_y = 0 \tag{6.1c}$$

where (u, v) are the zonal and meridional velocity components, f is the Coriolis parameter, τ^x is the zonal wind stress, h is the thickness of the accelerating layer, and p is the pressure (both τ and p have been divided by the density for simplicity of notation). The zonal jet that is the solution to (6.1) decays exponentially away from the equator with a meridional scale of the equatorial Rossby radius (typically 300 km). Away from the equator, meridional transport in (6.1a) is approximately Ekman convergence (for westerly winds), which feeds the accelerating zonal jet, and produces equatorial downwelling (6.1c) that provides the meridional pressure gradient to geostrophically balance the jet (6.1b). For a westerly wind burst magnitude of $7.5\,\mathrm{m\,s^{-1}}$ (e.g., Figure 6.2, top) the stress anomalies would be about $0.1\,\mathrm{N\,m^{-2}}$. Taking a 100 m layer thickness, the zonal acceleration at the equator indicated by (6.1a) is of the order of $10\,\mathrm{cm\,s^{-1}\,day^{-1}}$, comparable to observations under these conditions, and indicating that a very rapid current can be spun up within the timescale of a westerly wind burst (McPhaden *et al.*, 1988, 1992; Ralph *et al.*, 1997; Cronin *et al.*, 2000). Yoshida Jets appear to be a common and robust feature of the west Pacific under westerly wind bursts, and are frequently observed (McPhaden *et al.*, 1990; Delcroix *et al.*, 1993). Since the Yoshida balance does not consider zonal variability, the jet is assumed to occur everywhere under the wind; an unresolved question concerns the possible convergence at the east edge of the jet (Richardson *et al.*, 1999; Cronin and McPhaden, 2002; Lengaigne *et al.*, 2002).

Cronin *et al.* (2000) diagnosed the zonal momentum terms in the COARE region during March 1992–April 1994. Figure 6.5, from that paper, shows that near-surface zonal acceleration was nearly in phase with the wind and that the jet reaches maximum velocity within about 3 days from the peak of the wind. Examples of rapid acceleration of surface currents under westerly wind bursts are common (e.g., Figure 6.2). One model of the response to impulsive forcing simplified the situation by assuming linear, frictional dynamics in a homogeneous mixed layer above a sharp thermocline (McPhaden *et al.*, 1988). With switched-on zonal-wind forcing, their solution for the shear flow was a parabolic velocity profile accelerating in the direction of the wind, maximum at the surface and decaying to zero at the base of the mixed layer. With a vertical eddy viscosity estimated from observed shear profiles, the shear profile set up within several days.

The Yoshida Jet would accelerate without bound except that the zonal pressure gradient term, omitted from (6.1a), becomes important in days to weeks and results in equatorial Kelvin and Rossby waves being emitted from the edges of the wind patch (McCreary, 1985). (The meridional profile of the wind determines the mix of Kelvin and Rossby waves; see Richardson *et al.*, 1999.) The Kelvin wave part of the

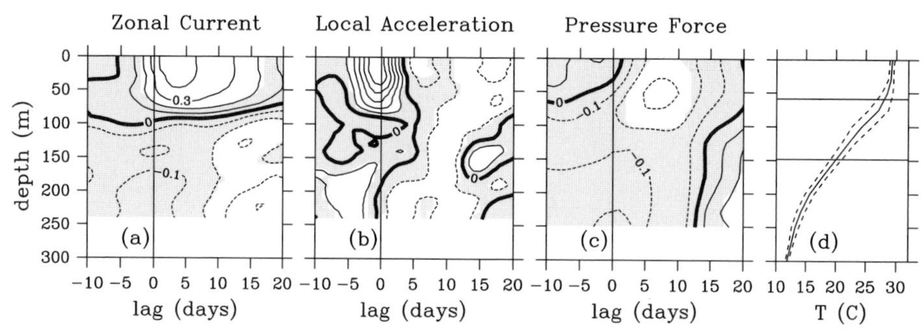

Figure 6.5. Lagged correlation between local zonal wind stress and (a) zonal current, (b) local zonal current acceleration, and (c) zonal pressure gradient force as a function of depth, at 0°, 156°E. A positive lag implies that wind stress variability occurs prior to the respective variable anomaly. (d) Mean temperature and its standard deviation envelope at 0°, 156°E. The two horizontal lines show the mean depths of the 28°C and 20°C isotherms, which define the surface, intermediate and EUC layers. Correlations that are not significant at the 90% confidence level are shaded. The contour interval is 0.1.
After Cronin *et al.* (2000).

response has no effect west of the wind patch but carries a downwelling signal to the east, such that the thermocline tilts down to the east under the (westerly) wind patch and is flat from its eastern edge to the back of the advancing wave. The Rossby part of the response has no role east of the patch but carries an upwelling signal westward. If a westerly wind remains steady (and ocean boundaries are unimportant), the fully-adjusted solution has a flat, upwelled thermocline to the west, a downward slope under the wind, and a flat, downwelled thermocline to the east. Under the wind patch itself, the vertically-integrated zonal pressure gradient comes to Sverdup balance with the wind stress, and the acceleration stops; in effect the waves carry the wind-input momentum away from the forcing region. The result is a downwind jet at the surface, decaying with depth, and a pressure-gradient-driven upwind current below, as the frictional influence declines with depth. For steady easterly winds, this two-layer structure describes the mean situation in the central Pacific, with a westward SEC at the surface, and EUC beneath. (For discussion of non-linearities associated with these circulations, see, among many others, Philander and Pacanowski, 1980; Johnson and McPhaden, 1993a, b; Johnson and Luther, 1994; Yu and Schopf, 1997; Zhang and Rothstein, 1998; Cronin *et al.*, 2000; Lengaigne *et al.*, 2002; Kessler *et al.*, 2003). The key questions for westerly wind bursts, therefore, are the timescale on which the waves establish the pressure gradient and how the vertical structure that allows *three* stacked jets to exist is set up.

The pressure gradient timescale depends primarily on the width of the wind patch compared to the propagation time of the waves. For typical westerly wind burst forcing with fetch of a few thousand kilometers, and first baroclinic mode Kelvin waves with speeds of 2–3 m s^{-1}, the pressure gradient setup takes about 10 days, which is borne out by observation (Figure 6.5(c), and see Cronin *et al.*, 2000).

Model experiments with idealized winds have shown a strong sensitivity of the ocean response to the wind fetch and zonal profile (Richardson *et al.*, 1999).

Yoshida dynamics alone can only set up shear of one sign and therefore only two stacked jets. The reversing jets observed under westerly wind bursts, however, demonstrate the possibility of a surface eastward Yoshida Jet, an eastward EUC in the center of the thermocline, and a westward flow (the reversing jet or SSWJ) in the weakly stratified upper thermocline between them (Figures 6.2 and 6.4). In fact, Figure 6.2 shows that a quite complicated vertical structure can occur with rapid wind changes; note that the zonal current at 156°E in September and early October, 1992, under easterly winds, has a surface westward current varying in phase with the wind, a subsurface eastward current near 80 m that is apparently driven by the shallow pressure gradient response to the easterlies, a remnant of a SSWJ at 130 m generated by the early September westerlies, and the EUC at 240 m below that. This suggests that the relatively diffuse west Pacific thermocline can support pressure gradient reversals in the vertical (Cronin *et al.*, 2000). Observations show examples of the EUC flowing nearly undisturbed during the occurrence of significant westerly winds with the formation of a Yoshida Jet and SSWJ lasting for several months (e.g., Figure 6.2, right-hand panels), and conversely of a reversed pressure gradient extending into the central thermocline and slowing the EUC as the SSWJ develops (McPhaden *et al.*, 1992; see Figure 6.2, left-hand panels). On average, the pressure gradient and zonal current at EUC-level are weakly correlated with local zonal winds with a lag of about 15 days (Cronin *et al.*, 2000; see also Figure 6.5(a, c)). One can imagine that the different responses depend sensitively on the preexisting stratification and current structure: in particular the thickness of the SEC and how far it extends into the thermocline, and whether the upper thermocline above the EUC can adjust to produce a pressure gradient to bring the Yoshida Jet to steady state, and thereby create a SSWJ, without involving the lower isotherms. Model experiments to try to isolate these effects have been performed by Zhang and Rothstein (1998) and Richardson *et al.* (1999).

This raises the deeper issue of what determines the vertical structure of ocean adjustment to time-varying winds. On the basin-scale at low frequency (say six months or more), the entire thermocline slope adjusts to Sverdrup balance with the wind stress (McPhaden and Taft, 1988). In the eastern Pacific, where the thermocline is sharp and shallow and the winds are relatively steady easterlies that provide, via upwelling, for quick communication of thermocline anomalies to the surface, there is little opportunity for a complex vertical structure to occur. But in the warm pool, where the upper layer can be more than 100 m thick, the thermocline can extend over 200 m or more, and the winds commonly change sign in a month or less, a more elaborate structure is possible. For example, Zhang (1997) noted that although intraseasonal temperature variability at 0°, 165°E is large down to at least 300 m, the signal is incoherent across about 75 m depth. The factors setting the vertical scales of these reversals presumably are related to the pre-existing stratification, but this is not well understood.

6.4 REMOTE SIGNATURES OF WIND-FORCED KELVIN WAVES

The propagation of equatorial Kelvin waves is so efficient at carrying wind-forced signals eastward along the equator that the first recognition of intraseasonal time-scales in the tropical Pacific Ocean was in sea level records along the coast of the Americas (Enfield and Lukas, 1983). Spillane *et al.* (1987) and Enfield (1987) documented coherent 30–70 day period coastal sea level variability from Peru to northern California. They quickly realized that nothing in the local winds could produce such a signal, and found a lag relation with west Pacific island sea levels that clearly showed the Kelvin wave propagation, at speeds of about 2.5 m s^{-1}. More recently, Hormazabal *et al.* (2002) made a similar diagnosis for sea level at 30°S on the Chilean coast (see also Clarke and Ahmed (1999), for an analysis of the role of the continental shelf in determining the phase speed of coastal propagation).

Since the TAO mooring array (Hayes *et al.*, 1991; McPhaden, 1995) has provided adequate temporal resolution, observation of the prominent intraseasonal Kelvin waves has become routine. The Kelvin wave due to intraseasonal westerly wind events is seen as a thermocline downwelling that commonly can be 50 m or more (Figure 6.6), well east of the wind itself, and accompanied by an eastward surge of surface current that can be as large as 1 m s^{-1}. Their effects on SST in the central and eastern Pacific have been noted many times. Under some conditions, SST change at 140°W can be dominated by the intraseasonal zonal advection due to west Pacific Kelvin waves (Kessler *et al.*, 1995). Vecchi and Harrison (2000) stratified westerly wind bursts by location and by the low-frequency background state of ENSO. They showed that, on average, the largest east Pacific SST effects were found when equatorial westerlies occurred with climatologically-normal SST, not during El Niño events, apparently because zonal advection is more efficient at changing SST when a large background zonal gradient exists. The westerly-wind-driven Kelvin waves can also remotely modulate eastern Pacific SST through lowering the thermocline (Figure 6.6) and changing the effect of background upwelling on SST, which Zhang (2001) argued was the dominant mechanism (see also Belamari *et al.*, 2003). Giese and Harrison (1991) suggested that another possible SST effect could be due to the passage of a downwelling Kelvin wave, with a meridional scale of 2°–3°, accelerating the EUC and thereby increasing the shear with the surrounding SEC. In their model, the resulting amplification of the tropical instability waves (see Section 6.7.2) resulted in an equatorward heat flux that was as large as the zonal advection warming.

The TAO moorings have also allowed the vertical structure of the intraseasonal Kelvin waves to be dissected and diagnosed. McPhaden and Taft (1988) used moorings at 140°W, 125°W and 110°W, where there is little intraseasonal wind forcing, to show that the principal intraseasonal signals in zonal current, temperature, and dynamic height had characteristics of a remotely-forced first baroclinic mode Kelvin wave, with speed about 2.1 m s^{-1}, and that this variability had an amplitude as large as that of the annual cycle. They also commented that the dominant intraseasonal period observed in these oceanic variables was 60–90 days, longer than the apparent MJO forcing (see Section 6.5 for further discussion of this

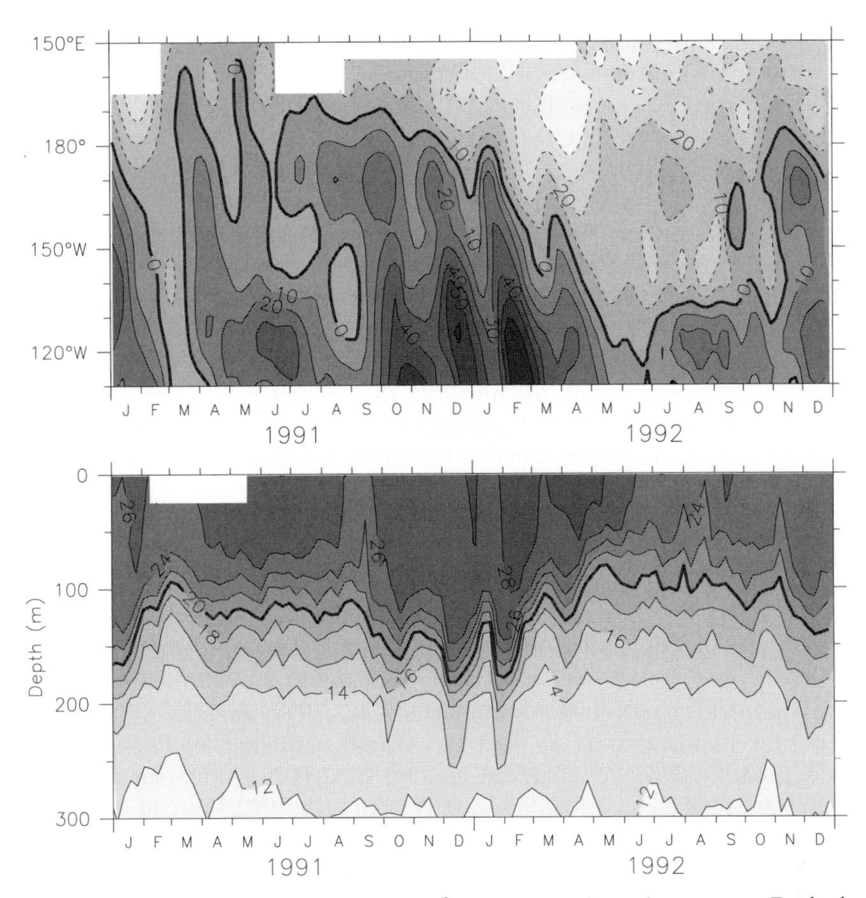

Figure 6.6. (*Top*) Anomalous depth of the 20°C isotherm along the equator. Dark shading indicates deep anomalies, with a contour interval of 10 m. The intraseasonal Kelvin waves are evident as tilted bands, especially during September 1991–February 1992. (*Bottom*) Temperature at 0°, 140°W. Dark shading indicates higher temperature, with a contour interval of 2°C. The 20°C isotherm is shown as the thick line. Kelvin waves arriving from the western Pacific produce the sharp downwelling events. For both panels, data comes from TAO moorings.

issue). Although their results suggested that the waves were approximately linear, the fact that the EUC speed is typically $1\,\mathrm{m\,s^{-1}}$ or more raised the question of mean current modifications of the wave modes. Johnson and McPhaden (1993b) analyzed time series of temperature and zonal current from moorings along 140°W and compared them with a meridionally-symmetric model that included an idealized mean flow comparable to the EUC, as well as a SEC centered near 3° latitude at the surface. Relatively little direct Doppler shifting was found, because the wave vertical scales are so much larger than those of the mean currents, and the main modifications to linear dynamics were the occurrence of a temperature amplitude minimum on the equator and an amplification of the wave zonal currents below the EUC core (see also Lengaigne *et al.* (2002) who emphasized that Kelvin advection of the mean EUC was responsible for this). Both these effects were due to wave vertical advection of the background temperature and current fields. Considering these studies, a linear diagnosis of the remotely-forced Kelvin waves seems to be first order appropriate.

Theory suggests that many vertical Kelvin modes would be excited by the observed wind forcing, and several studies have noted evidence for modal structures. Busalacchi and Cane (1985) showed that while both the first and second vertical modes are a significant contribution to sea level variability in the eastern Pacific, higher modes are not. Giese and Harrison (1990) found that in an OGCM with a realistically-sloping equatorial thermocline, the second baroclinic mode Kelvin wave's surface currents would be amplified relative to that of the first mode, and would be the dominant velocity signal at the South American coast. Kutsuwada and McPhaden (2002) pointed out that during El Niño events, when the thermocline is flatter than usual, this effect would be moderated and the first baroclinic mode more prominent. They also showed evidence of upward phase propagation in the free-wave region, suggesting the formation of a downward beam of energy (McCreary, 1984). Kindle and Phoebus (1995) found that a model including three modes gave a better simulation of sea level at the American coast. Cravatte *et al.* (2003) showed evidence of energy transfer from the first to the second baroclinic mode for intraseasonal Kelvin waves during 1992/1999. In light of these results suggesting the importance of at least the second baroclinic mode, the apparent success of single-active-layer (reduced gravity) models in simulating much of the observed wave-mediated variability (Metzger *et al.*, 1992; Wu *et al.*, 2000) is puzzling. Kessler and McPhaden (1995a) examined the signatures of the first four baroclinic modes in thermocline depth at 140°W and found that although in a strict modal decomposition two modes were needed, in fact reasonable choices of reduced gravity (thus single-mode) parameters gave a very similar solution, at least in the central Pacific not too far from the forcing region. The reason is that the typical choice for wave speed in the reduced gravity models ($c = 2.5\,\mathrm{m\,s^{-1}}$) is appropriate to a true first mode with about 250 m thickness (where $c^2 = g'h$, with g' the reduced gravity and h the layer thickness), but in fact these models are often taken to have h be a realistic thermocline depth of about 150 m. These choices therefore artificially pump up the mode-one amplitude, and thereby compensate to produce a fairly realistic representation of the total Kelvin signal (Kessler and McPhaden, 1995a).

Although Kelvin waves propagate non-dispersively with a simple velocity and thermocline depth anomaly structure that has u and h in phase, they can produce much more complex phasing of SST variability. Indeed, Kessler and McPhaden (1995a) had noted, but could not explain, the intraseasonal warming and cooling events that occurred nearly simultaneously over a wide longitude range during the onset and decay of the El Niño of 1991/1992. McPhaden (2002) interpreted this by showing that while Kelvin wave zonal advection dominates the intraseasonal SST balance in the central Pacific, vertical advection and entrainment are more important in the east where the thermocline is very shallow. If Kelvin wave vertical velocity is assumed to be due to thermocline motion ($w \approx dh/dt$), then upwelling leads the westward current anomaly by 1/4 of a cycle as the wave passes a point. Thus, the cooling due to upwelling also leads the cooling due to wave zonal advection of the mean SST gradient by 1/4 of a cycle. Depending on the relative importance of each of these processes to the SST balance at different longitudes, the phasing of SST due to intraseasonal Kelvin waves can appear to propagate in either direction or occur in phase. McPhaden (2002) showed that the growing dominance of vertical entrainment as the Kelvin wave propagates eastward led to the nearly simultaneous intraseasonal SST anomalies over a broad longitude range observed throughout the 1990s.

Although equatorial Kelvin waves can have arbitrary shape in (x, t), two factors combine to make the MJO fraction of ISV the dominant contribution to the oceanic Kelvin signal. First, since the ocean integrates the wind forcing along the Kelvin wave characteristics (Kessler *et al.*, 1995; Hendon *et al.*, 1998), organized, large-scale forcing is favored over more incoherent variability. Second, wind forcing that moves eastward will project more strongly onto the Kelvin mode (Weisberg and Tang, 1983), because it is partly resonant. Over the west Pacific warm pool, the MJO propagates east at a speed of about $5\,\mathrm{m\,s^{-1}}$ (Hendon and Salby, 1994; Shinoda *et al.*, 1998), which is comparable to the oceanic Kelvin speed (about $2.5\,\mathrm{m\,s^{-1}}$ for the first vertical mode). Hendon *et al.* (1998) noted that as the MJO speeds up east of the dateline, it gets ahead of the oceanic Kelvin wave; by about 130°W it is out of phase with the Kelvin current anomalies and thus serves to damp the wave.

In addition to the effects discussed in this section, intraseasonal Kelvin waves have also been related to the ENSO cycle and rectification mechanisms through a variety of processes. These will be discussed in Section 6.5.

6.5 EL NIÑO AND RECTIFICATION OF ISV

The question of a role for ISV, especially the MJO, in the ENSO cycle has been a hotly debated topic in the climate community (Zhang *et al.*, 2001), and no definitive resolution has thus far been reached (see Chapter 9). Although the intraseasonal signatures in the ocean during El Niños can be impressively large, comparable to the amplitude of the seasonal cycle or ENSO (e.g., Figure 6.6), a non-linear mechanism would be required to couple intraseasonal to lower frequencies, and this has been difficult to demonstrate. In addition, the usual indices of global MJO activity are uncorrelated with indices of the ENSO cycle (Slingo *et al.*, 1999; Hendon *et al.*, 1999), which has led some to argue that a systematic connection is unlikely. Nevertheless, the frequent observation of strong intraseasonal (especially MJO) variability in the western Pacific during the onset stage of recent El Niños (Gutzler, 1991; Kessler *et al.*, 1995; McPhaden, 1999; McPhaden and Yu, 1999; Zhang and Gottschalck, 2002) has generated a variety of speculation about this possibility. The spectacular failure of all the ENSO forecast models to predict the magnitude or rapid growth of the 1997/1998 El Niño, which occurred subsequent to a series of large MJO events in boreal winter/spring 1996/1997 (McPhaden, 1999; van

Oldenborgh, 2000; Barnston *et al.*, 2000), brought the problem to the fore. The question of the role of ISV in ENSO is part of a fundamental debate that revolves around the distinction between two views: ENSO seen as a quasi-cyclic mode of oscillation of the Pacific climate system (see Neelin *et al.* (1998) for a review), or as an initial value problem in which each El Niño is a largely independent event (Moore and Kleeman, 1999; Kessler, 2002b). In the first case, ISV is a source of noise that may contribute to irregularity of the cycle, but is not fundamental to it (Roulston and Neelin, 2000). In the second case an initiating perturbation external to ENSO itself is an essential element, and ISV could potentially provide it. (However, no one suggests that ISV *causes* the ENSO cycle itself, as is clear from the fact that coupled models without anything resembling the MJO develop fairly realistic ENSO cycles and statistics). Recently, theories have arisen that combine elements of these two viewpoints, arguing that the spatial characteristics of west Pacific westerlies associated with the MJO produce climate noise that is especially suited to influence a developing El Niño (Moore and Kleeman, 1999; Fedorov, 2002).

The occurrence of intraseasonal signatures in the ocean associated with El Niño was first noticed by Lukas *et al.* (1984), looking at central Pacific island sea levels, and others have followed using a variety of observed quantities (Enfield, 1987; McPhaden *et al.*, 1988; Kessler and McPhaden, 1995a, b; Kutsuwada and McPhaden, 2002; Zhang and Gottschalck, 2002). As described in the sections above, the ocean signatures include cooling under the strong winds and cloudiness of the west Pacific warm pool (e.g., Figure 6.3), and Kelvin-wave-mediated eastward advection and thermocline downwelling in the equatorial regions to the east.

There is no doubt that strong MJO activity is regularly seen during non-El Niño years, including its ocean signatures (Kessler *et al.*, 1995). Global interannual variability of the MJO is dominated by changes in the core region centered at 90°E, which are unrelated to ENSO (Hendon *et al.*, 1999; see also Figure 6.7, left-hand panel). Differences in MJO characteristics during El Niños have been noted, however. Several investigators have shown that MJO convection and surface zonal winds shift eastward during El Niño onset, from the far western Pacific to the eastern edge of the expanding warm pool (Gutzler, 1991; Fink and Speth, 1997; Hendon *et al.*, 1999; Kessler, 2001). Figure 6.7 shows that intraseasonal OLR in the warm pool region (150°E–180°) has large amplitude during El Niños that is not well correlated with the core intraseasonal OLR region over the Indian Ocean. During El Niños, ISV extends eastward (Figure 6.3), and its warm pool activity is in fact strongly correlated with the SOI ($r = -0.58$, see Figure 6.7, right-hand panel). Kessler (2001) showed that the eastward shift was not just incoherent ISV, which makes up perhaps half of the variance in this frequency band (Hendon *et al.*, 1999), but a systematic component of the organized MJO (note the large-scale events in Figure 6.3). The eastward shift is crucial to MJO/ENSO interaction because it can greatly increase the fetch of MJO winds over the Pacific (perhaps by a factor of two), thereby increasing the magnitude of the ocean signatures during those times.

The observed shift in the MJO envelope and its effect on the ocean was quantified by Zhang (2001) and Zhang and Gottschalck (2002), who suggested that an appropriate index could be based on the integral of MJO-filtered winds along

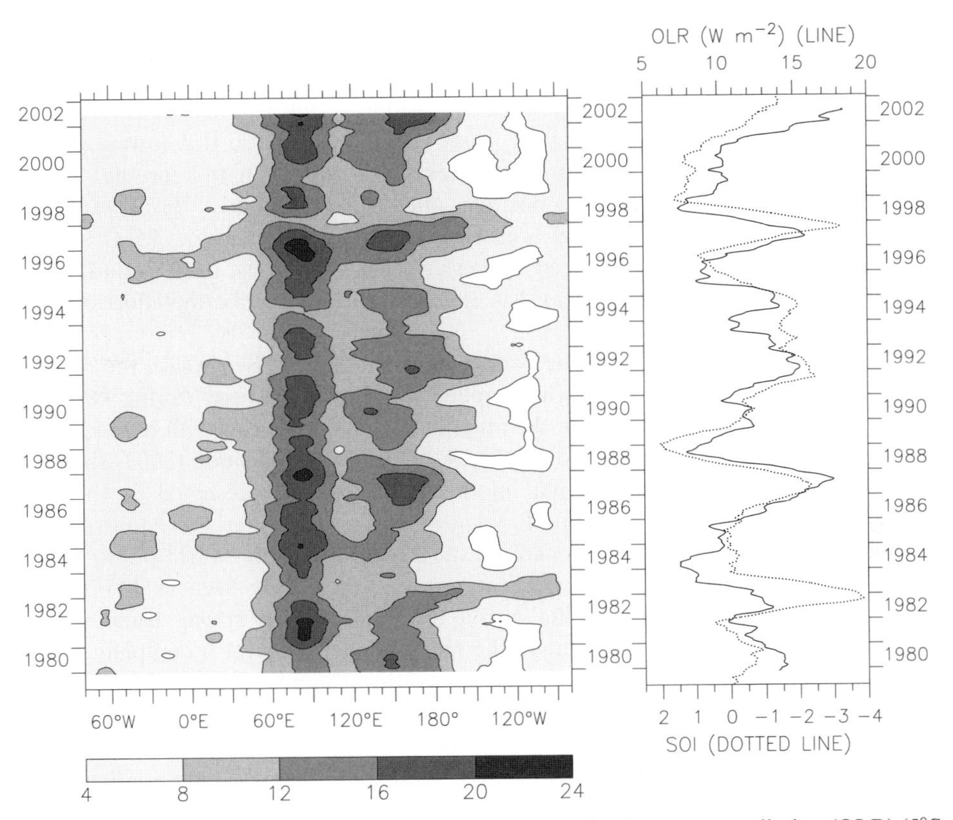

Figure 6.7. Interannual amplitude of intraseasonal outgoing long-wave radiation (OLR) (5°S–5°N) (W m^{-2}), defined as the one-year running standard deviation of intraseasonally-band-passed OLR. (*Left*) Amplitude in the global tropical strip, centered on the major region of variance at 100°E (the abscissa extends around the world, broken at the South American coast at 80°W). (*Right*) Time series of OLR amplitude averaged over the western Pacific (150°E–180°) (solid line, scale at top) in comparison with the SOI (dotted line, scale at bottom). Year ticks on each panel are at 1 January of each year, with year labels centered at mid-year. After Kessler (2001).

oceanic Kelvin wave characteristics. The index constructed in this way encompasses changes in the spatial pattern of the MJO as felt by ocean dynamics (though it does not consider changes in heat fluxes under the winds themselves). Zhang and Gottschalck (2002) used this technique to show that the MJO accounted for a significant fraction of interannual east Pacific SST variability, and that stronger El Niños (since 1980) were preceded by stronger MJOs. Although this work indicated a statistical relation between the intraseasonal and interannual frequencies, the rectifying mechanism still needs to be explained. If the MJO is simply an oscillation with zero mean, and the ocean feels this forcing linearly, there would be no interaction between frequencies. Several attempts have been made to elucidate such a mechanism.

One approach asks whether the occurrence of MJO events changes the background winds and heat fluxes over the Pacific. Ordinary statistical techniques used to extract MJO signatures from observations assume a linear separation between frequencies by band-pass filtering in some form to isolate the intraseasonal variance. The resulting zero mean time series are often taken to represent MJO anomalies with equal magnitude positive and negative phases. However, the realism of such representations has been questioned. If, for example, MJO events have systematically higher windspeed or westerly winds than the background, then anomalies defined to have zero mean will not adequately describe the effect on the ocean.

In the western Pacific where the background winds are often weak, the occurrence of large MJO wind oscillations implies stronger windspeed during *both* its easterly and westerly phases than in the absence of an MJO event, with correspondingly higher evaporation averaged over a cycle. Shinoda and Hendon (2002) showed that this process had an interannual modulation: MJOs represented in the US National Center for Environmental Prediction–National Center for Atmospheric Research (NCEP–NCAR) Reanalysis were more active over the warm pool during periods when the low-frequency winds were weak (probably because MJOs are restricted to the far western Pacific during La Niñas when strong trade winds extend westward), and thus on average the mean windspeed over a complete MJO cycle was enhanced by about $1\,\mathrm{m\,s^{-1}}$, with an averaged increased latent heat flux of about $23\,\mathrm{W\,m^{-2}}$. Zhang (1997) noted the difference in the effect of westerly anomalies on different backgrounds: during periods of mean easterlies (La Niña) a westerly wind burst represents a weakening of wind speed that will reduce latent heat fluxes, whereas on a westerly background, the same wind burst increases the wind speed and is a cooling term. Ocean models often use an *ad hoc* "gust factor" to represent disorganized small-scale wind speeds in the calculation of latent heat and mixing; this also tends to increase the wind speed produced by intraseasonal wind anomalies imposed in a model by limiting it on the low end. Kessler and Kleeman (2000) forced an ocean GCM with equal-amplitude easterly and westerly zonal wind anomalies (and a gust factor minimum of $4\,\mathrm{m\,s^{-1}}$) and found that the resulting higher windspeed produced SST cooling by about 0.6°C over an MJO cycle, compared to a climatological run. Although the gust factor influenced these results, SST cooling under stronger-than-climatology oscillating winds appears to be a robust feature of the MJO over the warm pool.

There are many other possibilities for rectifying interactions among the ocean responses to intraseasonal forcing, and these are just beginning to be explored. Waliser *et al.* (2003, 2004) forced an Indo-Pacific OGCM with realistic composite MJO and ISO anomalies and emphasized the potential for interaction between intraseasonal solar short-wave forcing and the ocean mixed layer. During the suppressed-convection phase of the MJO, positive short-wave anomalies occur in conjunction with low wind forcing; both of these act to stabilize and shoal the mixed layer and can produce SST warming. During the active-convection phase, the mixed layer deepens due to stronger winds and weaker solar heating; the result is that the cool anomaly is spread over a thicker layer and the negative SST change is not as large as the positive change of the opposite phase, so the rectified signal is a warming. This appeared to explain their results in the maritime continent region where ocean dynamical processes play a little role.

Some have explicitly argued that MJO winds do not in fact have zero mean: the MJO composites of Waliser *et al.* (2003, 2004) showed that the westerly phase of MJO winds averaged about 0.5 m s^{-1} stronger than the easterly phase over the warm pool. Raymond (2001) presented a model of the MJO in which the convective systems were associated with westerly wind bursts without a corresponding easterly anomaly; when there is no MJO there are no bursts (see also Clarke (1994) for a discussion of the preference for westerly winds under equatorial convection). In such a model, the occurrence of MJOs changes the mean winds, thus the intraseasonal events have a low-frequency component. It is difficult to objectively define from observations what the background winds would be "without the MJO", and therefore what is the net signature of the MJO, especially because of their frequent occurrence during El Niño onset phases, when the low-frequency winds are turning westerly. Does the occurrence of a particular background foster more or stronger MJOs? Or, conversely, does the chance occurrence of more MJOs add up to a different background? It appears that the answer to both questions is "yes", and that makes definition of the total effect of intraseasonal forcing a fuzzy concept. The best answer that can be given today is that the passage of an MJO across a large region of the west Pacific appears to be more likely during El Niño onset, when warming SST is spreading eastward, and the result of this passage is an increase in both westerly winds and wind speed over a wide zonal extent. We can now ask: how might these forcing changes interact with the coupled dynamics of the ENSO cycle?

One of the earliest attempts to quantify the effect of a short-term westerly event on the Pacific was by Latif *et al.* (1988). They forced a coupled GCM with a single, 30-day "westerly wind burst", with 10 m s^{-1} winds extending over 10°S–10°N, from the western boundary to 180°, after the model had achieved a stable climatology. Following the imposed burst, the coupled model was allowed to evolve freely. While the response of an uncoupled ocean model to such an event is short-lived, the coupled system developed long-term changes. An eastward shift of the area of warmest water to about 160°W led to an eastward shift of convection. The model atmosphere responded with persistent westerlies blowing into the convection, in a self-sustaining feedback which maintained the eastward-shifted SST and convection for more than a year. The Latif *et al.* (1988) experiment suggested that the coupled system is capable of rectifying short-term wind anomalies into a low-frequency change because the rapid response of the atmosphere to SST changes can reinforce the ocean anomalies before they have dissipated.

Around the same time, an experiment with a simple coupled model came to the opposite conclusion. Zebiak (1989) added intraseasonal noise to the model-generated zonal wind field in the Zebiak and Cane (1987; hereafter ZC) coupled model and found little impact on the evolution of the modelled ENSO cycle. The imposed ISV produced only a spread to the forecasts, not a systematic change in ENSO amplitude. Moore and Kleeman (2001) noted the insensitivity of the ZC model to perturbations in the western Pacific, and attributed it to the way atmospheric latent heating (that spurs the growth of convection) is treated over the warm pool where in reality small SST anomalies can produce a strong flowering of convection. This process is inhibited in the ZC model, and is a major difference from the coupled GCM of Latif *et al.* (1988). Another difference is that SST anomalies in ZC are closely tied to thermocline depth fluctuations, which is appropriate in the central and eastern Pacific but much less so in the west, where the background thermocline is very deep and surface fluxes dominate (McPhaden, 2002; see Section 6.2.2). Although the ZC model has shown notable success in forecasting the ENSO cycle, it is probably the wrong tool to investigate the effects of ISV.

A simple model of MJO rectification under zero mean winds was proposed by Kessler *et al.* (1995), based on a similar idea as Latif *et al.* (1988) discussed above: the atmosphere responds within days to SST changes by shifting the location of convection and associated westerlies, but the ocean's response to winds is lagged because it integrates the forcing. The rapid atmospheric shift is seen in Figure 6.3 as the intraseasonal winds following the maximum SST gradient eastward. Kessler *et al.* (1995) modeled this in highly idealized form by assuming that organized intraseasonal winds occur only over the warm pool, and that the wind fetch responds instantly to changes in the warm pool zonal width. Westerly winds generate Kelvin wave currents that advect the east edge of the warm pool eastward, and easterly winds do the opposite. As the warm pool width changes, so does the region of convection and the fetch of the oscillating winds: westerly winds increase the width and easterly winds decrease it. Thus, westerly winds increase their own fetch and easterlies decrease it, thus the ocean feels the eastward advection more strongly. The net effect of oscillating intraseasonal winds is to push the warm pool slowly eastward; in the idealized Kessler *et al.* (1995) model this was found to resemble the stepwise eastward expansion of warm SST seen during El Niños. The fact that the east edge of the warm pool is also a salinity front contributes an additional positive pressure gradient term that can enhance the eastward advection (Lengaigne *et al.,* 2002).

Kessler and Kleeman (2000) explored the consequences of the net latent heat cooling produced by the high windspeeds due to oscillating winds on a weak background. In an intermediate coupled model, slightly cool SST (presumed to have been generated by a series of MJOs) was imposed on the Pacific west of about 160°E, during the period just as the 1997/1998 El Niño was beginning. In fact, observations showed cooling of the far west Pacific at this time (Figure 6.3, see also McPhaden, 1999). Hindcasts of the 1997/1998 event were made with and without the imposed cooling. The control run produced a weak El Niño, typical of the forecasts that were made by many models before the event. The imposed-cooling run developed persistent westerlies blowing out of the cool western region; these increased the El Niño SST anomalies by about 30%, which improved the realism of the hindcast. They suggested that the MJO can thus act constructively on ENSO as a stochastic amplifier, and that the weather-like unpredictable nature of MJOs may make forecasting the amplitude of an oncoming El Niño event more difficult than predicting the occurrence of the event itself.

A different sort of rectifying process has been proposed to explain the perplexing discrepancy between the 40–50-day MJO signals observed in the atmosphere

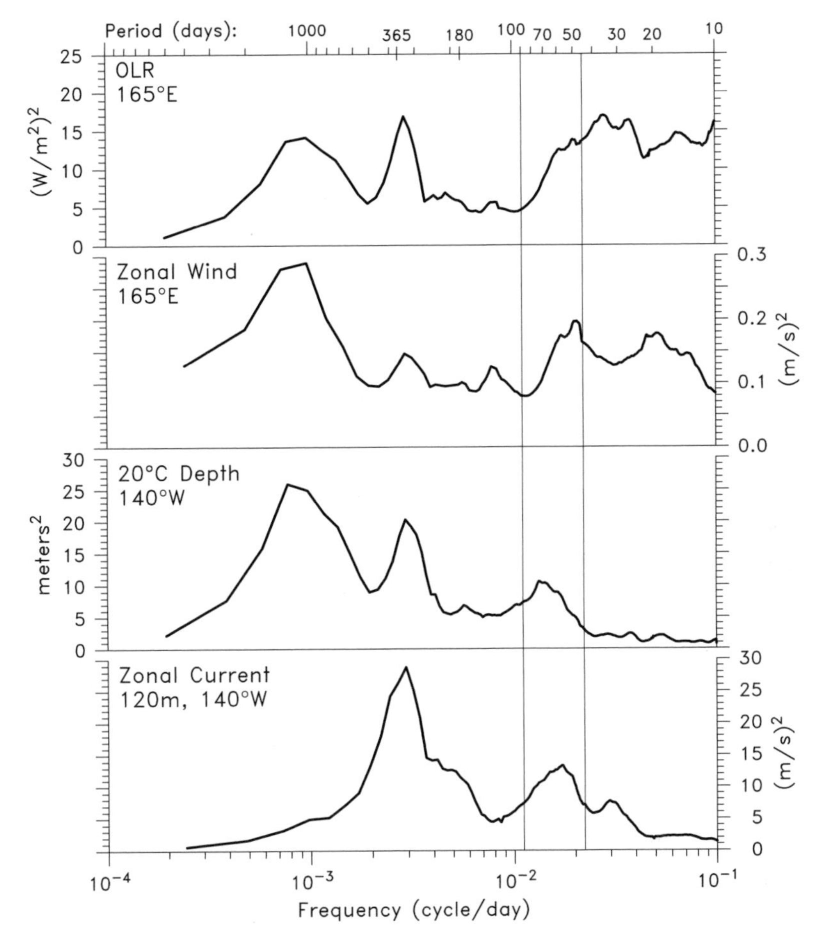

Figure 6.8. Variance-preserving spectra of OLR at 165°E, zonal wind at 165°E, 20°C depth at 140°W, and EUC speed at 140°W, 120°m depth, all at the equator. Each variable has a separate scale as indicated. The spectra are calculated for the 10-year period April 1983–April 1993 for all quantities except zonal wind, for which only the 7-years data, July 1986–July 1993, were available.
After Kessler *et al.* (1995).

(Madden and Julian, 1994) and the 60–70-day periods that dominate the ocean Kelvin wave response (Enfield, 1987; Kessler *et al.*, 1995, among others). Figure 6.8 shows variance-preserving spectra of two atmospheric quantities at 165°E (OLR, and zonal winds measured by the TAO buoy there), and two ocean quantities at 140°W (thermocline depth and zonal current at undercurrent level), which experience little local intraseasonal forcing, but strongly feel ISV through Kelvin wave propagation from the western Pacific. The intraseasonal variance of both the atmospheric quantities falls off sharply at periods longer than about 55 days, while the corresponding peak for both ocean quantities is clearly shifted to a lower frequency.

As noted above, the Kelvin wave amplitude east of a patch of oscillating zonal winds depends on the integral of the forcing along the wave characteristic. For steady winds, this is proportional to the time it takes the Kelvin wave to cross the patch, L/c, where L is the patch width and c is the wave speed (about $2.5\,\mathrm{m\,s^{-1}}$), but for oscillating winds the response is smaller because the winds may change sign while the wave is still traversing the patch. As the frequency of the wind increases to the point where the Kelvin wave crosses the patch in one period ($P = L/c$), the wave feels an equal amount of easterlies and westerlies, so the forcing integral cancels exactly and the response east of the patch falls to zero. Kessler *et al.* (1995) showed that for a 5,000 km width patch, there is a rapid fall-off in Kelvin wave amplitude between roughly 100-day and 30-day period oscillations, and suggested that this would account for the preference for lower intraseasonal frequencies in the ocean east of the warm pool (Figure 6.8). Hendon *et al.* (1998) improved on this crude fixed-patch model by considering the more realistic eastward propagation of MJO forcing over the warm pool, which moves at speeds similar to the Kelvin wave (Hendon and Salby, 1994). When the MJO speed equals the Kelvin wave speed, the forcing is resonant and the Kelvin amplitude is the same as for steady winds; in other cases it is less. For a realistic MJO wind, this maximum occurs at periods of about 70 days, and falls off very rapidly at higher frequencies. They conclude that the observed frequency offset between atmosphere and ocean is due to these linear Kelvin wave dynamics as a consequence of the spatial and temporal characteristics of the MJO winds.

6.6 ISV IN THE INDIAN OCEAN

The Indian Ocean is much more poorly observed than the Pacific; as a result much of the diagnosis has been done in models, often without adequate observational confirmation. Many hypotheses have been raised in model studies that cannot be fully substantiated and remain speculative. This situation is slowly being rectified through large-scale sampling (e.g., the Argo array of profiling floats, Gould *et al.* (2004)), and through regional programs (e.g., JASMINE, see Webster *et al.*, 2002). Schott and McCreary (2001) give a comprehensive review of the present state of knowledge of Indian Ocean circulation, focusing on the dynamics of the large-scale and low-frequency signals. Another useful overview is the textbook of Tomczak and Godfrey (1994), while Swallow (1983) reviews *in situ* observations of Indian Ocean eddies.

6.6.1 Differences between the Indian and Pacific Ocean warm pools and their consequences

The physical processes by which the Indian Ocean responds to intraseasonal forcing are the same as those of the Pacific, but differences in the background conditions have a large influence on the oceanic consequences. The shape of the Indian Ocean basin has an important effect because it is closed in the northern subtropics. In the Pacific (or Atlantic) equatorial Kelvin waves reflect along the eastern boundary to

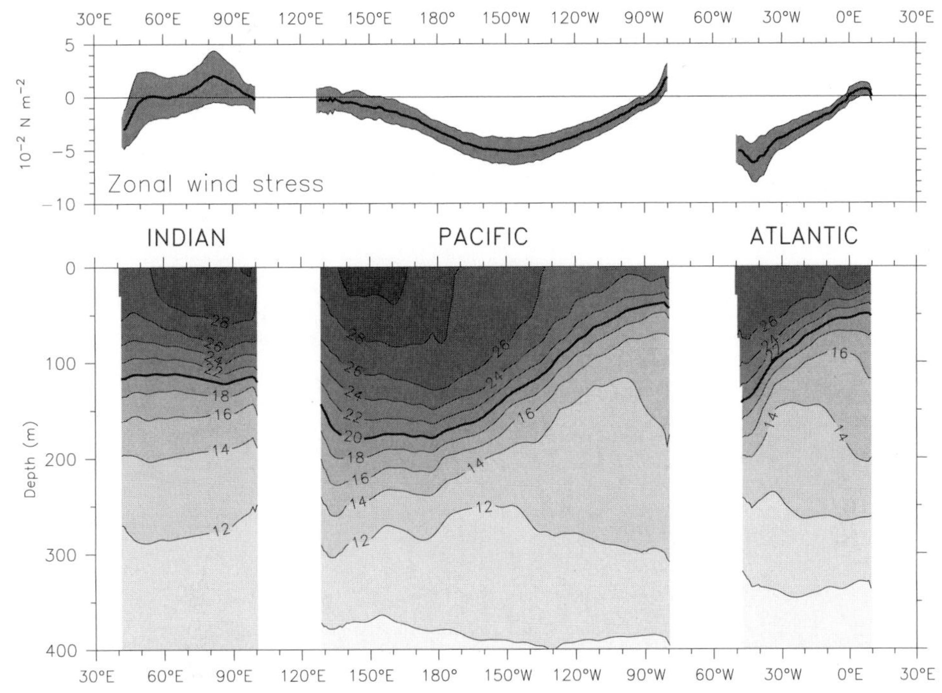

Figure 6.9. Mean zonal wind stress (*top*) and upper ocean temperature (*bottom*) along the equator. The winds are from the ERS scatterometer during 1992–2000, averaged over 5°S–5°N. The heavy line shows the mean, and the gray shading around it shows the standard deviation of the annual cycle. The ocean temperatures are from the Levitus (1994) *World Ocean Atlas*, with a contour interval of 2°C, and a supplemental contour at 29°C.

coastal signals that propagate poleward; as a result intraseasonal wind forcing becomes "lost" to the tropics in those basins. In the Indian Ocean, in contrast, coastal waves are directed into the Bay of Bengal (and from the Bay around the southern tip of India into the Arabian Sea), providing an important source of remote forcing to the off-equatorial tropics originating in equatorial winds (Potemra *et al.*, 1991; McCreary *et al.*, 1993; Schott *et al.*, 1994; Eigenheer and Quadfasel, 2000; Somayajulu *et al.*, 2003; Yu, 2003; Waliser *et al.*, 2004).

The fact that the climatological semi-annual wind forcing is much stronger than the mean winds in the Indian Ocean distinguishes it from the other basins that have permanent equatorial easterlies and thus a permanent zonally sloping thermocline and EUC, with their accompanying warm pool in the west and cold tongue due to upwelling in the east. The shallow east Pacific thermocline allows remotely forced thermocline depth changes to quickly and easily affect SST and thereby provide the potential for coupled interaction. Figure 6.9 shows the mean zonal thermocline slope associated with the mean easterlies in the Pacific and Atlantic, and the contrasting flat, deep thermocline of the Indian Ocean. Because of this profound difference in structure, the fluent communication between ocean dynamical processes and the

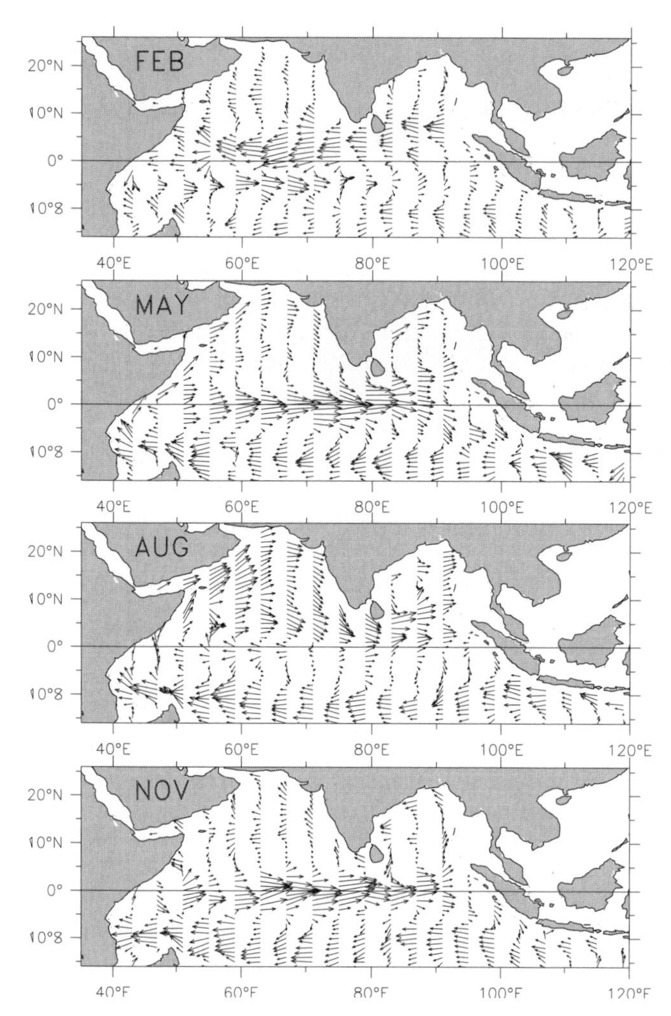

Figure 6.10. Seasonal cycle of Indian Ocean surface currents from historical ship-drift data (Mariano *et al.*, 1995). North of about 8°S the annual and semi-annual variation of most currents is much larger than the mean. The Wyrtki Jets are the equatorially-trapped eastward currents during May and November. The dramatic seasonal reversals of circulation in the Arabian Sea, the Bay of Bengal, and along the African Coast (Somali Current) are also evident.

atmosphere as occurs in the Pacific (Sections 6.4 and 6.5) is much more difficult to accomplish in the equatorial Indian Ocean. (However, an interannual "Indian Ocean Zonal Mode" has been proposed that depends on such changes in the narrow upwelling region close to the coast of Java; see Webster *et al.* (1999), Saji *et al.* (1999), Murtugudde *et al.* (2000), and Annamalai *et al.* (2003).)

The semi-annual equatorial zonal winds spin up eastward Yoshida Jet-like features in May and November (Wyrtki Jets; Figure 6.10) in response to westerly maxima during the monsoon transition seasons (Wyrtki, 1973; Reverdin, 1987; Han *et al.*, 1999). Although the climatological picture suggests two well-defined jets, observational (Reppin *et al.*, 1999) and modeling studies (Masson *et al.*, 2003) show that each semi-annual jet is broken up into oscillations with timescales of a month or less. As the zonal pressure gradient adjusts to the monsoon transition winds (see Section 6.3), semi-annual Kelvin waves are generated that contribute to the seasonally-reversing boundary currents in the Bay of Bengal (see references above), and by interacting with the locally-generated flow field there, contribute to its rich intraseasonal eddy field (Vinayachandran and Yamagata, 1998). This source of semi-annual remote forcing adds to the local intraseasonal forcing in boreal summer (Waliser *et al.*, 2004) to produce variability throughout the year, which may be one reason why the Bay of Bengal eddies do not appear to be strongly seasonally modulated (Somayajulu *et al.*, 2003).

In the central equatorial Pacific, meridional winds are weak compared to zonal winds, and the zonal winds are relatively uniform in latitude within 5°N–5°S. Thus the meridional circulation there is largely symmetric with a nodal point at the equator, and cross-equatorial oceanic heat transport occurs principally through mixing and small-scale processes (Blanke and Raynaud, 1997). In the Indian Ocean, by contrast, meridional winds are strong and seasonally reversing, and zonal winds are antisymmetric across the equator. This allows a significant mid-basin cross-equatorial (mean southward) Ekman mass transport (Miyama *et al.*, 2003) that balances the large cross-equatorial western boundary current mass transport driven by monsoon winds (the Somali Current, see Section 6.7.3); such a circulation has no counterpart in the other oceans. Similarly, in the west Pacific warm pool, the annual cycle of SST is damped by the tendency for convective cloudiness to increase with surface temperature (Ramanathan and Collins, 1991). In the North Indian Ocean, by contrast, the short-wave heat flux has a strong annual cycle because convection is absent during boreal spring. However, the annual cycle of SST there (east of the Somali Current region) is also small (Murtugudde and Busalacchi, 1999), comparable to that of the Pacific warm pool. Loschnigg and Webster (2000) suggested that this requires a seasonally-reversing oceanic cross-equatorial heat transport of the order of ± 1.5 PW, northward in winter and southward in summer, to maintain the SST (Figure 6.11). In their model, this transport was produced by a combination of western boundary and interior Ekman flows, driven by monsoon winds. It was strongly intraseasonally modulated. On the other hand, Waliser *et al.* (2004) found a similar seasonal but much smaller intraseasonal cross-equatorial transport oscillation in their ocean GCM and commented that the simplicity of the Loschnigg and Webster (2000) 2.5-layer ocean model might have led to an overestimate by neglecting the complex baroclinic variations.

The extreme rainfall and riverine input to the Bay of Bengal give it a surface salinity at least 1 psu fresher than the west Pacific warm pool (Bhat *et al.*, 2001) and low surface salinity extends across the equator in the eastern basin (Sprintall and Tomczak, 1992; Han *et al.*, 2001b). The resulting barrier layer is much stronger than in the Pacific warm pool (Section 6.2), especially in the western Bay (Shetye *et al.*,

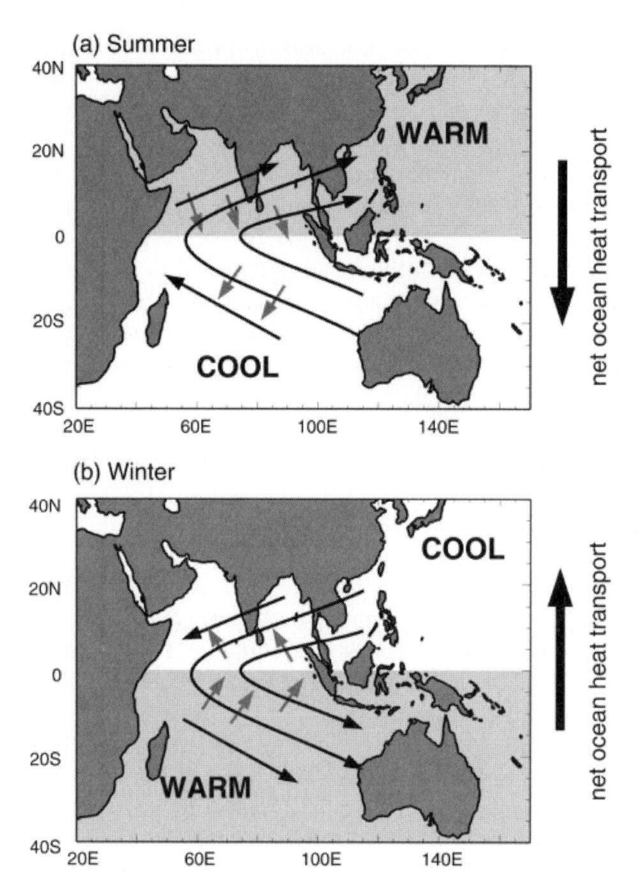

Figure 6.11. A regulatory model of the annual cycle of the Indian Ocean monsoon system depicted for (a) summer (June–September) and (b) winter (December–February). Curved black arrows denote the wind forced by the large-scale differential heating denoted by "warm" and "cool". The small gray arrows are the Ekman transport forced by the winds. The large vertical black arrows to the right show the net ocean heat transport that reverses between summer and winter. The net effect of the combined wind-forced ocean circulation is to transport heat to the winter hemisphere, thus modulating the SST differences between the hemispheres.
After Loschnigg and Webster (2000).

1996), and enlarges down to the equator most prominently in boreal fall (Masson *et al.*, 2002). As in the west Pacific warm pool, the barrier layer enhances the surface speed of the boreal fall Wyrtki Jet by trapping wind momentum in a thin surface layer (Section 6.2); Han *et al.* (1999) estimated this effect at $0.3\,\mathrm{m\,s^{-1}}$. In a model forced by observed precipitation, Masson *et al.* (2002, 2003) further suggested that advection of a subsurface salinity maximum by the Jet contributes to the intensification of the barrier layer in the eastern equatorial Indian Ocean.

The Indonesian Throughflow (ITF) exerts a fundamental control on the

Indo-Pacific warm pool; coupled model experiments suggest that it results in a warming of the Indian Ocean while cooling the Pacific and shifting the warm pool to the west (Schneider, 1998). Therefore, factors that influence ITF mass and property transport variability are of great interest. Velocities through the narrow ITF outflow straits can be significantly affected by Kelvin waves forced by equatorial winds and propagating along the Java coast at intraseasonal and semi-annual frequencies (Qiu *et al.,* 1999; Potemra *et al.*, 2002; Waliser *et al.*, 2003). These waves modulate sea level on the Indian Ocean ends of the straits and therefore change the along-strait pressure gradients; there may also be property effects due to changing the baroclinic structure of the outflows (Potemra *et al.*, 2003; Sprintall *et al.*, 2003). Qiu *et al.* (1999). Durland and Qiu (2003) also showed that intraseasonal Kelvin waves enter the Indonesian Seas at the Lombok Strait (a major outflow into the Indian Ocean) and modulate sea level in the Makassar Strait; this means that the straits further east (Timor and Omboi) are much less affected by Indian Ocean equatorial ISV. Although the intraseasonal equatorial forcing is a major influence on the velocity at Lombok, it is not yet clear whether this variability has a significant effect on the properties on the Indian Ocean side that would contribute to subsequent variability (Sprintall *et al.*, 2003).

It is worth noting that Qiu *et al.* (1999) also identified an ISO in the Celebes Sea that is unrelated to TISO winds. Their model results suggested that as the Mindanao Current retroflects into the Pacific at about 4°N, it sheds eddies into the Celebes Sea that closely match the gravest Rossby mode of this semi-enclosed basin, leading to resonance.

6.6.2 26-day oscillations in the western equatorial Indian Ocean

Meridional current oscillations concentrated on a narrow band of periods near 26 days were first observed by moorings in the western equatorial Indian Ocean in 1979–1980 (Luyten and Roemmich, 1982). There was very little zonal current signal in this band. Subsequently, similar oscillations were found in drifter tracks just north of the equator (Reverdin and Luyten, 1986), which raised the possibility that these were Tropical Instability Waves (TIW; see Section 6.7.2) similar to those in the Pacific and Atlantic. However, the close association of TIW with zonal current shear was not seen in the Indian Ocean. Models forced with smoothly-varying monthly winds were able to reproduce the 26-day waves, and showed that their dispersion properties (wavelength, westward phase propagation, and eastward group velocity) were consistent with Yanai wave kinematics (Kindle and Thompson, 1989). Tsai *et al.* (1992) used satellite SST to examine the spatial and temporal properties of the oscillations in this frequency band. These data confirmed the 26-day period, and its characteristics were consistent with westward-phase-propagating Yanai waves. The oscillations were found from 52°E to 60°E, about 1,000 km east of the coast. There have been two explanations proposed for the generation of these waves, which are apparently not forced by anything in the local winds. Moore and McCreary (1990) suggested that a periodic wind stress along the slanting western boundary could produce such Yanai waves propagating into the interior. However, several models (e.g., Kindle and Thompson, 1989) have produced 26-day Yanai waves when forced with climatological monthly winds alone. In these models, the waves were generated as an instability of the Somali Current system southern gyre (Schott and McCreary, 2001), a completely different mechanism than that which produces the TIW. It is not known why there is a preference for the apparently robust 26-day period.

6.6.3 Recent models of wind-forced ISV in the Indian Ocean

Using the detailed new satellite wind and SST products, modelers have begun to attempt simulations of Indian Ocean ISV forced with realistic winds and compared to realistic SSTs. Although many of the processes found are similar to those previously diagnosed in the Pacific, several studies have tried to disentangle the oceanic ISV due to TISO wind forcing vs. that due to internal instabilities.

Han *et al.* (2001a) noted the occurrence of two distinct peaks in simulated zonal currents in the central and eastern Indian Ocean, at 40–60 days and at 90 days. Comparing model runs with and without the intraseasonal winds showed that the 40–60-day signals were a predominantly linear ocean response to direct wind forcing. In the central and eastern basin, much of this variability was associated with organized, eastward-propagating MJO winds. The 90-day current peak, however, was significantly different from the forced linear solution. Part of the difference could be explained by a mechanism similar to that found in the Pacific by Kessler *et al.* (1995) and Hendon *et al.* (1998), in which the ratio between the period of the forcing and the time it takes a wave to cross the wind patch can lead to lower frequencies being preferentially felt by the ocean (see Section 6.5). However, in the Indian Ocean, reflected Rossby waves can be an additional influence because the distance from the region of strong intraseasonal winds and the eastern boundary is much shorter than it is in the Pacific. Han *et al.* (2001a) showed that a second baroclinic mode equatorial Kelvin/reflected Rossby wave is nearly resonant in the Indian basin when forced with 90-day period winds (also see Jensen, 1993), and thereby enhances the eastern ocean response at that period.

Sengupta *et al.* (2001) compared the results of an OGCM forced with full wind variability and with filtered seasonal cycle winds. Even when forced with smooth seasonal cycle winds, their model developed intraseasonal current variability in the western boundary region (see Section 6.7.3) and also south of Sri Lanka in the central basin, similar to observations (e.g., Figure 6.12). Tracing individual Kelvin waves and their Rossby reflections showed that the Sri Lanka ISV in this model was due to intraseasonal vortices generated when the eastern boundary Rossby reflections of the semi-annual Wyrtki Jet Kelvin waves meet the background eastward South Monsoon Current (Vinayachandran and Yamagata, 1998; Schott and McCreary, 2001). The fact that the Sri Lanka ISV in a complete-forcing run agreed in phase with observed velocities strongly suggested that despite the development of instabilities, these signals were predictable, and thus at least quasi-linear, responses to the winds.

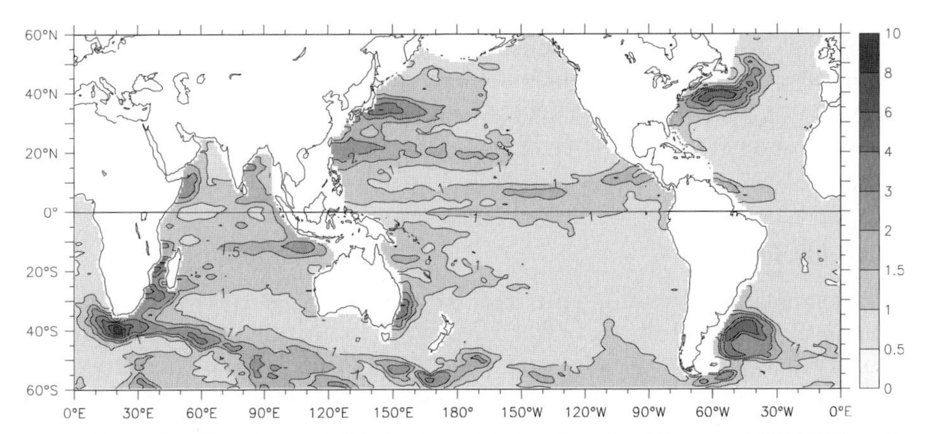

Figure 6.12. RMS of band-passed (35–85-day half power) sea level from the TOPEX/Poseidon satellite altimeter, for data during January 1992–July 2003. Dark shading indicates high sea level RMS, with a stretched contour interval (values indicated in the scale on the right).

6.7 OTHER INTRINSIC OCEANIC ISV

6.7.1 Global ISV

There is such a large variety of intrinsic ISV in the oceans, caused by many processes other than intraseasonal atmospheric forcing, that a review of the entire subject is well beyond the scope of this chapter. In addition, much of this variability is not germane to the principal thrust of this book. We will therefore focus on the most common signals that are likely to be intermingled with wind-forced ISV in the tropics, and that could therefore cause confusion in the interpretation of ocean observations connected with TISO. Other regional ISV signals will be discussed only briefly.

As an index of the occurrence of ISV in the global ocean, Figure 6.12 shows the RMS of intraseasonally band-passed sea level (sea surface height (SSH)) from the TOPEX/Poseidon altimeter (Fu *et al.*, 1994). It shows distinct regions with strong SSH ISV in many parts of the World Ocean; most of these are not associated with the tropical ISOs that are the principal subject of this book. Note that the equatorial region discussed above does not appear as a strong maximum of SSH ISV. That principally reflects the fact that small pressure gradients are more effective at driving currents near the equator because of the small value of the Coriolis parameter. Many investigators have studied altimetric SSH as an index of eddy variability, sometimes using it to estimate eddy kinetic energy through a geostrophic assumption which emphasizes the tropics (Stammer, 1997). Also note that the TOPEX altimeter does not sample small-scale very-near-coast signals very well, which is probably why the coastal Kelvin waves mentioned in Section 6.4 do not appear in Figure 6.12.

6.7.2 Non-TISO-forced ISV in the tropical Indo-Pacific

Two important intraseasonal phenomena that are not forced by TISO are observed in the tropical Pacific: the Tehuantepec and Papagayo eddies, which produce the bands of high ISV extending south-west from Central America in Figure 6.12, and the tropical instability waves, which are seen as the strip of SSH variability along 5°N. Although these two signals appear continuous in Figure 6.12, they are entirely separate phenomena (Giese *et al.*, 1994).

Central American eddies

The Tehuantepec eddies are generated by episodic winds blowing through the mountain pass at the Isthmus of Tehuantepec in southern Mexico (Chelton *et al.*, 2000b; Kessler, 2002a). High pressure behind winter cold fronts transitting North America causes a cross-mountain pressure gradient that funnels an intense wind jet through the pass and over the Pacific, on timescales of a few days (Hurd, 1929; Roden, 1961; Chelton *et al.,* 2000a). These winds produce locally strong mixing and SST fluctuations (Trasviña, 1995), and also generate a series of typically three to five anticyclonic (warm-core) eddies each winter that propagate westward approximately as free Rossby waves (Giese *et al.*, 1994), leading to the strip of high SSH variance in Figure 6.12. Individual eddies can be tracked in SSH along 11°N on occasion as far west as the dateline (Perigaud, 1990; Giese *et al.*, 1994; see also Figure 6.13). Ocean color is a useful technique for remotely sensing these eddies because their SST signal may be small, but the associated plankton blooms can still be evident (McClain *et al.*, 2002). While the forcing occurs on a timescale much shorter than intraseasonal, the fact that the wind events occur episodically several times each winter results in the apparent intraseasonal timescale. The observed preference for anticyclonic rotation was explained by McCreary *et al.* (1989): although eddies of both signs are generated by the wind jet (downwelling under the negative curl region on the right flank of the jet axis, upwelling on the left flank to the east), the high winds quickly mix away the upwelled thermocline of the cyclonic eddy.

Winds blowing through the lowlands of Nicaragua (known as Papagayo winds) are less variable than those through Tehuantepec, and not apparently associated with mid-latitude cold fronts, though they are still stronger in winter (Müller-Karger and Fuentes-Yaco, 2000; Chelton *et al.*, 2000a, 2000b). A different explanation for the eddies west of Nicaragua (seen as the southern strip of high intraseasonal variance west of Central America in Figure 6.12) was proposed by Hansen and Maul (1991), who estimated a propagation speed greater than that of free Rossby waves and diagnosed the eddies as strongly non-linear and similar to Gulf Stream rings. They suggested that the Papagayo eddies, which are also anticyclonic, could be due entirely to ocean dynamics, without any influence of the mountain-gap winds. Instead they point to conservation of potential vorticity as the North Equatorial Countercurrent (NECC), which is strongest in boreal winter, turns sharply northward when it meets the coast. They proposed that the anticyclonic relative vorticity gained in the northward flow results in eddy shedding. This hypothesis has been questioned (Giese *et al.*, 1994), and the possible difference in generation mechanism between the Papagayo and Tehuantepec eddies has not been resolved. Eddies are also seen west of a third wind jet through the lowlands of Panama

Color plates

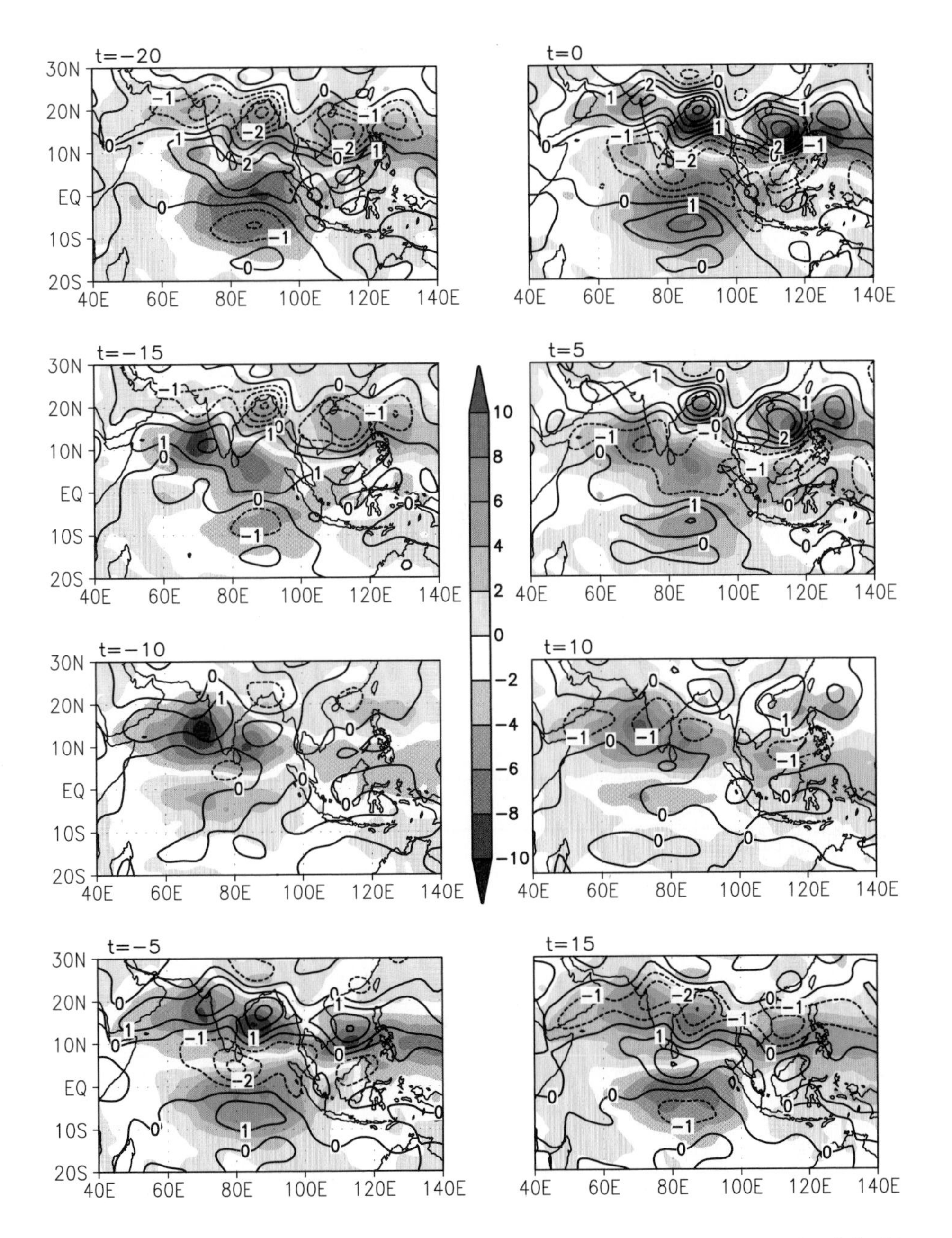

Figure 2.11. Evolution of convection and relative vorticity at 850 hPa over a cycle of the 30–60-day mode. Regressed 30–60-day filtered anomalies of OLR (in W m^{-2}; shaded) and relative vorticity of 850-hPa winds (contour) with respect to the same reference time series described in Figure 2.10 from a lag of 20 days ($t = -20$) to a lead of 15 days ($t = 15$). Solid (dashed) lines indicate positive (negative) relative vorticity, with a contour interval of $1 \times 10^{-6}\,\mathrm{s}^{-1}$, and with thick lines showing the zero contour.

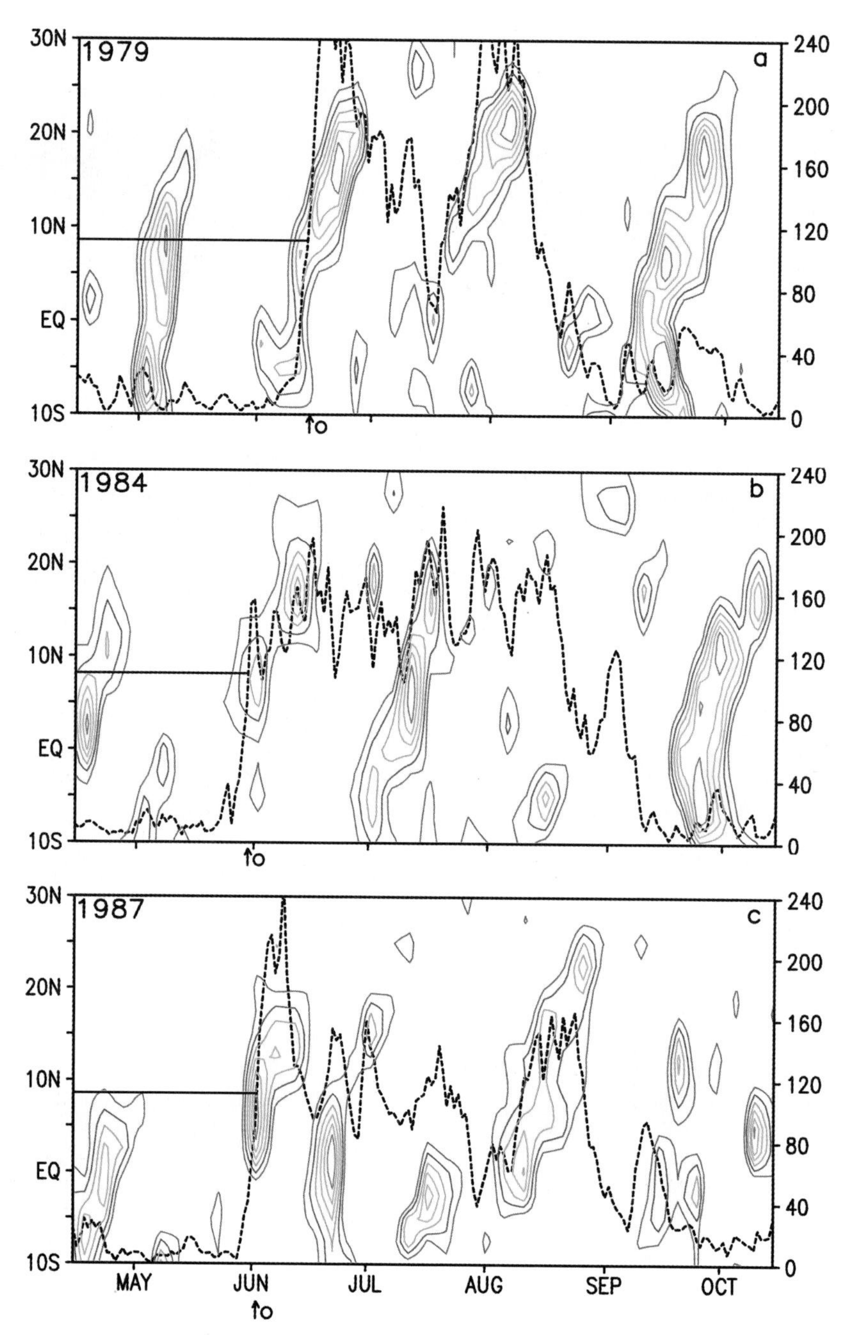

Figure 2.14. Time–latitude section of CMAP anomalies (unfiltered) averaged over 70°E–90°E. Only positive anomalies greater than $2\,\mathrm{mm\,day^{-1}}$ with a contour interval of $2\,\mathrm{mm\,day^{-1}}$ are plotted. Northward propagating wet spells can be seen. The dashed line shows the evolution of total K.E. of winds at 850 hPa averaged over the LLJ (55°E–65°E, 5°N–15°N). The scale for the K.E. $(\mathrm{m\,s^{-1}})^2$ is on the right. The latitude of Kerala (southern tip of India) is shown by the horizontal line. The monsoon onset over Kerala is shown by the arrow and roughly corresponds to K.E. exceeding $100(\mathrm{m\,s^{-1}})^2$ and precipitation exceeding $4\,\mathrm{mm\,day^{-1}}$.

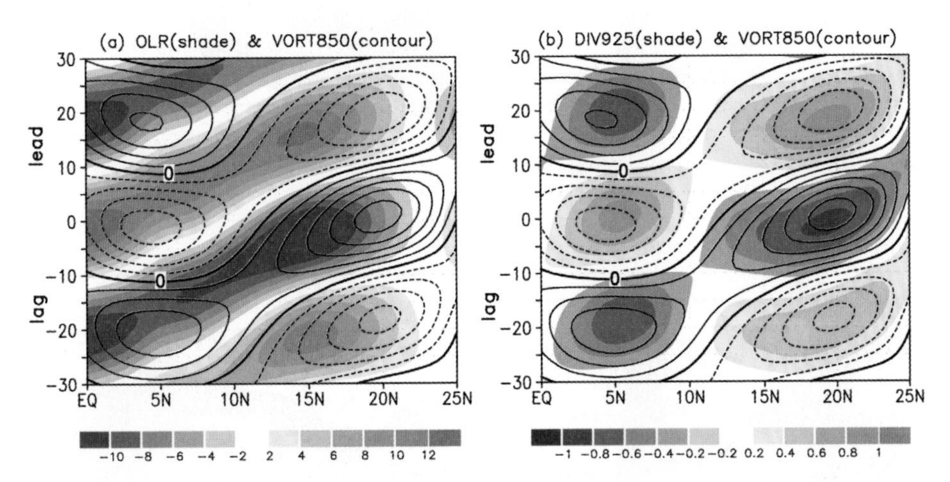

Figure 2.15. (a) Regressed 30–60-day filtered anomalies of OLR (shaded; $\mathrm{W\,m^{-2}}$) and 850-hPa relative vorticity (contour, positive solid and negative dashed, contour interval $1 \times 10^{-6}\,\mathrm{s^{-1}}$) with respect to the reference time series described in Figure 2.10 averaged over 80°E–90°E. (b) Regressed 30–60-day filtered anomalies of 850-hPa relative vorticity (contour, positive solid and negative dashed, contour interval $1 \times 10^{-6}\,\mathrm{s^{-1}}$) and divergence at 925 hPa (shaded; $10^{-6}\,\mathrm{s^{-1}}$) with respect to the same reference time series.

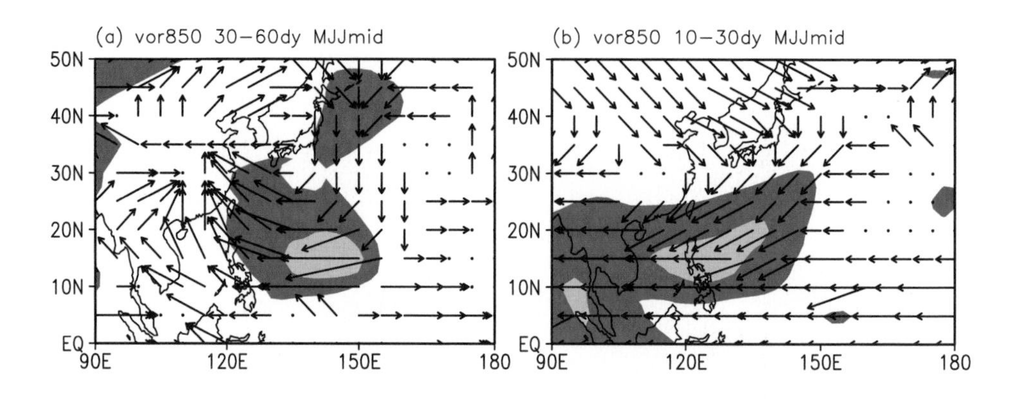

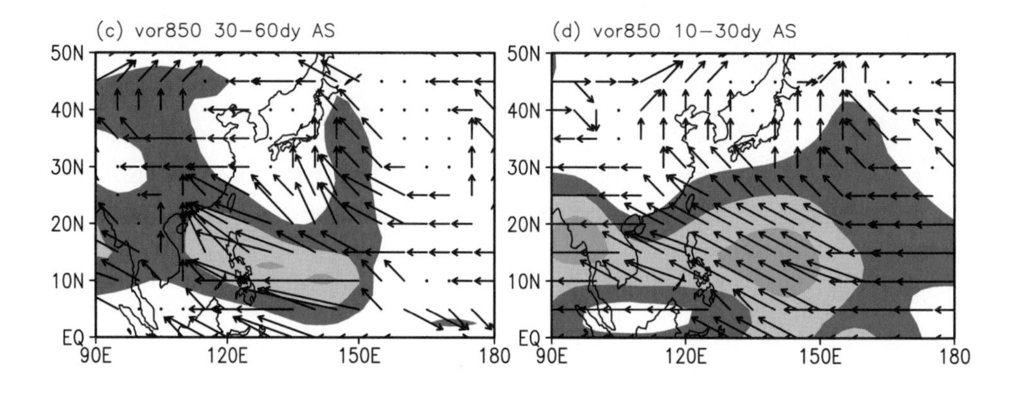

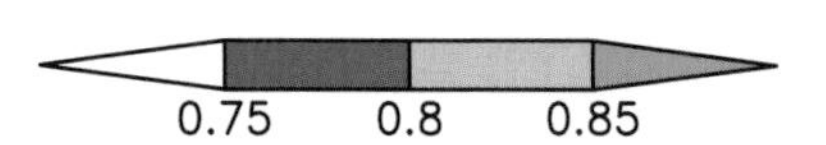

Figure 3.5. Propagation tendency vectors derived from the 5-day and 2-day lagged correlation maps for the 30–60-day (*left*) and 10–30-day (*right*) 850-hPa vorticity perturbations for (a, b) May to mid-July and (c, d) August to September. Shading represents the lagged correlation coefficients greater than 0.75 and the contour interval is 0.05. The vector denotes the direction and distance that an anomaly at the base point is most likely to travel in the next 5 or 2 days. The unit arrow denotes the length of a 3 grid-point distance (i.e., 7.5 degrees).

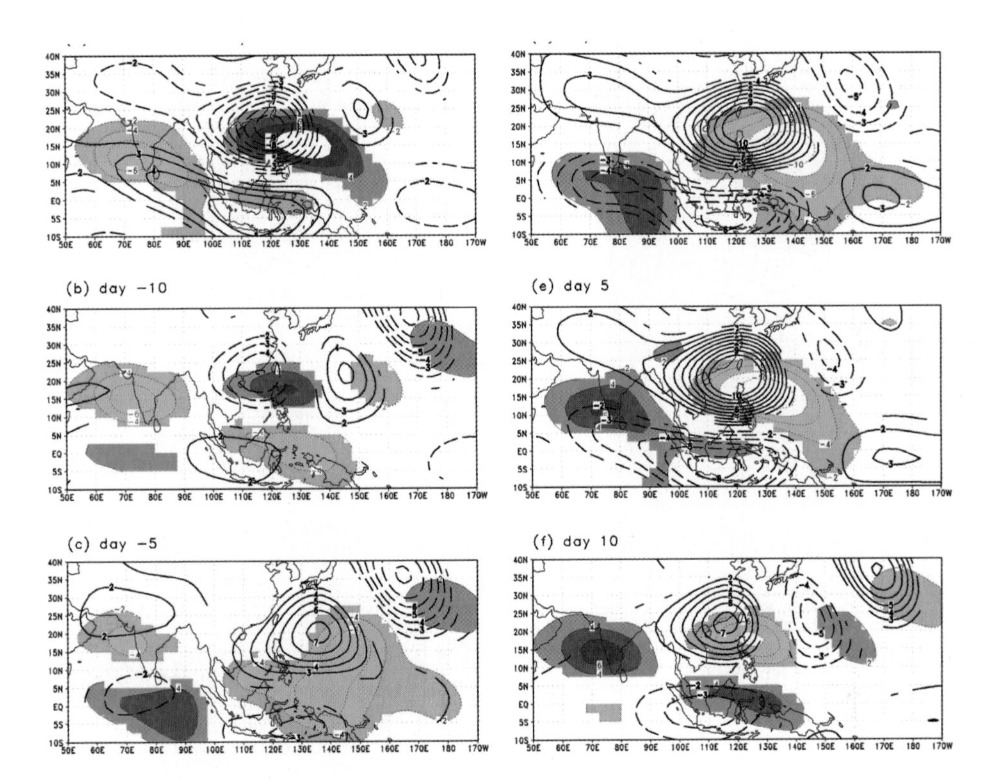

Figure 3.10. Evolution of the 30–60-day OLR and low-level circulation patterns in the western North Pacific adopted from Hsu and Weng (2001). Lagged regression coefficients between the OLR anomaly averaged at 120°E–160°E and 0–20°N and the OLR (shaded) and 850-hPa vorticity (contoured) at (a) day −15, (b) day −10, (c) day −5, (d) day 0, (e) day 5, and (f) day 10. Contour intervals are $2\,\mathrm{W\,m^{-2}}$ and $1 \times 10^{-7}\,\mathrm{s^{-1}}$ for the OLR and vorticity, respectively. Dark shading and solid lines indicate positive values, while light shading and dashed (and dotted) lines indicate negative values. The regression coefficients have been multiplied by one standard deviation of the OLR index and only those that are significant at the 0.05 level are plotted.

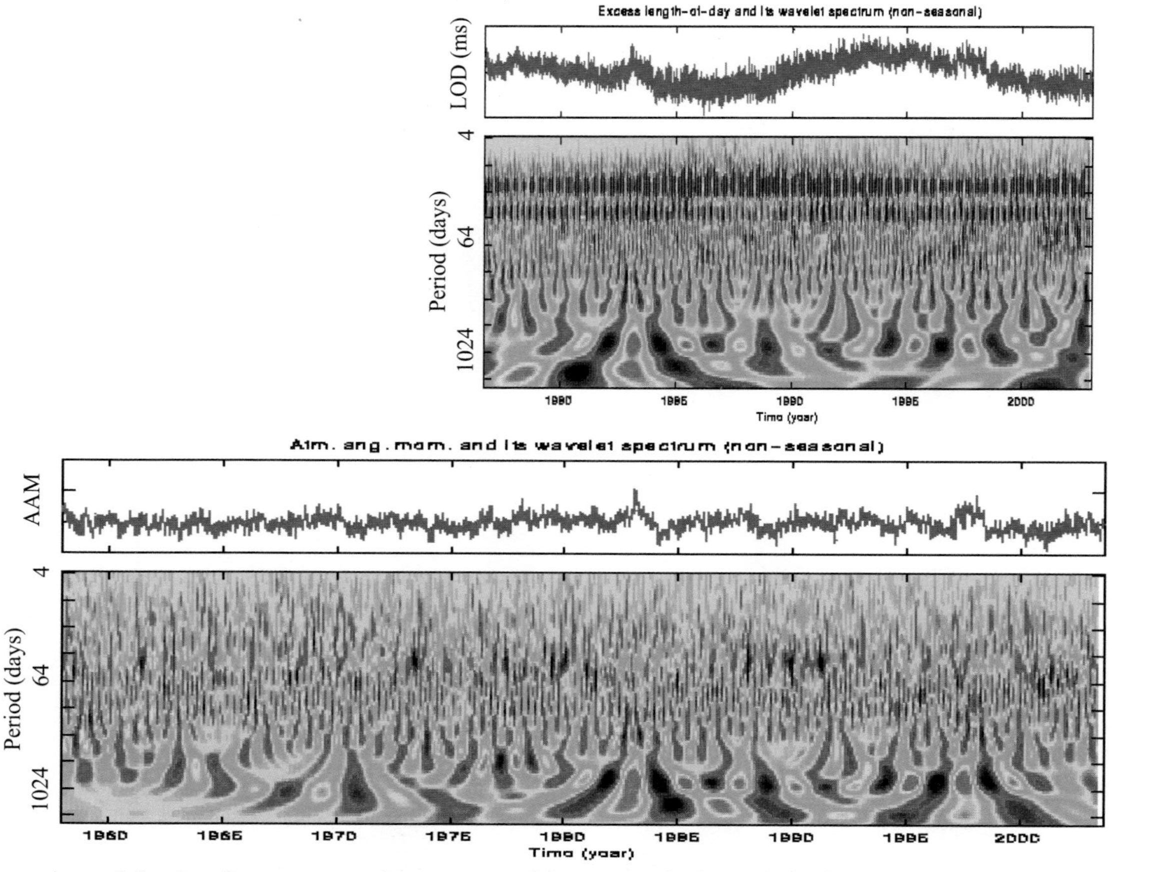

Figure 8.2. Comparison of the time-frequency wavelet spectrum of (*upper panel*) the geodetically observed ΔLOD time series; (*lower panel*) the computed AAM_z time series. The seasonal terms have been removed beforehand; the plot gives the real part (the cosine term of the wavelet transform) so that the positive–negative polarity of the undulations is shown. Note the similarity between the two spectra. The contour amplitude in both plots is normalized and dimensionless and hence only showing the relative strength (in fact the contour shade has been saturated on the high end so as to bring out the more moderate signals).

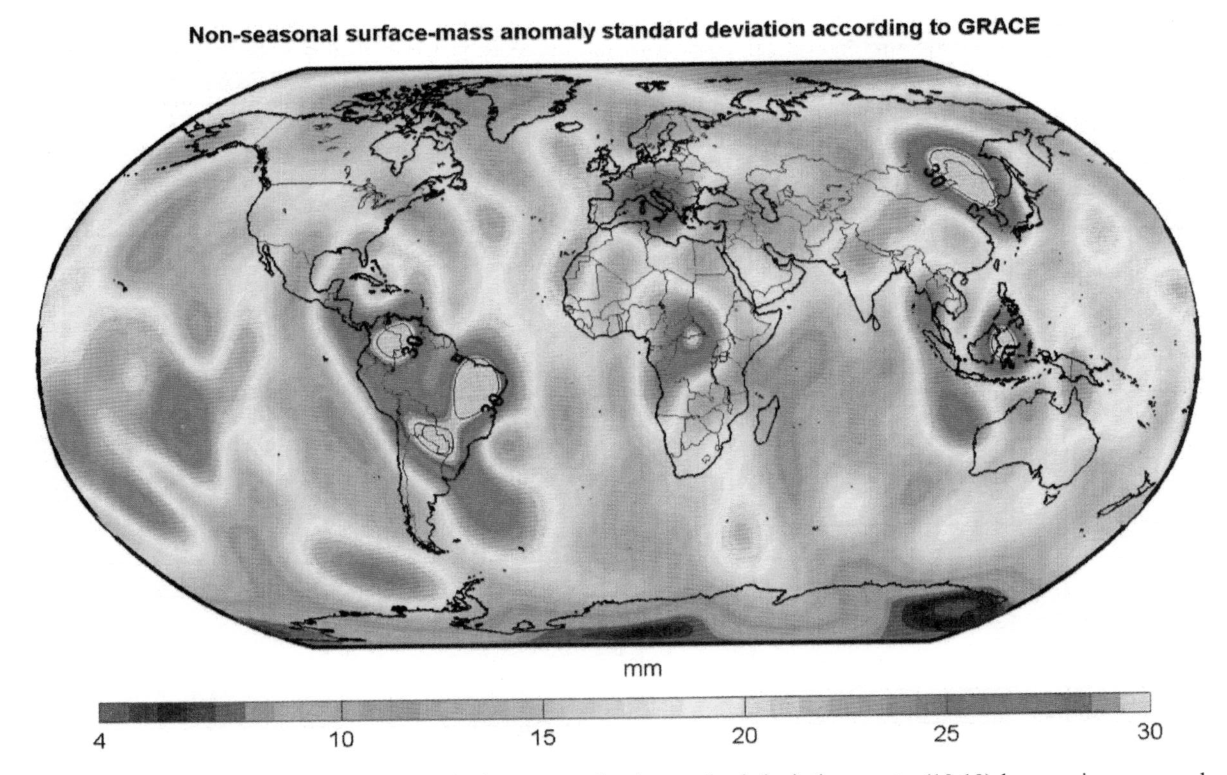

Figure 8.6. The ISV mass anomaly in terms of equivalent-water-depth standard deviation up to (10,10) harmonics averaged over a period between August 2002 and May 2004. These signals are first corrected for atmospheric mass redistribution (based on the ECMWF GCM model) and the oceanic contributions (according to a nominal barotropic ocean model), while the seasonal and linear trend signals are also removed. Note the moderate ISV signal level over the oceans, implying that the applied ocean model did a preliminary job in the modeling (and hence removing signals) of mass redistribution in the oceans up to harmonic degree 10.

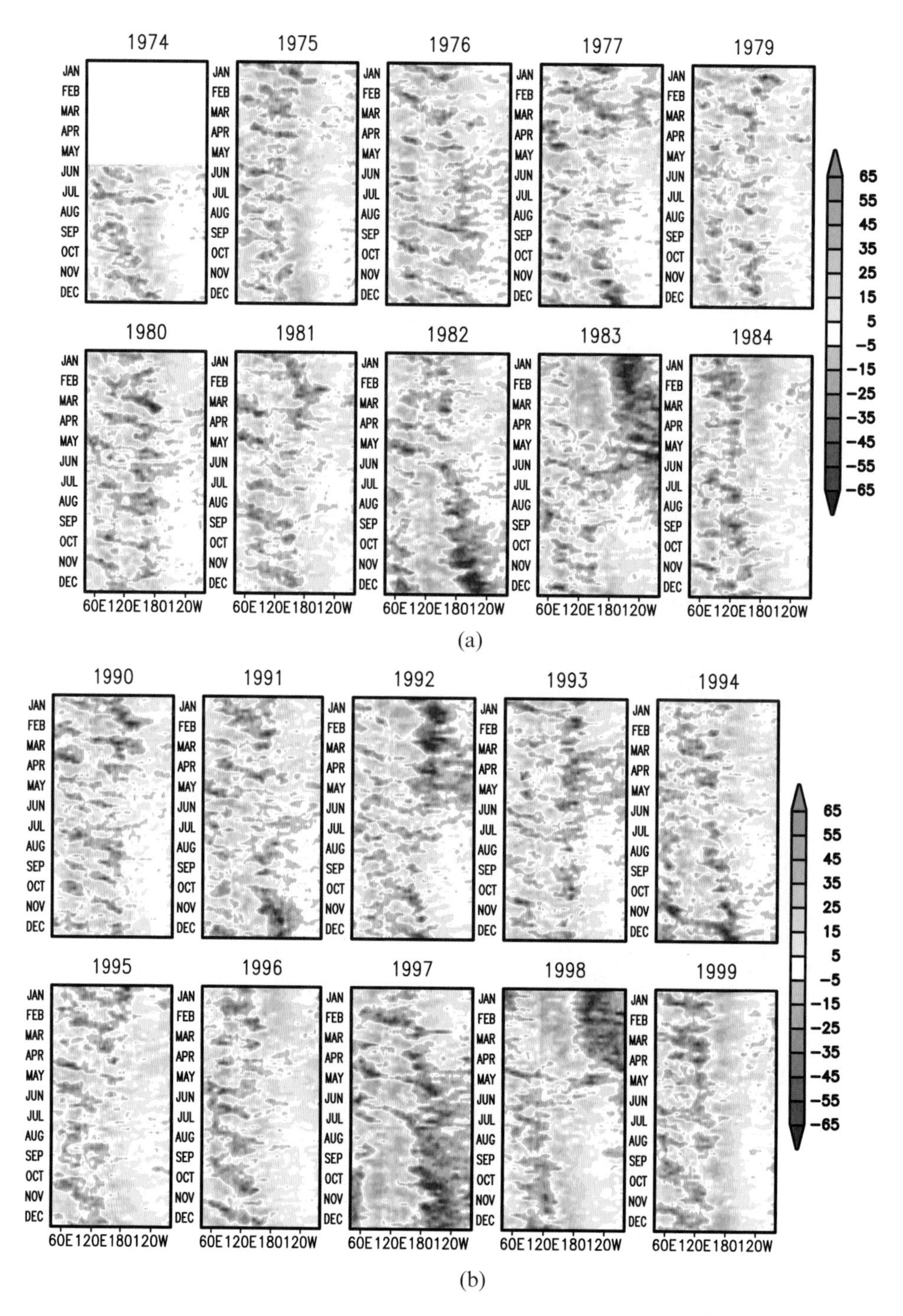

Figure 9.1. Time–longitude section of 5-day mean OLR ($\mathrm{Wm^{-2}}$) averaged between 5°S–5°N, for (a) 1974–1984 and (b) 1990–1999. Negative values (shaded blue) indicate enhanced deep convection, and positive values (shaded red), reduced deep convection.

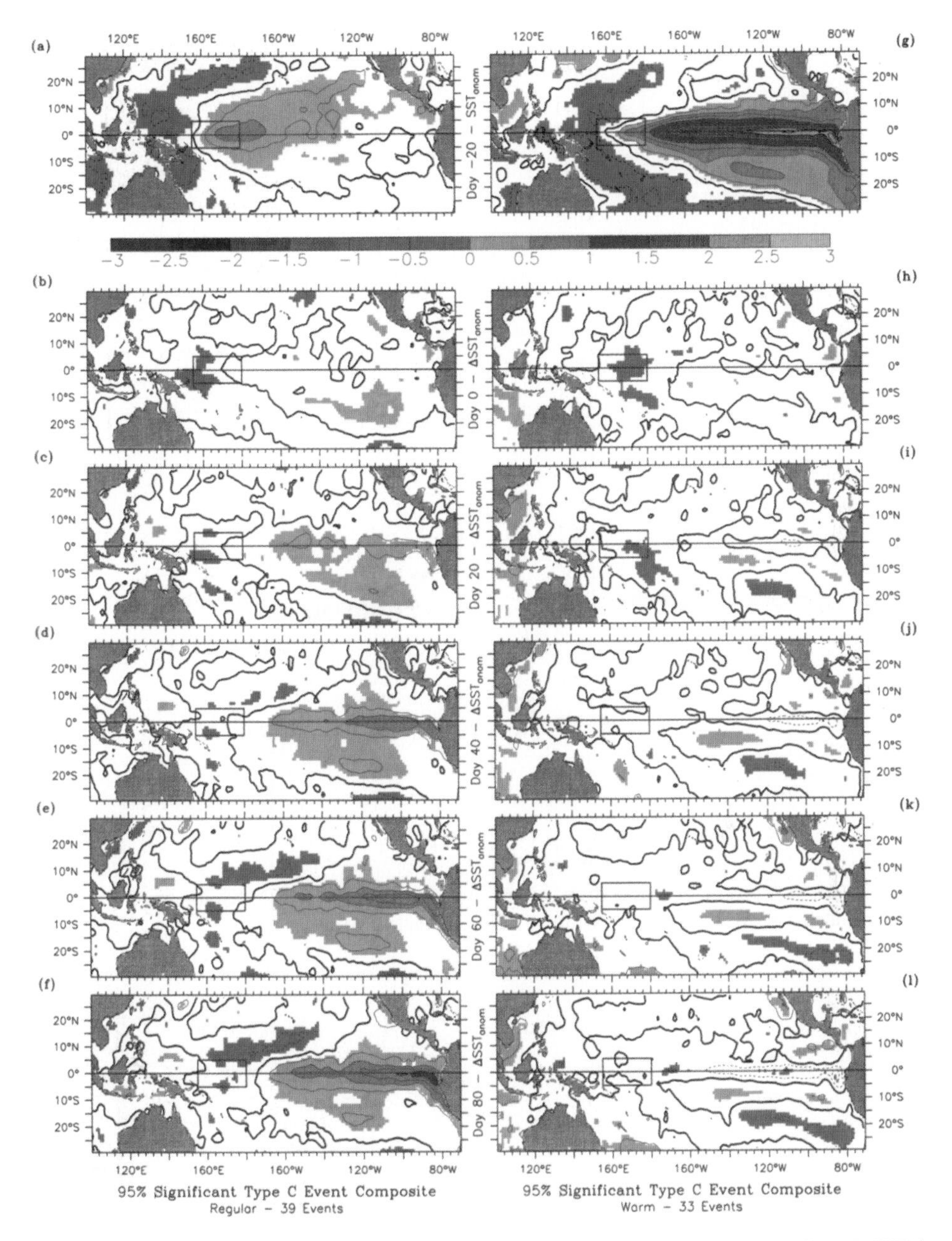

Figure 9.6. Composites of SSTA and changes of SSTA (ΔSSTA) from Day −20 for a WWE in the western Central Pacific (shown as square rectangle) for normal conditions (*left panels*): (a) Day (−20) SSTA, and ΔSSTA for (b) Day (0), (c) Day 20, (d) Day 40, (e) Day 60, and (f) Day 80. Same composites are shown in (g–l) except for warm ocean states. Values exceeding the 95% confidence level are color shaded. Contour interval is in 0.25°C. Shading contours are 0.5°C.

Adapted from Vecchi and Harrison (2000).

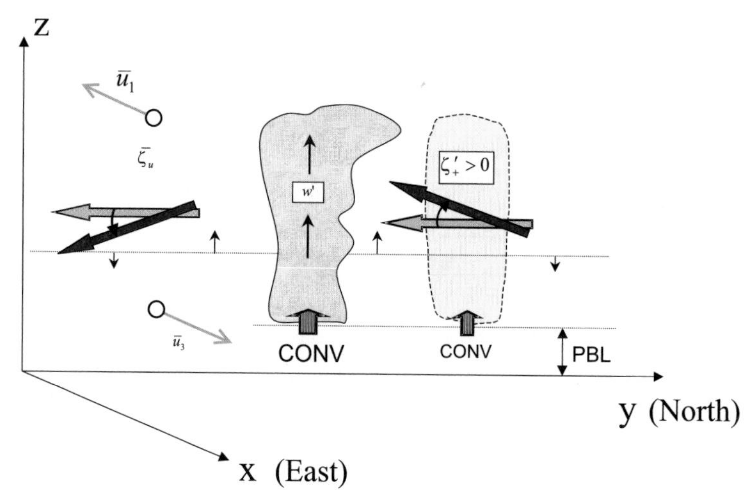

Figure 10.10. Schematic diagram showing the mechanism by which monsoon easterly vertical shear generates northward propagation of ISO convective anomalies.

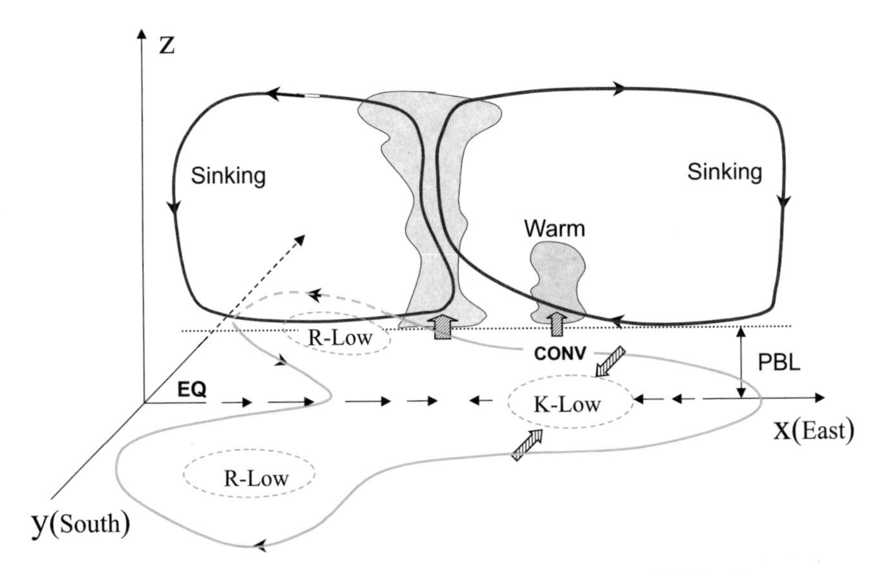

Figure 10.13. Schematic structure of the frictional CID mode, which is the counterpart of observed MJO mode. In the horizontal plane the "K-low" and "R-low" represents the low-pressure anomalies associated with the moist equatorial Kelvin and Rossby waves, respectively. Arrows indicate the wind directions. In the equatorial vertical plane the free-tropospheric wave circulation is highlighted. The wave-induced convergence is in phase with the major convection, whereas the frictional moisture convergence in the "K-low" region is ahead of the major convection due primarily to meridional wind convergence.

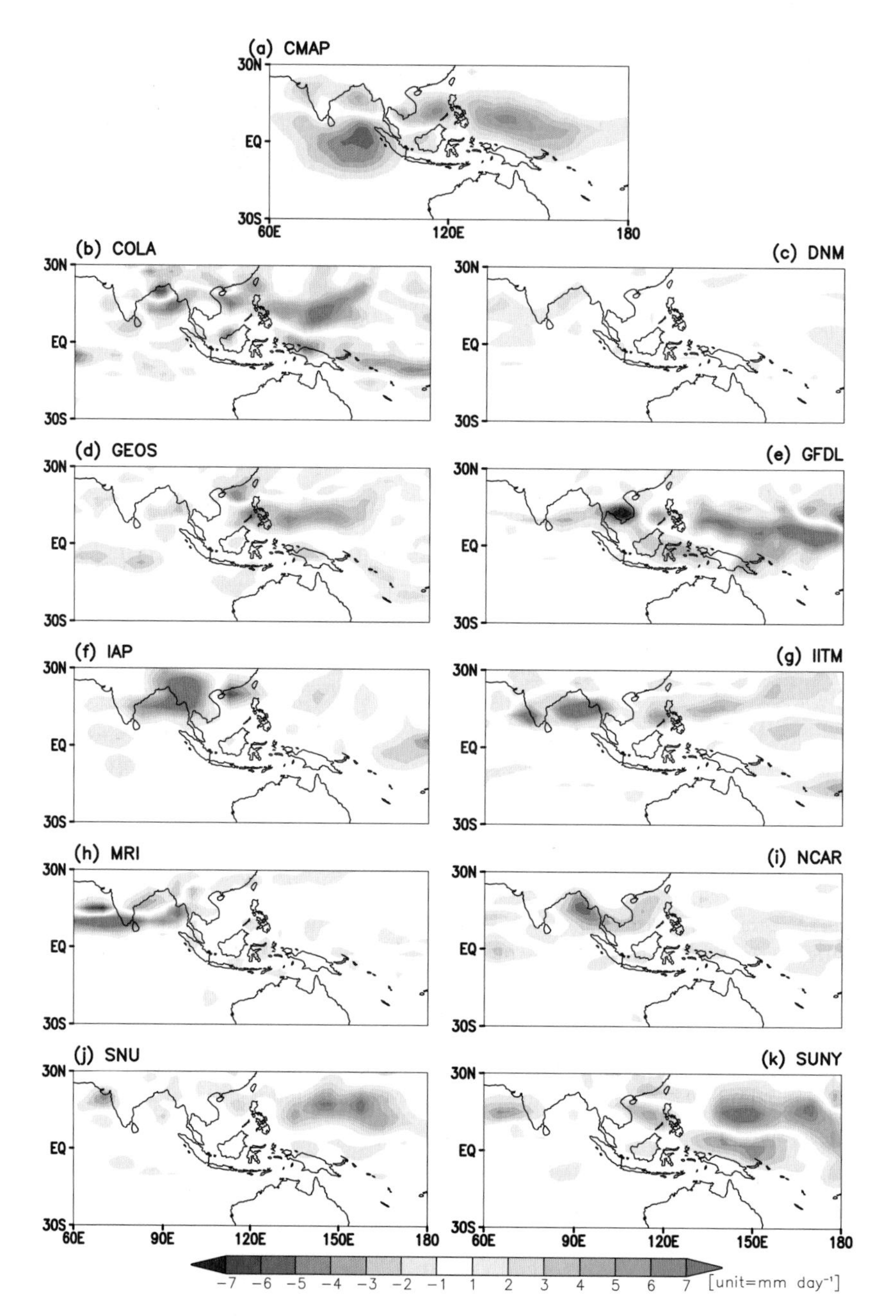

Figure 11.5. Composite BSISV rainfall for days 0 to approximately +5 based on identification using extended empirical orthogonal function analysis as per Waliser *et al.* (2003a). (a) Observations (CPC Merged Analysis of Precipitation (CMAP)). Atmospheric general circulation models forced with observed weekly SST – (b) COLA, (c) DNM, (d) GEOS, (e) GFDL, (f) IAP, (g) IITM, (h) MRI, (i) NCAR, (j) SNU, and (k) SUNY.

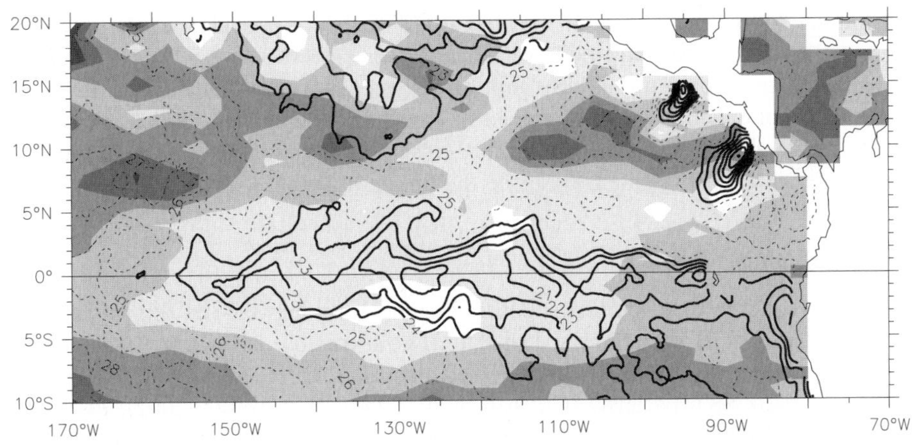

Figure 6.13. Example of the sea surface topography (shading) and temperature (contours) observed by satellite during January 2000, illustrating the signatures of the Central American eddies and of the tropical instability waves (Section 6.7.2). The eddies are visible as the four dark patches lined up along 8°N–12°N; these are warm-core (anticyclonic) vortices produced by episodic mountain-gap winds through the Central American Cordillera. They propagate west with a speed of about $15\,\mathrm{cm\,s^{-1}}$ ($400\,\mathrm{km\,month^{-1}}$). The direct SST effects of the winds are seen as the packed dark (cool) contours that indicate the location of the strongest winds at Tehuantepec and Papagayo. The dark SST contours along the equator show the cold tongue (a SST greater than 25°C is indicated by dashed contours). The cusps along the tight SST gradient on the north side of the tongue are the signature of tropical instability waves. These waves propagate west with a speed of about $1{,}000\,\mathrm{km\,month^{-1}}$.

(Müller-Karger and Fuentes-Yaco, 2000; note the third, weaker strip of ISV in Figure 6.12), but these have been studied even less.

The offshore passage of these eddies, however generated, was documented by Giese *et al.* (1994) using TOPEX altimetry, and by Hansen and Maul (1991) using surface drifter tracks. Confusingly, the Tehuantepec eddies appear to first move south-west and then follow the same track along 11°N as the Papagayo eddies westward from Central America (Giese *et al.*, 1994; see also Figure 6.12); it is not known why this behavior occurs. Their speed along 11°N is about $17\,\mathrm{cm\,s^{-1}}$, which is close to the theoretical first baroclinic mode Rossby wave speed at that latitude. The spatial scale of the eddies is a few hundred kilometers, and they tend to be zonally-elongated (Giese *et al.*, 1994; see also Figure 6.13). It is unlikely that the offshore eddies would interact with the ocean variability due to TISO forcing, because they are far enough from the equator to be well poleward of the Kelvin wave influenced region, and they die out east of the warm pool where the TISO signals have a wider meridional span. However, near the coast an interaction is possible, because TISO-forced equatorial Kelvin waves (Section 6.4) produce strong intraseasonal velocity and thermocline depth variability as their reflection leaves poleward-propagating coastal waves (note that the TISO Kelvin waves also display a preference for downwelling). An apparent coincidence between the arrival of TISO-origin Kelvin waves and the shedding of offshore-propagating eddies led to speculation that perhaps some of the Central American eddies were in fact triggered or modulated by TISO (B. Kessler, pers. commun., 1998), though the mechanism is unclear. Linear wave theory does not encompass the dynamics of coastal Kelvin wave propagation when a zonally-oriented coast is part of the picture, as occurs at the Isthmus of Tehuantepec; this problem has been glossed over (see the appendix to Kessler *et al.*, 2003).

Tropical Instability Waves

TIW were recognized as soon as satellites began observing SST, as the cusps in the sharp front near 2°N between the east Pacific cold tongue and warmer water along the North Equatorial Countercurrent (Legeckis, 1977). They are particularly obvious in their SST patterns (Chelton *et al.*, 2000c; Contreras, 2002), distorting the front into 1,000–1,500 km long cusp shapes (rounded to the south, pointed to the north; Figure 6.13). Since then, they have also been observed in satellite altimetry (Musman, 1989; Weidman *et al.*, 1999), surface drifter tracks (Hansen and Paul, 1984; Flament *et al.*, 1996; Baturin and Niiler, 1997), satellite ocean color (McClain *et al.*, 2002), and moored temperature and velocity time series (Halpern *et al.*, 1988; McPhaden, 1996), and are a robust and commonly-observed aspect of the eastern tropical Pacific (and Atlantic). In addition they are a ubiquitous feature of ocean GCMs (Cox, 1980; Philander *et al.*, 1986; Kessler *et al.*, 1998; Masina and Philander, 1999). With very large meridional velocity fluctuations, of the order of $\pm 50\,\mathrm{cm\,s^{-1}}$, TIW are a substantial source of noise that pose difficult aliasing problems in typically sparse ocean observations, even to sample the mean (Johnson *et al.*, 2001). TIW arise in the far eastern Pacific and propagate west with speeds of about 30–$60\,\mathrm{cm\,s^{-1}}$ (about 10° longitude per month), weakening west of about 150°W. It has not been clear how far west they penetrate, since as the SST front weakens to the west their SST signature that is easiest to observe dies away. Their frequency spans 20–35 days. A feature that has caused confusion is the apparent difference in frequency depending on the quantity being observed, with SST (whose signature is seen in the SST front near 2°N) showing a dominant period of about 25 days (Legeckis, 1977), whereas altimetric sea level (seen near 4°–6°N) appears to have a period near 35 days (Chelton *et al.*, 2003), and equatorial velocity a period near 21 days (Halpern *et al.*, 1988). These discrepancies may be the result of a fairly broadband instability shedding quasi-linear Yanai waves preferentially at certain frequencies (e.g., Weisberg *et al.*, 1979). Although the TIWs were first identified north of the equator, and their strongest signals continue to be found there, recent work has shown the existence of TIWs to the south (Chelton *et al.*, 2000c).

The principal mechanism producing the TIW is thought to be barotropic (shear) instability as first explained by Philander (1976, 1978). Since it is very difficult to diagnose energetics from sparse ocean observations, most of this work has been done in numerical models of various types (however, see Luther and Johnson (1990) and Qiao and Weisberg (1998) for observational analyses). The meridional shear in the equatorial Pacific is complex, with several strong oppositely-directed zonal currents in close proximity (Luther and Johnson, 1990; Johnson *et al.*, 2002), and there has been an evolution in thinking about this problem. The original Philander analysis concluded that the relevant shear was near 4°N between the westward SEC and the eastward NECC. More recent work points to the shear close to the equator between the SEC and the EUC; in addition the possibility of baroclinic instabilities associated either with the spreading isotherms around the EUC or with the sharp temperature front appears to be important as well (Yu *et al.*, 1995; Masina *et al.*, 1999). The sources of energy conversion driving the TIW remains an active area of research, and it is likely that different mechanisms come into play at different latitudes, perhaps explaining the multiple frequency structure seen. However, the fact that ocean GCMs of diverse types easily generate TIWs, whether forced with realistic or highly simplified winds, suggests that near-equatorial zonal shear is the dominant factor. The contribution to the variability by free-wave propagation out of the instability growth region is another current area of research. It has not been clear exactly how far east the TIWs begin, nor what is the initiating perturbation. All three currents involved are weaker in the far east (Johnson *et al.*, 2002; Lagerloef *et al.*, 1999) (but note that the Costa Rica Dome circulation may provide more shear east of 95°W; see Kessler, 2002a).

Because the TIWs depend on the background conditions, which vary seasonally and interannually, their low-frequency modulation is expected. The entire upper-equatorial circulation quickens when the winds are strongest in June–December: the westward SEC and eastward NECC are largest in boreal fall, as is equatorial upwelling (Kessler *et al.*, 1998). These conditions produce both the strongest meridional shears and sharpest temperature front, so it is not surprising to find that the TIWs appear in May or June and persist through the following February–March. Similarly, during El Niño events both the SEC and cold tongue weaken dramatically, and TIWs are found to be absent (Baturin and Niiler, 1997). It is possible that the TISO-generated Kelvin waves have an effect on TIWs, by increasing eastward current speeds on the equator. Giese and Harrison (1991) found that in their OGCM the resulting increase in meridional shear amplified the TIWs and produced a transient equatorward heat flux that was as large as the zonal advective warming due to the Kelvin wave.

Equatorward heat flux due to TIW mixing across the sharp SST front is a first-order term in the low-frequency east Pacific heat balance, as large as upwelling or surface fluxes (Hansen and Paul, 1984; Bryden and Brady, 1989; Kessler *et al.*, 1998; Swenson and Hansen, 1999; Wang and McPhaden, 1999). In the annual cycle it tends to counter the cooling due to wind-driven upwelling, which is also largest in June–December. The TIWs accomplish a substantial momentum flux as well, by mixing the shear between the SEC and EUC. The magnitude of mean zonal transport due to this poleward flux of eastward momentum is comparable to the wind-driven Sverdrup transport (Kessler *et al.*, 2003). Thus, the TIWs pose a substantial sampling challenge for observations, and a corresponding requirement for high temporal resolution in the diagnosis of ocean models, in order to properly represent these very large rectified signatures.

With the advent of high-resolution scatterometer winds, the cusp-like TIW SST pattern was found to be imprinted on the cross-equatorial south-easterlies of the eastern equatorial Pacific (Chelton *et al.*, 2001). This occurs because cold tongue SST stabilizes and decouples the atmospheric planetary boundary layer from the free atmosphere above, leading to a slow surface wind speed. As the south-easterlies blow across the distorted front and over warm SST, convection develops that mixes momentum downward and increases the windspeed. Chelton *et al.* (2001) showed that because the wind direction is quite consistent, this feedback produces a systematically positive curl and divergence signal along the SST front, raising the possibility of a coupled interaction that could further impact the ocean.

6.7.3 ISV outside the equatorial Indo-Pacific

Large intraseasonal signatures are prominent in several regions outside the equatorial Indo-Pacific that are unconnected to the TISO signals that form the main subject of this book. This large topic is sketched here very briefly, giving the bare picture with recent references for those who wish to delve more deeply.

ISV associated with western boundary currents

The poleward western boundary currents closing the Atlantic, Pacific, and South Indian Ocean subtropical gyres generate significant variability in the intraseasonal to interannual frequency bands (Figure 6.12). (The North Indian Ocean is discussed below.) ISV associated with western boundary currents has its origin in the fact that when these currents separate from the coast and are injected into the relatively quiescent open ocean, the strong jets become unstable and form meanders which may grow to become closed-core eddies or rings containing water pinched off from north or south of the jet axis. Bottom topography is an important influence on the path of the current and the regions of meander formation. Rings have scales of a few hundred kilometers, reach depths of more than one kilometer, and may persist for several years as identifiable water mass features. Each of these currents sheds typically three to ten rings each year, resulting in an apparent intraseasonal timescale. Once generated, the rings tend to drift westward with some characteristics of Rossby wave dynamics, though non-linear terms are usually significant, and they can also be swept eastward with the current or reabsorbed by it. Because the rings "contain" the water within them, substantial transport of heat and water properties can be accomplished by this drift. Such western boundary current ISV is seen in Figure 6.12 at the separation regions of the Gulf Stream and Brazil Current in the Atlantic, the Kuroshio and East Australia Current in the Pacific, and the Mozambique/Agulhas Current in the South Indian Ocean. Useful overviews of western boundary current structure and eddy generation mechanisms can be found in Olson (1991), Hogg and Johns (1995), and Ridgway and Dunn (2003). Recent analyses of eddies in the various regions include Qiu (2002), Ducet and LeTraon (2001), Tilburg *et al.* (2001), Goni and Wainer (2001), and Willson and Rees (2000).

The Mozambique/Agulhas Current system differs from the others because the South Indian Ocean subtropical gyre extends south of the Cape of Good Hope, and as a result the western boundary current cannot close the gyre within the Indian basin. According to linear Sverdrup theory, the western boundary current would extend across the South Atlantic to the coast of South America (Godfrey, 1989), but this is not realized because of the instability of the strong jet injected into the South Atlantic, combined with the background eastward flow around the Southern Ocean. The Agulhas Current retroflects, or turns around, developing an eddy at its western loop; this pinches off and the eddy usually drifts north-westward into the South Atlantic, accomplishing a large water property transport into that ocean, and producing the ISV maximum south of Africa in Figure 6.12. Agulhas eddies are discussed by Schouten *et al.* (2002) and references cited therein.

The Somali Current is distinguished from the other poleward western boundary currents by being present for only part of the year (see the May and August panels of Figure 6.10), a consequence of the fact that winds over the North Indian Ocean do not consist of permanent tropical trades and mid-latitude westerlies, and thus this basin does not have a permanent subtropical gyre. Nevertheless, during the intense summer monsoon winds, the western boundary current that arises along the Somali coast is comparable in speed and transport to the other major systems. ISV associated with the Somali Current is again due to instabilities as it turns offshore near 5°N, as well as further eddies near the tip of Somalia known as the Great Whirl and Socotra Eddy that are spun up with the boundary current (Figure 6.12). Schott and McCreary (2001) provide a comprehensive review of the Somali Current and its stationary and transient eddies.

ISV in the Southern Ocean

A strip of high ISV runs along the axis of the Antarctic Circumpolar Current (ACC; Figure 6.12), where meridional eddy fluxes have long been seen as a principal process of both its heat and momentum balance (Bryden, 1983). Eddy kinetic energy along this band is the most energetic in the world ocean. The poleward heat transport across the Southern Ocean required to maintain steady state is large, and because there are no meridional boundaries to support a gyre circulation, the eddies are the dominant mechanism by which this is accomplished (Gille, 2003). Zonal momentum from the strong and persistent westerly winds must be removed, presumably by vertical transport to be balanced by topographic drag, however, there has been considerable debate about whether eddy momentum fluxes accelerate or decelerate the mean flow (e.g., Morrow *et al.*, 1994; Hughes and Ash, 2001). The ACC appears to be broken up into interleaved jets that flow along quasi-permanent temperature and salinity fronts, some of which are evident at the surface (Belkin and Gordon, 1996). This complex structure tends to be anchored to topography, and much of the eddy variance is generated as the strong, nearly barotropic current flows over bottom features (Morrow *et al.*, 1992). Regions of intense eddy generation are visible in Figure 6.12, and include the ridges south of New Zealand, Drake Passage, and the mid-ocean ridges in the central Indian and Pacific Oceans.

Because of the extreme difficulty of working in these distant and dangerous waters, much of the early work on Southern Ocean eddies took place in Drake Passage where a decades-long observational program took place (Bryden, 1983), but since satellite altimetry has allowed the production of maps of SSH variability, statistics of the global picture have become possible (Morrow *et al.*, 1992, 1994; Gille and Kelly, 1996). Although the large-scale wind forcing might be expected to produce a large-scale response, Southern Ocean SSH variance is dominated by scales less than 1,000 km. Timescales were found to be about 34 days (Gille and Kelly, 1996). It is not known what produces this frequency structure.

ISV due to baroclinic instability

Currents on the westward limb (equatorial side) of the subtropical gyres (i.e., the North and South Equatorial Currents) in all the oceans typically present conditions favorable to the development of baroclinic instability. At the surface, especially in winter, temperatures are cooler with increasing latitude, but at thermocline level temperatures become warmer moving poleward into the deep bowls of the gyres. The resulting poleward tilt of the gyres with depth implies a change of sign in the vertical of the meridional gradient of potential vorticity, which is a necessary (though not sufficient) condition for such instabilities to grow (see Pedlosky, 1987, for a review of these dynamics). Linear stability analysis suggests that such disturbances should have the preferred size of 2π times the first baroclinic mode Rossby radius (a few hundred kilometers in the subtropics) and disperse as baroclinic Rossby waves; these characteristics will produce an intraseasonal timescale in subtropical time series (Qiu, 1999). Two such regions stand out in the high-passed SSH variability map of Figure 6.12: in the eastern Indian Ocean along 12°S, and in the western Pacific Ocean along 20°N. In both cases, although these are regions potentially influenced by intraseasonal wind and heat flux forcing, the prominent ISV maxima have been shown to be due to internal oceanic phenomena, unconnected to TISO.

The band of high ISV roughly along 12°S in the Indian Ocean (Figure 6.12) occurs along the axis of the westward SEC, which carries low-salinity Pacific water from the ITF into the central Indian Ocean. Feng and Wijffels (2002) showed that the high-variance signal shown in Figure 6.12 has a strong seasonal modulation, being larger by a factor of about two in July–September when the SEC itself is largest (as is the ITF which partly feeds it). They noted that this seasonality is inconsistent with the ISV forced by equatorial winds, which brings strong ISV to the Indonesian coast via Kelvin wave propagation, but peaks in January–June (see Section 6.6). Instead, the ISV observed along 12°S was found to be an internally-generated baroclinic instability of the SEC itself. Feng and Wijffels (2002) showed propagating sea level features, arising near 115°E, with a period of 40–80 days, a length-scale of 100–150 km, and a westward phase speed of 15–19 cm s^{-1}, consistent with the first-mode baroclinic Rossby wave speed in this region, and concluded that the observed ISV maximum was produced as an internal instability in the ocean.

A band of high ISV extends along 20°N from Hawaii to Taiwan (Figure 6.12), following the path of the eastward Subtropical Countercurrent (STCC) in the North Pacific. The STCC overlies the subsurface part of the westward North Equatorial Current (NEC) which extends poleward to the center of the gyre at thermocline level. As in the Indian Ocean signal discussed above, the 20°N ISV has a pronounced seasonal modulation, with its maximum in boreal spring about 50% larger than its boreal fall minimum (Qiu, 1999). Qiu showed that as the vertical shear between the STCC and the NEC increases in boreal winter, the vertical-meridional structure becomes baroclinically unstable and eddies develop. Roemmich and Gilson (2001) used temperature profiles along a ship track roughly at 20°N to show that their subsurface structure produced an overturning circulation and substantial poleward heat flux. The eddies propagate westward as baroclinic Rossby waves with a phase speed of about $10\,\mathrm{cm\,s^{-1}}$ and wavelength of about 500 km, thus a period of about 60 days. Although there is substantial intraseasonal forcing in this region, it occurs primarily in boreal summer, and the ISV maximum seen in Figure 6.12 appears to be due to internal ocean processes.

6.8 CONCLUSION

The literature reviewed here has shown that tropical intraseasonal forcing leads to a wide complex of dynamic and thermodynamic effects in the ocean, some of which have the potential to produce subsequent effects on the evolution of the tropical climate system. A principal driver of present enthusiasm for tropical oceanic ISV has been the possibility of oceanic rectification leading to a connection between the MJO and the ENSO cycle, but the recent progress in observing and modeling the eastern Indian Ocean and the development of the Asian monsoon suggests that some of the same mechanisms might be operating there as well.

The MJO/ENSO question revolves around the still-unquantified net effect of intraseasonally-oscillating forcing on the west Pacific warm pool. As reviewed in Sections 6.2 and 6.3, it is clear that the results of TISO heat, moisture, and wind forcing, profoundly affect the character and composition of the west Pacific warm pool, producing its commonly stacked velocity structure with several layers of reversing jets and its frequent salinity-stratified barrier layer. These dynamic and thermodynamic consequences are tied together because the result of precipitation is to enhance the response to wind forcing by concentrating it in a thin surface layer. While the possibilities for rectification are rife, the quantification of suggested mechanisms is hindered by the difficulty of modeling these processes, which depend sensitively on the mixed layer depth and the vertical structure of momentum mixing, among the least believable aspects of present-generation OGCMs. Indeed, similar models come to opposite conclusions about even the sign of some of the important rectified terms (e.g., the sign of the rectified current in the model of Kessler and Kleeman, 2000, compared to that of Waliser *et al.*, 2003). In addition, the paucity of salinity observations means that the spatial structure of the mixed and barrier layers is barely known.

A quantitative description of Indian Ocean ISV is only recently being realized through programs such as JASMINE (Webster *et al.*, 2002) and the maintenance of moored time series in the Bay of Bengal. Since the role of salinity stratification is likely to be even more intense than in the Pacific, we can expect that complex vertical structures, with the consequent creation of oceanic memory of intraseasonal forcing and resulting potential for feedback and rectification will turn out to be important in the evolution of Indian Ocean climate as well.

6.9 REFERENCES

Anderson, S. P., R. A. Weller, and R. B. Lukas (1996) Surface buoyancy forcing and the mixed layer of the western equatorial Pacific warm pool: Observations and 1S model results. *J. Climate*, **9**, 3056–3085.

Annamalai, H., R. Murtugudde, J. Potemra, S. P. Xie, P. Liu, and B. Wang (2003) Coupled dynamics over the Indian Ocean: Spring initiation of the Zonal Mode. *Deep-Sea Res. II*, **50**, 2305–2330.

Barnston, A. G., Y. He, and D. A. Unger (2000) A forecast product that maximizes utility for state-of-the-art seasonal climate prediction. *Bull. Amer. Meteorol. Soc.*, **81**, 1271–1290.

Baturin, N. G. and P. P. Niiler (1997) Effects of instability waves in the mixed layer of the equatorial Pacific. *J. Geophys. Res.*, **102**, 21771–21793.

Belamari, S., J.-L. Redelsperger, and M. Pontaud (2003) Dynamic role of a westerly wind burst in triggering an equatorial Pacific warm event. *J. Climate*, **16**, 1869–1890.

Belkin, I. M. and A. L. Gordon (1996) Southern Ocean fronts from the Greenwich Meridian to Tasmania. *J. Geophys. Res. Oceans*, **101**, 3675–3696.

Bhat, G. S., S. Gadgil, P. V. H. Kumar, S. R. Kalsi, P. Madhusoodanan, V. S. N. Murty, C. V. K. P. Rao, V. R. Babu, L. V. G. Rao, and R. R. Raos (2001) BOBMEX: The Bay of Bengal monsoon experiment. *Bull. Amer. Meteorol. Soc.,* **82**, 2217–2243.

Blanke, B. and S. Raynaud (1997) Kinematics of the Pacific equatorial undercurrent: An Eulerian and Lagrangian approach from GCM results. *J. Phys. Oceanogr.*, **27**, 1038–1053.

Bryden, H. L. (1983) The Southern Ocean. In: A. R. Robinson (ed.), *Eddies in Marine Science*, Springer-Verlag, New York, pp. 265–277.

Bryden, H. L. and E. C. Brady (1989) Eddy momentum and heat fluxes and their effects on the circulation of the equatorial Pacific Ocean. *J. Mar. Res.*, **47**, 55–79.

Busalacchi, A. J. and M. A. Cane (1985) Hindcasts of sea level variations during the 1982–83 El Niño. *J. Phys. Oceanogr.*, **15**, 213–221.

Chelton, D. B., S. K. Esbensen, M. G. Schlax, N. Thum, M. H. Freilich, F. J. Wentz, C. L. Gentemann, M. J. McPhaden, and P. S. Schopf (2001) Observations of coupling between surface wind stress and sea surface temperature in the eastern tropical Pacific. *J. Climate*, **14**, 1479–1498.

Chelton, D. B., M. H. Freilich, and S. K. Esbensen (2000a) Satellite observations of the wind jets off the Pacific coast of Central America. Part I: Case studies and statistical characteristics. *Mon. Wea. Rev.*, **128**, 1993–2018.

Chelton, D. B., M. H. Freilich, and S. K. Esbensen (2000b) Satellite observations of the wind jets off the Pacific coast of Central America. Part II: Regional relationships and dynamical considerations. *Mon. Wea. Rev.*, **128**, 2019–2043.

Chelton, D. B., F. J. Wentz, C. L. Gentemann, R. A. deSzoeke, and M. G. Schlax (2000c) Satellite microwave SST observation of transequatorial Tropical Instability Waves. *Geophys. Res. Lett.*, **27**, 1239–1242.

Chelton, D. B., M. G. Schlax, J. M. Lyman, and G. C. Johnson (2003) Equatorially trapped Rossby waves in the presence of meridionally sheared baroclinic flow in the Pacific Ocean. *Prog. Oceanogr.*, **56**, 323–380.

Clarke, A. J. (1994) Why are surface equatorial ENSO winds anomalously westerly under anomalous large-scale convection? *J. Climate*, **7**, 1623–1627.

Clarke, A. J. and R. Ahmed (1999) Dynamics of remotely forced intraseasonal oscillations off the western coast of South America. *J. Phys. Oceanogr.*, **29**, 240–258.

Contreras, R. F. (2002) Long-term observations of tropical instability waves. *J. Phys. Oceanogr.*, **32**, 2715–2722.

Cox, M. D. (1980) Generation and propagation of 30-day waves in a numerical model of the Pacific. *J. Phys. Oceanogr.*, **10**, 1168–1186.

Cravatte, S., J. Picaut, and G. Eldin (2003) Second and first baroclinic modes in the equatorial Pacific at intraseasonal timescales. *J. Geophys. Res.*, **108**, 3226, doi:10.1029/2002JC001511.

Cronin, M. F. and M. J. McPhaden (1997) The upper ocean heat balance in the western equatorial Pacific warm pool during September–December 1992. *J. Geophys. Res.*, **102**, 8533–8553.

Cronin, M. F. and M. J. McPhaden (2002) Barrier layer formation during westerly wind bursts. *J. Geophys. Res.*, **107**, 8020, doi:10.1029/2001JC001171.

Cronin, M. F., M. J. McPhaden, and R. H. Weisberg (2000) Wind-forced reversing jets in the western equatorial Pacific. *J. Phys. Oceanogr.*, **30**, 657–676.

Delcroix, T., G. Eldin, M. J. McPhaden, and A. Morliere (1993) Effects of westerly wind bursts upon the western equatorial Pacific Ocean, February–April 1991. *J. Geophys. Res.*, **98**, 16379–16385.

Ducet, N. and P. Y. LeTraon (2001) A comparison of surface eddy kinetic energy and Reynolds stresses in the Gulf Stream and the Kuroshio Current systems from merged TOPEX/Poseidon and ERS-1/2 altimetric data. *J. Geophys. Res.*, **106**, 16603–16662.

Durland, T. S. and B. Qiu (2003) Transmission of subinertial Kelvin waves through a strait. *J. Phys. Oceanogr.*, **33**, 1337–1350.

Eigenheer, A. and D. Quadfasel (2000) Seasonal variability of the Bay of Bengal circulation inferred from TOPEX/Poseidon altimetry. *J. Geophys. Res.*, **105**, 3243–3252.

Eldin, G., T. Delcroix, C. Henin, K. Richards, Y. duPenhoat, J. Picaut, and P. Rual (1994) Large-scale current and thermohaline structures along 156° during the COARE intensive observing period. *Geophys. Res. Lett.*, **21**, 2681–2684.

Enfield, D. B. (1987) The intraseasonal oscillation in eastern Pacific sea levels: How is it forced? *J. Phys. Oceanogr.*, **17**, 1860–1876.

Enfield, D. B. and R. Lukas (1983) Low-frequency sea level variability along the South American coast in 1981–83. *Trop. Ocean-Atmos. Newslett.*, **28**, 2–4.

Fedorov, A. V. (2002) The response of the coupled tropical ocean–atmosphere to westerly wind bursts. *Quart. J. Royal Meteorol. Soc.*, **128**, 1–23.

Feng, M. and S. Wijffels (2002) Intraseasonal variability in the South Equatorial Current of the east Indian Ocean. *J. Phys. Oceanogr.*, **32**, 265–277.

Fink, A. and P. Speth (1997) Some potential forcing mechanisms of the year-to-year variability of the tropical convection and its intraseasonal (25–70-day) variability. *Int. J. Climatol.*, **17**, 1513–1534.

Flament, P. J., S. C. Kennan, R. A. Knox, P. P. Niiler, and R. L. Bernstein (1996) The three-dimensional structure of an upper-ocean vortex in the tropical Pacific Ocean. *Nature*, **383**, 610–613.

Fu, L.-L., E. J. Christensen, C. A. Yamarone, M. Lefebvre, Y. Menard, M. Dorrer, and P. Escudier (1994) TOPEX/POSEIDON mission overview. *J. Geophys. Res.*, **99**, 24369–24382.

Giese, B. S. and D. E. Harrison (1990) Aspects of the Kelvin wave response to episodic wind forcing. *J. Geophys. Res. Oceans*, **95**, 7289–7312.

Giese, B. S. and D. E. Harrison (1991) Eastern equatorial Pacific response to three composite westerly wind types. *J. Geophys. Res.*, **96**, 3239–3248.

Giese, B. S., J. A. Carton, and L. J. Holl (1994) Sea level variability in the eastern Pacific as observed by TOPEX and Tropical Ocean Global–Atmosphere Tropical Atmosphere–Ocean experiment. *J. Geophys. Res.*, **99**, 24739–24748.

Gille, S. T. (2003) Float observation of the Southern Ocean. Part II: Eddy fluxes. *J. Phys. Oceanogr.*, **33**, 1182–1196.

Gille, S. T. and K. A. Kelly (1996) Scales of spatial and temporal variability in the Southern Ocean. *J. Geophys. Res.*, **101**, 8759–8773.

Godfrey, J. S. (1989) A Sverdrup model of the depth-integrated flow for the world ocean, allowing for island circulations. *Geophys. Astrophys. Fluid Dyn.*, **45**, 89–112.

Godfrey, J. S., R. A. Houze, R. H. Johnson, R. Lukas, J.-L. Redelsperger, A. Sumi, and R. Weller (1998) Coupled Ocean-Atmosphere Response Experiment (COARE): An interim report. *J. Geophys. Res.*, **103**, 14395–14450.

Goni, G. J. and I. Wainer (2001) Investigation of the Brazil Current front variability from altimeter data. *J. Geophys. Res.*, **106**, 31117–31128.

Gould, J., D. Roemmich, S. Wijffels, H. Freeland, M. Ignaszewsky, X. Jianping, S. Pouliquen, Y. Desaubies, U. Send, and K. Radhakrishman (2004) Argo profiling floats bring new era of in situ ocean observations. *EOS*, **85**, 179–191.

Gutzler, D. S. (1991) Interannual fluctuations of intraseasonal variance of near-equatorial zonal winds. *J. Geophys. Res.*, **96**, 3173–3185.

Halpern, D., R. A. Knox, and D. S. Luther (1988) Observations of 20-day period meridional current oscillations in the upper ocean along the Pacific equator. *J. Phys. Oceanogr.*, **18**, 1514–1534.

Han, W., J. P. McCreary, D. L. T. Anderson, and A. J. Mariano (1999) On the dynamics of the eastward surface jets in the equatorial Indian Ocean. *J. Phys. Oceanogr.*, **29**, 2191–2209.

Han, W., D. M. Lawrence, and P. J. Webster (2001a) Dynamical response of equatorial Indian Ocean to intraseasonal winds: Zonal flow. *Geophys. Res. Lett.*, **28**, 4215–4218.

Han, W., J. P. McCreary, and K. E. Kohler (2001b) Influence of precipitation minus evaporation and Bay of Bengal rivers on dynamics, thermodynamics, and mixed layer physics in the upper Indian Ocean. *J. Geophys. Res.*, **106**, 6895–6916.

Hansen, D. V. and G. A. Maul (1991) Anticyclonic current rings in the eastern tropical Pacific Ocean. *J. Geophys. Res.*, **96**, 6965–6979.

Hansen, D. V. and C. A. Paul (1984) Genesis and effect of long waves in the equatorial Pacific. *J. Geophys. Res.*, **89**, 10431–10440.

Hayes, S. P., L. J. Mangum, J. Picaut, A. Sumi, and K. Takeuchi (1991) TOGA-TAO: A moored array for real-time measurements in the tropical Pacific Ocean. *Bull. Amer. Meteorol. Soc.*, **72**, 339–347.

Hendon, H. H. and M. L. Salby (1994) The life cycle of the Madden–Julian Oscillation. *J. Atmos. Sci.*, **51**, 2225–2231.

Hendon, H. H., B. Liebmann, and J. D. Glick (1998) Oceanic Kelvin waves and the Madden–Julian Oscillation. *J. Atmos. Sci.*, **55**, 88–101.

Hendon, H. H., C. Zhang, and J. D. Glick (1999) Interannual variability of the Madden–Julian Oscillation during austral summer. *J. Climate*, **12**, 2358–2550.

Hisard, P., J. Merle, and B. Voituriez (1970) The equatorial undercurrent observed at 170°E in March and April 1967. *J. Mar. Res.*, **28**, 281–303.

Hogg, N. G. and W. E. Johns (1995) Western boundary currents. *Rev. Geophys.*, **33**, 1311–1334.

Hormazabal, S., G. Shaffer, J. Letelier, and O. Ulloa (2001) Local and remote forcing of sea surface temperature in the coastal upwelling system off Chile. *J. Geophys. Res. Oceans*, **106**(C8), 16657–16671.

Hormazabal, S., G. Shaffer, and O. Pizarro (2002) Tropical Pacific control of intraseasonal oscillations off Chile by way of oceanic and atmospheric pathways. *Geophys. Res. Lett.*, **29**, Art 1081.

Hughes, C. W. and E. R. Ash (2001) Eddy forcing of the mean flow in the Southern Ocean. *J. Geophys. Res.*, **106**, 2713–2722.

Hurd, W. E. (1929) Northers of the Gulf of Tehuantepec. *Mon. Wea. Rev.*, **57**, 192–194.

Jensen, T. G. (1993) Equatorial variability and resonance in a wind-driven Indian Ocean model. *J. Geophys. Res.*, **98**, 22533–22552.

Johnson, E. S. and D. S. Luther (1994) Mean zonal momentum balance in the upper and central equatorial Pacific Ocean. *J. Geophys. Res.*, **99**, 7689–7705.

Johnson, E. S. and M. J. McPhaden (1993a) Structure of intraseasonal Kelvin waves in the equatorial Pacific Ocean. *J. Phys. Oceanogr.*, **23**, 608–625.

Johnson, E. S. and M. J. McPhaden (1993b) Effects of a 3-dimensional mean flow on intraseasonal Kelvin waves in the equatorial Pacific Ocean. *J. Geophys. Res. Oceans*, **98**, 10185–10194.

Johnson, G. C., M. J. McPhaden, and E. Firing (2001) Equatorial Pacific Ocean horizontal velocity, divergence and upwelling. *J. Phys. Oceanogr.*, **31**, 839–849.

Johnson, G. C., B. M. Sloyan, W. S. Kessler, and K. E. McTaggart (2002) Direct measurements of upper ocean currents and water properties across the equatorial Pacific during the 1990s. *Prog. Oceanogr.*, **52**, 31–61.

Kessler, W. S. (2001) EOF representations of the Madden–Julian Oscillation and its connection with ENSO. *J. Climate*, **14**, 3055–3061.

Kessler, W. S. (2002a) Mean three-dimensional circulation in the northeast tropical Pacific. *J. Phys. Oceanogr.*, **32**, 2457–2471.

Kessler, W. S. (2002b) Is ENSO a cycle or a series of events? *Geophys. Res. Lett.*, **29**, 2125, doi:1029/2002GL015924.

Kessler, W. S. and R. Kleeman (2000) Rectification of the Madden–Julian Oscillation into the ENSO cycle. *J. Climate*, **13**, 3560–3575.

Kessler, W. S. and M. J. McPhaden (1995a) Oceanic equatorial waves and the 1991–93 El Niño. *J. Climate*, **8**, 1757–1774.

Kessler, W. S. and M. J. McPhaden (1995b) The 1991–93 El Niño in the central Pacific. *Deep-Sea Res., II*, **42**, 295–333.

Kessler, W. S., G. C. Johnson, and D. W. Moore (2003) Sverdrup and nonlinear dynamics of the Pacific equatorial currents. *J. Phys. Oceanogr.*, **33**, 994–1008.

Kessler, W. S., M. J. McPhaden, and K. M. Weickmann (1995) Forcing of intraseasonal Kelvin waves in the equatorial Pacific. *J. Geophys. Res.*, **100**, 10613–10631.

Kessler, W. S., L. M. Rothstein, and D. Chen (1998) The annual cycle of SST in the eastern tropical Pacific, diagnosed in an ocean GCM. *J. Climate*, **11**, 777–799.

Kindle, J. C. and P. A. Phoebus (1995) The ocean response to operational wind bursts during the 1991–92 El Niño. *J. Geophys. Res.*, **100**, 4803–4920.

Kindle, J. C. and J. D. Thompson (1989) The 26- and 50-day oscillations in the western Indian Ocean: Model results. *J. Geophys. Res.*, **94**, 4721–4736.

Kutsuwasa, K. and M. J. McPhaden (2002) Intraseasonal variations in the upper equatorial Pacific Ocean prior to and during the 1997–98 El Niño. *J. Phys. Oceanogr.*, **32**(4), 1133–1149.

Lagerloef, G. S. E., G. T. Mitchum, R. B. Lukas, and P. P. Niiler (1999) Tropical Pacific near-surface currents estimated from altimeter, wind and drifter data. *J. Geophys. Res.*, **104**, 23313–23326.

Latif, M., J. Biercamp, and H. von Storch (1988) The response of a coupled ocean–atmosphere general circulation model to wind bursts. *J. Atmos. Sci.*, **45**, 964–979.

Legeckis, R. (1977) Long waves in the eastern equatorial ocean: A view from a geostationary satellite. *Science*, **197**, 1179–1181.

Lengaigne, M., J. P. Boulanger, C. Menkes, S. Masson, G. Madec, and P. Delecluse (2002) Ocean response to the March 1997 westerly wind event. *J. Geophys. Res. (O)*, **107**(C12): art. no. 8015.

Levitus, S. (1994) Climatological Atlas of the World Ocean, NOAA Prof. Pap. No. 13. US Government Printing Office.

Loschnigg, J. and P. J. Webster (2000) A coupled ocean-atmosphere system of SST modulation for the Indian Ocean. *J. Climate*, **13**, 3342–3360.

Lukas, R. and E. Lindstrom (1991) The mixed layer of the western equatorial Pacific. *J. Geophys. Res.*, **96**, 3343–3357.

Lukas, R., S. P. Hayes, and K. Wyrtki (1984) Equatorial sea level response during the 1982–1983 El Niño. *J. Geophys. Res.*, **89**, 10425–10430.

Luther, D. S. and E. S. Johnson (1990) Eddy energetics in the upper equatorial Pacific during the Hawaii-to-Tahiti Shuttle Experiment. *J. Phys. Oceanogr.*, **20**, 913–944.

Luyten, J. R. and D. H. Roemmich (1982) Equatorial currents at semi-annual period in the Indian Ocean. *J. Phys. Oceanogr.*, **12**, 406–413.

Madden, R. A. and P. A. Julian (1994) Observations of the 40–50-day tropical oscillation – A review. *Mon. Wea. Rev.*, **122**, 814–837.

Mariano, A. J., E. H. Ryan, B. D. Perkins, and S. Smithers (1995) The mariano global surface velocity analysis 1.0. U.S. Coast Guard Technical Report, CG-D-34-95.

Masina, S. and S. G. H. Philander (1999) An analysis of tropical instability waves in a numerical model of the Pacific Ocean. Part 1: Spatial variability of the waves. *J. Geophys. Res.*, **104**, 29613–29635.

Masina, S., S. G. H. Philander, and A. B. G. Bush (1999) An analysis of tropical instability waves in a numerical model of the Pacific. Ocean. Part 2: Generation and energetics of the waves. *J. Geophys. Res.*, **104**, 29637–29661.

Masson, S., P. Delecluse, J.-P. Boulanger, and C. Menkes (2002) A model study of the seasonal variability and formation mechanisms of the barrier layer in the eastern equatorial Indian Ocean. *J. Geophys. Res.*, **107**, 8017, doi:10.1029/2001JC000832.

Masson, S., P. Delecluse, and J.-P. Boulanger (2003) Impacts of salinity on the eastern Indian Ocean during the termination of the fall Wyrtki Jet. *J. Geophys. Res.*, **108**, doi:10.1029/2001JC00083.

McClain, C. R., J. R. Christian, S. R. Signorinin, M. R. Lewis, I. Asanuma, D. Turk, and C. Dupouy-Douchement (2002) Satellite ocean color observations of the tropical Pacific Ocean. *Deep-Sea Res. II*, **49**, 2533–2560.

McCreary, J. P. (1984) Equatorial beams. *J. Mar. Res.*, **42**, 395–430.

McCreary, J. P. (1985) Modeling equatorial ocean circulation. *Ann. Rev. Fluid Mech.*, **17**, 359–409.

McCreary, J. P., P. K. Kundu, and R. L. Molinari (1993) A numerical investigation of dynamics, thermodynamics and mixed-layer processes in the Indian Ocean. *Prog. Oceanogr.*, **31**, 181–244.

McCreary, J. P., H. S. Lee, and D. B. Enfield (1989) Response of the coastal ocean to strong offshore winds: With application to circulations in the Gulf of Tehuantepec and Papagayo. *J. Mar. Res.*, **47**, 81–109.

McPhaden, M. J. (1995) The Tropical Atmosphere–Ocean array is completed. *Bull. Amer. Meteorol. Soc.*, **76**, 739–741.

McPhaden, M. J. (1996) Monthly period oscillations in the Pacific North Equatorial Countercurrent. *J. Geophys. Res.*, **101**, 6337–6360.

McPhaden, M. J. (1999) Genesis and evolution of the 1997–98 El Niño. *Science*, **283**, 950–954.

McPhaden, M. J. (2002) Mixed layer temperature balance on intraseasonal timescales in the equatorial Pacific Ocean. *J. Climate*, **15**, 2632–2647.

McPhaden, M. J. and S. P. Hayes (1991) On the variability of winds, sea surface temperature and surface layer heat content in the western equatorial Pacific. *J. Geophys. Res.*, **96**(suppl.), 3331–3342.

McPhaden, M. J. and B. A. Taft (1988) Dynamics of seasonal and intraseasonal variability in the eastern equatorial Pacific. *J. Phys. Oceanogr.*, **18**, 1713–1732.

McPhaden, M. J., H. P. Freitag, S. P. Hayes, B. A. Taft, Z. Chen, and K. Wyrtki (1988) The response of the equatorial Pacific Ocean to a westerly wind burst in May 1986. *J. Geophys. Res.*, **93**, 10589–10603.

McPhaden, M. J., S. P. Hayes, L. J. Mangum, and J. M. Toole (1990) Variability in the western equatorial Pacific Ocean during the 1986–87 El Niño/Southern Oscillation event. *J. Phys. Oceanogr.*, **20**, 190–208.

McPhaden, M. J., F. Bahr, Y. duPenhoat, E. Firing, S. P. Hayes, P. P. Niiler, P. L. Richardson, and J. M. Toole (1992) The response of the western equatorial Pacific Ocean to westerly wind bursts during Novermber 1989 to January 1990. *J. Geophys. Res.*, **97**, 14289–14303.

McPhaden, M. J. and X. Yu (1999) Equatorial waves and the 1997–98 El Niño. *Geophys. Res. Lett.*, **26**, 2961–2964.

Metzger, E. J., H. E. Hurlburt, J. C. Kindle, Z. Sirkes, and J. M. Pringle (1992) Hindcasting of wind-driven anomalies using a reduced gravity global ocean model. *Mar. Tech. Soc. J.*, **26**, 23–32.

Meyers, G., J.-R. Donguy, and R. Reed (1986) Evaporative cooling of the western equatorial Pacific. *Nature*, **312**, 258–260.

Miyama, T., J. P. McCreary, T. G. Jensen, J. Loschnigg, S. Godfrey, and A. Ishida (2003) Structure and dynamics of the Indian Ocean cross-equatorial cell. *Deep-Sea Res., II*, **50**, 2023–2047.

Moore, A. M. and R. Kleeman (1999) Stochastic forcing of ENSO by the intraseasonal oscillation. *J. Climate*, **12**, 1199–1220.

Moore, A. M. and R. Kleeman (2001) The differences between the optimal perturbations of coupled models of ENSO. *J. Climate*, **14**, 138–163.

Moore, D. W. and J. P. McCreary (1990) Excitation of intermediate-frequency equatorial waves at a western ocean boundary: With application to observations from the Indian Ocean. *J. Geophys. Res.*, **95**, 5219–5231.

Morrow, R., J. Church, R. Coleman, D. Chelton, and N. White (1992) Eddy momentum flux and its contribution to the Southern Ocean momentum balance. *Nature*, **357**, 482–484.

Morrow, R., R. Coleman, J. Church, and D. Chelton (1994) Surface eddy momentum flux and velocity variances in the Southern Ocean from Geosat altimetry. *J. Phys. Oceanogr.*, **24**, 2050–2071.

Müller-Karger, F. E. and C. Fuentes-Yaco (2000) Characteristics of wind-generated rings in the eastern tropical Pacific Ocean. *J. Geophys. Res.*, **105**, 1271–1284.

Murtugudde, R. and A. J. Busalacchi (1999) Interannual variability of the dynamics and thermodynamics of the tropical Indian Ocean. *J. Climate*, **12**, 2300–2326.

Murtugudde, R., J. P. McCreary, and A. J. Busalacchi (2000) Oceanic processes associated with anomalous events in the Indian Ocean with relevance to 1997–1998. *J. Geophys. Res.*, **105**, 3295–3306.

Musman, S. (1989) Sea height wave form in equatorial waves and its interpretation. *J. Geophys. Res.*, **94**, 3303–3309.

Neelin, J. D., D. S. Battisti, A. C. Hirst, F.-F. Jin, Y. Wakata, T. Yamagata, and S. E. Zebiak (1998) ENSO theory. *J. Geophys. Res.*, **103**, 14261–14290.

Olson, D. B. (1991) Rings in the ocean. *Ann. Rev. Earth Planet Sci.*, **19**, 283–311.

Pedlosky, J. (1987) *Geophysical Fluid Dynamics* (2nd edition). Springer-Verlag, New York. 710 pp.

Perigaud, C. (1990) Sea level oscillations observed with Geosat along the two shear fronts of the Pacific North Equatorial Countercurrent. *J. Geophys. Res.*, **95**, 7239–7248.

Philander, S. G. H. (1976) Instabilities of zonal equatorial currents. *J. Geophys. Res.*, **81**, 3725–3735.

Philander, S. G. H. (1978) Instabilities of zonal equatorial currents, 2. *J. Geophys. Res.*, **83**, 3679–3682.

Philander, S. G. H. and R. C. Pacanowski (1980) The generation of equatorial currents. *J. Geophys. Res.*, **85**, 1123–1136.

Philander, S. G. H., W. J. Hurlin, and R. C. Pacanowski (1986) Properties of long equatorial waves in models of the seasonal cycle in the tropical Atlantic and Pacific Oceans. *J. Geophys. Res.*, **91**, 14207–14211.

Potemra, J. T., S. L. Hautala, J. Sprintall, and W. Pandoe (2002) Interaction between the Indonesian Seas and the Indian Ocean in observations and numerical models. *J. Phys. Oceanogr.*, **32**, 1838–1854.

Potemra, J. T., S. L. Hautala, and J. Sprintall (2003) Vertical structure of the Indonesian Throughflow in a large-scale model. *Deep-Sea Res. II*, **50**, 2143–2161.

Potemra, J. T., M. E. Luther, and J. J. O'Brien (1991) The seasonal circulation of the upper ocean in the Bay of Bengal. *J. Geophys. Res.*, **96**, 12667–12683.

Qiao, L. and R. H. Weisberg (1998) Tropical instability wave energetics: Observations from the Tropical Instability Wave Experiment. *J. Phys. Oceanogr.*, **28**, 345–360.

Qiu, B. (1999) Seasonal eddy field modulation of the North Pacific Subtropical Countercurrent: TOPEX/Poseidon observations and theory. *J. Phys. Oceanogr.*, **29**, 2471–2486.

Qiu, B. (2002) The Kuroshio Extension system: Its large-scale variability and role in the midlatitude ocean–atmosphere interaction. *J. Oceanogr.*, **58**, 57–75.

Qiu, B., M. Mao, and Y. Kashino (1999) Intraseasonal variability in the Indo-Pacific Throughflow and the regions surrounding the Indonesian Seas. *J. Phys. Oceanogr.*, **29**, 1599–1618.

Ralph, E. A., K. Bi, P. P. Niiler, and Y. duPenhoat (1997) A Lagrangian description of the western equatorial Pacific response to the wind burst of December 1992: Heat advection in the warm pool. *J. Climate*, **10**, 1706–1721.

Ramanathan, V. and W. Collins (1991) Thermodynamic regulation of ocean warming by cirrus clouds deduced from observations of the 1987 El Niño. *Nature*, **351**, 27–32.

Raymond, D. J. (2001) A new model of the Madden–Julian Oscillation. *J. Atmos. Sci.*, **58**, 2807–2819.

Reppin, J., F. A. Schott, J. Fischer, and D. Quadfasel (1999) Equatorial currents and transports in the upper central Indian Ocean: Annual cycle and interannual variability. *J. Geophys. Res.*, **104**, 15495–15514.

Reverdin, G. (1987) The upper equatorial Indian Ocean: The climatological seasonal cycle. *J. Phys. Oceanogr.*, **17**, 903–927.

Reverdin, G. and J. Luyten (1986) Near-surface meanders in the equatorial Indian Ocean. *J. Phys. Oceanogr.*, **16**, 1088–1100.

Richardson, R. A., I. G. Ginnis, and L. M. Rothstein (1999) A numerical investigation of the local ocean response to westerly wind burst forcing in the western equatorial Pacific. *J. Phys. Oceanogr.*, **29**, 1334–1352.

Ridgway, K. R. and J. R. Dunn (2003) Mesoscale structure of the mean East Australian Current System and its relationship with topography. *Prog. Oceanogr.*, **56**, 189–222.

Roden, G. I. (1961) On the wind-driven circulation in the Gulf of Tehuantepec and its effect on surface temperatures. *Geofis. Int.*, **1**, 55–72.

Roemmich, D. and J. Gilson (2001) Eddy transport of heat and thermocline waters in the North Pacific: A key to interannual/decadal climate variability? *J. Phys. Oceanogr.*, **31**, 675–687.

Roemmich, D., M. Morris, W. R. Young, and J. R. Donguy (1994) Fresh equatorial jets. *J. Phys. Oceanogr.*, **24**, 540–558.

Roulston, M. S. and J. D. Neelin (2000) The reponse of an ENSO model to climate noise, weather noise and intraseasonal forcing. *Geophys. Res. Lett.*, **27**, 3723–3726.

Saji, N. H., B. N. Goswami, P. N. Vinayachandran, and T. Yamagata (1999) A dipole mode in the tropical Indian Ocean. *Nature*, **401**, 360–363.

Schiller, A. and J. S. Godfrey (2003) Indian Ocean intraseasonal variability in an ocean general circulation model. *J. Climate*, **16**, 21–39.

Schneider, N. (1998) The Indonesian Throughflow and the global climate system. *J. Climate*, **11**, 676–689.

Schott, F. A. and J. P. McCreary (2001) The monsoon circulation of the Indian Ocean. *Prog. Oceanogr.*, **51**, 1–123.

Schott, F., J. Reppin, and J. Fischer (1994) Currents and transports of the Monsoon Current south of Sri Lanka. *J. Geophys. Res.*, **99**, 25127–25141.

Schouten, M. W., W. P. M. de Ruijter, and P. J. van Leeuwen (2002) Upstream control of Agulhas ring shedding. *J. Geophys. Res.*, **107**, doi:10.1029/2001JC000804.

Sengupta, D., R. Senan, and B. N. Goswami (2001) Origin of intraseasonal variability of circulation in the tropical central Indian Ocean. *Geophys. Res. Lett.* **28**, 1267–1270.

Shetye, S. R., A. D. Gouveia, D. Shankar, S. S. C. Shenoi, P. N. Vinayachandran, D. Sundar, G. S. Michael, and G. Nampoothiri (1996) Hydrography and circulation in the western Bay of Bengal during the northeast monsoon. *J. Geophys. Res.*, **101**, 14011–14025.

Shinoda, T. and H. H. Hendon (1998) Mixed layer modeling of intraseasonal variability in the tropical western Pacific and Indian Oceans. *J. Climate*, **11**, 2668–2685.

Shinoda, T. and H. H. Hendon (2001) Upper-ocean heat budget in response to the Madden–Julian oscillation in the western equatorial Pacific. *J. Climate* **14**(21), 4147–4165.

Shinoda, T. and H. H. Hendon (2002) Rectified wind forcing and latent heat flux produced by the Madden–Julian Oscillation. *J. Climate*, **15**(23), 3500–3508.

Shinoda, T., H. H. Hendon, and J. D. Glick (1998) Intraseasonal variability of surface fluxes and sea surface temperature in the tropical western Pacific and Indian Oceans. *J. Climate*, **11**, 1685–1702.

Slingo, J. M., D. P. Rowell, K. R. Sperber, and F. Nortley (1999) On the predictability of the inter annual behavior of the Madden–Julian Oscillation and its relationship with El Niño. *Quart. J. Roy. Met. Soc.*, **125**, 583–609.

Somayajulu, Y. K., V. S. N. Murty, and Y. V. B. Sarma (2003) Seasonal and interannual variability of surface circulation in the bay of Bengal from TOPEX/Poseidon altimetry. *Deep-Sea Res. II*, **50**, 867–880.

Spillane, M. C., D. B. Enfield, and J. S. Allen (1987) Intraseasonal oscillations in sea level along the west coast of the Americas. *J. Phys. Oceanogr.*, **17**, 313–325.

Sprintall, J. and M. J. McPhaden (1994) Surface layer variations observed in multiyear time series measurements from the western equatorial Pacific. *J. Geophys. Res.*, **99**, 963–979.

Sprintall, J., J. T. Potemra, S. L. Hautala, N. A Bray, and W. W. Pandoe (2003) Temperature and salinity variability in the exit passages of the Indonesian Throughflow. *Deep-Sea Res., II*, **50**, 2183–2204.

Sprintall, J. and M. Tomczak (1992) Evidence of the barrier layer in the surface layer of the tropics. *J. Geophys. Res.*, **97**, 7305–7316.

Stammer, D. (1997) Global characteristics of ocean variability estimated from regional TOPEX/Poseidon altimeter measurements. *J. Phys. Oceanogr.*, **27**, 1743–1769.

Stommel, H. M. (1960) Wind-drift near the equator. *Deep-Sea Res.*, **6**, 298–302.

Swallow, J. C. (1983) Eddies in the Indian Ocean. In: A. R. Robinson (ed.), *Eddies in Marine Science*, Springer-Verlag, New York, pp. 200–218.

Swenson, M. S. and D. V. Hansen (1999) Tropical Pacific Ocean mixed layer heat budget: The Pacific cold tongue. *J. Phys. Oceanogr.*, **29**, 69–82.

Tilberg, C. E., H. E. Hurlburt. J. J. O'Brien, and J. F. Shriver (2001) The dynamics of the East Australian Current system: The Tasman Front, the East Auckland Current, and the East Cape Current. *J. Phys. Oceanogr.*, **31**, 2917–2943.

Tilburg C. E., H. E. Hurlburt, J. J. O'Brien, and J. F. Shriver (2002) The dynamics of the East Australian Current system: The Tasman Front, the East Auckland Current, and the East Cape Current. *J. Phys. Oceanogr.*, **31**, 2917–2943.

Tomczak, M. and J. S. Godfrey (1994) *Regional Oceanography: An Introduction*. Pergamon Press, Oxford, 422 pp.

Trasviña, A. (1995) Offshore wind forcing in the Gulf of Tehuantepec, Mexico: The asymmetric circulation. *J. Geophys. Res.*, **96**, 12599–12618.

Tsai, P. T. H., J. J. O'Brien, and M. E. Luther (1992) The 26-day oscillation observed in satellite sea surface temperature measurements in the equatorial western Indian Ocean. *J. Geophys. Res.*, **97**, 9605–9618.

van Oldenborgh, G. J. (2000) What caused the onset of the 1997–98 El Niño? *Mon. Wea. Rev.*, **128**, 2601–2607.

Vecchi, G. A. and D. E. Harrison (2000) Tropical Pacific sea surface temperature anomalies, El Niño, and westerly wind bursts. *J. Climate*, **13**, 1814–1830.

Vinayanchandran, P. N. and T. Yamagata (1998) Monsoon response of the sea around Sri Lanka: Generation of thermal domes and anticyclonic vortices. *J. Phys. Oceanogr.*, **28**, 1946–1960.

Waliser, D. E., R. Murtugudde, and L. E. Lucas (2003) Indo-Pacific ocean response to atmospheric intraseasonal variability. Part I: Austral summer and the Madden–Julian Oscillation. *J. Geophys. Res.*, **108**, 3160, doi:10.1029/2003JC001620.

Waliser, D. E., R. Murtugudde, and L. E. Lucas (2004) Indo-Pacific ocean response to atmospheric intraseasonal variability. Part 2: Boreal summer and the intraseasonal oscillation. *J. Geophys. Res.*, **109**, C03030, doi:10.1029/2003JC002002.

Wang, W. and M. J. McPhaden (1999) The surface-layer heat budget in the equatorial Pacific Ocean. Part I: Mean seasonal cycle. *J. Phys. Oceanogr.*, **29**, 1812–1831.

Webster, P. J. and R. Lukas (1992) TOGA-COARE: The coupled ocean–atmosphere response experiment. *Bull. Amer. Meteor. Soc.*, **73**, 1377–1416.

Webster, P. J., E. F. Bradley, C. W. Fairall, J. S. Godfrey, P. Hacker, R. A. Houze, R. Lukas, Y. Serra, J. M. Hummon, T. D. M. Lawrence *et al.* (2002) The JASMINE pilot study. *Bull. Amer. Meteorol. Soc.*, **83**, 1603–1630.

Webster, P. J., A. M. Moore, J. P. Loschnigg, and R. R. Leben (1999) Coupled ocean–atmosphere dynamics in the Indian Ocean during 1997–98. *Nature*, **401**, 356–360.

Weidman, P. D., D. L. Mickler, B. Dayyani, and G. H. Born (1999) Analysis of Legeckis eddies in the near-equatorial Pacific. *J. Geophys. Res.*, **104**, 7865–7887.

Weisberg, R. H., A. M. Horigan, and C. Colin (1979) Equatorially-trapped variability in the equatorial Atlantic. *J. Mar. Res.*, **37**, 67–86.

Weisberg, R. H. and T. Y. Tang (1983) Equatorial ocean response to growing and moving wind systems, with application to the Atlantic. *J. Mar. Res.*, **41**, 461–486.

Weisberg, R. H. and C. Wang (1997) Slow variability in the equatorial west-central Pacific in relation to ENSO. *J. Climate*, **10**, 1998–2017.

Weller, R. A. and S. P. Anderson (1996) Surface meteorology and air–sea fluxes in the western equatorial Pacific warm pool during the TOGA coupled ocean–atmosphere response experiment. *J. Climate*, **9**, 1959–1990.

Willson, H. R. and N. W. Rees (2000) Classification of mesoscale features in the Brazil–Falkland Current confluence zone. *Prog. Oceanogr.*, **45**, 415–426.

Wu, L., Z. Liu, and H. E. Hurlburt (2000) Kelvin wave and Rossby wave interaction in the extratropical–tropical Pacific. *Geophys. Res. Lett.*, **27**, 1259–1262.

Wyrtki, K. (1973) An equatorial jet in the Indian Ocean. *Science*, **181**, 262–264.

Yoshida, K. (1959) A theory of the Cromwell Current (the equatorial undercurrent) and of equatorial upwelling. *J. Oceanogr. Soc. Jap.*, **15**, 159–170.

Yu, L. S. (2003) Variability of the depth of the 20°C isotherm along 6°N in the Bay of Bengal: Its response to remote and local forcing and its relation to satellite SSH variability. *Deep-Sea Res.-II*, **50**, 2285–2304.

Yu, Z. J., J. P. McCreary, and J. A. Proehl (1995) Meridional asymmetry and energetics of tropical instability waves. *J. Phys. Oceanogr.*, **25**, 2997–3007.

Yu, Z. J. and P. S. Schopf (1997) Vertical eddy mixing in the tropica upper ocean: Its influence on zonal currents. *J. Phys. Oceanogr.*, **27**, 1447–1458.

Zebiak, S. E. (1989) On the 30–60 day oscillation and the prediction of El Niño. *J. Climate*, **2**, 1381–1387.

Zebiak, S. E. and M. A. Cane (1987) A model El Niño-Southern Oscillation. *Mon. Wea. Rev.*, **115**, 2262–2278.

Zhang, C. (1996) Atmospheric intraseasonal variability at the surface in the tropical western Pacific Ocean. *J. Atmos. Sci.*, **53**, 739–758.

Zhang, C. (1997) Intraseasonal variability of the upper-ocean thermal structure observed at 0° and 165°E. *J. Climate*, **10**, 3077–3092.

Zhang, C. (2001) Intraseasonal perturbations in sea surface temperatures of the equatorial eastern Pacific and their association with the Madden–Julian Oscillation. *J. Climate*, **14**(6), 1309–1322.

Zhang, C. D. and J. Gottschalck (2002) SST anomalies of ENSO and the Madden–Julian oscillation in the equatorial Pacific. *J. Climate*, **15**, 2429–2445.

Zhang, C. and M. J. McPhaden (2000) Intraseasonal surface cooling in the equatorial western Pacific. *J. Climate*, **13**, 2261–2276.

Zhang, C., H. H. Hendon, W. S. Kessler, and A. J. Rosati (2001) A workshop on the MJO and ENSO: Meeting summary. *Bull. Amer. Meteorol. Soc.*, **82**, 971–976.

Zhang, K. Q. and L. M. Rothstein (1998) Modeling the oceanic response to westerly wind bursts in the western equatorial Pacific. *J. Phys. Oceanogr.*, **28**, 2227–2249.

7

Air–sea interaction

Harry Hendon

7.1 INTRODUCTION

Air–sea interaction associated with tropical intraseasonal variability (ISV) and, particularly, the Madden–Julian Oscillation (MJO) is of interest for three reasons. First, variations of the air–sea fluxes of heat and moisture may be fundamental to mechanisms of tropical ISV. For instance, air–sea interaction may promote the slow eastward propagation of the MJO and its northward propagation in the Indian summer monsoon. Besides playing a critical role for the interplay between convection and dynamics, surface fluxes of heat, moisture, and momentum drive sea surface temperature (SST) perturbations that may feedback to the surface fluxes and ultimately to the atmospheric dynamics, thus, for instance, contributing to the growth of the MJO. Second, the episodic variations of surface momentum, heat, and freshwater fluxes driven by atmospheric ISV may play a role in the maintenance and low-frequency variability of the warm pool in the tropical Indian and Pacific Oceans. For example, the MJO induces transports in the equatorial west Pacific that act in the mean to remove about the same amount of heat from the warm pool as is provided by the mean surface heat flux (Ralph *et al.*, 1997). From the opposite perspective of the ocean driving the atmosphere, interannual variations of SST in the warm pool may also drive interannual variations in MJO activity, which may bear on the ability to predict seasonal variations of MJO activity. Third, the MJO forces surface currents that drive SST variations at the eastern edge of the warm pool (e.g., Kessler *et al.*, 1995). Kelvin waves are also efficiently excited by the MJO (e.g., Hendon *et al.*, 1998), which radiate into the eastern Pacific where they can perturb the SST (e.g., Giese and Harrison, 1991; Zhang, 2001; McPhaden, 2002). These intraseasonal SST variations may lead to a rectified coupled-response, which plays a role in the evolution of the El Niño Southern Oscillation (ENSO) (e.g., Bergman *et al.*, 2001; Zhang and Gottschalck, 2002).

W. K. M. Lau and D. E. Waliser (eds), *Intraseasonal Variability in the Atmosphere–Ocean Climate System*.

The focus in this chapter is on observations of air–sea interaction that may be relevant to the mechanism and variability of the MJO. The interactions relevant to the evolution of ENSO are only briefly discussed but are covered in detail in Chapters 6 and 9. Diagnostic studies of the response in the upper ocean to intraseasonal surface flux variations, which provides insight into the processes that control the associated SST variability in the warm pool, are also covered. Theory and modeling studies of the role of air–sea interaction for the mechanism for the MJO are touched upon briefly but are treated in detail in Chapters 10 and 11.

7.2 AIR–SEA FLUXES FOR THE EASTWARD MJO

The evolution of surface heat, moisture, and momentum fluxes associated with the eastward propagating MJO has been described in the western Pacific using satellite and in situ observations (e.g., Zhang, 1996; Lau and Sui, 1997; Zhang and McPhaden, 2000) and across the entire warm pool using global analyses (e.g., Flatau *et al.*, 1997; Hendon and Glick, 1997; Jones *et al.*, 1998; Shinoda *et al.*, 1998; Woolnough *et al.*, 2000). The basic structure of the fluxes is schematically illustrated in Figure 7.1. The abscissa (in kilometers) spans one-half the wavelength of the MJO, which is an equivalent duration of ~ 30 days at a given point as the MJO systematically propagates eastward. This schematic is typical for the MJO (with some minor shifts in phasing as discussed below) across the entire warm pool of the Indian and western Pacific Oceans.

In the convectively active phase, which has a zonal extent of about 8,000 km and meridional width of about 2,000 km, increased deep convection is associated with increased cloud cover, increased rainfall, and decreased surface insolation. Enhanced deep convection slightly leads (by ~ 1 week) enhanced surface westerlies. In the warm pool region where the basic state wind is weak westerly, these anomalous westerlies act to increase the surface windspeed, hence increasing the flux of latent and sensible heat. In the convectively suppressed phase, which has a slightly larger zonal extent (and longer duration) than the active phase, reduced westerlies act to reduce the windspeed, thus reducing the latent and sensible heat flux. Decreased convection is also associated with decreased cloud cover and increased surface insolation.

The observed phase lag (~ 1 week) of latent heat flux with respect to enhanced convection is counter to that assumed in some simple "quasi-equlibrium" models of the MJO, whereby enhanced latent heat flux to the east of enhanced convection acts to intensify convection to the east (e.g., Emanuel, 1987). Such models presume an easterly basic state so that easterly anomalies in advance of enhanced convection act to increase the windspeed and latent heat flux. While such simple theory is at odds with the observed phase lag of the latent heat flux with respect to convection, a positive impact of wind-induced latent heat flux variations has been demonstrated in some models (e.g., Neelin *et al.*, 1987; Lin *et al.*, 2000; Raymond, 2001; Colón *et al.*, 2002).

Magnitudes of the fluxes at the extrema of a large MJO event are indicated in Figure 7.1. The sensible heat flux anomaly is not shown, as it is about 1/10th the size

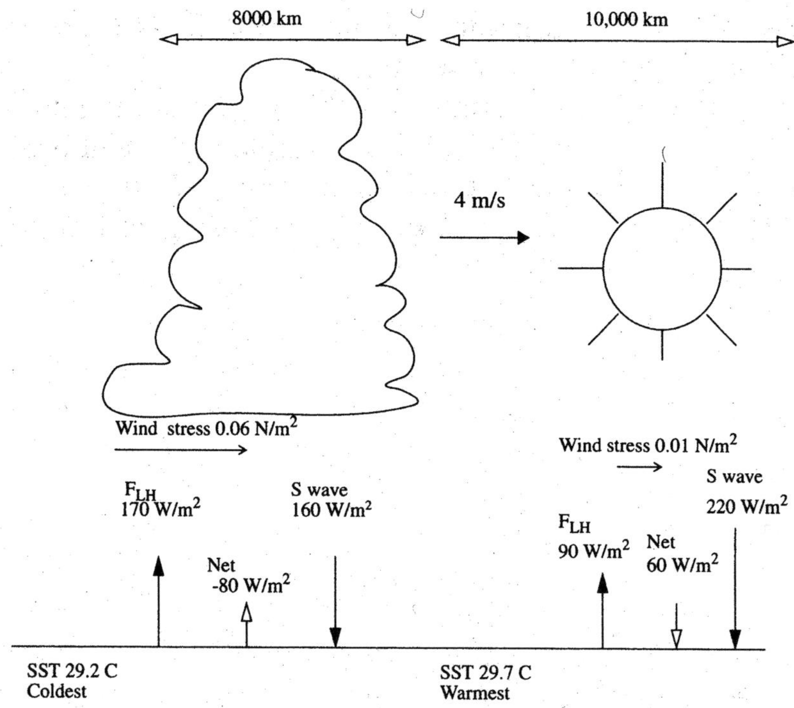

Figure 7.1. Schematic diagram showing magnitude and phase relationship relative to the convective anomaly of the surface fluxes and SST variations produced by the canonical MJO. The asymmetric zonal scale of the cloudy-windy and suppressed-calm phases and eastward phase speed ($4\,\mathrm{m\,s^{-1}}$) of the joint atmosphere–ocean disturbance across the warm pool are indicated. Typical extrema of surface fluxes and SST over the life cycle of the MJO are shown for the western Pacific.
From Shinoda *et al.* (1998).

of the latent heat flux anomaly and has similar phasing. The net surface long-wave radiation (also not shown) tends to oppose the surface insolation anomaly but is much smaller. Hence, the surface heat flux variation is dominated by the insolation and latent heat flux variations.

Enhanced convection and associated enhanced cloud cover typically reduce surface insolation by about 20–40 $\mathrm{W\,m^{-2}}$. Shortly thereafter ($\sim$7 day lag), the latent heat flux, in association with enhanced surface westerlies, peaks with similar magnitude. This near collocation implies that the latent heat flux and insolation anomaly act together to produce a peak surface cooling of 40–80 $\mathrm{W\,m^{-2}}$, which propagates coherently eastward across the equatorial warm pool in conjunction with the convection and surface westerly anomalies. In the suppressed convective phase, insolation is increased by 10–30 $\mathrm{W\,m^{-2}}$ and the reduced surface westerlies act to decrease the latent heat flux by 10–30 $\mathrm{W\,m^{-2}}$. Together the insolation and latent heat flux anomalies produce a peak surface warming of about 20–60 $\mathrm{W\,m^{-2}}$ during the suppressed phase. Note that in this analysis, the peak amplitude of the warming during the suppressed convective phase is weaker than the cooling during the enhanced convective phase. However, the suppressed phase is of longer duration (greater zonal extent) than the enhanced phase. Hence, averaged over a life cycle the net heat flux is weakly positive into the warm pool.

As the convective anomaly approaches the dateline, induced easterly anomalies east of the convection act to increase the windspeed and latent heat flux (e.g., Hendon and Glick, 1997; Woolnough *et al.*, 2000) because the mean (trade) winds are easterly in the eastern Pacific. Because enhanced latent heat flux now occurs east of enhanced convection (i.e., in a region of enhanced insolation) and the convective anomalies weaken as the MJO propagates into the eastern Pacific (hence producing weaker surface insolation anomalies), the surface heat flux perturbation is generally small east of the dateline. However, during northern summer, the MJO perturbs convection in the Intertropical Convergence Zone (ITCZ) north of the equator and coherent surface heat flux perturbations of similar phasing and magnitude to those in the warm pool are observed in the eastern Pacific (e.g., Maloney and Kiehl, 2002).

The MJO also perturbs the freshwater and surface momentum fluxes. Over the warm pool, Shinoda *et al.* (1998) estimate rainfall to increase by 5–7 $\mathrm{mm\,day^{-1}}$ during the enhanced convective phase and decrease by a similar amount during the calm-suppressed convective phase. Similarly, westerly stress increases to about $0.05\,\mathrm{N\,m^{-2}}$ during the westerly-convective phase and decreases to about $0.01\,\mathrm{N\,m^{-2}}$ during the suppressed convective phase. A hallmark of the MJO is that the maximum (westerly) stress nearly coincides with the highest flux of freshwater into the ocean. Hence, their individual influences on the buoyancy forcing of the warm pool mixed layer will tend to cancel (e.g., Anderson *et al.*, 1996; Zhang and Anderson, 2003).

The phasing and relative magnitude of the fluxes in Figure 7.1 are observed to vary systematically as the MJO traverses the Indian and western Pacific Oceans. When convection is developing in the Indian Ocean, the convective anomaly tends to be near the node of the surface zonal wind anomaly (e.g., Hendon and Salby, 1994), which is consistent with a Gill-type response to equatorially-symmetric diabatic heating. As the convection moves into the western Pacific, the surface zonal wind anomaly shifts eastward relative to the convection so that anomalous westerlies blow entirely through the region of anomalous convection. Hendon and Salby (1994) interpreted this changing phase as indicative of the evolution of the energetics of the MJO through its lifecycle. However, a simple dynamical explanation is that the convective anomaly tends to be equatorially-centered in the Indian Ocean but shifts off the equator into the South Pacific Convergence Zone (SPCZ) in the west Pacific. Hence, the phasing of the surface zonal wind relative to convection in the Indian Ocean is consistent with an equatorial symmetric heat source. In the western Pacific, however, the phasing of the surface zonal winds is consistent with an equatorial asymmetric heat source displaced into the southern hemisphere.

The changing phase of the surface zonal winds relative to the convective anomaly implies that the phasing of the surface fluxes will change as the active convective phase propagates eastward from the Indian Ocean. In the western

Pacific, the surface insolation anomaly will be more in phase with the latent heat flux anomaly because the surface westerlies coincide with increased convection. But, in the Indian Ocean, there will be less cooperation. Shinoda *et al.* (1998) further diagnosed the latent heat flux anomaly to contribute less to the total surface heat flux anomaly in the Indian Ocean. Hence, the surface heat flux variation in the Indian Ocean is dominated by the insolation variation whereas the latent heat and insolation anomalies make equal contributions in the western Pacific.

7.3 AIR–SEA FLUXES ASSOCIATED WITH NORTHWARD PROPAGATION IN THE INDIAN SUMMER MONSOON

During boreal summer, the eastward propagating MJO is notably weaker (e.g., Salby and Hendon, 1994), while the dominant intraseasonal mode exhibits pronounced northward propagation ($\sim 1^\circ$ lat day^{-1}) from the equatorial Indian Ocean into the Indian monsoon (e.g., Sikka and Gadgil, 1980; Wang and Rui, 1990; Kemball-Cook and Wang, 2001). The dominant period is shorter as well (35 days as compared to 50 days during winter). There is some debate as to whether northward propagation occurs independent of the eastward propagating MJO along the equator (e.g., Lawrence and Webster, 2002; Jiang *et al.*, 2004). Nonetheless, the northward propagating events during boreal summer significantly perturb the surface fluxes of heat, moisture, and momentum. The associated coherent SST anomalies are indicative of robust air–sea interaction (e.g., Bhat *et al.*, 2001; Sengupta and Ravichandran, 2001; Vecchi and Harrison, 2002; Webster *et al.*, 2002). While dynamical mechanisms for the poleward propagation away from the equator, which involve emitted Rossby waves, have been suggested (Chapter 10), air–sea interaction appears to foster the northward propagation in much the same fashion as it fosters eastward propagation along the equator (e.g., Fu *et al.*, 2003).

Figure 7.2 displays the structure of the air–sea fluxes for the typical northward propagating intraseasonal oscillation (ISO) in the Indian Ocean sector during boreal summer. The abscissa spans about 20–25° latitude, representative of a section of the Indian Ocean from the equator to 20–25°N. Equivalently, a span of about 20 days is displayed at a given point in the Indian Ocean as the oscillation propagates to the north.

To the north of the developing convective anomaly, reduced convection acts to increase the surface insolation. In the suppressed region, anomalous easterlies act on the westerly basic state of the summer monsoon, thereby reducing the windspeed and latent heat flux. During the convective phase, ennhanced convection acts to decrease insolation, while anomalous westerlies act on the westerly basic state to increase the windspeed and latent heat flux. Typical magnitudes of these intraseasonal variations are $\sim 5\,\mathrm{m\,s^{-1}}$ for the zonal wind, and $\sim 25\,\mathrm{W\,m^{-2}}$ for the latent heat flux and insolation. These anomalies can increase by a factor of 3 to 4 for individual events (e.g., Webster *et al.*, 2002).

As for the eastward propagating MJO, the maximum latent heat flux anomaly occurs slightly after the maximum convection, which again brings into question the

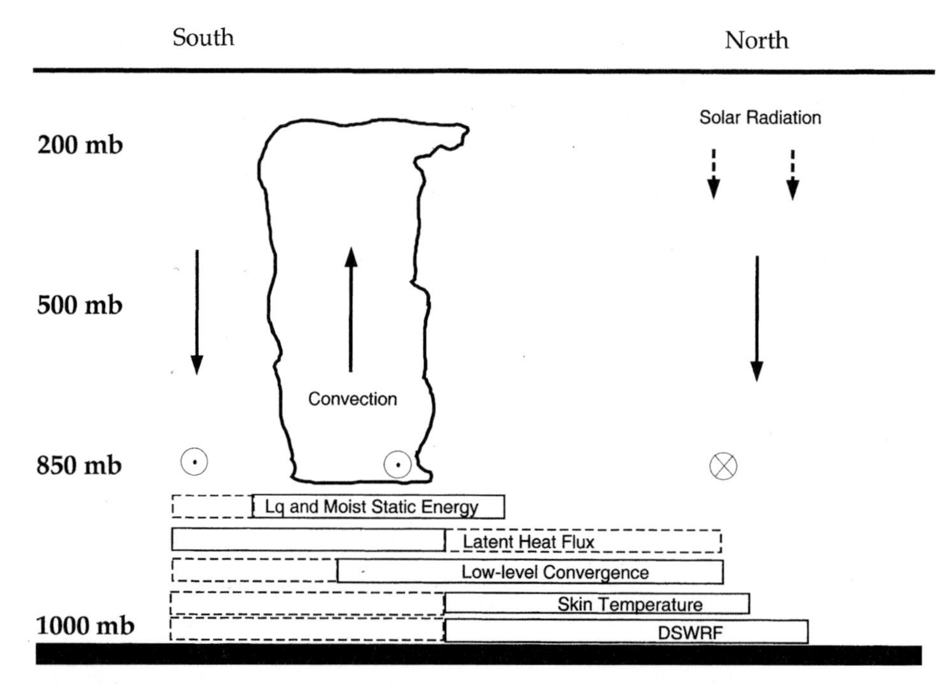

Figure 7.2. Schematic of air–sea interaction in the northward propagation of convective anomalies associated with the ISO during boreal summer in the Indian and western Pacific Oceans. Dark vertical lines indicate the mid-troposphere vertical velocity anomaly. The cloud indicates deep precipitating convection. The boxes represent the approximate locations of anomalies relative to the convection. The solid box indicates a positive anomaly, and the dashed box indicates a negative anomaly. Circles indicate direction of 850-mb zonal wind anomaly with the $\odot(\oplus)$ representing easterlies (westerlies).
From Kemball-Cook and Wang (2001).

relevance of simple "quasi-equilibrium" theories for explaining the intraseasonal behavior during boreal summer. The insolation anomaly slightly leads the latent heat flux anomaly, but they still add constructively to perturb the net surface heat flux, which can result in significant SST perturbations.

7.4 SST VARIABILITY

For the ocean to play an active role in the dynamics of the eastward propagating MJO and the northward propagating variability in the Indian monsoon, the surface flux variations must induce a SST variation. Krishnamurti *et al.* (1988), motivated by the need to explain the long timescale of the MJO, were the first to examine SST variability associated with the MJO. Using data from the First GARP Global Experiment (FGGE) year, they showed that intraseasonal (30–60-day period) SST

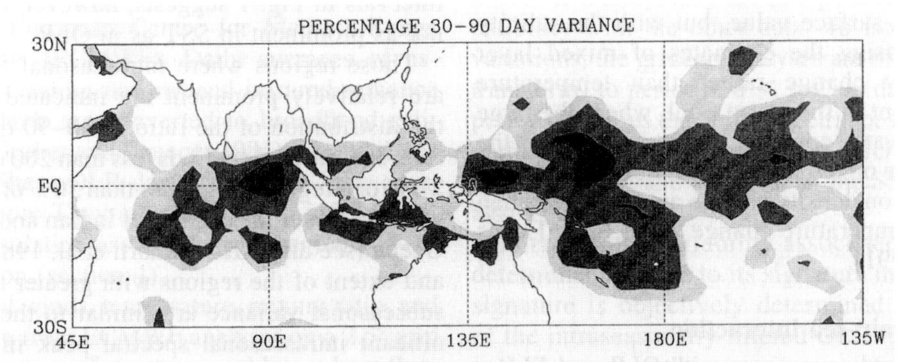

Figure 7.3. Ratio (percentage) of intraseasonal (30–90-day period) SST variance to the total sub-seasonal (10–200-day period) SST variance for the period 1 July, 1986–30 June, 1993. Shading levels begin at 45%, 50%, and 60%.
From Hendon and Glick (1997).

variability was most prominent across the equatorial Indian and western Pacific Oceans and had temporal phasing indicative of the atmosphere forcing the ocean on the intraseasonal timescale (i.e., minimum SST lagged maximum surface westerlies by $\sim 1/4$ of a cycle). Hendon and Glick (1997), using seven years of weekly SST analyses from Reynolds and Smith (1994), confirmed their results. Figure 7.3 displays the ratio (as a percentage) of intraseasonal (30–90-day period) to sub-seasonal (10–200-day period) SST variance. The regions in the Indian and western Pacific with a ratio greater than 50% are also regions where SST variability exhibits a significant spectral peak near 60 days (Zhang, 1996; Hendon and Glick, 1997). These regions are also where the signal in convection and surface winds associated with the MJO is the strongest (e.g., Salby and Hendon, 1994). Note that the ISV in the northern Indian Ocean may be underestimated in the satellite-based SST data used to create Figure 7.3 due to retrieval problems in regions of persistent precipitating convection (Harrison and Vecchi, 2001; Sengupta and Ravichandran, 2001).

For the eastward propagating MJO, warm SST anomalies develop after passage of the calm-suppressed phase, when the surface heat flux is most positive into the ocean, and cold SST anomalies develop after passage of the windy-convective phase, when the surface heat flux is most negative (Figure 7.1). The typical SST anomaly has an amplitude $\sim 1/3$ K (Flatau *et al.*, 1997; Hendon and Glick, 1997; Zhang, 1997; Shinoda *et al.*, 1998; Woolnough *et al.*, 2000; Kemball-Cook and Wang, 2001), although strong MJO events often produce SST swings of more than 1 K in the western equatorial Pacific (e.g., Weller and Anderson, 1996).

Pronounced SST anomalies in the Indian Ocean are also observed during boreal summer when the ISO is propagating northward. To the north of the developing convective anomaly in the equatorial Indian Ocean, clear skies (enhanced insolation) and reduced latent heat flux (easterly anomalies) act to warm the Arabian Sea and Bay of Bengal (e.g., Kimball-Cook and Wang, 2001; Vecchi and Harrison, 2002). During the convective phase, increased cloud cover reduces insolation and anomalous westerlies act to increase the windspeed and latent heat flux, thereby producing surface cooling. Hence, a warm SST anomaly leads the northward propagating convective anomaly by 1–2 weeks (1/4 of a cycle) and a cold SST anomaly follows the convective anomaly by a similar lag. The magnitude of these SST anomalies, especially in the Bay of Bengal (1–3 K), can be much larger than for the near-equatorial SST anomalies associated with the eastward propagating MJO (e.g., Sengutpa and Ravichandran, 2001). The freshness of the mixed layer in the Bay of Bengal, where mean precipitation and river runoff is high, results in a shallower and more stably stratified mixed layer with a deeper barrier layer than in the western equatorial Pacific (e.g., Bhat *et al.*, 2001; Sengupta and Ravichandran, 2001; Webster *et al.*, 2002). Hence, the mixed layer in the Bay of Bengal remains shallow in the presence of stronger winds and is more sensitive to the intraseasonal surface heat flux variations.

7.5 MECHANISMS OF SST VARIABILITY

Understanding how SST and the upper ocean mixed layer vary intraseasonally is important both for successful coupled simulation and for validation of air–sea interaction theories of ISV. The focus in this section is on mechanisms of near-equatorial SST variability associated with the eastward MJO. Much less work has been done on the mechanisms of SST variability in the northern Indian Ocean associated with northward propagating ISOs during boreal summer, probably due to lack of quality observations of the upper ocean and surface meteorology. However, recent field campaigns (e.g., JASMINE and BOBMEX) and new satellite SST products (e.g., Vecchi and Harrison, 2002) have revealed some similarities and differences with the behavior of the near-equatorial warm pool, which are commented on at the end of this section.

The near-equatorial warm pool is a region of weak winds and horizontal SST gradient, deep thermocline (~ 100 m), and a shallow (~ 25 m) fresh (stable) mixed layer, which overlays a deeper, more saline isothermal layer (e.g., Lukas and Lindstrom, 1991). The layer between the halocline, which typically defines the base of the mixed layer, and the thermocline is referred to as the barrier layer, because it effectively shields the mixed layer from colder sub-thermocline water. These conditions of weak horizontal temperature gradient and strong vertical stability mean that 1-D processes primarily govern the intraseasonal SST variations in the warm pool. The relatively weak wind fluctuations produced by the MJO are typically not sufficient to mix through the barrier layer. Hence the SST variation associated with the MJO is primarily accounted for by surface heat flux variation (Anderson *et al.*, 1996; Flatau *et al.*, 1997; Hendon and Glick, 1997; Lau and Sui, 1997; Jones *et al.*, 1998; Shinoda and Hendon, 1998; Woolnough *et al.*, 2000; Schiller and Godfrey, 2002; Zhang and Anderson, 2003). The largest heat flux out of the ocean occurs at about the time of maximum convection (Figure 7.1), resulting in a minimum SST about 1/4 of a cycle later (about two weeks for the typical MJO event). Similarly, the largest heat flux into the ocean occurs at the time of most suppressed convection, resulting in a warmest SST about 1/4 of a cycle prior to onset of deep convection.

Detailed observations in the western equatorial Pacific during the Tropical Ocean Global Atmosphere–Coupled Ocean Atmosphere Response Experiment (TOGA–COARE) (Webster and Lukas, 1992), however, suggest a complex evolution of the upper ocean through the lifecycle of the MJO. TOGA–COARE ran from November 1992–February 1993. The experiment was fortunate in that two major MJO events traversed the Intensive Flux Array (IFA) located in the equatorial western Pacific (roughly spanning 2°N–2°S, 152°S–157°E) in December 1992 and January 1993 (e.g., Gutzler *et al.*, 1994). Hourly observations of surface fluxes and SST at the IMET mooring (located at 1.45°S, 156°E; Weller and Anderson, 1996) are shown in Figure 7.4 (Anderson *et al.*, 1996). During the suppressed convective phase, SST gradually warms in the presence of a positive surface heat flux and weak surface wind. A pronounced diurnal cycle of SST is evident. During the convective phase, when the winds strengthen and become westerly, the surface heat flux is negative and SST rapidly cools with an absence of a diurnal cycle. The westerlies at this time are seen to be highly variable, which is a hallmark of the so-called westery wind bursts that tend to occur within the large-scale envelope of enhanced convection during the MJO (e.g., Hendon and Liebmann, 1994). These wind bursts could potentially play an important role in the evolution of the warm pool mixed layer because the wind stirring of the mixed layer does not simply increase linearly with the wind speed.

High-resolution modeling of the mixed layer indicates that 1-D processes govern this SST behavior provided that accurate surface fluxes are utilized (Figure 7.4; see also Shinoda and Hendon, 1998). However, for this TOGA–COARE period at the Improved Meteorology (IMET) site, a cumulative systematic bias is apparent (Figure 7.4), which is only accounted for by considering horizontal and vertical advection (e.g., Cronin and McPhaden, 1997; Feng *et al.*, 1998; Feng *et al.*, 2000). Except within the oceanic equatorial radius of deformation ($\sim$2–3° latitude; e.g., Ralph *et al.*, 1997; Feng *et al.*, 2000; Shinoda and Hendon, 2001; Schiller and Godfrey, 2003), horizontal and vertical advection appear not to be coherent on the scale of the MJO. Hence, advection does not play a systematic role in governing the meridionally broad SST variation associated with the MJO (Shinoda and Hendon, 1998, 2001). While advective processes are clearly important at some places for individual events, they depend critically on the initial state of the upper ocean (e.g., pre-existing anomalous horizontal temperature gradient) and the atmospheric noise (the non-spatially coherent circulation). Thus, the magnitude and sign of the advective tendencies varies significantly from one MJO event to another. Systematic advection in the near equatorial belt is returned to shortly.

The mixed layer depth does vary systematically over the lifecycle of the MJO (Hendon and Glick, 1997; Shinoda and Hendon, 1998), which will affect the sensitivity of the mixed layer to surface heat flux variations and the amount of short-wave radiation that penetrates through the mixed layer. Mixed layer deepening is also indicative of entrainment, which may act to additionally cool or warm the mixed layer depending on the stratification.

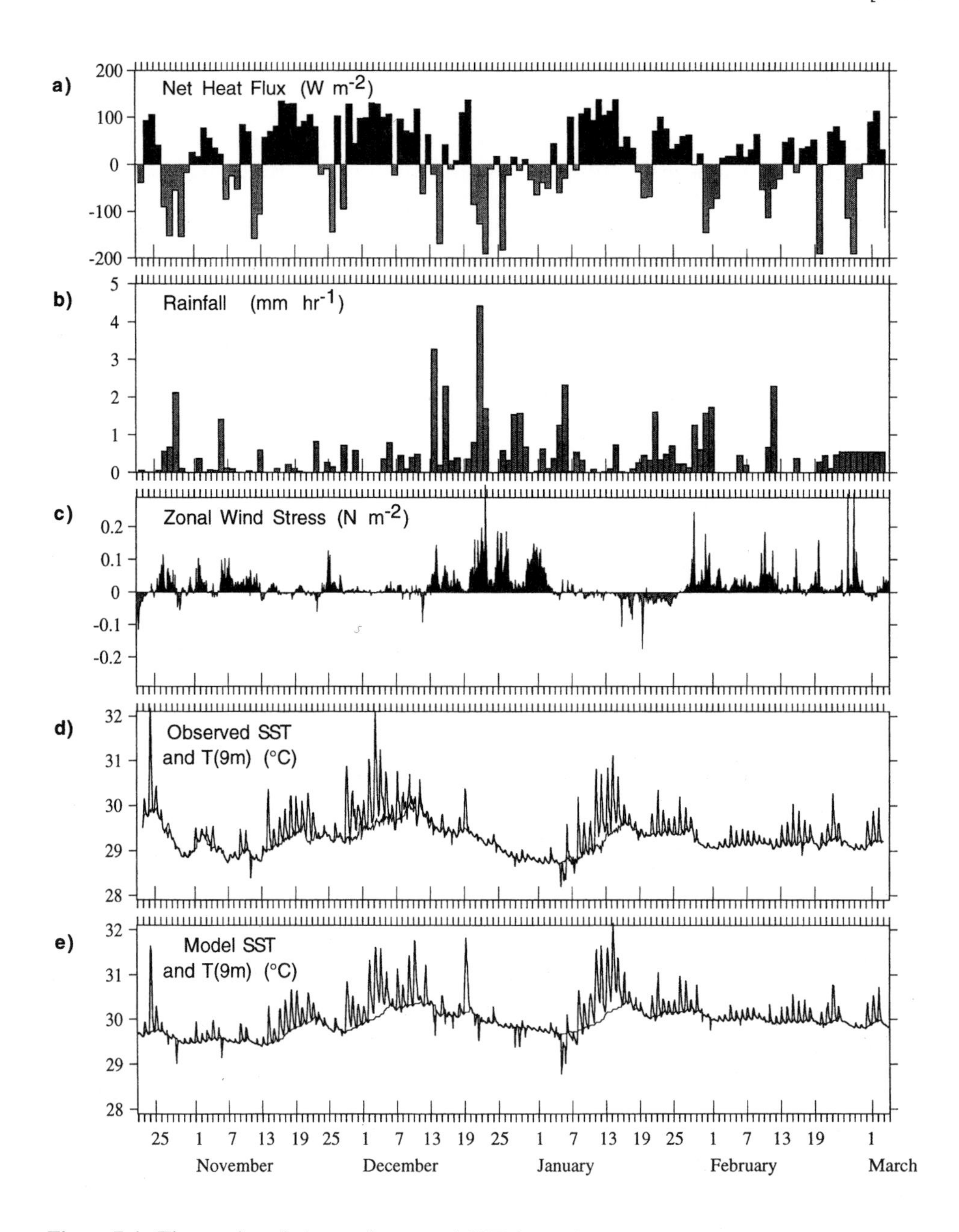

Figure 7.4. Time series of air–sea fluxes and SST from the WHOI–IMET mooring located in the IFA during TOGA–COARE (25 October, 1992–1 March, 1993). The net heat flux and rainfall were averaged over 24 hours. The wind stress and observed SST were smoothed over 2 hours. The modeled SST was estimated using a 1-D mixed-layer model forced with the observed surface fluxes from the IMET mooring.
From Anderson *et al.* (1996).

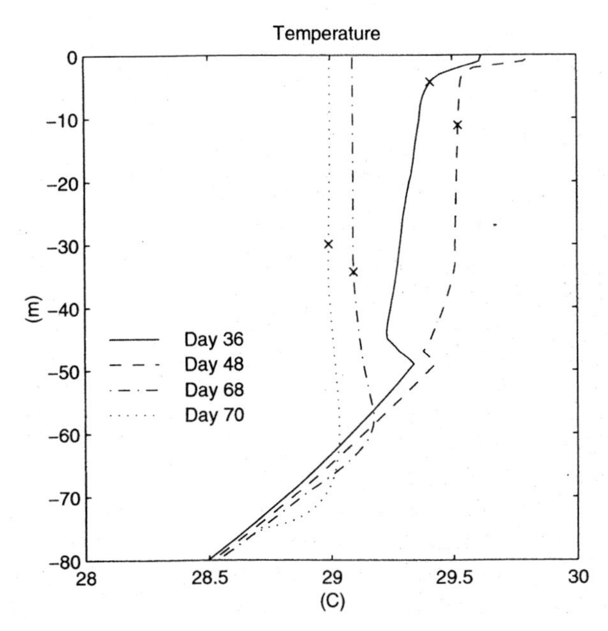

Figure 7.5. Daily averages temperature profile simulated in the equatorial western Pacific with a 1-D mixed layer model driven with observed surface fluxes for the MJO event that traversed the IFA in late December 1992 (see Figure 7.4). The crosses indicate the mixed layer depth each day. Day 36 corresponds to 26 November, 1992 (suppressed convective phase), and day 70 corresponds to 30 December, 1992 (post convective phase).
From Shinoda and Hendon (1998).

The typical evolution of the ocean mixed layer from the suppressed-calm phase to the convective-windy phase of the MJO is illustrated in Figure 7.5. During the calm-clear phase of the MJO (day 36–48), which corresponds to 26 November–8 December, 1992, during COARE, a shallow mixed layer forms above a deep barrier layer in response to a positive surface heat flux and small turbulent mixing. A strong diurnal cycle of mixed layer depth and temperature develops (Figure 7.6), with night-time convection acting to spread the intense daytime warming to a deeper layer. The shallowness of the mixed layer during this warming phase (<10 m) also results in significant ($>10\,\mathrm{W\,m^{-2}}$) penetration of short-wave radiation through the base of the mixed layer. Absorption of this radiation below the mixed layer results in a stable inversion owing to the freshness of the mixed layer (Lukas and Lindstrom, 1991; Sprintall and Tomczak, 1992; Anderson *et al.*, 1996; Shinoda and Hendon, 1998; Schiller and Godfrey, 2002; Zhang and Anderson, 2003).

During the cloudy-windy phase (days 68–70, which corresponds to 26–28 December, 1992), the surface heat flux is negative, the mixed layer deepens, and the diurnal cycle of radiation is weak. The negative surface heat flux initially cools the mixed layer more than the sub-mixed-layer and the mixed layer begins to deepen, eroding the barrier layer. Entrainment of the sub-mixed-layer water into the mixed

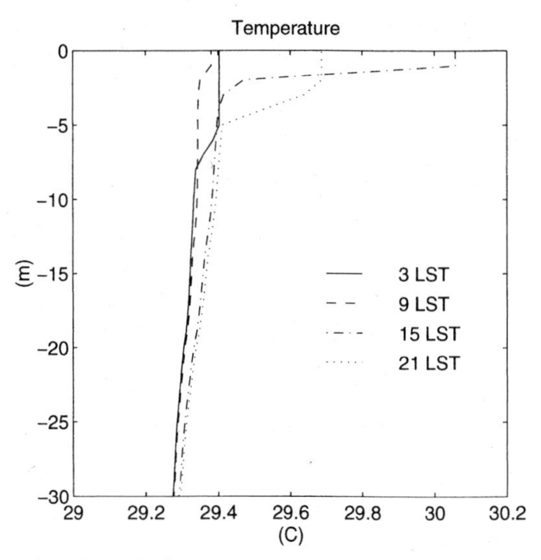

Figure 7.6. Diurnal cycle of vertical temperature profile simulated during the suppressed convective phase of the MJO in the equatorial western Pacific on 26 November, 1992.
From Shinoda and Hendon (1998).

layer initially acts to warm the mixed layer because of the slight (but stable) temperature inversion in the barrier layer. Thus, the entrainment heat flux into the mixed layer tends to be out of phase with the surface heat flux (Shinoda and Hendon, 1998); diurnal entrainment (night-time convection) during the surface warming phase cools the mixed layer, while entrainment of warm sub-mixed-layer water during the surface cooling phase tends to warm the mixed layer.

In the experiments of Shinoda and Hendon (1998), cold sub-thermocline water was never entrained into the mixed layer during the windy-convective phase because of the weakness of the large-scale MJO-induced winds, the strong stratification at the thermocline in the initial conditions, and the development of a barrier layer during the calm-clear phase. However, entrainment of sub-thermocline water into the mixed layer is observed to occur in the western Pacific during some strong westerly wind bursts (e.g., Feng *et al.*, 1998). The results of Shinoda and Hendon (1998) suggest that over the large spatial scale of the typical MJO, entrainment cooling during the cloudy-windy phase of the MJO is not systematic but may be important during some intense events.

The phasing of the mixed layer depth (deepest during the cloudy-windy phase and shallowest during the calm-clear phase) does result in enhanced sensitivity of the mixed layer temperature to surface heat flux forcing during the warming phase, and reduced sensitivity during the cooling phase. The enhanced sensitivity during the warming phase is partially offset by an increased amount of short-wave radiation that penetrates through the base of the shallow mixed layer. Still, if the mean mixed depth is used to simulate the MJO-induced SST variation, then the cooling phase is over-predicted and the warming phase is under-predicted by 25–50% (Shinoda and Hendon, 1998).

Lukas and Lindstrom (1991) and Anderson *et al.* (1996) suggest that the intraseasonal variation of freshwater flux may also play a role in governing the mixed layer evolution over the course of the MJO. They envision that intense precipitation prior to the windy-cooling phase would generate a very stable fresh mixed layer, hence preventing mixing of cold water into the mixed layer during the windy-cooling phase. While such behavior has been observed at individual locations throughout the warm pool (e.g., Sprintall and Tomczak, 1992), the phasing of the freshwater flux relative to the surface cooling and maximum surface windspeed for the MJO (Figure 7.1) suggests that it is not an important process for the typical MJO event. That is, the bulk of the rainfall during the MJO tends to fall during the windiest period, thus preventing formation of a shallow fresh layer (e.g., Zhang and Anderson, 2003).

Shinoda and Hendon (1998) showed the lack of sensitivity to the freshwater flux variation associated with the MJO by driving the mixed layer model with composite surface fluxes from 10 well-defined MJO events. Two simulations were conducted. The control run used the full surface fluxes for each MJO event, while the other run prescribed the freshwater flux to be held constant at its mean value. Very little difference in mixed layer temperature is predicted. Hence, while the mean freshwater flux is critical for maintaining the freshness and stratification of the warm pool (e.g., Lukas and Lindstrom, 1991; Anderson *et al.*, 1996), the intraseasonal variation of freshwater flux over the lifecycle of the MJO appears not to be systematically important. Zhang and Anderson (2003) point out that accurate simulation in climate models of the phasing of rainfall and winds for the MJO is challenging. Hence, coupled models may erroneously simulate sensitivity to the freshwater flux simply by simulating an MJO with an unrealistic phasing of rainfall relative to the wind speed maximum.

Inclusion of the diurnal cycle of insolation, on the other hand, does have a large systematic impact on the evolution of the mixed layer temperature. Figure 7.7 shows the daily mean mixed layer temperature predicted at the IMET mooring during TOGA–COARE for simulations using hourly surface fluxes and daily mean surface fluxes. During the calm-clear phases of the MJO, when the diurnal cycle of mixed layer temperature is large, inclusion of the diurnal cycle of insolation increases the daily mean mixed layer temperature by 0.2–0.5 K. More heat is absorbed in a shallower mixed layer during daylight hours when the insolation and mixed layer depth vary diurnally, even though night-time convection tends to spread this heat out over a deeper layer. Hence, the mean SST during the calm phase increases. The amplitude of the intraseasonal variation of SST over the lifecycle of the MJO also increases by 20–30% (see also Schiller and Godfrey, 2002). Impact of this amplified intraseasonal SST variation on the atmosphere and specifically on the dynamics of the MJO has yet to be determined. On the other hand, the diurnal cycle of SST during the suppressed phase has been postulated to drive shallow convection in the afternoon that acts to progressively moisten the lower troposphere, setting up

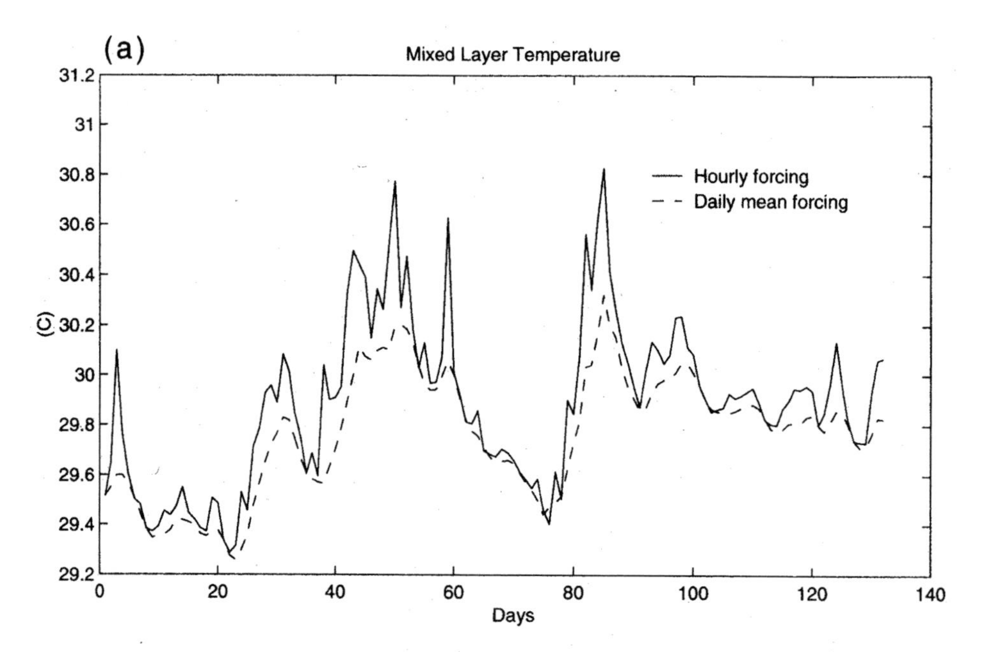

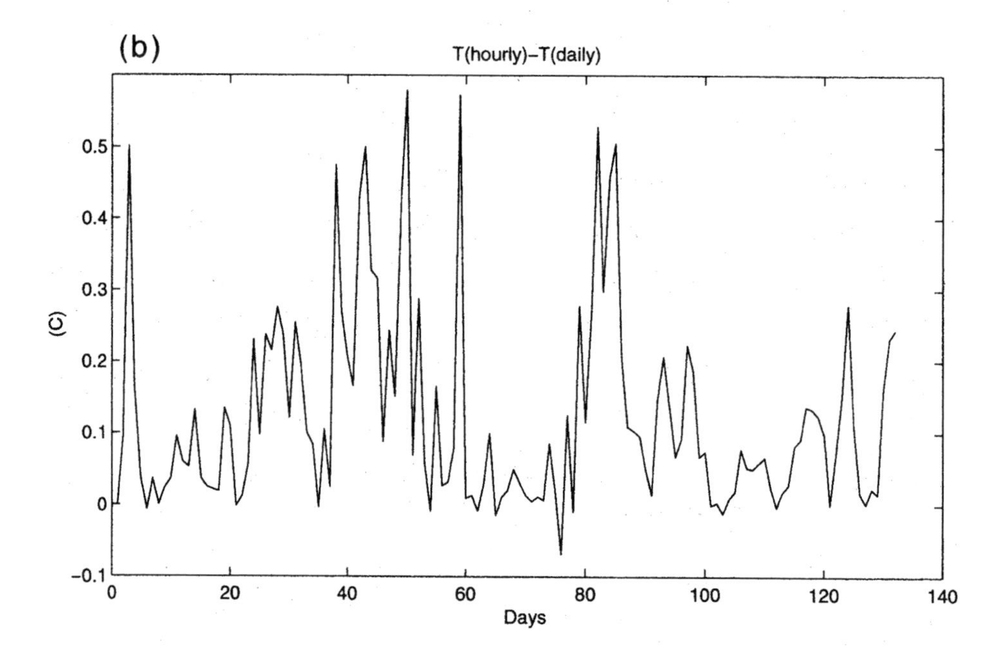

Figure 7.7. (a) Predicted SST at the IMET site for 22 October, 1992–2 March, 1993 using hourly surface fluxes from the IMET mooring data (thick curve) and daily mean fluxes (dashed curve). (b) The SST difference between the two experiments shown in (a).
From Shinoda and Hendon (1998).

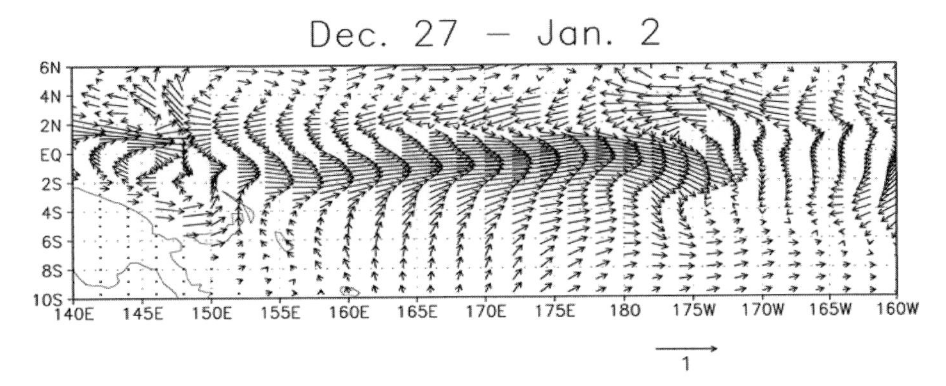

Figure 7.8. Surface currents simulated for the week beginning 27 December, 1992, using an OGCM driven with observed surface fluxes. The maximum velocity vector has a magnitude of $1\,\mathrm{m\,s^{-1}}$.
From Shinoda and Hendon (2001).

conditions that are favorable for the onset of organized convection within the MJO (Godfrey *et al.*, 1998).

The broad-scale surface wind perturbations associated with the MJO, though typically weak ($<5\,\mathrm{m\,s^{-1}}$), do act to rapidly spin up a near-equatorial surface current (e.g., McPhaden *et al.*, 1988; Maes *et al.*, 1998; Cronin *et al.*, 2000; Shinoda and Hendon, 2001). These currents may then advect both the mean zonal SST gradient on the eastern edge of the warm pool (which may be relevant to the initiation of El Niño as discussed in Chapters 6 and 9) and the anomalous zonal SST gradient in the warm pool induced by the MJO via the surface heat flux variations (Ralph *et al.*, 1997). An example of such a narrow equatorial jet (i.e., Yoshida, 1959) was observed with drifters during TOGA–COARE (Ralph *et al.*, 1997). This strong eastward current developed in response to the surface westerlies associated with the MJO event that traversed the western Pacific in late December 1992. The narrow, meridionally convergent jet is well simulated by an OGCM forced with observed surface stress for this period (Figure 7.8).

Ralph *et al.* (1997) demonstrate that this jet acted to advect eastward the anomalous cold SST that developed in the equatorial western Pacific in response to the MJO-induced surface cooling. They further diagnosed that, in the long-term mean, such non-linear advection acts to cool the equatorial western Pacific warm pool at just the rate that it is warmed by the net surface heat flux ($\sim 10\,\mathrm{W\,m^{-2}}$). Hence, episodic non-linear advection generated by the MJO helps to maintain the mean state of the warm pool. Shinoda and Hendon (2001) qualitatively confirmed this result by driving an OGCM with observed intraseasonal surface fluxes. But, the simulated mean cooling produced by the MJO was underestimated, presumably due to difficulties in predicting subtle changes to the weak zonal SST gradient.

Much less work has been done on the mechanisms of SST variability in the northern Indian Ocean associated with northward propagating intraseasonal activity during boreal summer. Analysis of limited observations in the Bay of

Bengal reveal a near-quadrature relationship between net surface heat flux variations associated with ISOs in convection and SST, consistent with the notion that the SST variation is primarily driven by the surface heat flux variation (Sengupta and Ravichandran, 2001). Taking into account the variation of mixed layer depth and penetration of short-wave radiation, Sengupta and Ravichandran (2001) showed qualitative agreement between the surface heat flux variation and observed intraseasonal SST tendency. Despite the fact that mean winds are strong ($\sim 10\,\mathrm{m\,s^{-1}}$), abundance of rainfall and runoff in the region maintains a stable mixed layer upon a deep barrier layer, which prevents mixing of cold sub-thermocline water into the mixed layer (Bhat *et al.*, 2001). On the other hand, model results from Loschnigg and Webster (2000) suggest the oceanic poleward heat transport integrated across the entire Indian Ocean varies in association with the poleward propagating intraseasonal events. Their result implies that the ISO drives changes in the wind-driven circulation of the Indian Ocean, but it is not clear how or whether these changes in heat transport relate to the associated SST changes.

7.6 SST–ATMOSPHERE FEEDBACK

One consequence of the induced intraseasonal SST variations is that the surface fluxes will be modified. The induced SST anomaly associated with the MJO (Figure 7.1) has been diagnosed to cause a slight eastward shift of the latent heat flux anomaly relative to the wind speed anomaly because of the temperature effect on the surface saturation humidity (e.g., Zhang, 1996; Hendon and Glick, 1997). This eastward shift causes the maximum latent heat flux to be more collocated with the maximum convection, rather than lagging it by a few days. This subtle phase difference has important ramifications for the mechanism of the eastward propagating MJO (e.g., Raymond, 2001). However, the quality of current global surface flux and humidity data sets prohibits a more definitive description at this time.

The induced SST anomaly will also act to reduce the sensible and latent heat fluxes driven by the MJO (Hendon and Glick, 1997; Shinoda *et al.*, 1998). That is, SST will warm in response to reduced latent heat flux in the suppressed phase, but a warmer SST will act to increase the latent heat flux. The opposite happens in the suppressed phase. Shinoda *et al.* (1998) estimate that the induced SST anomaly acts to reduce the amplitude of the latent heat flux by 10–15%. How this reduction may affect the dynamics of the MJO is not obvious, but in some atmospheric models coupled to shallow slab mixed layers (so that a large SST anomaly is induced) the MJO amplitude is diminished because the wind–evaporation feedback is effectively eliminated (E. Maloney, pers. commun., 2004).

The phasing of the SST anomalies relative to the convective anomaly, whereby warm SST precedes enhanced convection by ~1–2 weeks and cold SST precedes suppressed convection by a similar lag, is also suggestive of a positive feedback. Various mechanisms for a positive feedback of the SST perturbations onto the eastward propagating MJO have been proposed. Wang and Xie (1998), Waliser *et al.* (1999), Kemball-Cook *et al.* (2002), and Maloney and Kiehl (2002) propose that the warm SST in advance of the convective anomaly hydrostatically acts to lower the surface pressure, thereby promoting more moisture convergence in advance of the convective phase, thus helping to destabilize the MJO. This thermodynamically driven enhancement is estimated to directly account for about a 10% reduction in surface pressure and a similar increase in surface convergence. However, due to feedbacks between moisture convergence, diabatic heating, and circulation, the total effect of this slight pressure reduction could be much larger. In the Waliser *et al.* (1999) and Kemball-Cook *et al.* (2002) studies, the SST-induced surface convergence acts to amplify a pre-existing MJO mode in the models. On the other hand, in the Wang and Xie (1998) study, the SST-induced convergence acts to destabilize the MJO-mode for conditions where the model otherwise does not support an MJO.

Flatau *et al.* (1997), Hendon and Glick (1997), and Lau and Sui (1997) propose that the warm SST anomaly directly destabilizes the atmospheric column in advance of the convection by acting to raise the surface equivalent potential temperature (or moist static energy). At 30°C, a 0.5°C swing in SST is associated with a $0.8\,\mathrm{g\,kg^{-1}}$ change in saturation mixing ratio or about a 3°C change in saturation equivalent potential temperature. In the Flatau *et al.* (1997) model, where the SST tendency was simply parameterized to be negative (cool) when the surface winds were westerly (which encompasses enhanced windspeed and reduced insolation due to enhanced convection) and positive (warm) when the surface winds were easterly (which encompasses reduced windspeed and enhanced insolation due to reduced convection), coupling with the SST acted to amplify, slow down, and better organize an existing MJO-like mode.

Judging the realism of these coupled model studies is difficult. In the Flatau *et al.* (1997) study, the induced SST anomalies are 3–4 times bigger than observed. Though coupling did dramatically improve the intraseasonal organization of convection and its eastward propagation, the impact of the induced SST anomalies may be overestimated. Waliser *et al.* (1999), using a slab-mixed layer coupled to a GCM, and Kemball-Cook *et al.* (2002), using an intermediate ocean model coupled to a GCM, report more positive impacts whereby coupling slowed down and strengthened an existing MJO-like mode. While more realistic intraseasonal SST anomalies were simulated (amplitudes $\sim 0.1\,\mathrm{K}$), it is difficult to assess the robustness of these results based on short (10–15-year) integrations due to large internal variability of the MJO in the absence of coupling. However, coupling is certainly not a panacea for simulation of the MJO. Hendon (2000) found little impact of coupling in a GCM that produces a robust MJO-like mode when uncoupled. The uncoupled mode in this particular model appears to be promoted by evaporation–wind feedback (Neelin *et al.*, 1987), with enhanced evaporation unrealistically occurring in the easterlies to the east of the convective anomaly. As a result, the latent heat flux and insolation anomalies do not constructively add to perturb the net surface heat flux, thus little SST anomaly is generated when coupled. The conclusion from these GCM studies, then, is that the coupling at best improves the simulation of the MJO, but only if it is reasonably simulated in the uncoupled models.

The near quadrature relationship between northward propagating convection during boreal summer and SST suggests that the ocean is primarily being forced by the atmosphere, similar to the situation for the eastward propagating MJO. But, as for the eastward MJO events, the warm SST anomaly in advance of the poleward propagating convective anomaly acts to hydrostatically decrease surface pressure, thereby increasing surface convergence and humidity, which acts to destabilize the atmosphere in advance of the convection (e.g., Kemball-Cook *et al.*, 2002; Fu *et al.*, 2003). While northward propagation of the MJO is simulated in some models to occur in the absence of air–sea coupling, inclusion of coupling increases the amplitude of the northward propagating events by ~50% (Fu *et al.*, 2003). In fact, the impact of air–sea coupling on northward propagation in the Indian summer monsoon appears to be more robust than that on eastward propagation of the MJO along the equator.

7.7 IMPACT OF SLOW SST VARIATIONS ON MJO ACTIVITY

Interaction of the MJO with the ocean has been discussed from the perspective of the atmosphere driving the ocean with the possibility of a coupled feedback. One-way interaction, where by the slowly varying SST influences the behavior of the MJO, does play an important role for the seasonal and interannual variation of MJO activity. The seasonal variation of MJO activity, with the strongest eastward propagation occurring near vernal equinox (e.g., Salby and Hendon, 1994), is partially explained by the seasonal cycle of SST and mean convection (e.g., Li and Wang, 1994; Salby *et al.*, 1994). The most zonally extensive warm pool and region of active convection develops after the austral solstice, but is displaced into the southern hemisphere. The strongest MJO activity tends to occur at this time because near-equatorial convection can develop in the Indian Ocean and propagate furthest east into the Pacific. At the autumnal equinox, intraseasonal convection is observed to peak again in the Indian Ocean (Salby and Hendon, 1994), but because SSTs have not yet warmed in the equatorial western Pacific, MJO activity tends to not be as strong as near the vernal equinox.

Large year-to-year variability in the strength of MJO activity is also observed (e.g., Salby and Hendon, 1994; Hendon *et al.*, 1999; Slingo *et al.*, 1999). This variability, at least at that time of year when the MJO is strongest as measured by globally integrated indices, is not clearly connected to interannual variations of the SST. For instance Hendon *et al.* (1999) and Slingo *et al.* (1999) found little relationship between year-to-year variations of the level of global MJO activity and SST variations, and, in particular, those associated with ENSO. A possible exception is an apparent tendency for global MJO activity to weaken at the peak of the strongest El Niño events. Figure 7.9 (top panel), which is an updated version of the plot shown in Hendon *et al.* (1999), confirms the lack of a simple relationship between MJO activity in austral summer (the season when MJO activity is strongest) and ENSO. This result has been confirmed with GCM studies where observed SST is prescribed for a number of years (e.g., Slingo *et al.*, 1999; Gualdi *et al.*, 1999; Waliser *et al.*, 2001). However, simulation of the MJO in these models is far from perfect. Thus,

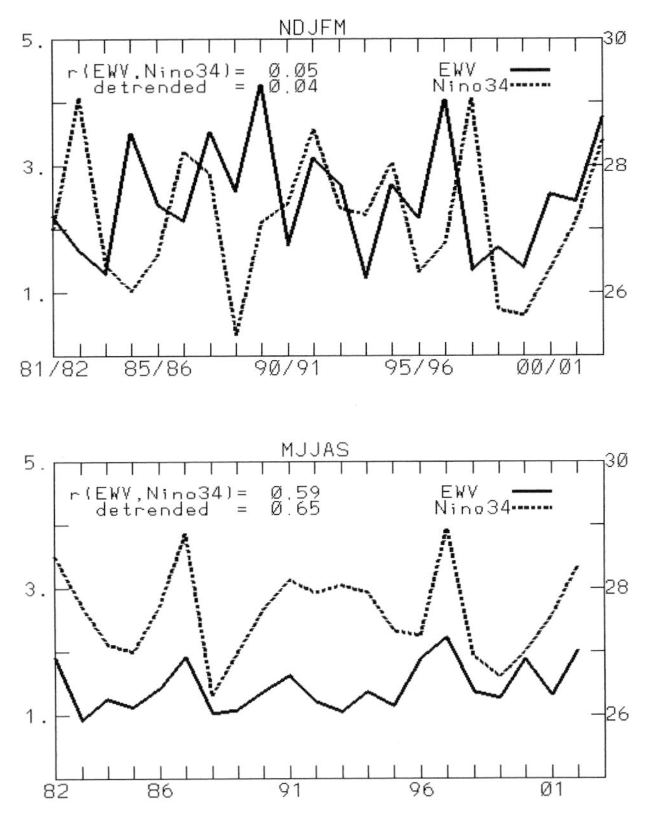

Figure 7.9. Time series of tropically averaged (15°S–15°N) outgoing long-wave radiation (OLR) variance filtered to eastward wave numbers 1–4 and periods of 30–90 days. (*top panel*) The five month period November–March. (*bottom panel*) May–September. Also shown is the Nino34 SST index for the same 5-month periods. Simultaneous correlations, with and without linear trend, are indicated.
Filtered OLR data were kindly supplied by M. Wheeler, BMRC.

there is ample scope for further study of seasonal predictability of the MJO once reliable simulations of the MJO are achieved.

On the other hand, a significant relationship between global MJO activity in boreal summer and ENSO is apparent (Figure 7.9 (bottom panel)). Despite the fact that this occurs at the time of year when the MJO is weakest along the equator, it is during this time of the year when MJO activity may have its greatest impact on the evolution of the coupled system in the Pacific (e.g., Fedorov, 2002). A simple explanation for the enhanced MJO activity in boreal summer during developing El Niño events is that the total SST field then is similar to the normal SST field during late boreal winter. In particular, the cold tongue in the central and eastern Pacific, which normally acts to reduce eastward propagation along the equator during boreal summer (e.g., Li and Wang, 1994) is absent during a warm event. Thus, MJO activity can propagate further east and maintain its strength longer. The lack of a relationship between the strength of global MJO activity and ENSO during boreal winter results because, as warm events peak in boreal winter, anomalous convection then often shifts so far east in the Pacific that suppressed conditions develop in the western Pacific. Thus, the MJO experiences less of a fetch upon which to grow.

MJO activity, along with all other intraseasonally varying convective activity, does shift eastward in the Pacific during warm events and contracts westward during cold events (Gutzler, 1991; Anyamba and Weare, 1995; Fink and Speth, 1997; Hendon *et al.*, 1999, Kessler, 2001). This eastward shift appears to simply reflect the eastward shift of warm SST as El Niño develops: warming SST provides the basis for MJO activity to develop further eastward. Kessler (2001) points out that this apparently diagnostic relationship may still be supportive of an active role of the MJO in the development of El Niño, especially if the MJO produces a rectified response in the western Pacific (e.g., Shinoda and Hendon, 2002), where the coupled system is especially sensitive (e.g., Kessler and Kleeman, 2000; see also Chapters 6 and 9).

7.8 CONCLUDING REMARKS

This review has focussed on observations of air–sea interaction associated with eastward and northward propagation of the MJO. Specifically, air–sea interactions that are relevant to the mechanism and variability of the MJO were addressed. Two areas of future research will shed more light onto the role of air–sea interaction for ISV throughout the global tropics. First, improved global data sets from new observing platforms and refined reanalysis of surface fluxes, SST, and rainfall, will allow more comprehensive study of the delicate interplay of convection, SST, and oceanic and atmospheric circulation. Second, improved simulation of tropical ISV in climate models will allow careful study of the mechanism and importance of active coupling on intraseasonal timescales.

7.9 ACKNOWLEDGMENT

While this chapter is purportedly a comprehensive review of tropical intraseasonal air–sea interaction, the perspective is a personal one. I thank S. Anderson, S. Kemball-Cook, and the American Meteorological Society for allowing me to reproduce Figures 7.2 and 7.4. Comments on an earlier version of this chapter by C. Zhang and the editors led to improvements in the presentation and scope.

7.10 REFERENCES

Anderson, S. P., R. A. Weller, and R. Lukas (1996) Surface buoyancy forcing and the mixed layer in the western Pacific warm pool: Observation and 1D model results. *J. Climate*, **9**, 3056–3085.

Anyamba, E. K. and B. C. Weare (1995) Temporal variability of the 40–50 day oscillation in tropical convection. *Int. J. Climatol.*, **15**, 379–402.

Bergman, J. W., H. H. Hendon, and K. M. Weickmann (2001) Intraseasonal variation of west Pacific convection at the onset of the 1997–98 El Niño. *J. Climate.*, **14**, 1702–1719.

Bhat, G. S., S. Gadgil, P. U. Hareesh Kumar, S. R. Kalse, P. Madhusoodanan, V. S. N. Muriz, C. U. K. Prasada Rao, V. Ramosh Babu, L. U. G. Rao., R. R. Rao, *et al.* (2001) BOBMEX: The Bay of Bengal Monsoon Experiment. *Bull. Amer. Meteor. Soc.*, **82**, 2217–2243.

Colón, E., J. Lindesay, and M. J. Suarez (2002) The impact of surface flux-and circulation-driven feedbacks on simulated Madden–Julian Oscillations. *J. Climate*, **15**, 624–641.

Cronin, M. F., M. J. McPhaden, and R. H. Weisberg (2000) Wind-forced reversing jets in the western equatorial Pacific. *J. Phys. Ocean.*, **30**, 657–676.

Cronin, M. F. and M. J. McPhaden (1997) The upper ocean heat balance in the western equatorial Pacific warm pool during September–December 1992. *J. Geophys. Res.*, **102**, 8533–8553.

Emanuel, K. A. (1987) An air–sea interaction model of intraseasonal oscillations in the Tropics. *J. Atmos. Sci.*, **44**, 2324–2340.

Fedorov, A. V. (2002) The response of the coupled tropical ocean–atmosphere to westerly wind bursts. *Quart. J. Roy. Meteor. Soc.*, **128**, 1–23.

Feng, M., P. Hacker, and R. Lukas (1998) Upper ocean heat and salt balances in response to a westerly wind burst in the western equatorial Pacific during TOGA COARE. *J. Geophys. Res.*, **103**, 10289–10311.

Feng, M., R. Lukas, P. Hacker, R. A. Weller, and S. P. Anderson (2000) Upper-ocean heat and salt balances in the western equatorial Pacific in response to the intraseasonal oscillation during TOGA COARE. *J. Climate*, **13**, 2409–2427.

Fink, A. and P. Speth (1997) Some potential forcing mechanisms of the year-to-year variability of the tropical convection and its intraseasonal (25–70 day) variability. *Int. J. Climatol.*, **17**, 1513–1534.

Flatau, M., P. J. Flatau, P. Phoebus, and P. P. Niiler (1997) The feedback between equatorial convection and local radiative and evaporative processes: The implications for intraseasonal oscillations. *J. Atmos. Sci.*, **54**, 2373–2386.

Fu, X., B. Wang, T. Li, and J. P. McCreary (2003) Coupling between northward propagating intraseasonal oscillations and sea surface temperature in the Indian Ocean. *J. Atmos. Sci.*, **60**, 1733–1753.

Giese, B. S. and D. E. Harison (1991) Eastern equatorial Pacific response to three composite westerly wind types. *J. Geophys. Res.*, **96**(suppl.), 3239–3248.

Godfrey, J. S., R. A. Houze, R. H. Johnson, R. Lukas, J. L. Redelsperger, A. Sumi, and R. Weller (1998) Coupled Ocean–Atmosphere Response Experiment (COARE): An interim report. *J. Geophys. Res.*, **103C**, 14395–14450.

Gualdi, S., A. Navarra, and G. Tinarelli (1999) The interannual variability of the Madden–Julian Oscillation in an ensemble of GCM simulations. *Climate Dyn.*, **15**, 643–658.

Gutzler, D. S. (1991) Interannual fluctuations of intraseasonal variance of near-equatorial zonal winds. *J. Geophys. Res.*, **96**, 3173–3185.

Gutzler, D. S., G. N. Kiladis, G. A. Meehl, K. M. Weickmann, and M. Wheeler (1994) The global climate of December 1992–February 1993. Part II: Large-scale variability across the tropical western Pacific during TOGA COARE. *J. Climate*, **7**, 1606–1622.

Harrison, D. E. and G. A. Vecchi (2001) January 1999 Indian Ocean cooling event. *Geophys. Res. Lett.*, **28**, 3717–3720.

Hendon, H. H. (2000) Impact of air–sea coupling on the Madden–Julian oscillation in a general circulation model. *J. Atmos. Sci.*, **57**, 3939–3952.
Hendon, H. H., C. Zhang, and J. D. Glick (1999) Interannual variability of the Madden–Julian Oscillation during austral summer. *J. Climate.*, **12**, 2538–2550.
Hendon, H. H., B. Leibmann, and J. Glick (1998) Oceanic Kelvin waves and Madden–Julian Oscillation. *J. Atmos. Sci.*, **55**, 88–101.
Hendon, H. H., and J. Glick (1997) Intraseasonal air–sea interaction in the tropical Indian and Pacific Oceans. *J. Climate*, **10**, 647–661.
Hendon, H. H. and B. Liebmann (1994) Organization of convection within the Madden–Julian Oscillation. *J. Geophys. Res.*, **99**, 8073–8083.
Hendon, H. H. and M. L. Salby (1994) The life cycle of the Madden–Julian Oscillation. *J. Atmos. Sci.*, **51**, 2225–2237.
Jiang, X., T. Li, and B. Wang (2004) Strutures and mechanisms of the northward propagating boreal summer intraseasonal oscillation. *J. Climate*, **17**, 1022–1039.
Jones, C., D. E. Waliser, and C. Gautier (1998) The influence of the Madden–Julian oscillation on ocean surface heat fluxes and sea surface temperature. *J. Climate*, **11**, 1057–1072.
Kemball-Cook, S., B. Wang, and X. Fu (2002) Simulation of the intraseasonal oscillation in ECHAM-4 model: The impact of coupling with an ocean model. *J. Atmos. Sci.*, **59**, 1433–1453.
Kemball-Cook, S. and B. Wang (2001) Equatorial waves and air–sea interaction in the boreal summer intraseasonal oscillation. *J. Climate*, **14**, 2923–2942.
Kessler, W.S. (2001) EOF representations of the Madden–Julian Oscillation and its connection with ENSO. *J. Climate*, **14**, 3055–3061.
Kessler, W. S. and R. Kleeman (2000) Rectification of the Madden–Julian Oscillation into the ENSO cycle. *J. Climate*, **13**, 3560–3575.
Kessler, W. S., M. J. McPhaden, and K. M. Weickmann (1995) Forcing of intraseasonal Kelvin waves in the equatorial Pacific. *J. Geophys. Res.*, **100**, 10613–10631.
Krishnamurti, T. N., D. K. Oosterhof, and A. V. Metha (1988) Air–sea interaction on the timescale of 30–50 days. *J. Atmos. Sci.*, **45**, 1304–1322.
Lau, K. M. and C.-H. Sui (1997) Mechanisms of short-term sea surface temperature regulation: Observations from TOGA COARE. *J. Climate*, **10**, 465–472.
Lawrence, D. M. and P. J. Webster (2002) The boreal summer intraseasonal oscillation: Relationship between northward and eastward movement of convection. *J. Atmos. Sci.*, **59**, 1593–1606.
Li, T. and B. Wang (1994) The influence of sea surface temperature on the tropical intraseasonal oscillation: A numerical study. *Mon. Wea. Rev.*, **122**, 2349–2362.
Lin, J. W.-B., J. D. Neelin, and N. Zeng (2000) Maintenance of tropical intraseasonal variability: Impact of evaporation–wind feedback and midlatitude storms. *J. Atmos. Sci.*, **57**, 2793–2823.
Loschnigg, J. and P. J. Webster (2000) A coupled-atmosphere system of SST modulation for the Indian Ocean. *J. Climate*, **13**, 3342–3360.
Lukas, R., and E. Lindstrom (1991) The mixed layer of the western equatorial Pacific ocean. *J. Geophys. Res.*, **96**(suppl.), 3343–3357.
Maes, C., P. Delecluse, and G. Madec (1998) Impact of westerly wind bursts on the warm pool of the TOGA COARE domain in an OGCM. *Climate Dyn.*, **14**, 55–70.
Maloney, E. D. and J. T. Kiehl (2002) MJO-related SST variations over the tropical Eastern Pacific during Northern Hemisphere summer. *J. Climate*, **15**, 675–689.
McPhaden, M. J. (2002) Mixed layer temperature balance on intraseasonal timescales in the equatorial Pacific Ocean. *J. Climate*, **15**, 2632–2647.

McPhaden, M. J., H. P. Fretiag, S. P. Hayes, B. A. Taft, Z. Chen, and K. Wyrtki (1988) The response of the equatorial Pacific Ocean to a westerly wind burst in May 1986. *J. Geophys., Res.*, **93**, 10589–10603.

Neelin, J. D., I. M. Held, and K. H. Cook (1987) Evaporation-wind feedback and low-frequency variability in the tropical atmosphere. *J. Atmos. Sci.*, **44**, 2341–2348.

Ralph, E. A., K. Bi, and P. P. Niiler (1997) A Lagrangian description of the western equatorial Pacific response to the wind burst of December 1992. *J. Climate*, **10**, 1706–1721.

Raymond, D. J. (2001) A new model of the Madden–Julian Oscillation. *J. Atmos. Sci.*, **58**, 2807–2819.

Reynolds, R. W. and T. M. Smith (1994) Improved global sea surface temperature analyses using optimum interpolation. *J. Climate*, **7**, 929–948.

Salby, M. L. and H. H. Hendon (1994) Intraseasonal behavior of clouds, temperature, and winds in the Tropics. *J. Atmos. Sci.*, **51**, 2207–2224.

Salby, M. L., G. Rolando, and H. H. Hendon (1994) Planetary-scale circulations in the presence of climatological and wave-induced heating. *J. Atmos. Sci.*, **51**, 2344–2367.

Sengupta, D. and M. Ravichandran (2001) Oscillations of Bay of Bengal sea surface temperature during the 1998 summer monsoon. *Geophys. Res. Letter*, **28**, 2033–2036.

Schiller, A. and J. S. Godfrey (2003) Indian Ocean intraseasonal variability in an ocean general circulation model. *J. Climate*, **16**, 21–39.

Shinoda, T. and H. H. Hendon (2002) Rectified wind forcing and latent heat flux produced by the Madden–Julian Oscillation. *J. Climate*, **15**, 3500–3508.

Shinoda, T. and H. H. Hendon (2001) Upper-ocean heat budget in response to the Madden–Julian Oscillation in the western equatorial Pacific. *J. Climate*, **14**, 4147–4165.

Shinoda, T. and H. H. Hendon (1998) Mixed layer modeling of intraseasonal variability in the tropical western Pacific and Indian Oceans. *J. Climate*, **11**, 2668–2685.

Shinoda, T., H. H. Hendon, and J. Glick (1998) Intraseasonal variability of surface fluxes and sea surface temperature in the tropical western Pacific and Indian Oceans. *J. Climate*, **11**, 1685–1702.

Sikka, D. R. and S. Gadgil (1980) On the maximum cloud zone and the ITCZ over Indian longitudes during the southwest monsoon. *Mon. Wea. Rev.*, **108**, 1840–1853.

Slingo, J. M., D. P. Rowell, K. R. Sperber, and F. Nortley (1999) On the predictability of the interannual behaviour of the Madden–Julian Oscillation and its relationship to El Niño. *Quart. J. Roy. Meteor. Soc.*, **125**, 583–609.

Sprintall, J. and M. Tomczak (1992) Evidence of the barrier layer in the surface layer of the tropics. *J. Geophys. Res.*, **97**, 7305–7316.

Vecchi, G. A., and D. E. Harrison (2002) Monsoon breaks and subseasonal sea surface temperature variability in the Bay of Bengal. *J. Climate*, **15**, 1485–1493.

Waliser, D. E., Z. Zhang, K. M. Lau, and J.-H. Kim (2001) Interannual sea surface temperature variability and the predictability of tropical intraseasonal variability. *J. Atmos. Sci.*, **58**, 2596–2615.

Waliser, D. E., K.-M. Lau, and J. H. Kim (1999) The influence of coupled sea surface temperatures on the Madden–Julian Oscillation: A model perturbation experiment. *J. Atmos. Sci.*, **56**, 333–358.

Wang, B. and H. Rui (1990) Synoptic climatology of transient tropical intraseasonal convection anomalies: 1975–1985. *Meteor. Atmos. Phys.*, **44**, 43–61.

Wang, B. and X. Xie (1998) Coupled modes of the warm pool climate system. Part I: The role of air–sea interaction in maintaining Madden–Julian Oscillation. *J. Climate*, **11**, 2116–2135.

Webster, P. J., E. F. Bradley, C. W. Fairall, J. S. Godfrey, P. Hacker, R. A. Houze, R. Lukas, Y. Serra, J. M. Hummon, T. D. M. Lawrence, *et al.* (2002) The JASMINE pilot study. *Bull. Amer. Meteor. Soc.*, **83**, 1603–1630.

Webster, P. J. and R. Lukas (1992) TOGA COARE: The Coupled Ocean–Atmosphere Response Experiment. *Bull. Amer. Meteor. Soc.*, **73**, 1377–1416.

Weller, R. A. and S. P. Anderson (1996) Surface meteorology and air–sea fluxes in the western equatorial Pacific warm pool during the TOGA coupled ocean–atmosphere response experiment. *J. Climate*, **9**, 1959–1990.

Woolnough, S. J., J. M. Slingo, and B. J. Hoskins (2000) The relationship between convection and sea surface temperature on intraseasonal timescales. *J. Climate*, **13**, 2086–2104.

Yoshida, K. (1959) A theory of the Cromwell current and of the equatorial upwelling—An interpretation in a similarity to a coastal circulation. *J. Oceanogr. Soc. Jap.*, **15**, 159–170.

Zhang, C. (2001) Intraseasonal perturbations in sea surface temperatures of the equatorial eastern Pacific and their association with the Madden–Julian Oscillation. *J. Climate*, **14**, 1309–1322.

Zhang, C. (1997) Intraseasonal variability of the upper-ocean thermal structure observed at 0° and 165°E. *J. Climate*, **10**, 3077–3092.

Zhang, C. (1996) Atmospheric intraseasonal variability at the surface in the tropical western Pacific Ocean. *J. Atmos. Sci.*, **53**, 739–758.

Zhang, C. and S. P. Anderson (2003) Sensitivity of intraseasonal perturbations in SST to the structure of the MJO. *J. Atmos. Sci.*, **60**, 2196–2207.

Zhang, C. and J. Gottschalck (2002) SST anomalies of ENSO and the Madden–Julian Oscillation in the equatorial Pacific. *J. Climate*, **15**, 2429–2445.

Zhang, C. and M. J. McPhaden (2000) Intraseasonal surface cooling in the equatorial western Pacific. *J. Climate*, **13**, 2261–2276.

8

Mass, momentum, and geodynamics

Benjamin F. Chao and David A. Salstein

8.1 INTRODUCTION

While other chapters of this book describe the meteorological intraseasonal variability (ISV) phenomena in the atmosphere–ocean system, and examine the possible causes of the ISV or the dynamic interactions between the meteorological components that are involved, in this chapter we will study certain global *geodynamic* effects that relate to the *mass transport* associated with the ISV, which for the most part occur in the atmosphere–ocean system. In particular, we will discuss the angular momentum variability of the atmosphere and its influences on Earth's rotation; we also visit the associated mass-induced gravity variations. The observations of these can help improve our understanding of, and our modeling capability for, atmospheric–oceanic circulations in general, including those related to the ISV.

The transports of mass and energy are key processes that determine the dynamics of our Earth system. Mass transports occur in all the *geophysical fluid* components of the Earth system – the atmosphere, hydrosphere, cryosphere, biosphere, lithosphere, and the deep interior of mantle and cores, on a wide range of temporal and spatial scales. In the atmosphere, the variations in position and strength of the pressure systems, as found on weather maps, as well as in the wind systems indicate that different masses of air move around the planet; dynamical fluctuations result as part of this atmospheric general circulation. Mass transport also occurs in the oceans, resulting mainly from forcing by the surface wind, pressure, thermohaline fluxes, and tides, but also as internal ocean modes. Numerical models of these circulations allow the response of the atmosphere and oceans to these processes to be investigated in detail. The mass near the Earth's land surface in the cryosphere and hydrological reservoirs varies due to natural as well as anthropogenic influences. Hydrological modeling, in combination with observations, can supplement the atmospheric and ocean models to determine the global mass

W. K. M. Lau and D. E. Waliser (eds), *Intraseasonal Variability in the Atmosphere–Ocean Climate System*.

variability. These mass transport components may all have strong ISV signals, and studying them is undoubtedly a most interdisciplinary field in Earth sciences.

However, mass transport has not received due attention. For example, many current ocean general circulation models (GCMs) do not conserve correctly the water mass, but instead conserve water volume under the Boussinesq approximation (e.g., Da Szoeke and Samelson, 2002), although some modern models have investigated aspects of mass conservation. Although atmospheric GCMs largely conserve dry air mass, including many that are used daily for weather forecasting, like the ocean models, not all do. A most critical piece in the global water mass puzzle, the large-scale land hydrological and cryospheric water mass cycle budget, that is constantly under direct exchange with the atmosphere, remains least known (e.g., IPCC, 2001).

There are, of course, mass transports that occur in the interior geophysical fluids as well (e.g., in the mantle associated with the post-glacial rebound and tectonic movements, and in the fluid core under the forcing of the geodynamo). These occur, however, on markedly different timescales than ISV.

As mass moves or redistributes in the geophysical fluids, three distinct geodynamic effects occur as a consequence (e.g., Chao *et al.*, 2000):

(i) *Temporal variation in the solid Earth's rotation.* By virtual of the conservation of angular momentum of the closed Earth system, the angular momentum variation associated with any geophysical mass transport has to be balanced by the same but opposite change in the angular momentum of the rest of the Earth, essentially the solid-Earth. In this sense, the former, or more accurately the combined sum of all such angular momentum changes, manifests itself in the solid-Earth's rotation variation. As will be discussed below, the mass transport imparts angular momentum not only by virtue of the actual motion of the mass, but also by the resultant change in mass distribution, and hence in the inertia tensor, of the geophysical fluid. Note also that the rotation of the Earth is a 3-D quantity; all 3 components vary as a result of the angular momentum exchange.

(ii) *Temporal variations in the Earth's gravity field.* The redistributed mass in question slightly modifies the global gravitational attraction force, as dictated by Newton's gravitational law, thereby changing the gravity field in both time and space. This time-variable gravity signal, coming from all geophysical fluids, is a superimposition onto the background *static* gravity field.

(iii) *Temporal variation in the solid Earth's center of mass (CM).* The conservation of linear momentum dictates a motion in the CM of the solid Earth resulting from a net shift of the CM of the fluid envelope, in such a way that the CM of the whole Earth (the "geocenter") remains stationary (or, more accurately, stays in its orbit about the Sun according to celestial mechanics.)

In this chapter we will mainly discuss the Earth's rotation variation (i), which is the most mature subject of the three as far as observation and modeling is concerned.

We shall however give rudimentary discussion of the time-variable gravity (ii), in anticipation of its exciting future prospects, and to a lesser extent the geocenter motion (iii) as its observations are presently in their infancy.

Although relatively minute – typically no larger than 1–10 parts per billion in relative terms, the aforementioned geodynamic signals have become observable through very precise space-geodetic techniques (e.g., AGU, 1993). For over three decades, space geodetic measurement precision has improved at the rate of one order of magnitude per decade (something of a space-geodetic "Moore's law", see, e.g., Chao, 2003). Today the observations come from several advanced geodetic techniques, whose developments have been primarily led by NASA while incorporating an extensive international cooperation. The earliest such technique was based on ranging by pulsed laser to the retro-reflectors placed on the moon by the Apollo Mission circa 1970, though it was soon supplemented and now all but replaced since the late 1970s by similar ranging to retro-reflectors on board the now dozen or so geodetic satellites in very stable orbits at various altitudes and inclinations; the technique is known as satellite laser ranging (SLR). Another space-geodetic technique that began in the early 1980s, very-long-baseline interferometry (VLBI), uses signals from remote radio sources in the universe for geometrical determination of the precise baseline length between pairs of radiotelescopes. Earth's orientation at a given instant in relation to inertial space can be derived. Since the 1990s the Global Positioning System (GPS) came on the scene using a whole constellation of artificial satellites as a reference system, within which the position of places on the Earth's surface can be precisely determined "anywhere, anytime". Somewhat more recently, the French Doppler Orbitography and Radio positioning Integrated by Satellite (DORIS) system uses a tracking system providing range-rate measurements of signals from a dense network of ground-based radio beacons. Together these observing systems define precisely our celestial and terrestrial reference frames (among many other observables), and more importantly for our present study, the link between the two reference systems. This link, primarily simply a uniform daily Earth rotation, also contains the slight variations in the rotation, which is the main target of our study. The origin of the terrestrial reference frame is defined relative to the geocenter. In addition, satellites' dynamic orbit determinations from SLR are used to determine the Earth's gravity field, and have been precise enough to detect temporal *variations* in gravity (e.g., Cox and Chao, 2002).

The variability in Earth rotation, geocenter, and gravity parameters represents the sum total of contributions from all mass transports taking place in the geophysical fluids, as mentioned above. In this sense, the space geodesy system has become an integrating, effective, and unique tool for remote sensing of global mass transports, subject naturally to its precision/accuracy and temporal/spatial resolutions. With respect to the ISV in the meteorological system, we examine only a focussed component within an otherwise extensive and interdisciplinary subject. Specifically, we emphasize the interactions and connections of ISV with Earth-rotation parameters and related geodynamic signals which occur as a result of the ISVs themselves.

8.2 ANGULAR MOMENTUM VARIATIONS AND EARTH ROTATION

The distribution of atmospheric mass and its changing motions have been known for quite some time. The basic structure of the zonal general circulation, namely the westerly winds in the middle latitudes and easterlies in the low and high latitudes, were well appreciated in the 19th century, as was the meridional circulation, featuring the direct Hadley circulation of rising air in the tropics and descending air further polewards. More structure was noted when it was determined that the mean cells in the middle latitude are indirect, known as Ferrel cells, are indirect. Both zonal mean motion and eddy motions are responsible for transporting momentum meridionally. Using the network of radiosondes, the annual and interannual signals in these large-scale circulation systems were synthesized; a summary may be noted in Peixoto and Oort (1992), although they only treated the axial component of the atmospheric angular momentum. Similar results were obtained from the various operational meteorological series as well as reanalyses.

Remarkably, the atmosphere changes its overall atmospheric angular momentum (AAM) substantially, especially with season. Between northern hemisphere winter and summer, the relative AAM changes by a factor of two, with the strong winter winds at the jet stream level of the winter hemisphere being particularly strong. Whereas the westerly winds, with the maxima in the jets at levels near 200 hPa in the subtropics and middle latitudes of both hemispheres, carry most of the axial AAM, variations on interannual, seasonal, and intraseasonal scales are distributed somewhat differently. For the ISV, Rosen and Salstein (1983) noted that they are more evenly distributed in the lower latitudes of the hemispheres.

An important question arose early on as to how the changes in AAM can occur so quickly and dramatically. Physically, angular momentum in a system can only be changed only by torques upon that element, and so the search for torques of the terrestrial AAM led to two particular mechanisms: that from normal pressure forces against topography, and that from tangential friction along the boundary. We discuss these torques and their implications later in the chapter. For ISV in particular, detailed studies of the angular momentum-torques on such timescales have been performed, and related to other elements. Lau *et al.* (1989) and Kang and Lau (1990) considered elements of how the already known ISVs could impact the global angular momentum, whose contributions did peak in the upper troposphere, and they linked the variations to the effect of heating, dynamics, and convection, particularly over the tropical Pacific, with concomitant transfers of angular momentum across the ocean boundary as frictional torques. With the transfer of angular momentum from the atmosphere to the solid Earth, in the form of the Earth's rotation, it is important to understand concepts of its rotational variability, and how it is measured.

As stated above, the Earth rotation, linking the so-called terrestrial and celestial reference frames, is simply a uniform daily rotation of the Earth plus some very minute variations. As with any rotation in our 3-D world, Earth's rotation can be represented as a 3-D vector. (Note that the tensor dimension of the rotation vector in an n-D space is $n(n-1)/2$.) The departure of the Earth's rotation from a uniform rotation in space can thus be conveniently divided into two parts: (i) the magnitude of the (1-D) axial component along the mean rotation z-axis that pierces the surface at the mean North and South Poles, determining the *length-of-day* variation (ΔLOD); and (ii) the (2-D) equatorial component in the x–y plane, giving the orientation of the rotational axis as seen on the Earth known as the *polar motion*.

SLR and VLBI, and more recently GPS, have been the major techniques relied upon within the group of techniques measuring the Earth's rotation. The data set we use here is an optimal combination from these techniques, as described by Gross *et al.* (1998a). Sub-milliarcsecond precision (1 milliarcsecond, or mas, corresponds to about 3 centimeters if projected onto the Earth's surface) is now routinely achieved in quasi-daily measurements.

8.2.1 Length-of-day variation and axial angular momentum

As stated, Earth's rotation varies as a result of angular momentum variations in the geophysical fluids. Among these, the AAM is a fundamental and important quantity in the atmospheric circulation system; in addition to exchanges internal to the atmosphere, its total amount is constantly changing, so that angular momentum is transferred to and from the solid Earth, modifying the (solid) Earth's rotation. Let us first focus on the (1-D) axial component – the AAM_z on the atmospheric side of the balance, and ΔLOD on the geodetic side where the z-axis be the axis connecting the mean North and South Poles. (A modern perspective of the complete LOD variations can be found in, e.g., Chao, 2003.) The formulation by which one computes AAM (or any angular momentum due to mass transports in the geophysical fluids) was clearly laid out in the seminal work by Munk and MacDonald (1960), and later brought up to date with increasingly more appropriate Earth parameters by, e.g., Barnes *et al.* (1983) and Eubanks (1993).

In general, when moving, mass of a geophysical fluid produces an angular momentum variation that consists of two parts – one part due to the actual motion of that mass relative to the Earth, plus another part due to the change in the inertia tensor as a consequence of the mass redistribution (participating with the planet in its solid body rotation). For the axial z-component of AAM, we have, approximately:

$$\text{Motion term for AAM}_z \text{ (or the ``wind term'')} = \frac{1.00a^3}{C\omega g} \iiint u \cos\theta \, d\Omega \, dp \tag{8.1a}$$

$$\text{Mass term of AAM}_z \text{ (or the ``pressure term'')} = \frac{0.70a^4}{Cg} \iint p \cos^2\theta \, d\Omega \tag{8.1b}$$

Physically, (8.1a) is analogous to what happens to a log floating on water as a lumberjack "races" (and balances himself) on it; (8.1b) is analogous to the spinning skater phenomenon – where a skater speeds up her spinning as she draws her arms closer to her body and slows down as she does the opposite.

In the equations u and p are the (zonal) westerly-wind and surface-pressure fields, respectively; a, ω, and g are the mean radius, mean rotation rate, and mean gravitational acceleration of the Earth, respectively; Ω is an abbreviation for

(θ, λ) = (latitude, longitude), where $d\Omega = \cos\theta\, d\theta\, d\lambda$ is the surface element. The dp can be converted into a vertical distance element dh by the hydrostatic relation $dp = -\rho g\, dh$, where ρ is the air density. The multiplicative coefficients in front of (8.1a) and (8.1b) account for some subtle but important effects including the Earth's elastic yielding under surface loading. C is the mean axial moment of inertia of the mantle; the cores are excluded in C because it is approximately true that they do not participate in the mantle's rotational variation on ISV timescales (e.g., Barnes *et al.*, 1983). In other words we assume zero core–mantle coupling strength on such timescales that are much shorter than the decadal and longer scales associated with core–mantle coupling. Note that for simplicity we have normalized the AAM terms with respect to Earth parameters, so that they are in non-dimensional "*excitation*" units relative to the length of the mean solar day, 86,400 s. The resultant change in LOD due to AAM_z is:

$$AAM_z \text{ (sum of 8.1a + 8.1b)} = -\Delta LOD/LOD. \tag{8.2}$$

Note that an increase in AAM_z slows the spinning planet, causing LOD to increase. In absolute units, a change of 1 millisecond (ms) of ΔLOD is equivalent to 15 milliarcseconds (mas) in Earth orientation or ~45 cm on the Earth's surface, the rotation angle and distance subtended in that time, which in turn corresponds to a transfer of $5.95 \times 10^{25}\,\text{kg}\,\text{m}^2\,\text{s}^{-1}$ of AAM.

Not all LOD changes come from AAM_z of course (e.g., Chao, 2003). Period-specific tidal signals in the ΔLOD data do not behave according to relation (8.2), because they are a result of external lunar and solar torques and resultant internal interactions (which is in itself a set of very complex phenomena). For example, the long-period tidal signals (such as the fortnightly and monthly tides) in ΔLOD arise from the solid-Earth tidal deformations and, to a lesser extent, the ocean tides, which on the other hand are the main causes of diurnal and semidiurnal ΔLOD. For the seasonal (annual + semiannual) signals, though their majority is related to AAM. A considerable amount comes from tides as well as non-atmospheric mass transports. Oceanic angular momentum and land hydrological angular momentum are examples of the latter, which would have their own expressions similar to (8.1). There is also a large, slow (decadal) fluctuation in ΔLOD resulting primarily from core angular momentum variation, and a secular increase in LOD due to "tidal braking" where the Moon slows down the Earth's rotation via tidal energy dissipation, while the distance between the two bodies increases – a phenomenon long discussed since George Darwin in the 19th century (e.g., Cartwright, 1999); these, too, will not be discussed here.

It should be mentioned that the evaluation of the mass term of AAM is subject to an important uncertainty related to the ocean's behavior, namely the inverted-barometer (IB) assumption. The IB assumption stipulates that the world ocean adjusts its level instantaneously and isostatically to the overlying atmospheric pressure variations. Dynamically, the net effect is that the ocean simply smears out the atmospheric pressure variation evenly over the entire ocean area. An idealization which is a good approximation to reality especially on timescales longer than ten days or so such as with ISV, the IB effect leads to a significantly reduced

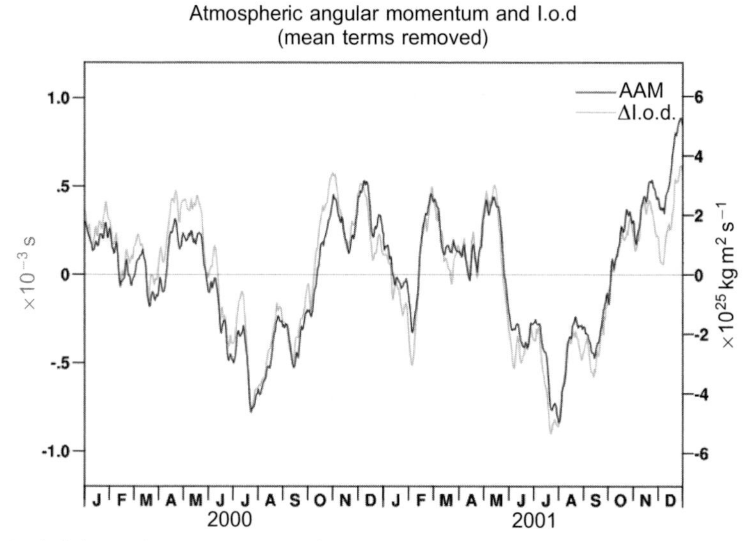

Figure 8.1. Axial angular momentum of the atmosphere AAM_z (black) computed from the NCEP–NCAR. Reanalysis, compared with the LOD (gray), from a combined geodetic solution, in equivalent units for 2 years. The atmospheric excitation AAM_z is a sum of the wind terms and the pressure terms with the IB effect assumed for the ocean. Mean values have been removed. Note the very good agreement of the two series; the discrepancies signify non-atmospheric contributions to the ΔLOD. Besides the seasonal signature a very prominent ISV is noted.

variability in the mass (pressure) terms for ΔLOD, and for that matter, for polar motion, time-variable gravity, and geocenter motion.

Among the first detailed results concerning the AAM_z–ΔLOD relationship were those of Hide *et al.* (1980) and Langley *et al.* (1981). The linear relationship between the two quantities was later firmly demonstrated by Rosen and Salstein (1983) and a number of subsequent investigations in the following two decades, based on progressively improved AAM estimation and LOD observation, which, reassuringly, led to ever improving correlation between them (e.g., Eubanks, 1993; Dickey, 1993). The linear relationship was observed over a wide range of timescales from interannual, to intraseasonal, to as short as a few days (Rosen *et al.*, 1990; Dickey *et al.*, 1992), approaching the limiting resolving power of modern LOD data.

Thus, the close relation between the zonal atmospheric circulation system and ΔLOD on a range of timescales between a few days and years (Figure 8.1) became very evident, once high-quality high-temporal resolution data became available. This was owing to the following facts: (i) ΔLOD on timescales of several days to interannual is dominated, amongst all contributing sources, by AAM_z; (ii) AAM_z in turn is dominated by the contribution made by the motion term determined by the zonal u-wind field (8.1a); (iii) the zonal u-wind is the dominant feature in the atmospheric circulation field; and (iv) the mass term (8.1b) contributes relatively less, so its uncertainly regarding "IB vs. non-IB" is only of secondary influence. Similar close relationships, though, were more difficult to establish about the other geodetic observables, e.g., polar motion, time-variable gravity, or geocenter motion – as it turned out, those corresponding relationships are more elusive and were only later made possible via better observations, analysis, and improvements in atmospheric models.

Today, the AAM_z is routinely computed according to (8.1) at weather prediction centers such as the U.S. National Centers for Environmental Prediction (NCEP) and the European Centre for Medium-range Weather Forecasts (ECMWF), using output from their respective atmospheric analyses, typically at 6-hour intervals. The mass terms are evaluated both ways: with and without the IB assumption. These data sets are available from the Special Bureau for Atmosphere of the International Earth Rotation and Reference Systems Service's Global Geophysical Fluids Center (Salstein *et al.*, 1993, 2001; Chao *et al.*, 2000). An example is given in Figure 8.1 showing a 2-year span of excellent agreement between ΔLOD and AAM_z, including the seasonal signature and the clearly strong and contemporaneous intraseasonal variations. Again, the small differences suggest the existence of non-AAM contributions to ΔLOD, given small errors in the data sets. In fact, such remaining signals have been linked to variations in the oceanic angular momentum (e.g., Marcus *et al.*, 1998; Johnson *et al.*, 1999).

Langley *et al.* (1981) specifically examined the near-50 day (ISV) oscillation in ΔLOD, making a connection to the AAM_z possibly associated with the Madden–Julian Oscillation (MJO; Madden and Julian, 1971, 1972) and the discovery of LOD variability with similar timescales (Feissel and Gambis, 1980). A series of investigations has since examined the ISV in both AAM_z and ΔLOD, notably those of Anderson and Rosen (1983) relating the oscillation to convection in the central Pacific, Weikmann and Sardeshmukh (1994), and Weickmann *et al.* (1997) who broadened the investigation to consider activity in the Indian Ocean as well, and tied the dynamics of angular momentum to wave trains across the Pacific Ocean. In fact, even the confinement of an intraseasonal oscillation to the tropical area was being questioned, with the suggestion that there are angular momentum signatures from an independent extra-tropical oscillation (Dickey *et al.*, 1991). Lau *et al.* (1989) studied the relation of AAM_z and the outgoing long-wave radiation (OLR) in terms of tropical ISV modes. Gutzler and Ponte (1990) examined the coherences among tropical zonal winds, near-equatorial sea level, AAM_z and ΔLOD, all of which exhibit a broad ISV including MJO. Rosen *et al.* (1991) separated the variability in AAM_z in terms of frequency bands and geographic zonal bands, and found in particular that the ISV results mostly from behavior in the tropics and subtropics. Itoh (1994) further specified the subtropics as the source region of the AAM_z ISV. Hendon (1995) confirmed that the 50-day peak in LOD is associated with active phases of the MJO. Marcus *et al.* (2001) reported correlation of ISV in AAM_z and ΔLOD with the El Niño Southern Oscillation (ENSO), but a lack of direct relationship between MJO and ENSO.

For a fuller diagnostic analysis revealing the temporal dependence of the strength of different frequency signals we can decompose the time series using a time–frequency wavelet spectrum. This special technique basically shows a "contour" of spectral power, using a running short-wave packet ("wavelet") of a given shape to pick out the power within any given (narrow) frequency band over a given (short) time period (c.f., Chao and Naito, 1995). Naturally such a spectrum is subject to limited resolution in both frequency and time, but they provide an effective overview to a time series that has broadband spectral content. Thus, in the case of ISV signals with varying frequency–time characteristics, one will be able to see its full evolution.

To produce the wavelet spectrum of the ΔLOD time series (Figure 8.2, upper panel), the strongest signals, i.e., the seasonal terms (annual and semi-annual), have been removed beforehand by subtracting least-squares fits in order to reveal other signals of interest. Clearly seen, though not the subject under discussion here, are the long-period tidal terms (mostly the fortnightly M_f at 13.66 days and the monthly M_m at 27.55 days, modulated at the 18.6-year period related to the tide at the lunar precession period, as well as semi-annual tidal periods as expected). Here we only plot the real part (the cosine term of the wavelet transform) so that the positive–negative polarity of the undulations is shown. It is interesting to note the long, decadal-scale undulations in period exhibited by the strongest (sequence with shading contrast) ISV signals (e.g., the strongest sub-seasonal signals that occur in the last part of the record, around 2001–2002). In comparison, Figure 8.2 (lower panel) (see color section) shows the computed non-seasonal AAM_z. Close examination of the ISV portion of the two spectra reveals the remarkably good agreement, corroborating what was evident in the time-domain comparison (Figure 8.1). Also evident is the agreement in the interannual band, demonstrating that interannual ΔLOD is caused by the changing AAM_z (primarily in the u-wind fields) related to the important ENSO and QBO (Quasi-biennial Oscillation) signals (e.g., Chao, 1989); the alternating signs around 1980–1985 and 1995–2000 are such examples.

8.2.2 Polar motion excitation and equatorial angular momentum

Next we shall study the 2-D equatorial x–y components–AAM_{xy} (in the equatorial plane orthogonal to AAM_z) on the atmospheric side and polar motion, a rotational wobble of the Earth, on the geodetic side. The direction to which the Earth's rotational axis points (near the mean poles) varies due to a number of astronomical and geophysical processes. When observed from space the absolute variations, similar to those of a spinning (and gyrating) top, are known as the *nutation* (including the familiar astronomical precession). On the other hand, the *polar motion* is that of the same rotation axis direction but relative to an observer sitting on Earth and hence rotating with the terrestrial reference frame. The observer sees the polar motion even though the absolute momentum of the entire planet is conserved. An analogy of this motion is that felt by an out on a poorly-thrown, wobbling frisbee. The important fact relevant to us here is that, in contrast to the nutation that magnifies the external astronomical influences, the polar motion magnifies the *internal geophysical* influences such as angular momentum exchanges among geophysical fluids.

Unlike the z-component that is dominated by the atmospheric zonal wind field, the x–y components of the atmospheric dynamics have more subtle interplay, and they have commanded somewhat less attention by atmospheric scientists, though the modes that cause polar motion have been noted in idealized atmospheric models (Feldstein, 2003). Corresponding to (8.1), the expressions for the x–y component of (the normalized, non-dimensional) AAM are approximately (Munk and MacDonald, 1960; Barnes *et al.*, 1983):

$$\text{Motion term of } \mathrm{AAM}_{xy} = \frac{-1.43a^3}{(C-A)\omega g}\iiint (u\sin\theta + iv)e^{i\lambda}\,d\Omega\,dp \tag{8.3a}$$

$$\text{Mass term of } \mathrm{AAM}_{xy} = \frac{-1.00a^4}{(C-A)g}\iint p\cos\theta\sin\theta\, e^{i\lambda}\,d\Omega \tag{8.3b}$$

where C is the mantle's principal momentum of inertia as in (8.1), and A is its mean moment of inertia pointing in the equatorial plane.

Note that now the motion term involves v, the meridional wind field, as well as u. The pressure term dominates the wind term for polar motion excitation. This term results from the (degree = 2, order = 1) term in spherical harmonics favoring zonal wave number 1, and waves with largest amplitudes and opposite phases in the two middle latitude regions of each hemisphere. Such uneven mass distributions, in either latitude or longitude, cause imbalances of mass at opposite parts of the globe that are excitations for polar motion. In terms of our spinning skater analogy, if she draws in her two arms in an asymmetric way (e.g., one arm higher than the other), she would wobble while continuing spinning. Loading near the equator and the poles (0, 90°N, and 90°S latitude) creates no polar motion, as these latitudes are the nodes of the (2,1) excitation term; hence strong ISV across the globe need occur in the middle latitudes to have an appreciable effect on polar motion.

In contrast to (8.1), here the *difference* between the axial and equatorial moment of inertia, $C - A$, comes into play. The relative difference $(C - A)/C$, about 1 part in 300, represents the Earth's oblateness, which in turn is a result of the Earth rotation. It is the stabilizing factor that keeps the polar motion in check and prevents the Earth's axis from, say, undergoing a disastrous tumbling. In fact, modified by a factor related to the elasticity of the Earth and the participation/non-participation of the cores and the ocean, it is responsible for the Earth's resonance oscillation, known as the Chandler wobble, at the natural period of $p_c = 434$ days. As a damped oscillator, the Chandler wobble also has a natural damping factor, or quality factor Q_c, estimated to be upwards from 50 (e.g., Furuya and Chao, 1996).

The observed polar motion is a relatively largely prograde motion (counter-clockwise if viewed from above the North Pole) of the rotation axis around the nominal North Pole, with an amplitude of several meters consisting mainly of a forced annual wobble plus an excited Chandler wobble. The equation of motion that relates the AAM_{xy} to its excited polar motion P, expressed in radian, is:

$$\mathrm{AAM}_{xy}\ (\text{sum of } 8.3\text{a} + 8.3\text{b}) = P - \frac{1}{i\omega_c}\frac{d}{dt}P \tag{8.4}$$

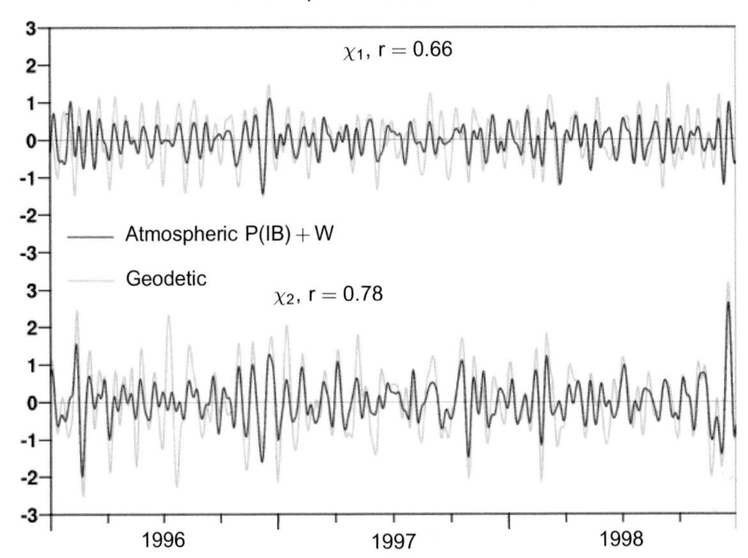

Figure 8.3. Comparison of the excitation function for polar motion: black is the equatorial AAM_{xy} computed from the NCEP–NCAR Reanalysis system, gray is from a deconvolution of geodetically-observed polar motion data from the free Chandler resonance, for the x and y components. The atmospheric excitation AAM_{xy} is a sum of the wind terms and the pressure terms with the IB effect assumed for the ocean. Units are $\times 10^{-7}$ radian (1 radian $= 2.06 \times 10^8$ mas). Here the series are filtered in the ISV band, and have a relatively good phase relationship at that timescale. Correlation coefficients, r, are indicated.

where the Chandler frequency $\omega_c \equiv 2\pi(1 + i/2Q_c)/p_c$, given as a complex number to allow for the damping. In (8.3) and (8.4), the real and imaginary parts represent the x-component along the Greenwich meridian and the y-component along the 90°E longitude.

Equation (8.4) describes an excited 2-D linear, resonance system, with excitation function on the left-hand side in the form of AAM_{xy}. When solved, the polar motion P is the temporal convolution of the polar motion excitation function AAM_{xy} with the Earth's resonance as the free Chandler wobble. Conversely, the right-hand side operation of (8.4) represents the deconvolution of the observed polar motion P from the free Chandler wobble. The result, often expressed as a χ-function (see Figure 8.3), is then the observed *excitation* of the polar motion. It consists of a broadband signal (including ISV), plus strong seasonal (annual + semi-annual) terms mostly of atmospheric origin, and a long-term secular drift primarily due to the post-glacial rebound of the solid Earth resulting from the unloading of the ice sheets since the last glacial age 10,000 years ago. The broadband signal in χ is thus directly comparable to the non-seasonal AAM_{xy}.

Figure 8.3 shows, for a selected few years, a comparison of such polar motion excitations in the ISV band; here we have removed the seasonal and secular terms similarly to the earlier ΔLOD example, as they are outside the temporal scale of our present interest. Again, a good agreement is evident; the broad ISV-band correlation coefficients are as high as 0.66 for the x-component and 0.78 for the y-component.

Such a correlation was reported first by Eubanks *et al.* (1988) and then by a number of other studies. It may be noted, however, that this agreement is not as good as for the axial case for ΔLOD. This difference relates to the different physical mechanisms at work. For example, unlike for ΔLOD, the mass (pressure) term has a larger contribution than the motion (wind) term, hence the IB effect that reduces the effective surface pressure excitations introduces a larger uncertainty in the evaluation of AAM_{xy}. In the axial case, forced mostly by westerly winds, the prevailing zonal circulation of the atmosphere projects strongly onto the excitation of ΔLOD. Regarding the agreement in the atmospheric and geodetic polar motion signals, the amplitude of the meteorological signal exceeds that of the geodetic when the IB is not taken into account, but is too small when it is included (as shown in Figure 8.3); hence a state somewhere in between the non-IB and the full IB effect appears to be likely closer to reality. Also, non-atmospheric sources prove to be relatively important in contributing their own χ, especially the oceanic angular momentum, as demonstrated by Ponte *et al.* (1998), Johnson *et al.* (1999) for example; see also the review by Gross *et al.* (2003).

Understanding the meteorological origins for the polar motion variability is instructive. The variance of the excitation term has its greatest power in regions of variability that often strongly feature fluctuating low pressures, like the North Pacific, North Atlantic, and regions of the southern oceans (Salstein and Rosen, 1989). Though the atmospheric excitation for polar motion from pressure variations over oceanic regions are largely reduced by the IB effect on ISV timescales, over the continents the strong semi-permanent high-pressure regions fluctuate, mainly over Siberia and secondarily over North America, impacting polar motion (Nastula and Salstein, 1999).

Some further insight into ISV can be acquired from the power spectrum of the observed polar motion excitation χ, given in Figure 8.4. Now that the input time series in this case is complex-valued, both positive and negative frequencies are meaningful – the positive frequency refers to the circularly prograde component (i.e., the same direction as the prevailing polar motion), while the negative frequency refers to the retrograde component. Here one sees a broadband, red spectrum which shows a very distinctive asymmetry between the positive (prograde) and negative (retrograde) frequencies all across the broad ISV band. Thus, a considerably larger partition of the AAM_{xy} power resides over the retrograde band, with periods longer than a few days, than in the opposite, prograde direction. This asymmetry has been demonstrated to originate from AAM_{xy}, as it disappears once the AAM_{xy} is subtracted from the time series (Gross *et al.*, 1998b). Superimposed are the long-period peak (at the central, approximately zero frequency), four seasonal peaks (prograde + retrograde for annual + semi-annual, next to the central peak, with the strongest one being the prograde annual term), and a few isolated tidal peaks (e.g., at monthly and fortnightly periods due to ocean tides); all are on timescales quite distant from the ISV scales of interest here.

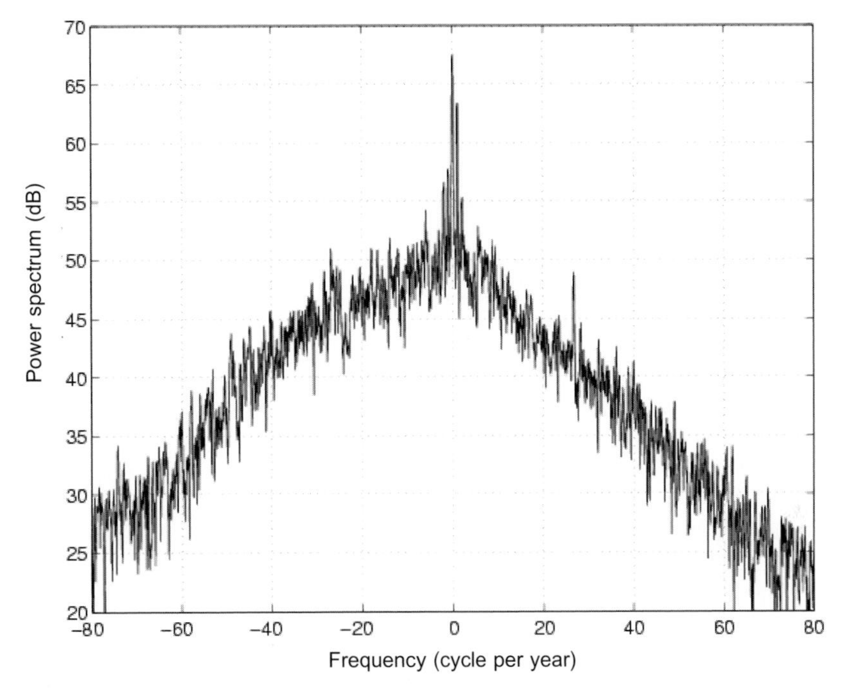

Figure 8.4. The power spectrum of the polar motion excitation obtained from a deconvolution of geodetically-observed polar motion from the free Chandler resonance, 1976–2003. The positive frequency refers to the circularly prograde component, the negative frequency the retrograde component. Note the distinctive asymmetry in the spectrum where considerably higher power resides in the negative than the positive frequencies all across the broad ISV band.

8.2.3 Angular momentum and torques

The transfer of angular momentum between the solid Earth and its fluid envelope is accomplished dynamically by torques from forces acting on the fluid–solid Earth interfaces. The torque vector, of course, is the cross-product of the force and radius vectors. The formulation of the relation between angular momentum and torques can be found in Munk and MacDonald (1960) and Wahr (1982). Basically, in treating Earth rotation problems, one can rely solely on the conservation of angular momentum and equate the opposite changes in the AAM and that of the solid Earth without any prior knowledge of the actual torques that actually transfer the angular momentum. On the other hand, one can endeavor to model and compute the appropriate torques, which in principle should effect the said transfer of angular momentum in exactly the same way (and conserve the total angular momentum) as found in the angular-momentum approach. The angular-momentum and torque approaches are dynamically equivalent, but they have different formulations and face different data availability and uncertainty issues. In particular, the torque approach is less of an "exact science" at present because of an insufficient

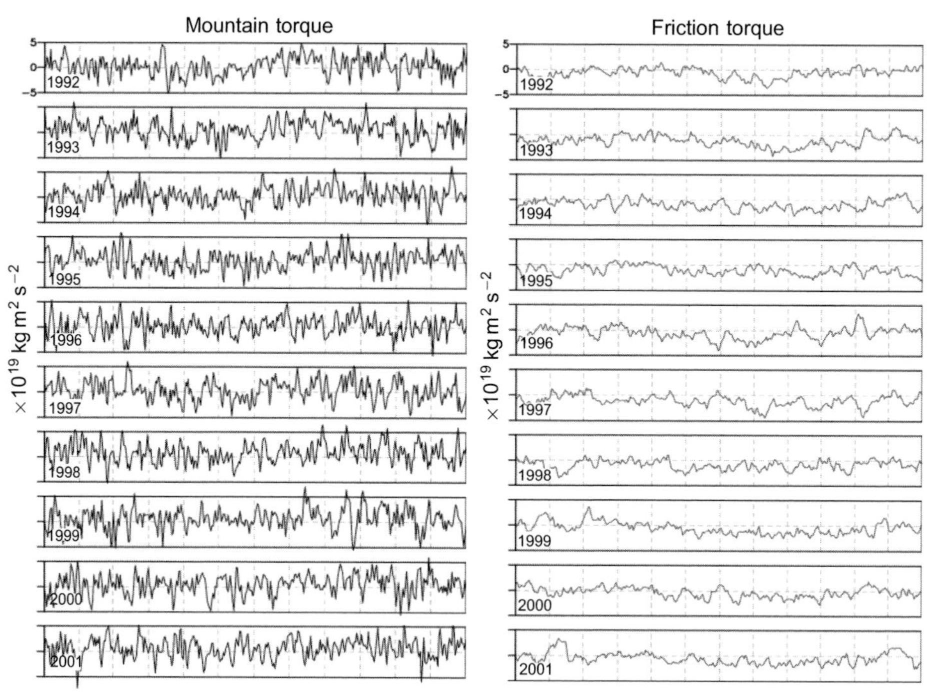

Figure 8.5. Two different atmospheric torques exerted upon the solid Earth in the axial component related to ΔLOD, including the normal mountain torque and the tangential friction torque. Note the prominence of the highest frequencies in the mountain torques, though for the ISV timescales the mountain and friction torques have comparable power.

knowledge about torque mechanisms and hence larger uncertainties in their modeling.

In the case of the atmosphere and ocean, there are two major types of torques, the study of which dates back to White (1949) and Newton (1971). The first is a frictional torque, which is derived from a tangential force in the form of wind stress over land and ocean surfaces, and ocean bottom drag; the second is a pressure torque based on a normal force acting against the topography. At atmosphere–land and ocean–land boundaries, the pressure torque is known as the mountain torque and the continental torque, respectively. For the atmosphere, they are calculated by models within a data assimilation scheme (e.g., Madden and Speth, 1995). Similarly, oceanic torques affecting Earth's rotation were reviewed and computed from models by Fujita *et al.* (2002).

Fundamentally, actions occurring at smaller than the grid scale of most general circulation models generate an additional torque, based on the gravity-wave drag. It has been adapted as an additional torque, though it is small over the timescales considered here. Such gravity waves may be viewed as presenting an accumulation of the effect of horizontal pressure gradients; when described in a model, though, they resemble frictional torques. We plot in Figure 8.5 the two larger torques, the mountain and friction torques, for the axial case, based on the NCEP Reanalysis system. It may be noted that the mountain torque has power at high frequencies related to synoptic activity of storm and fair-weather systems crossing mountainous terrain (Iskenderian and Salstein, 1998), leading to the pressure gradients mentioned above. On such rapid, synoptic scales, friction torque has low-temporal variability, as prevailing winds near the surface do not change so markedly on those timescales. A power spectrum in time (not shown) reveals that around the ISV timescales the friction and mountain torques are of comparable magnitudes. Madden (1987) has indicated that friction torque from anomalies to the east of areas of convection may be responsible for interactions on the 40–50-day timescales.

In principle there is an additional, gravitational torque acting at distance on density anomalies between the fluid–solid Earth (de Viron *et al.*, 1999). The mass, and hence gravity, anomaly of a mountain range would cause an additional force on the atmospheric mass above it that would project onto a torque. Variable atmospheric mass surrounding the mountain would have an uneven effect, yielding a residual force in a particular direction. This effect has an extremely small magnitude in the axial case and so is not plotted in Figure 8.5. However, for signals in the equatorial plane that excite polar motion, the gravitational torque related to the oblate bulge of the Earth is important because of the large latitude-dependent mass anomaly that is the Earth's oblateness. This gravitational torque counterbalances to a considerable extent a pressure "mountain" torque obtained when considering the equatorial bulge as a huge mountain on the Earth (Wahr, 1982; Bell, 1994). Such gravitational torques, however, have relatively small variability on the ISV timescales, even in the equatorial polar-motion related direction. On ISV timescales, nevertheless, the net bulge torque dominates the atmospheric effect on polar motion (Marcus *et al.*, 2004).

8.3 TIME-VARIABLE GRAVITY

Little was known about the Earth's global gravity field beyond a nominal value of g until 1957 when the Sputnik satellites ushered in the space age. From space, the type of data used to solve the gravity "anomaly" is the actual satellite orbit itself, determined by tracking from ground stations. Putting a gravimeter into orbit is fruitless as far as measuring gravity is concerned, because the gravimeter would always be in a free fall, experiencing weightlessness along with the satellite body itself. Deep down this is a consequence of the equivalence principle between gravitational mass and inertia mass and hence between gravity and acceleration, a unique property of gravity according to general relativity. This situation makes gravity much different from (and its space measurement so much harder than) other physical properties that are conducive to direct remote sensing from space. Nevertheless, the orbit of a satellite does contain information about the gravity field through which the satellite traverses – the overall mean elliptic orbit gives the main spherically-symmetric (the "monopole") part, whereas the departures therefrom signify the gravity *anomalies*. With enough different satellite orbits, the complete gravity field can in principle be determined.

Over the years, the tracking of more and more satellites began to provide just such data, and ever higher spatial degree harmonic components of the gravity field were recovered (higher harmonics mean shorter "wavelengths" or higher spatial resolution, see below). Satisfying the Laplace equation, the gravity potential field U has the closed-form solution, customarily expressed as (e.g., Kaula, 1966):

$$U(r, \theta, \phi) = \frac{GM}{a} \sum_{l=0}^{\infty} \sum_{m=0}^{l} \left(\frac{a}{r}\right)^{l+1} P_{lm}(\cos\theta)(C_{lm}\cos m\lambda + S_{lm}\sin m\lambda) \tag{8.5}$$

where M is the Earth's mass, and P_{lm} is the 4π-normalized Legendre function. The coefficients C_{lm} and S_{lm} are known as the (normalized) harmonic *Stokes coefficient* of degree $l(= 0, 1, 2, \ldots, \infty)$ and order $m(= 0, 1, 2, \ldots, l)$; they constitute a basis set of functions which at a given truncation specify a gravity model. On the other hand, from Newton's gravitational law and the so-called multipole expansion of the field U (e.g., Jackson, 1975), one realizes that the Stokes coefficients are simply normalized multipoles of the density distribution of the body (e.g., Chao, 1994):

$$C_{lm}(t) + iS_{lm}(t) = \frac{1}{(2l+1)Ma^l} \iiint \varrho(\mathbf{r}; t) r^l Y_{lm}(\Omega)\, dV \tag{8.6}$$

where $Y_{lm}(\Omega) = P_{lm}(\cos\theta)\exp(im\lambda)$. When there is mass redistribution $\Delta\varrho(\mathbf{r}; t)$ in the body, the external gravity field would change accordingly. In particular, for mass redistributions $\Delta\sigma(\Omega; t)$ that can be considered as happening approximately on a spherical surface $r = a$, (8.6) reduces from a volume integral to a surface integral:

$$\Delta C_{lm}(t) + i\Delta S_{lm}(t) = \frac{a^2}{(2l+1)M} \iint \Delta\sigma(\Omega; t) Y_{lm}(\Omega)\, d\Omega \tag{8.7}$$

Again, for the loading of the atmospheric mass over ocean, $\Delta\sigma(\Omega; t)$ is modified greatly by the oceanic IB effect. For the atmosphere (or oceans), $\Delta\sigma(\Omega; t)$ can be easily converted from the surface (or ocean bottom) pressure field assuming hydrostatic equilibrium: $p = g\Delta\sigma$. Similarly as with the angular momentum, one employs the pressure data sets provided by weather centers such as NCEP or ECMWF.

As with Earth rotation, contributions to the time-variable gravity (8.7) come from mass transports in all geophysical fluids. Therefore, ultimately our study pertains to all mass redistribution as a function of space and time. So in principle, variations of gravity would be related to ISV if masses are redistributed at that timescale. In Section 8.2, the Earth rotation variation represents a globally integrated angular momentum quantity; the little spatial resolution that exists is afforded by the kernel function in (8.1) and (8.3) during the integration (e.g., the Legendre functions of degree 2 in the case of the mass terms). Note that the Stokes coefficients of degree 2, order 1 (2,1), is identical to the mass term of the polar motion excitation (apart from a constant factor), and so is the $(2, 0)$ coefficient to the mass term of the ΔLOD excitation. In comparing gravity to Earth rotation studies in understanding interactions between the atmosphere and solid Earth, the major difference, and a fundamental advantage, is the fact that the gravity observation carries with it spatial resolution, growing with the harmonic degree afforded by the observations. If the maximum harmonic degree that can be supported by the available data is L, then the nominal spatial resolution is the Earth's circumference (40,000 km) divided by $2L$.

The precise SLR techniques have detected relatively minute but important time-variable gravity signals on the largest spatial scales; variations longer than monthly can now be clearly identified. Many studies have made clear connection of these signals to atmospheric and oceanic mass redistributions (e.g., Chao and Au, 1991; Johnson *et al.*, 2001). In another example, Cox and Chao (2002) reported and speculated on possible meteorological causes for a positive anomaly in the Earth's oblateness, corresponding to the negative of the (2, 0) Stokes coefficient, during 1998–2001, superimposed on the well-known secular decrease of the oblateness due primarily to the post-glacial rebound. However, the SLR data are intrinsically limited in resolution: monthly sampling may not be very useful in studying ISV in mass redistribution, and the geographical resolution associated with low-degree harmonics (say, below degree 4 or 5) is rather low.

On the other hand, the space gravity missions of CHAMP (launched in July 2000) and especially GRACE (launched in March 2002, with an expected lifetime of over 5 years) are yielding gravity information at much higher precision and geographical resolution than the earlier SLR-based information. For example, GRACE promises to be able to detect sub-centimeter level water-layer-equivalent mass changes over an area of a few hundred kilometers across every month or so (Tapley *et al.*, 2004; Wahr *et al.*, 1998, 2004).

At monthly and longer time resolutions, both the classic SLR and the new GRACE time-variable gravity data are yet barely capable of resolving many interesting ISV signals. In the near future, though, GRACE solutions at time intervals as short as 10 days will become available, through trade-offs of temporal resolution with spatial resolution and precision.

An indicative example of what GRACE can elucidate about intraseasonal mass redistribution is given in Figure 8.6 (see color section). From these GRACE observations, the effect of the atmospheric mass redistribution, including its high temporal resolution, based on the ECMWF analyses is removed, as well as oceanic contributions according to a nominal barotropic ocean model (see Ali and Zlotnicki, 2003). This process is needed to de-alias the signal as observed by GRACE at shorter timescales, and leaves an approximately monthly time resolution for the gravity signals. The map contours in Figure 8.6 (see color section) show the strength of residual intraseasonal mass anomaly in terms of equivalent-water-depth standard deviation after removal of any remaining seasonal signals and linear trends. The map is composed of the superposition of up to (10,10) harmonics (hence a geographical resolution of about 2,000 km) and is averaged over 22 monthly solutions spanning the period between April 2002 and July 2004. The most prominent signals with the strongest ISV are land hydrology. In particular, the moderate ISV signal level over the oceans at face value implies that the applied ocean model did a preliminary job in modeling (and hence removing signals) of mass redistribution in the oceans up to harmonic degree 10. However, more complete scientific results await further and more detailed analyses to higher degrees based on longer and more accurate observations.

8.4 GEOCENTER MOTION

The degree $l = 0$ term in (8.5) corresponds to the gravity generated by the spherically symmetric Earth, or conceivably the "monopole" as if the total Earth mass were collapsed into a singular point. The next higher components, the three $l = 1$ terms or the dipoles, correspond to the position vector of the geocenter, or the CM of the Earth system. This vector would of course be identically zero if the coordinate origin is always fixed at the geocenter. In practice, however, observatories fixed on the surface of the solid Earth, tracking the dynamic satellites, define a slightly different reference system, where the CM of the solid Earth deviates slightly from the geocenter; these slight shifts lead to so-called geocenter motion. This apparent motion is a result of the mass shifts in the moving geophysical fluid, as dictated by the conservation of linear momentum. Its observed amplitude is of the order of 1 cm, smaller than the size of a cherry, with strong seasonality as well as ISV.

Just as for any other Stokes coefficients (8.7), the geocenter motion due to the atmosphere in particular, ISV or otherwise, can be readily evaluated given the global data sets from, say, NCEP or ECMWF. Together with additional contributions from the oceans and land hydrology (if similarly evaluated via global data or model outputs), they should be able to explain the actual geocenter motion (e.g., Chen *et al.*, 1999). The space-geodetic observation of the geocenter motion is currently in its infancy; the solutions sometimes have systematic differences between those determined by different techniques as large as the signal level (IERS, 1999). However, the three components of the geocenter motion in principle contain as much (independent) information about global mass redistribution as the three components of the Earth rotation variation. Monitoring the geocenter thus promises great utility in studying global mass changes, as in fact has already been demonstrated by Blewitt *et al.* (2001).

8.5 CONCLUSIONS

The conservation of angular momentum is the physical principle that governs the influence of AAM on the Earth's rotation. This is of course true for any angular momentum variation in any geophysical fluid on any timescale. The practical significance for knowing the Earth's rotation is of great interest because of navigational needs, both for near-Earth and for interplanetary spacecraft applications, for timekeeping concerns, and for understanding the geophysical balances of the planet. Besides practical applications in monitoring and predicting Earth rotation variations, the scientific significance of the study of this relationship as presented above is several-fold: at a root level, we have a quantitative "confirmation" of the validity of the principle of angular momentum conservation on planetary scales; but more importantly, to the extent permitted by the error bounds of the data, we have a quantitative confirmation of the validity of the Earth rotation measurement as well as the validity of the calculated angular momentum based on GCM outputs. Conversely, the discrepancies as a function of frequency and time contain information about angular momentum variations from other secondary sources not already accounted for, whether meteorological, geophysical, or even anthropogenic (as in the case of major diversions of water mass, say, in building dams). In addition, they can lead to quantitative improvements of the models and the geophysical formulations employed in the angular momentum calculations, thereby providing useful constraints leading to a better understanding of mass transports in the geophysical fluids.

The same argument is valid with respect to other global geodynamic observables due to mass transports in the geophysical fluids. In this sense, Earth rotation variations, time-variable gravity, and geocenter motion all contain information about ISV in particular. An important and unique feature is that these observables pertain to the mass, a quantity not readily derived by remote sensing observations otherwise.

Nevertheless, these geodynamic observables have intrinsic limitations due to their integrating nature, despite the triumphant progress seen in the increasingly accurate space-geodetic techniques in the last 2–3 decades. For one thing, the geodynamic effects sense the sum total of all mass transports, not discriminating the meteorological/geophysical sources. To effectively distinguish the processes, geographical identification comes into play. While, as integrated global quantities, the Earth rotation and geocenter observations have little spatial resolution to speak of, a reasonably high resolution afforded by the time-variable gravity provides a powerful identification of the actual sources at any given time. Such is the case shown in Figure 8.6 (see color section), where the knowledge of time-variable gravity can have dramatic effects on monitoring various mass transport processes.

Another key consideration related to the usefulness of geodetic techniques as applied in the research of ISV, is the timescale and temporal resolution of observations. Meteorological and geophysical processes are typically characterized by distinct timescales, but there are exceptions, such as a number of geophysical/climatic changes that are broadband phenomena having timescales upwards from decadal to secular (e.g., ice sheet dynamics, post-glacial rebound, core flows). The ISV, of one such resolution, has been noted in geodetic signals and appears simply linked to the MJO discussed in other chapters. For the broad range of timescales, however, other than tidal variations which have specific and well-defined periods, the only significant sources are meteorological and some oceanic forcing as discussed here. Relative to ISV, the Earth rotation and geocenter observations can (and do) have reasonably high temporal resolutions, whereas the time-variable gravity currently has temporal resolutions inadequate for ISV but is potentially capable of reaching higher resolutions. There are challenges and at the same time exciting prospects ahead of us in interpreting and making optimal applications of the geodynamic observations in the study of ISV in the Earth system.

8.6 ACKNOWLEDGMENTS

Supported by NASA's Solid Earth Sciences programs, the space-geodesy techniques that resulted in the Earth rotation, gravity, and geocenter data have been developed over decades by a truly international effort. The geophysical fluids data used in this chapter are provided by IERS' Global Geophysical Fluids Center, in particular its Special Bureaus for the Atmosphere and for the Oceans. The Earth rotation data were kindly provided by Richard Gross; Figure 8.6 (see color section) was made available by David Garcia from newly released GRACE data, courtesy of the joint USA-Germay Project of GRACE. The work is also supported by grants ATM0002688 and ATM0429975 from the US National Science Foundation.

8.7 REFERENCES

AGU (1993) *Contributions of Space Geodesy to Geodynamics: Technology, Geodynamics Series 25*, D. E. Smith and D. L. Turcotte (eds.). Amer. Geophys. Union, Washington, D.C.

Ali, A. H. and Zlotnicki, V. (2003) Quality of wind stress fields measured by the skill of a barotropic ocean model: Importance of stability of the Marine Atmospheric Boundary Layer. *Geophys. Res. Lett.*, **30**(3): Art. #1129, February 11.

Anderson, J. R. and R. D. Rosen (1983) The latitude–height structure of the 40–50 day variations in the atmospheric angular momentum. *J. Atmos. Sci.*, **40**, 1584–1591.

Barnes, R. T. H., R. Hide, A. A. White, and C. A. Wilson (1983) Atmospheric angular momentum fluctuations, length-of-day changes and polar motion. *Proc. Roy. Soc. Lond. A*, **387**, 31–73.

Bell, M. J. (1994) Oscillations in the equatorial components of the atmosphere's angular momentum and torques on the earth's bulge. *Q. J. R. Meteorol. Soc.*, **120**, 195–213.

Blewitt, G., D. Lavallee, P. Clarke, and K. Nurutdinov (2001) A new global mode of Earth deformation: Seasonal cycle detected. *Science*, **294**, 2342–2345.

Cartwright, D. E. (1999) *Tides: A Scientific History*, Cambridge University Press, Cambridge, UK.

Chao, B. F. (1989) Length-of-day variations caused by El Niño-Southern Oscillation and Quasi-Biennial Oscillation. *Science*, **243**, 923–925.

Chao, B. F. and A. Y. Au (1991) Temporal variation of Earth's zonal gravitational field caused by atmospheric mass redistribution: 1980–1988. *J. Geophys. Res.*, **96**, 6569–6575.

Chao, B. F. (1994) The Geoid and Earth Rotation. In: P. Vanicek and N. Christou (eds), *Geophysical Interpretations of Geoid*. CRC Press, Boca Raton.

Chao, B. F. and I. Naito (1995) Wavelet analysis provides a new tool for studying Earth's rotation. *EOS, Trans. Amer. Geophys. Union*, **76**, 161–165.

Chao, B. F., V. Dehant, R. S. Gross, R. D. Ray, D. A. Salstein, M. M. Watkins, and C. R. Wilson (2000) Space geodesy monitors mass transports in global geophysical fluids. *EOS, Trans. Amer. Geophys. Union*, **81**, 247–250.

Chao, B. F. (2003) Geodesy is not just for static measurements any more. *EOS, Trans. Amer. Geophys. Union*, **84**, 145–156.

Chen, J. L., C. R. Wilson, R. J. Eanes, and R. S. Nerem (1999) Geophysical interpretation of observed geocenter variations. *J. Geophy. Res.*, **104**, 2683–2690.

Cox, C. M. and B. F. Chao (2002) Detection of a large-scale mass redistribution in the terrestrial system since 1998. *Science*, **297**, 831–833.

Da Szoeke, R. A. and R. M. Samelson (2002) The duality between the Boussinesq and non-Boussinesq hydrostatic equations of motion. *J. Phys. Oceanogr.*, **32**, 2194–2303.

de Viron, O., C. Bizouard, D. A. Salstein, and V. Dehant (1999) Atmospheric torque on the Earth and comparison with atmospheric angular momentum variations. *J. Geophys. Res.*, **104**, 4861–4875.

Dickey, J. O., M. Ghil, and S. L. Marcus (1991) Extratropical aspects of the 30–60 day oscillation in length-of-day and atmospheric angular momentum. *J. Geophys. Res.*, **96**, 22643–22658.

Dickey, J. O., S. L. Marcus, J. A. Steppe, and R. Hide (1992) The Earth's angular momentum budget on subseasonal timescales. *Science*, **255**, 321–324.

Dickey J. O. (1993) Atmospheric excitation of the Earth's rotation: Progress and prospects via space geodesy. In: D. E. Smith and D. L. Turcott (eds), *Contributions of Space Geodesy to Geodynamics: Earth Dynamics*, AGU, Washington, D.C., pp. 55–70.

Eubanks, T. M., J. A. Steppe, J. O. Dickey, R. D. Rosen, and D. A. Salstein (1988) Causes of Rapid Motions of the Earth's Pole. *Nature*, **334**, 115–119.

Eubanks, T. M. (1993) Variations in the orientation of the Earth. In: D. E. Smith and D. L. Turcott (eds), *Contributions of Space Geodesy to Geodynamics: Earth Dynamics*. AGU, Washington, D.C., pp. 1–54.

Feissel, M. and D. Gambis (1980) La mise en evidence de variations rapides de la duree du jour. *Compt. Rendus Hebdomadaires Sceances Acad. Sci.*, Ser. B, **291**, 271–273.

Feldstein, S. B. (2003) The dynamics associated with equatorial atmospheric angular momentum in an aquaplanet GCM. *J. Atmos. Sci.*, **60**, 1822–1834.

Fujita, M., B. F. Chao, B. V. Sanchez, and T. J. Johnson (2002) Oceanic torques on solid Earth and their effects on Earth rotation: Results from the Parallel Ocean Climate Model. *J. Geophys. Res.*, **107**(B8), doi:10.1029/2001JB000339.

Furuya, M. and B. F. Chao (1996) Estimation of period and *Q* of the Chandler wobble. *Geophys. J. Int.*, **127**, 693–702.

Gross, R. S., T. M. Eubanks, J. A. Steppe, A. P. Freedman, J. O. Dickey, and T. F. Runge (1998a) A Kalman-filter-based approach to combining independent Earth-orientation series. *J. Geodesy*, **72**, 215–235.

Gross, R. S., B. F. Chao, and S. D. Desai (1998b) Effect of long-period ocean tides on the Earth's polar motion. *Prog. in Oceanography*, **40**, 385–397.

Gross, R. S., I. Fukumori, and D. Menemenlis (2003) Atmospheric and oceanic excitation of the Earth's wobbles during 1980–2000. *J. Geophys. Res.*, **108**, No. B8, 2370, doi:10.1029/2002JB002143.

Gutzler, D. S. and R. M. Ponte (1990) Exchange of momentum, among atmosphere, ocean, and solid Earth associated with the Madden–Julian Oscillation. *J. Geophys. Res.*, **95**, 18679–18686.

Hendon, H. H. (1995) Length of day changes associated with the Madden–Julian oscillation. *J. Atmos. Sci.*, **52**, 2373–2383.

Hide, R., N. T. Birch, L. V. Morrison, D. J. Shea, and A. A. White (1980) Atmospheric angular momentum fluctuations and changes in the length of the day. *Nature*, **286**, 114–117.

IERS (1999) IERS analysis campaign to investigate motions of the geocenter, IERS Tech. Note #25, edited by J. Ray.

IPCC (2001) A contribution of Working Groups I, II and III to the Third Assessment Report of the Intergovernmental Panel on Climate Change (R. T. Watson and the Core Writing Team (eds)). Cambridge University Press, Cambridge, UK, pp. 398.

Iskenderian, H. and D. A. Salstein (1998) Regional sources of mountain torque variability and high-frequency fluctuations in atmospheric angular momentum. *Mon. Wea. Rev.*, **126**, 1681–1694.

Itoh, H. (1994) Variations of atmospheric angular momentum associated with intra-seasonal oscillations forced by zonally moving prescribed heating. *J. Geophys. Res.*, **99**, 12981–12998.

Jackson, J. D. (1975) *Classical Electrodynamics* (2nd ed.). Wiley, New York.

Jault, D. and J. L. Le Mouel (1989) The topographic torque associated with the tangentially geostrophic motion at the core surface and inferences on the flow inside the core. *Geophys. Astrophys. Fluid Dynamics*, **48**, 273–296.

Johnson, T. J., C. R. Wilson, and B. F. Chao (1999) Oceanic angular momentum variability estimated from the Parallel Ocean Climate Model, 1988–1998. *J. Geophys. Res.*, **104**, 25183–25196.

Johnson, T. J., C. R. Wilson, and B. F. Chao (2001) Non-tidal oceanic contributions to gravitational field changes: Predictions of the Parallel Ocean Climate Model. *J. Geophys. Res.*, **106**, 11315–11334.

Kang, I.-S. and K. M. Lau (1990) Evolution of tropical circulation anomalies associated with 30–60 day oscillation of globally averaged angular momentum during northern summer. *J. Meteorol. Soc. Jap.*, **68**, 237–249.

Kaula, W. M. (1966) *Theory of Satellite Geodesy*. Blaisdell Publishing Co., Waltham, M.A.

Langley, R. B., R. W. King, I. I. Shapiro, R. D. Rosen, and D. A. Salstein (1981) Atmospheric angular momentum and length of day: A common fluctuation with a period near 50 days. *Nature*, **294**, 730–733.

Lau, K. M., I. S. Kang, and P. J. Sheu (1989) Principal modes of intraseasonal variations in atmospheric angular momentum and tropical convection. *J. Geophys. Res.*, **94**, 6319–6332.

Madden, R. A. and P. R. Julian (1971) Description of a 40–50 day oscillation in the zonal wind in the tropical Pacific. *J. Atmos. Sci.*, **28**, 702–708.

Madden, R. A. and P. R. Julian (1972) Description of global-scale circulation cells in the tropics with a 40–50 day period. *J. Atmos. Sci.*, **29**, 1109–1123.

Madden, R. A. (1987) Relationships between changes in the length of day and the 40–50 day oscillation in the tropics. *J. Geophys. Res.*, **92**, 8391–8399.

Madden, R. A. and P. Speth (1995) Estimates of atmospheric angular momentum, friction, and mountain torque during 1987–1988. *J. Atmos. Sci.*, **52**, 3681–3694.

Marcus, S. L., Y. Chao, J. O. Dickey, and P. Gegout (1998) Detection and modeling of nontidal oceanic effects on Earth's rotation rate. *Science*, **281**, 1656–1659.

Marcus, S. L., J. O. Dickey, and O. de Viron (2001) Links between intraseasonal (extended MJO) and ENSO timescales: Insights via geodetic and atmospheric analysis. *Geophys. Res. Lett.*, **28**, 3465–3458.

Marcus, S. L., O. de Viron, and J. O. Dickey (2004) Atmospheric contributions to Earth nutation: Geodetic constraints and limitations of the torque approach. *J. Atmos. Sci.*, **61**, 352–356.

Munk, W. H., and G. J. F. MacDonald (1960) *The Rotation of the Earth*. Cambridge University Press, New York.

Nastula, J. and D. A. Salstein (1999) Regional atmospheric angular momentum contributions to polar motion excitation. *J. Geophys. Res.*, **104**, 7347–7358.

Newton, C. W. (1971) Mountain torques in the global angular momentum balance. *J. Atmos. Sci.*, **28**, 623–628.

Peixoto, J. P. and A. H. Oort (1992) *Physics of Climate*. American Institute of Physics, New York.

Ponte, R. M., D. Stammer, and J. Marshall (1998) Oceanic signals in observed motions of the Earth's pole of rotation. *Nature*, **391**, 476–479.

Rosen, R. and D. Salstein (1983) Variations in atmospheric angular momentum on global and regional scales and the length of day. *J. Geophys. Res.*, **88**, 5451–5470.

Rosen, R. D., D. A. Salstein, and T. M. Wood (1990) Discrepancies in the Earth–Atmospheric Angular Momentum Budget. *J. Geophys. Res.*, **95**, 265–279.

Rosen, R. D., D. A. Salstein, and T. M. Wood (1991) Zonal contributions to global momentum variations on intraseasonal through interannual time scales. *J. Geophys. Res.*, **96**, 5145–5151.

Salstein, D. A., D. M. Kann, A. J. Miller, and R. D. Rosen (1993) The sub-bureau for Atmospheric Angular Momentum of the International Earth Rotation Service (IERS): A meteorological data center with geodetic applications. *Bull. Am. Met. Soc.*, **74**, 67–80.

Salstein, D. A. and R. D. Rosen (1989) Regional contributions to the atmospheric excitation of rapid polar motion. *J. Geophys. Res.*, **94**, 9971–9978.

Salstein, D. A., O. de Viron, M. Yseboodt, and V. Dehant (2001) High-frequency geophysical fluid modeling necessary to understand Earth rotation variability. *EOS*, **82**, 237–238.

Tapley, B. D. (2002) The GRACE mission: Status and performance assessment. American Geophysical Union fall meeting, San Francisco.

Tapely, B. D., S. Battadpur, M. Watkins, and C. Reigber (2004). The Gravity and Climate Recovery Experiment: Mission overview and early results. *Geophys. Res. Lett.*, **31**, L09607, doi:10.1029/2004GL019920.

Wahr, J. M. (1982) The effects of the atmosphere and oceans on the Earth's wobble. Part I: Theory. *Geophys. J. R. Astron. Soc.*, **70**, 349–372.

Wahr, J., M. Molenaar, and F. Bryan (1998) Time variability of the Earth's gravity field: Hydrological and oceanic effects and their possible detection using GRACE. *J. Geophys. Res.*, **103**, 30205–30230.

Wahr, J., S. Swenson, V. Zlotnicki, and I. Velocogna (2004) Time-variable gravity from GRACE: First results. *Geophys. Res. Lett.*, **31**, L11501, doi:10.1029/2004GL019779.

Weickmann, K. and P. Sardeshmukh (1994) The atmospheric angular momentum budget associated with a Madden–Julian Oscillation. *J. Atmos. Sci.*, **51**, 3194–3204.

Weickmann, K., G. Kiladis, and P. Sardeshmukh (1997) The dynamics of intraseasonal atmospheric angular momentum oscillations. *J. Atmos. Sci.*, **54**, 1445–1461.

White, R. M. (1949) The role of the mountains in the angular momentum balance of the atmosphere. *J. Meteor.*, **6**, 353–355.

9

El Niño Southern Oscillation connection

William K. M. Lau

9.1 INTRODUCTION

The Madden–Julian Oscillation (MJO) is the most pronounced signal in tropical intraseasonal (20–90 days) variability, and the El Niño Southern Oscillation (ENSO) is the most dominant interannual climate phenomenon in the tropical ocean–atmosphere system. Both MJO and ENSO involve major shifts in tropical convection, large-scale circulation, and weather patterns around the world. The hypothesis that the MJO and ENSO may be intrinsically linked was first proposed in the mid-1980s by Lau (1985a, b). Subsequently, many observational and modeling studies have appeared in the literature debating the merits of the hypothesis. Today, while the MJO–ENSO connection is still a topic of active research (Zhang *et al.*, 2001), knowledge gained from better understanding of the causes and evolution of the MJO and ENSO has been incorporated into long-range weather forecasting and climate prediction schemes, resulting in improved forecasts not only in the tropics but also in many extra-tropical regions. A key factor in the prediction improvement is the recognition that the MJO and ENSO events do not act independently, but may interact with each other to provide the long-term pre-conditions, and the short-term fine tuning needed for better skill in long-range (> months) predictions. How can the MJO and ENSO – two phenomena with widely separate timescales, be physically linked and interact with each other? This critical question and related issues will be addressed in this chapter.

Here, as in other chapters of this book, the MJO is considered in the context of tropical intraseasonal variability (TISV). In Section 9.2, a historical perspective of the TISV–ENSO connection will be presented. This is followed in Sections 9.3–9.6 by discussions, in terms of four major phases, about the development and evolution of the paradigm of the TISV–ENSO relationship, from 1980 to the present-day. Note that the phases have much overlap, and do not strictly follow a chronologic order.

W. K. M. Lau and D. E. Waliser (eds), *Intraseasonal Variability in the Atmosphere–Ocean Climate System.*

- *Phase-1*: this is the embryonic stage beginning in the early 1980s consisting of mostly observational studies based on emerging satellite data and operational analyses, with somewhat sketchy descriptions and rudimentary ideas. This stage also included observational and theoretical work, documenting the hierarchical structure of the MJO, westerly wind bursts, and coupling instability of the tropical ocean–atmosphere.
- *Phase-2*: this is the exploratory stage from the late 1980s to mid-1990s, represented by a large number of more in-depth observational and modeling studies, focusing on the understanding of possible mechanisms. These studies generally did not find strong statistical evidence of a MJO–ENSO link, but yet could not exclude the possible relevance of such a connection.
- *Phase-3*: this phase encompassed many case studies of El Niño and La Niña, including the oceanic remote response to TISV forcings carried out in the mid-1990s to early 2000s. The strong TISV forcings associated with the onset and demise of the 1997/1998 El Niño lended definitive support to, and rejuvenated the debate on the MJO–ENSO connection. This period also covers work on the development of a more comprehensive and dynamical framework, including the concept of optimal stochastic forcings, and rectification of the MJO into ENSO cycles.
- *Phase-4*: this phase covers recent developments, including new observational insights, and many studies focused on realistic simulation of the MJO in atmospheric models and coupled climate models, and effects of TISV and MJO on the predictability of ENSO.

Studies conducted in Phase-2 and 3 have considerable overlap with material covered in Chapters 6 and 7 of this book. Here, we shall only concentrate on the aspects relevant to the TISV–ENSO connection and refer details of TISV to those chapters. For Phase-4, we also include results from recent observations to shed new light on the TISV–ENSO relationship. The chapter ends with a discussion in Section 9.7 on the role of TISV in ENSO predictability.

9.2 AN HISTORICAL PERSPECTIVE

The MJO, originally known as the 40–50-day oscillation of the tropical atmosphere was first reported in two seminal papers by Madden and Julian (1971, 1972; see Chapter 1). However, the importance of the discovery was largely left unnoticed for almost a decade. It was not until the early 1980s, when the phenomenon was "rediscovered" almost serendipitously and its paramount importance was recognized, thanks to the convergence of several major events in the field of atmospheric and oceanic sciences.

The first was the launch of the First GARP Global Experiment (FGGE) in 1978/1979, which was the first global-scale observation and field program undertaken by the international meteorological and oceanographic communities in an attempt to provide a complete description of the general circulation of the atmosphere with the objective to improve weather prediction (McGovern and Fleming 1977; Kalnay *et al.*, 1981). One of the special observing periods in FGGE was the Monsoon Experiment (MONEX), in which regional field campaigns were conducted to provide better observations of the Asian summer and winter monsoons. FGGE and MONEX led to the discovery of pronounced intraseasonal oscillations (ISOs), with quasi-biweekly and 30–60-day timescales in the global tropics and in the monsoon regions (Lorenc, 1984; Krishnamurti and Subramaya, 1983; Krishnamurti *et al.*, 1985). These studies led some scientists to believe in the early 1980s that the Asian monsoon was responsible for exciting the MJO (Yasunari, 1982; Murakami *et al.*, 1984; Murakami and Nakazawa, 1985). Today, it is recognized that the MJO is an intrinsic phenomenon of the tropical atmosphere, but subject to modulation by air–sea interaction, as well as regional land and oceanic processes associated with the Asian–Australian monsoon.

The second major event was the abrupt occurrence of the major El Niño of 1982/1983. In view of the influential work of Wyrtki (1975, 1979), who postulated a sea-level rise in the western Pacific as a pre-condition for the onset of El Niño, everyone was caught by surprise by the sudden onset of the 1982/1983 El Niño, which occurred without a major build-up phase (Cane, 1983; Gill and Rasmusson, 1983). Many scientists came to realize that they did not understand the mechanisms of El Niño at all, and that more observations and better models were needed. The frustration at being fooled by nature had provided the impetus for many scientists to start looking for better explanations of El Niño, including coupled instability of the tropical ocean–atmosphere and the role of atmospheric transients. This impetus led to the establishment of the very successful Tropical Ocean Global Atmosphere (TOGA) program in 1985–1995 for monitoring and improving the prediction capability of ENSO (National Research Council Report, 1996).

The third event that influenced the underlying ideas of a MJO–ENSO connection was the commencement of the modern satellite era in the late 1970s, coupled with the development of the Global Telecommunication System (GTS) enabling operational weather services around the world to use and produce satellite products for weather forecasting. Indeed, the recognition of the importance of using and sharing satellite information for weather forecasting was the driving force behind the World Weather Watch (WWW), a component of the FGGE. One of the first and most important satellite products used for climate research was the Advanced Very High Resolution Radiometer (AVHRR) product on the NOAA TIROS satellite since 1974. By the early 1980s, almost 10 years of daily AVHRR data were archived and available to the research community. The most widely used AVHRR product was the outgoing long-wave radiation (OLR), which provided a broadband measure of the total flux of long-wave radiation loss to space at the top of the atmosphere (Gruber and Krueger, 1984). Deep convective clouds in the tropics have cold cloud tops, and therefore have a low value of OLR. Typically a value of less than 200 W m^{-2} signals the presence of deep convection in the tropics (Waliser *et al.*, 1993). Because of this simple, but unique property of OLR, it has been widely used as a proxy for deep convection over the tropical oceans, where the background long-wave radiation from low clouds or from the ocean surface is much higher (>220 W m^{-2}). In the early to mid-1980s, a series of observational papers on the MJO (at the time still referred to as the 40–50-day or 30–60-day oscillation) using OLR and wind analyses from the US National Meterological Center (NMC) appeared (e.g., Lau and Chan, 1983a, b, 1985, 1986a, b; Weickmann 1983; Weickmann *et al.*, 1985; Murakami and Nakazawa, 1985; Knutson and Weickmann, 1987). These papers helped to bring the MJO to the attention of the scientific community.

9.3 *PHASE-1*: THE EMBRYONIC STAGE

The basic concept of an MJO–ENSO connection was first proposed in Lau (1985a, b). Noting the similar spatial scales, but vastly different time scales of variability associated with the MJO and ENSO, Lau proposed a theoretical framework in which three basic elements (unstable air–sea interactions, the seasonal cycle, and stochastic forcing from high-frequency transients such as MJO) were identified to be crucial factors leading to the ENSO cycle and its long-term behavior. Using a simple non-linear oscillator model, Lau demonstrated that the interplay of the aforementioned processes could produce many salient features of the variability of ENSO including the 2–5 years recurrence interval, auto-correlation and spectral characteristics, as well as phase-locking with the seasonal cycle. In a series of subsequent papers, the basic paradigm was bolstered by further observational and theoretical studies (Lau and Chan, 1986, 1988; Lau and Shen, 1988; Hirst and Lau, 1990).

9.3.1 OLR time–longitude sections

Figure 9.1(a) (see color section) shows the time–longitude section, also known as a Hovmöller diagram, of 7-day mean OLR along the equator covering the entire Indo-Pacific basin, reproduced from Lau and Chan (1986a) for the period 1974–1984. Data were missing and never recovered for 1978. This figure provided for the first time a vivid depiction of the episodic onset of an El Niño, juxtaposed against a backdrop of a space–time evolution of deep convection associated with the MJO year after year. It provided convincing evidence to the intrinsic nature of the MJO in the tropical atmosphere, and its possible transformation during ENSO. Notable features in Figure 9.1(a) included the almost continuous streams of eastward propagation of convective pulses from the western Indian Ocean to the central Pacific. During the El Niño of 1982/1983, deep convection began to shift eastward to the eastern Pacific in June 1982, and remained active there until June 1983, when convection abruptly ceased and the coupled system began to enter the La Niña phase. Lau and Chan also noted the increased MJO activity in the Indian Ocean and western Pacific several months prior to the onset of the 1982/1983 El Niño. Interestingly, Figure 9.1(a) suggested that the MJO was present all year round, with a stronger signal (along the equator) during spring and fall, but weakest signal in the northern summer. This helped dispel the notion that the Asian summer monsoon was the cause of the MJO. Based on observations such as these, a number of authors went on to develop basic theories of atmospheric low-frequency oscillations arising from moisture convergence and latent heating feedback from tropical convection in an aqua-planet all-ocean covered earth) (Hyashi and Sumi, 1986; Lau and Peng, 1987; Chang and Lim, 1988; Wang and Rui, 1990; Hendon, 1988). The readers are referred to Chapter 10 for discussions of theories of the MJO.

Figure 9.1(b) (see color section) shows a time–longitude section similar to that used in Lau and Chan (1986a), but for the period 1990–1999. Two El Niño's occurred during the period in 1991/1992 and in 1997/1998. The latter has been referred to as the El Niño of the century, because of its exceptional strength and impact on weather and climate worldwide. In normal years, pronounced MJO convection signals were more or less confined to the Indian Ocean and western Pacific warm pool (>28°C) of the tropical western Pacific. Near the dateline, the MJO convective activities appear to stop just at the eastern extreme of the warm pool, except during El Niño, when the warm pool expands to the eastern Pacific, setting up conditions for deep convection there. Compared to the period 1974–1984, the general similarity in variability associated with MJO and ENSO is striking. Even more remarkable was the similarity of the MJO propagation, ENSO evolution, during the two decades, as if nature had a memory to repeat itself through the entire MJO–ENSO evolution, even after more than 10 years. The increased MJO activity in the Indian Ocean and western Pacific during the first part of 1997, and other definitive observations of oceanic Kelvin-wave signals from the TOGA–tropical atmosphere and ocean (TAO) array, suggested strong MJO impacts on the onset of 1997/1998 El Niño and provided the impetus for renewed interest in studies of the MJO–ENSO connection (see discussion in Section 9.5).

9.3.2 Seasonality

The seasonal cycle was recognized as an important factor in the interaction between MJO and ENSO, because of the strong phase locking of the ENSO to the annual cycle (Lau, 1985a, b). Lau and Chan (1988) showed that the tropical averaged root-mean-square fluctuation of OLR associated with the annual cycle is about 1.5 times that of the TISV, and that is about 3 times stronger than the interannual component. Figure 9.2 shows the amplitude and spatial distribution of TISV of OLR in different seasons. The TISV centers of activity stretch from the Indian Ocean to the western Pacific, with an obvious minimum over the maritime continent. The reason for the minimum is not clear, but may be related to the effects of topography which inhibits the large-scale organization of deep convection, or to strong land heating (cooling) during the day (night) which tends to favor strong diurnal variability over low-frequency variability. Also, the maritime continent is frequently impacted by mid-latitude and subtropical disturbances such as cold air intrusion from the East Asian continent, which inhibits the development of deep convection (Chang and Lau, 1980; Lau *et al.*, 1983). The TISV is strongest and most extensive in DJF, with a pronounced signal in the southern tropics between the equator and 20°S. In MAM, the MJO appears more symmetric about the equator, with centers of activity well

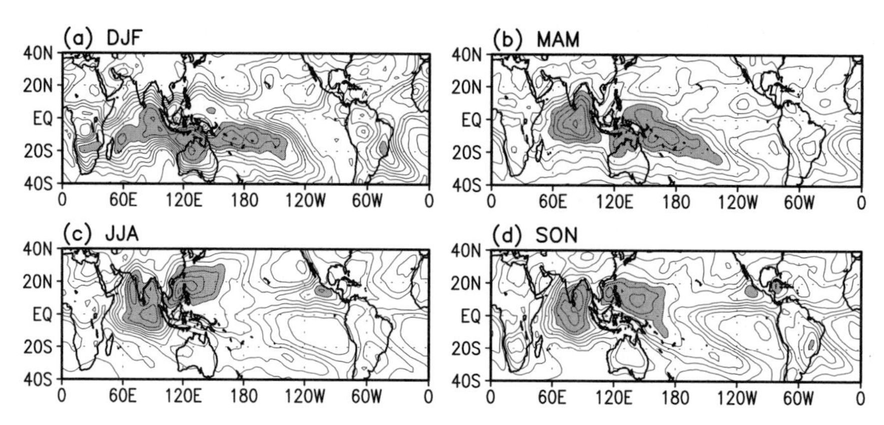

Figure 9.2. Spatial distribution of variance of 20–70-day band-passed OLR for the four seasons: (a) DJF, (b) MAM, (c) JJA, and (d) SON. Unit is in $(\mathrm{W\,m^{-2}})^2$. Regions of deep convection are highlighted.

separated between the Indian Ocean and the western Pacific. During JJA, the TISV activity shifts to the Asian monsoon region, with centers around the eastern Arabian Sea, the Bay of Bengal, the South China Sea, and the subtropical western Pacific. The SON pattern is similar to JJA, except for signs of TISV activity in the South Pacific Convergence Zone (SPCZ) and the South Atlantic Convergence Zone (SACZ) in the southern hemisphere. As will be discussed in Section 9.5, in terms of remote forcing and possible triggering of the onset of ENSO anomalies by surface winds, the TISV in MAM may be most effective through excitation of oceanic Kelvin waves in the equatorial oceanic wave guide which is within approximately 2° of the equator (Geise and Harrison, 1990; Hendon *et al.*, 1998; Harrison and Schopf, 1984). However, not all TISV in MAM will lead to the onset of El Niño. To increase the probability of triggering a full-blown warm event, pre-conditioning (e.g., increased heat content in the tropical Pacific), in the preceding winter season may be required.

9.3.3 Supercloud clusters

In addition to coherent eastward propagation, the MJO is associated with a hierarchy of different scales of atmospheric motion, which may provide stochastic forcing of El Niño. Nakazawa (1988) showed that westward propagating high-frequency convective systems are embedded in the eastward propagating MJO envelope, along the equator, to form a large-scale organized convective complex known as a "supercloud cluster" (SCC). The SCC constitutes the eastward propagating convective envelope of the MJO (see Chapter 1 and Figure 1.4). Lau *et al.* (1991) found that the substructures within the SCC possessed a hierarchy of spatial and temporal scales ranging from a few tens to hundreds and thousands of kilometers, with multiple periodicities from diurnal, to 2–3 days and 10–15 days. Associated with the SCC substructures are fluctuations of surface westerly winds over a variety of timescales over the western and central Pacific. The westerly wind fluctuations, and related twin-cyclonic activities, belong to the phenomenon known as westerly wind bursts (Keen, 1982; Harrison, 1984). Lau *et al.* (1989) proposed a unified dynamical framework for studying the mechanisms of SSC, MJO, twin-cyclones, westerly wind bursts, and ENSO, in terms of atmospheric moist dynamics and coupled air–sea interactions over the tropical oceans. Today, the term "westerly wind event" (WWE) is used to refer to a broad spectrum TISV in surface westerly wind fluctuations from days to weeks over the western and central equatorial Pacific with no preferred periodicity. Discussions of the relationship of WWEs and sea surface temperature (SST) variations, and WWEs as a stochastic forcing of the tropical coupled ocean–atmosphere will follow in Sections 9.4 and 9.5.

9.3.4 Early modeling framework

During the early 1980s, simple coupled ocean–atmosphere models of El Niño began to appear (Lau, 1981; Zebiak, 1982; McCreary and Anderson, 1984; Gill, 1984). Cane and Zebiak (1985), and Zebiak and Cane (1987) pioneered the development of an intermediate model of the El Niño coupling a dynamical ocean to a steady-state atmosphere, with convergence feedback. Lau and Shen (1988) and Hirst and Lau (1990) reasoned that besides providing stochastic forcing, the MJO may fundamentally contribute to El Niño. They argued that to study the effect of MJOs, and WWE forcings on the coupled system, it is necessary to include interactive moist convection in coupled models, as opposed to steady-state atmosphere or "slave-atmosphere" coupled models. For the atmosphere, they used the following shallow-water system:

$$U_t - \beta y V + D_m U = \theta_x \tag{9.1}$$

$$V_t + \beta y U + D_m V = \theta_y \tag{9.2}$$

$$\theta_t - c_a^2(U_x + V_y) + D_T\theta = \lambda c_a^2 P \tag{9.3}$$

$$mq_o(U_x + V_y) = E - P \tag{9.4}$$

where the subscripts t, x, and y denote differentiation with respect to time and space; U and V are the zonal and meridional perturbation velocity of the lower atmosphere; and θ is the perturbation tropospheric potential temperature. Equation (9.3) is scaled by the factor gH_a/θ_a with H_a and θ_a representing the mean depth and the mean potential temperature of the lower troposphere respectively; c_a is the equivalent phase speed of the shallow-water system; and λ is a dimensional factor relating to the densities of air, water, heat capacity of air, and c_a. D_m and D_T represent the damping coefficients in momentum and temperature, respectively. The perturbation precipitation rate P is related to the background moisture q_o, the precipitation efficiency m, and the surface evaporation E by (9.4). Lau and Shen (1988) used the following expression for the evaporation:

$$E = -\alpha_a U + \alpha_o T \tag{9.5}$$

where T is the SST anomaly. The wind coupling coefficient, α_a is derived from a linearized form of the bulk aerodynamic formula for surface momentum, and is positive for mean easterlies and negative for mean westerlies. The SST coupling coefficient α_o is always positive. The basic-state moisture q_o is related to the basic-state SST through a standard formula relating saturation moisture to temperature. Equations (9.1–9.4) for the tropical atmosphere are coupled through surface latent heat and momentum fluxes to a similar system of equations for the tropical ocean (Hirst, 1986; Hirst and Lau, 1990):

$$u_t - \beta y v - h_x = \gamma U \tag{9.6}$$

$$v_t + \beta y u - h_y = \gamma V \tag{9.7}$$

$$h_t - c_o^2(u_x + v_y) = 0 \tag{9.8}$$

$$T_t + T_x^* u - K_T h + d_o T = 0 \tag{9.9}$$

where u and v are the eastward and northward perturbation currents; γ is the coefficient for wind stress; h is the dynamic height perturbation of the upper ocean; c_o is the oceanic gravity wave speed parameter; T^* is the basic-state SST of the ocean; K_T is a coefficient governing the SST changes associated with thermocline depth variations; and d_o is a thermal dissipation of the upper ocean.

Lau and Shen (1988) found that the presence of interactive moisture (i.e., condensation–convergence feedback in the tropical atmosphere), causes atmospheric motions to slow down through the reduction of the effective moist static stability of the lower troposphere. This influence becomes increasingly strong as the saturated moisture content of the lower troposphere increases in warmer SSTs, especially when the damping is weak. When atmospheric waves are slowed to the timescales commensurate with tropical intraseasonal oscillations (TISOs), due to reduced moist static stability from interaction between dynamics with moist convection, ocean–atmosphere coupling can destablize both the TISO and the low-frequency (interannual) modes. As a special solution for coupled Kelvin waves ($V = v = 0$) in equations (9.1–9.9), two basic unstable modes can be identified in the intermediate coupled moist atmosphere–ocean system (i.e., an advective mode and an upwelling mode) (see Figure 9.3). The advective mode stems from the destablization of atmospheric waves, identifiable as MJO's, by air–sea interaction and east–west SST advection by anomalous zonal currents in the equatorial wave guide. This mode is characterized by eastward propagation with the region of deep convection found to the west of the anomalous SST maximum (Figure 9.3(a)). It slows down and becomes increasingly unstable over warmer background SST, due to enhanced condensation–convergence feedback. Lau and Shen suggested that this mode may be responsible for the initial growth of the 1982/1983 El Niño. The upwelling mode arises from destablization of oceanic Kelvin waves by air–sea interaction through oceanic upwelling and moisture convergence feedback in the atmosphere (Figure 9.3(b)). The upwelling mode corresponds to the unstable coupled mode for "slave-atmosphere" models of El Niño (Lau, 1981; Philander *et al.*, 1984; Cane and Zebiak, 1985; Hirst, 1986; and many others). This mode has no east–west displacement between SST and deep convection, and is stationary. The inclusion of evaporative-wind feedback in the presence of mean surface easterly wind causes the upwelling

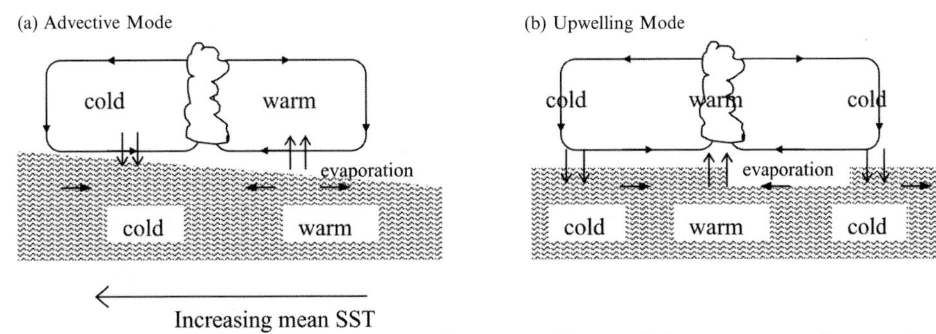

Figure 9.3. Schematic showing structure of two unstable coupled ocean–atmosphere modes: (a) advective mode and (b) upwelling mode.
From Lau and Shen (1988).

mode to move eastward (Emanuel, 1988). The evaporative-wind feedback may have been operative during the amplifying stage of the 1982/1983 El Niño. Hirst and Lau (1990) further analyzed the coupled modes, including contributions from Rossby waves, and confirmed the importance of non-equilibrium atmospheric dynamics in ocean–atmospheric interaction that is capable of amplifying ISV of internal atmospheric origin, through evaporation–wind feedback.

9.4 *PHASE-2*: THE EXPLORATORY STAGE

This stage is characterized by a large number of in-depth diagnostic and modeling studies in attempts to establish a long-term statistical, and more physically based, TISV–ENSO connection. Establishing a long-term statistical relationship between MJO and ENSO proved to be elusive. One indisputable feature is that the MJO and associated convection are strongly modulated by the ENSO cycles. The MJO wind signals extend eastward to the equatorial eastern Pacific as the mean convection and the warm water spread eastward along the equator during El Niño (Gill and Rasmusson, 1983; Lau and Chan, 1988; Weickmann, 1991; Schrage *et al.*, 1999). Yet over the extreme western Pacific and the Indian Ocean, there is no noticeable overall change in the amplitudes of MJO convection and wind activities during El Niño (Fink and Speth, 1997). A number of subsequent studies find that statistically the overall interannual variability (IAV) of MJO activity is not related to ENSO cycles (Slingo *et al.*, 1999; Hendon *et al.*, 1999; Waliser *et al.*, 2001). Some studies showed that introducing TISV as white-noise forcings into intermediate coupled models did not significantly alter the El Niño response of the model (Zebiak, 1989). In general, results from SST-forced atmospheric general circulation models suggested that the reproducibility of the IAV of MJO was poor (Gauldi *et al.*, 1999; Slingo *et al.*, 1999). These results have often been taken to mean that the MJO and ENSO do not interact in a substantial way.

It is important to note that the MJO–ENSO interaction may also be dependent on the use of definitions of MJO activities. The most common method of defining a MJO is to decompose band-passed (e.g., 20–70 days) OLR or wind fields (e.g., 200-mb velocity potential) into empirical orthogonal functions (EOFs). Typically, the first two eigenmodes which appear as a quadrature pair in both space and time can be interpreted as signals of coherent eastward propagation of the MJO. Compositing or regression analyses of large-scale fields of wind, temperature, and moisture are then carried out with respect to an index of MJO activity from the first two principal components (Knutson and Weickman, 1987; Lau and Chan, 1988; Zhang and Hendon, 1997). Kessler (2001) suggested the use of the third EOF to capture signals over the central Pacific where interaction of the MJO and ENSO was likely to occur. In the following, we provide a description of MJO–ENSO interaction based on one of the aforementioned popular definitions. In Section 9.6, we introduce a new index specially designed to study MJO–ENSO interactions.

9.4.1 MJO and ENSO interactions

A sense of the modulation of the MJO by ENSO, and possible MJO–ENSO interaction can be gleaned from an EOF analysis of pentad (5-day running mean) OLR. The analysis is conducted separately for (i) normal, (ii) La Niña (cold), and (iii) El Niño (warm) states of the tropical Pacific, defined respectively as periods in which the the area-averaged monthly SST in the Niño-3 region (5°N–5°S; 150°W–90°W) (i) has an absolute value less than 0.5°C, (ii) is less than 0.5°C, and (iii) is greater than 0.5°C. The pentads are defined with respect to the climatological seasonal cycle at each grid point, which has been removed before the EOF analysis. The spatial distributions of the dominant EOFs for the three categories, representing the eastward propagating components of the MJO are shown in Figure 9.4. For normal (Figure 9.4(a, b)) and cold events (Figure 9.4(c, d)), these components are represented by the first two EOF's, which together explain up to 6.8% and 7.1% of the total anomalous variance respectively. For warm events (Figure 9.4(e, f)), the propagating modes are represented by the second and third EOFs, which together explain 6.4% of the total variance. The first EOF for warm events, which will be described separately later, represents the MJO–ENSO interaction mode, which explains in its own right 6.1% of the total anomalous variance.

For all three categories, a pair of east–west dipoles in anomalous convection associated with eastward translating centers of action of the MJO from the Indian Ocean to the central Pacific is quite clear. The spatial distributions of the dipoles are quite similar between the normal and the cold states, except that the dipoles appear to be more compressed over the western Indo-Pacific region, and weakened over the central Pacific during cold events. These features are consistent with the westward extension of the cold tongue in the equatorial eastern Pacific during La Niña. During warm events (Figure 9.4(e, f)), while still maintaining the dipole structure over the western Indo-Pacific, the OLR anomalies expand eastward over the equatorial eastern Pacific and the Atlantic, consistent with the spread of warm water over these regions. Also noticed is an east–west dipole structure over the Indian Ocean

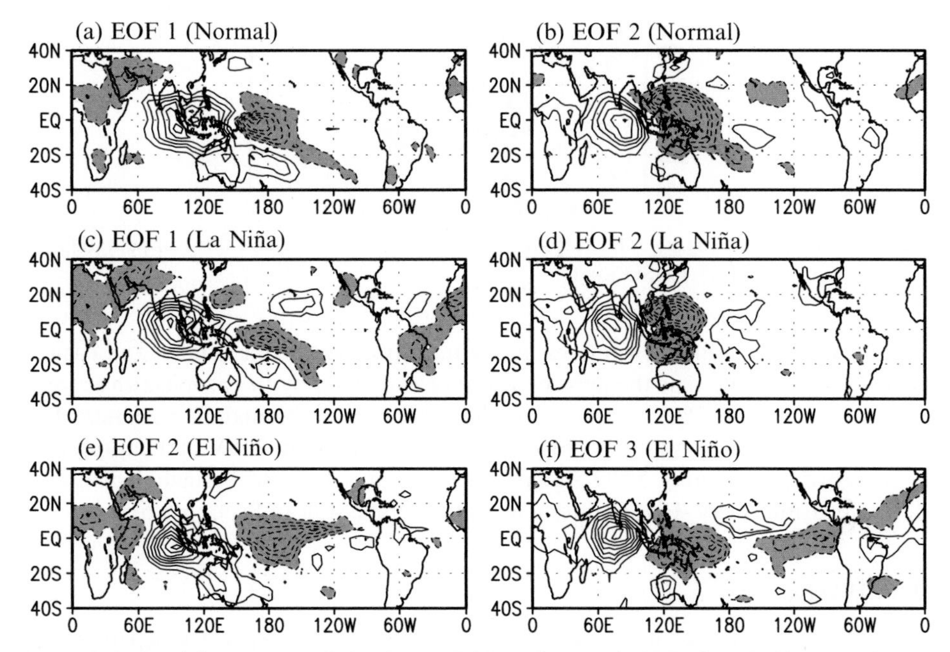

Figure 9.4. Spatial patterns of dominant EOFs of pentad OLR for: (a, b) normal state, (c, d) La Niña state, and (e, f) El Niño state, for the period 1979–1999. Units are non-dimensional.

(Figure 9.4(e)), not found in normal or cold states. However, there appears to be no noticeable difference in the magnitude of the variance of the MJO signal over the Indo-Pacific region between the normal, warm, and cold oceanic states, implying that the intensity of the MJO is somewhat independent of the interannual variation of SST. However, this does not imply that air–sea interaction is not important on the timescale of the MJO. In fact, there is now increasing evidence from observations and modeling studies that the structure and propagation of MJOs are strongly modulated by air–sea interaction (Waliser *et al.*, 1999; Lau and Sui, 1997; Flatau *et al.*, 1997; Wang and Xie, 1998; Fu *et al.*, 2003).

Returning to the MJO–ENSO mode (first EOF mode) for warm events (Figure 9.5), we find that this mode has a pattern over the central Pacific and the eastern Pacific similar to those found during ENSO (Gill and Rasmusson, 1983). The MJO–ENSO mode is unique in the sense that it possesses both interannual and intraseasonal timescales. It is not found in the normal, nor cold events, and represents an extension of the MJO signal when the tropical Pacific is in its warm state. Strictly speaking this mode should not be considered as a part of the intrinsic MJO mode, as it does not exist under normal or cold ocean states. The above description is in agreement with previous results indicating that the amplitude of the MJO activity over the Indian Ocean and the western Pacific is relatively independent of the state of the ocean. An alternate interpretation is that the MJO–ENSO connection does not

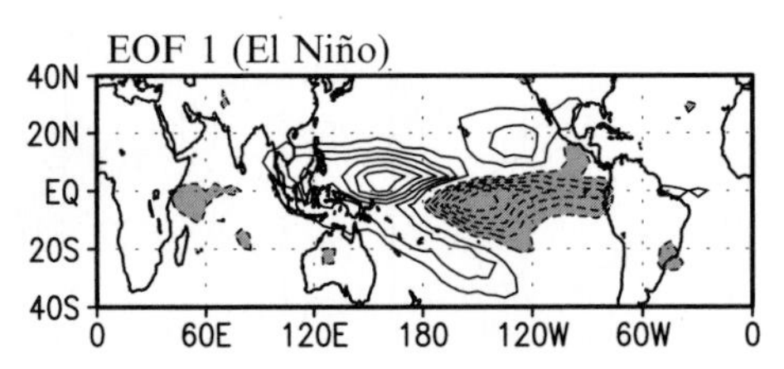

Figure 9.5. Same as in Figure 9.4, except for the spatial distribution of EOF 1 of pentad OLR, showing the mixed MJO–ENSO mode.

occur all the time, and when it does, it is concentrated over the central and eastern tropical Pacific. Note that the percent variance explained by the propagating modes is relatively small (6–7%) compared to the total pentad anomalous variance. This means that there are substantial portions of the variability in TISV, not captured by the eastward propagating MJO signals and that may have to be included in considering possible TISV–ENSO interactions (see discussion in Section 9.6).

9.4.2 WWEs

Westerly wind anomalies in the western and central Pacific, regardless whether they stem from the eastward propagating MJO, or from other sources such as the Asian monsoons or cold air intrusion from the extra-tropics (Lau *et al.*, 1983; Yu and Rienecker, 1998), are important agents for remote forcing of ENSO. Hence, a more direct way of examining the TISV–ENSO connection is through impacts of the WWEs on SST. WWEs are defined here as surface wind fluctuations in the western and central Pacific, between 110°E–170°W, within 15° latitude on either side of the equator. They can occur singly or in succession, typically with zonal wind anomalies spanning 30°–40° longitude, and lasting from 7–10 days (Harrison, 1984; Luther and Harrison, 1984; Luther *et al.*, 1983; Harrison and Geise, 1988; Harrison and Vecchi, 1997). As discussed in Section 9.3, a portion of the WWE signals may be associated with the hierarchical supercloud structure of the MJO. A large number of observations and modeling studies have shown that WWEs can induce downwelling oceanic Kelvin waves which propagate eastward along the equator, induce zonal SST advection, suppressed upwelling along the equator, and possibly trigger the onset of El Niño (Harrison and Schopf, 1984; Harrison and Geise, 1988; Giese and Harrison, 1990; Kindel and Phoebus, 1995; Kessler *et al.*, 1996; Belarmari *et al.*, 2003; and many others). Meyers *et al.* (1986) showed that there was substantial cooling of the western Pacific Ocean induced by WWEs during the 1982/1983 El Niño. Kessler and Kleeman (2000) suggested that the cooling of the western Pacific associated with MJO passage can alter the east–west SST and hence pressure gradients across the entire Pacific, thus providing a rectifying effect on ENSO (see also Chapter 6).

Vecchi and Harrison (2000) conducted a comprehensive study of the relationship between WWEs and SST variation. They found that in the absence of WWEs, SST perturbations in the equatorial central and eastern Pacific tend to relax back to climatology. While WWEs west of the dateline lead to local surface cooling in the western Pacific (Meyers *et al.*, 1986), those east of the dateline spawn warming in the equatorial wave guide when the ocean is in a normal state. When the Pacific Ocean is in a warm state, equatorial WWEs may be important in sustaining the warming of the equatorial eastern Pacific. Figure 9.6 shows the composites of initial SST anomalies (SSTAs) and changes in SSTAs at 20-day intervals, before and after WWEs over the western central Pacific, during normal and warm states of the ocean respectively. The changes in SSTAs are with respect to the initial SSTAs. In the normal state, 20 days prior to strong WWEs (Figure 9.6(a), see color section), SST is above normal in the central Pacific and below normal in the western Pacific in large areas of tropics and the extra-tropics. The most pronounced feature is the warming that develops over the south-eastern Pacific and the equatorial wave guide starting from Days 10–20 (Figure 9.6(c)). The warming spreads to the coast of South America by Day 80 (Figure 9.6(f)), and reaches an amplitude of 1.0°–1.5°C. The WWEs produce local cooling in the western Pacific to the extent of 0.25–0.5°C. In the warm state, changes in the SSTA subsequent to the WWEs are relatively small compared to the normal state. WWE-induced local cooling is found over the western Pacific at Days 0–20 (Figure 9.6(e)). Enhanced warming in the wave-guide region, and cooling in the south-eastern Pacific are found starting at Day 20, but all the induced SST changes are less than 0.25°C. Hence, the impact of WWEs appear to be strongly dependent on the ambient state of the ocean–atmosphere system. The above results suggest that WWE may be instrumental in transitioning the tropical ocean–atmosphere from a normal to a warm state, by providing initial warming in the equatorial wave guide and then in the coastal region of South America. The result also illustrates the importance of the initial warming of the central Pacific, even in the normal category, suggesting that an El Niño may already be underway, in order for the WWE wind forcing to be effective.

9.5 *PHASE-3*: ENSO CASE STUDIES

In the mid-1990s through to early 2000, there were many case studies of El Niño documenting the impact of WWEs and excitation of oceanic Kelvin waves which were believed to have triggered the onset of El Niño (e.g., Kindle and Phoebus, 1995; Kessler *et al.*, 1996; McPhaden, 1999; McPhaden *et al.*, 1988; Yu and Reinecker, 1998; Kutsuwada and McPhadan, 2002). Nakazawa (2000) documented various atmospheric conditions associated with the MJO and tropical cyclone activities during the 1997/1998 El Niño. Indeed, definitive observations of atmospheric MJO signals, and oceanic Kelvin wave responses from satellites and the TOGA–TAO moorings, prior to the onset of the 1997/1998 El Niño re-invigorated the debate on the role of the MJO in possibly leading to the abrupt onset and termination of ENSO cycles (McPhaden and Yu, 1999; Takayabu *et al.*, 1999).

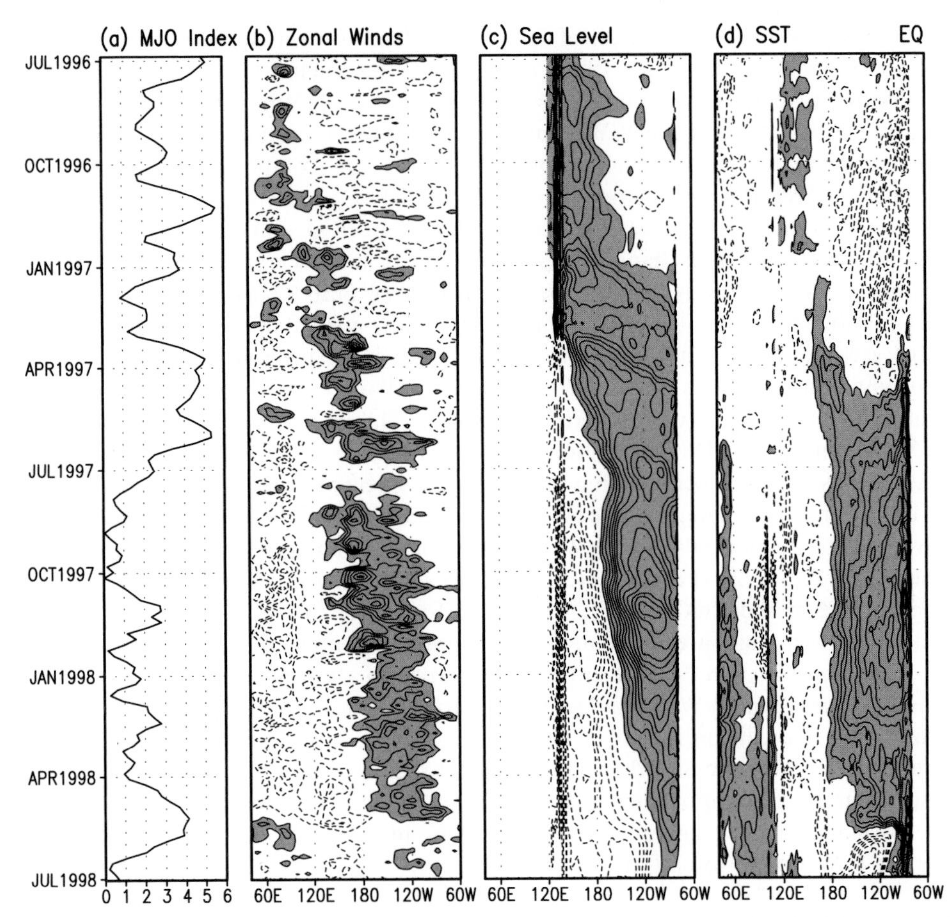

Figure 9.7. Space–time evolution of oceanic–atmospheric variables associated with onset and termination of the 1997/1998 El Niño: (a) Time series of a MJO index (see Section 9.6), and equatorial time–longitude sections of anomalies spanning the Indian Ocean and the Pacific, for (b) pentad 850-mb zonal wind ($m s^{-1}$), (c) weekly sea level height (cm), and (d) weekly SST (°C). Anomalies are defined with respect to the period from July 1996–July 1998. Contour interval for wind is $0.5 m s^{-1}$, for sea level is cm, and for SST is 0.2°C.

9.5.1 The 1997/1998 El Niño

The 1997/1998 El Niño was the strongest on record (McPhaden, 1999). It began in the spring of 1997 when SST in the central and eastern Pacific rose rapidly (Figure 9.7(d)) in conjunction with the appearance of extensive surface westerly wind anomalies signaling the collapse of the trade winds in the equatorial central and eastern Pacific (Figure 9.7(b)). The El Niño was terminated even more abruptly than the onset in May 1998, with an unprecedented 8°C drop in SST in a 30-day period (Figure 9.7(d)). Almost a year prior to the onset, there were pronounced WWEs over the Indian Ocean and the western Pacific (60°E–160°E), and a build

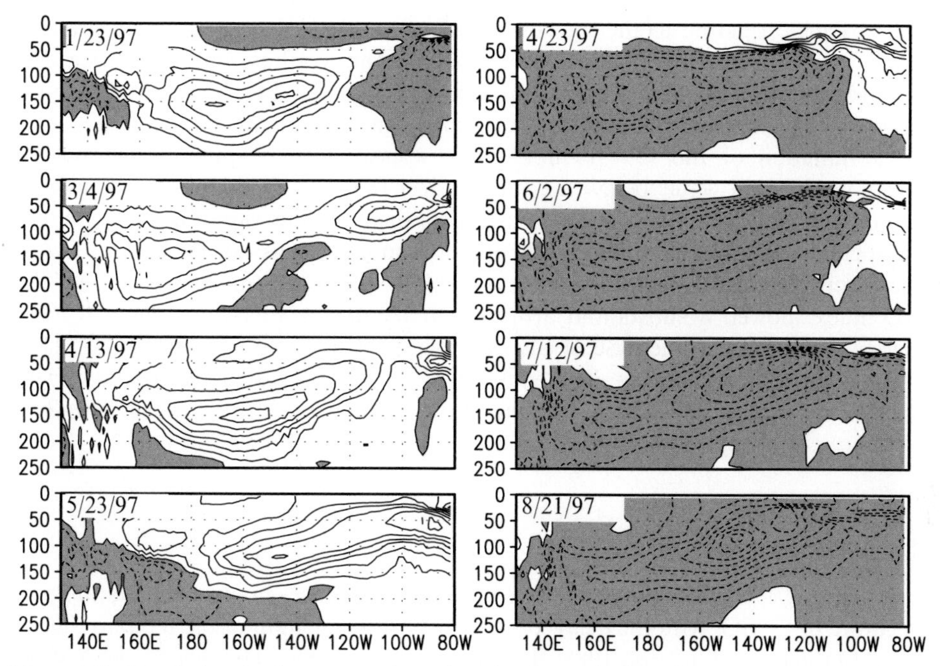

Figure 9.8. Depth–longitude cross sections showing the evolution of water temperature during the onset (*left panels*) and the termination (*right panels*) phases of the 1997/1998 El Niño. Contour interval is 0.5°C.
Data from NCEP Ocean Data Assimilation (Ji *et al.*, 1995).

up of the sea level in the western Pacific (Figure 9.7(c)). The increased WWEs prior to the El Niño onset was also captured in an index of MJO based on the variability of the 850-mb streamfunction (Figure 9.7(a); see definition in Section 9.6). During this build-up phase, the eastern Pacific cold tongue was well developed with noticeable signals of westward propagating tropical instability waves in the equatorial eastern Pacific (Figure 9.7(d), see Chapter 6). In the Indo-Pacific region, three major MJO signals, occurring in October–November 1996, December–January 1997, and March–April 1997 can be identified in the 850-mb wind, propagating from the Indian Ocean to the western and central Pacific. Figure 9.7(c) shows clearly the Kelvin-wave signal in sea level anomalies propagating across the entire Pacific to the coast of South America, excited by the last two MJOs in December–January and in March–April. The downwelling Kelvin waves led to rapid deepening of the thermocline, and abrupt warming in the eastern Pacific, caused by a combination of warm advection and the cessation of upwelling (McPhaden, 1999). As shown in Figure 9.8 (left panels), warm water was already well developed in the central Pacific below the surface in January 1997. The strong MJO/WWE forced the warm water to migrate eastward and upward to the eastern Pacific, reaching the coast of South America in early March. By mid-April, the result of the MJO wind forcing was to trigger a transition of the entire Pacific from the cold to the warm state. During the peak of the El Niño in January 1998, easterly wind anomalies emerged over the far western Pacific, leading to thermocline shoaling, sea level depression, and development of cold subsurface water. The shoaling and cooling progressed steadily eastward and upward from April through June, as the El Niño ran its course (Figure 9.8, right panels), in a manner similar to a reversal of the onset phase. The termination of the 1997/1998 El Niño appeared to be pre-destined by the appearance of cold water below the surface over the western and central Pacific in the boreal spring of 1998, and was accomplished by the rapid shoaling of the thermocline which appears first in the open ocean of the eastern Pacific in May to June. From July to August, the transition to La Niña was complete. Note that the MJO index showed a substantial drop in activity during El Niño (Figure 9.7(a)). However it appears to increase in activity, due to an easterly impulse of the MJO, just prior to the termination. Takayabu *et al.* (1999) suggested the easterly MJO impulse may lead to the sudden shoaling of the thermocline and return of upwelling and hence the abrupt SST drop, terminating the 1997/1998 El Niño.

9.5.2 Stochastic forcings

Since the 1982/1983 El Niño, there have been great advances in our understanding of the nature of ENSO, with the delayed oscillator as the linchpin of the modern theory of ENSO (Suarz and Schopf, 1988; Battisti and Hirst, 1989). The theory of ENSO is outside the scope of this chapter. For a review of ENSO dynamics, the readers are referred to Neelin *et al.* (1998) and Latif *et al.* (1994). However, the delayed oscillator cannot explain the quasi-irregularity of the ENSO cycles, its timing, and relationship to ISV. Toward the latter part of the 1990s, the idea of stochastic forcing of ENSO by TISV was gaining momentum (Penland and Sardeshmukh, 1995; Penland, 1996; Kleeman and Moore, 1997). Moore and Kleeman (1999) formulated a generalized dynamical framework to examine the spatial structure of the noise forcing which is most effective in forcing ENSO variability in the coupled ocean–atmosphere system.

They proposed the concept of a "stochastic optimal" defined as the most effective spatial structure that would give rise to the fastest growth of El Niño from stochastic forcings. Their results showed that the spatial structure of the surface-heating functions and surface wind stress of the stochastic optimal from an intermediate coupled ocean–atmosphere model are similar to the east–west dipole structure in OLR and wind stress associated with the MJO (Lau and Chan, 1985, 1988; Hendon *et al.*, 1999). Experiments with various stochastic forcings due to single or multiple WWEs suggested that: (a) when forced with white noise, the onset of an ENSO-like large amplitude perturbation is most favored when the forcing has the structure of the stochastic optimal – the optimal stochastic forcings can effectively trigger ENSO, or disrupt developing ENSO episodes; (b) the history of the noise forcing and its integral effects are important in amplifying or restricting the growth of the ENSO mode in the coupled ocean–atmosphere system; and (c) the effectiveness of the stochastic forcing is dependent on the phase of the seasonal cycle, and the evolving state of the ocean and atmosphere.

These results are in agreement with the observations that groups of MJOs and associated WWEs may have contributed to development of El Niño, such as the 1997/1998 event. If stochastic forcings are not applied at the right phase of the seasonal cycle, or the right phase of the El Niño event itself, their impacts may be minimal. However, because ENSO is an intrinsic mode in the tropical coupled ocean–atmosphere, El Niños and La Niñas may be generated even without WWE perturbations. These characteristics of forcings by stochastic optimal are consistent with those originally proposed by Lau (1985a, b) and Lau and Chan (1986a, b, 1988).

9.6 *PHASE-4*: RECENT DEVELOPMENT

At present, the TISV–ENSO relationship is still a subject of intense debate. While better observations of oceanic Kelvin waves have certainly provided strong evidence of the importance of TISV, particularly the MJO, in triggering of the 1997/1998 El Niño, there are still many uncertainties in determining under what conditions the relationship will or will not hold (Bergman *et al.*, 2001). Zhang *et al.* (2002) proposed an index for measuring Kelvin-wave forcings by wind stress associated with MJO and found evidence that strong Kelvin-wave forcing in the western Pacific precedes greater SSTAs in the equatorial Pacific by 6–12 months during 1980–1999. However, no such evidence can be found for the period 1950–1979. While some studies show substantial impact of stochastic forcings (Moore and Kleeman, 1999; Penland and Sardesmukh, 1995), others have suggested that power in the ISV range cannot be effectively channeled from sub-annual frequencies to the low frequencies associated with ENSO, but rather more likely through the reddening of the frequency spectrum through SST processes outside the tropics (Roulston and Neelin, 2000). Further progress in better understanding the MJO–ENSO connection is hampered by the inability of models to produce the realistic structure and statistics of the MJOs (Slingo *et al.*, 1997; Maloney and Hartmann, 2000; Sperber *et al.*, 1997; Waliser *et al.*, 2003). For an assessement of the modeling capabililty for the MJO, the readers are referred to Chapter 11.

Motivated by the need to assess the scientific controversy, and to improve models of MJO and ENSO for climate prediction, an MJO–ENSO workshop was convened at the National Oceanic and Atmospheric Administration (NOAA), Geophysical Fluid Dynamics Laboratory (GFDL), Princeton, New Jersey, in March 2000. Over 70 scientists from 8 countries attended the workshop. During the workshop, scientists discussed a wide range of topics from the dynamics and air–sea interaction associated with MJOs, oceanic response to westerly wind bursts, stochastic forcings of ENSO, MJO predictability, regional manifestations of MJOs, and impacts on ENSO prediction skills (Zhang *et al.*, 2001). Workshop participants noted the lack of skill of state-of-the-art climate models to simulate MJO/ISV and its possible interactions with ENSO in coupled ocean–atmosphere models. They summarized the broad spectrum of opinions, into three competing hypotheses:

Hypothesis I. ISV, as with other weather systems, provides a source of irregularity of ENSO and limits ENSO predictability. ISV is no different from noise forcing, and inherent chaotic behavior in the coupled system, and ISV is not required for ENSO prediction.

Hypothesis II. Influence of ISV on ENSO is unique and distinct from other weather systems. The timing and strength of an ENSO event may be sensitive to ISV forcing, and alters the phase of ENSO cycles. MJO/ISV is a necessary component of the coupled ocean–atmosphere system. Better understanding of the mechanisms of MJO/ISV interaction may lead to improved prediction.

Hypothesis III. Stochastic forcings of the MJO/ISV are essential for maintaining ENSO variability. This implies that the coupled ocean–atmosphere is damped, and ENSO prediction with long lead time will be extremely difficult since it is dependent only on stochastic forcings.

Further discussion of these hypotheses with regard to ENSO predictability will be presented in Section 9.7. Suffice it to point out here that there is not yet definitive observational evidence nor modeling studies to confirm or reject any one of the hypotheses. As noted in the workshop, the debate on the MJO–ENSO relationship was partly aggravated by the lack of clear statistical evidence of such a relationship.

One of the problems in previous studies of the MJO–ENSO relationship may have arisen from too much focus on the MJO in one season, either boreal winter or summer. Yet, MJO wind forcings are likely to be imposed on the ocean throughout the year. Oceanic Kelvin waves forced by the MJO are likely to be most effective when the forcing is confined to the equatorial waveguide, and symmetric with respect to the equator during the spring and fall, but less so during boreal winter or summer. Furthermore, the MJO has pronounced signals not only in the tropics, but also in the extra-tropics. Some studies have suggested the possible forcing of MJO from the extra-tropics (Lau *et al.*, 1983; Hsu, 1996). Hence, identifying the full spectrum of variability associated with the MJO, including all seasons, tropical and extra-tropical variability, and separating modes of ISV with respect to SST forcings and responses, is essential in understanding the MJO–ENSO relationship.

9.6.1 A new ISO index

In this section, we re-examine the MJO–ENSO relationship, by identifying dominant space–time modes of ISOs in the tropics and extra-tropics, and their possible separate roles in triggering or responding to ENSO. Using 50 years of the US National Center for Environmental Prediction (NCEP) wind reanalysis, we have computed the space–time extended empirical orthogonal functions (EEOF) of 20–70-day filtered 850-mb streamfunction to identify dominant modes of ISV all year round. The EEOF analysis, also known as multi-channel Singular Decomposition (M-SSA), is a powerful technique for identifying temporal and spatial structures in large-scale geophysical fields. Oscillatory modes are represented as a pair of eigenvectors approximately in quadrature (Vautard and Ghil, 1989). The rotational component of the wind in the lower atmosphere is used, because it is closely linked not only to surface wind forcing in the tropics, but also extra-tropical cyclone–anticyclone development, both of which may be linked to ENSO development. The first two EEOF modes describe an eastward, circum-global propagating

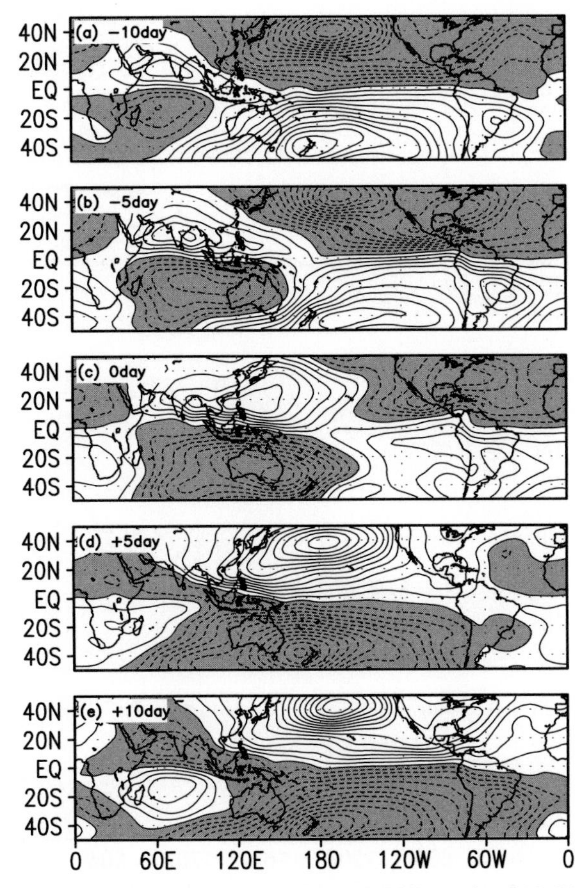

Figure 9.9. Space–time structure of the first dominant EEOF mode of 20–70-day band-passed 850-mb streamfunction, representing the eastward propagating component of the ISO: EEOF 1 for (a) −10 days, (b) −5 days, (c) center day, (d) +5 days, and (e) +10 days. The data period is 1956–1999. Unit is non-dimensional.

planetary pattern with equatorial surface wind build up and propagation from the Indian Ocean to the eastern Pacific, in tandem with the development of large-scale cyclonic and anticyclonic wind circulation patters over the North and South Pacific (Figure 9.9, EEOF 1 only; see also Weickmann *et al.*, 1985). The pattern is only a part of the large-scale circulation variability within the 20–70-day window, which we shall refer to as the eastward propagating mode (EPM). The third and fourth modes depict a quasi-stationary but oscillatory component, associated with wave signals emanating from the Indo-Pacific region along the subtropical jet streams of East Asia and northern Australia, to the subtropics and extra-tropics of both hemispheres (Figure 9.10, EEOF 3 only). The centers of action appear to be anchored over the eastern South Indian Ocean, the eastern South Pacific, the North Atlantic, and the North Pacific. Hereafter, these modes will be referred to as the quasi-stationary

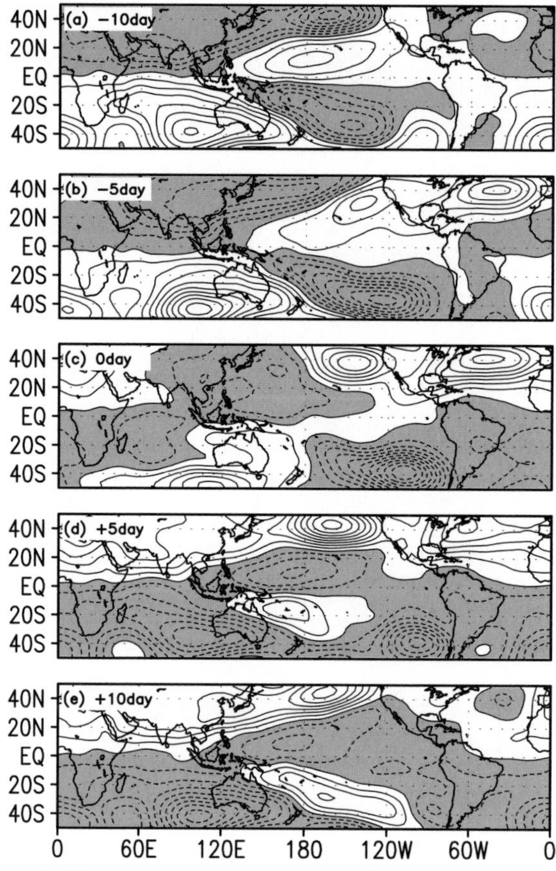

Figure 9.10. Same as in Figure 9.9, except for the third EEOF mode, representing the quasi-stationary component of the ISO signal.

mode (QSM). EEOF 2 and EEOF 4 (not shown) have spatial structures similar to but shifted approximately one-quarter of a wavelength relative to EEOF 1 and EEOF 3 respectively.

Figure 9.11 shows the comparison of the time series of a new ISO index, formed by the square root of the sum of the squares of the two principal components for EPM (EEOF 1 and EEOF 2), with several commonly used indices of the MJO (Slingo *et al.*, 1999; Hendon *et al.*, 1999). It can be seen that the new index is significantly correlated with those derived using the 200-mb velocity potential and streamfunction (Figure 9.11(c, d)), as well as using OLR (Figure 9.11(e)). The correlation with the 200-mb velocity potential index (Figure 9.11(c)) is the highest, and lowest with the zonal wind index (Figure 9.11(b)). Because the velocity potential index is dominated by the first two EEOFs (>73% variance), it contains almost exclusively tropical signals associated with the divergent wind components and

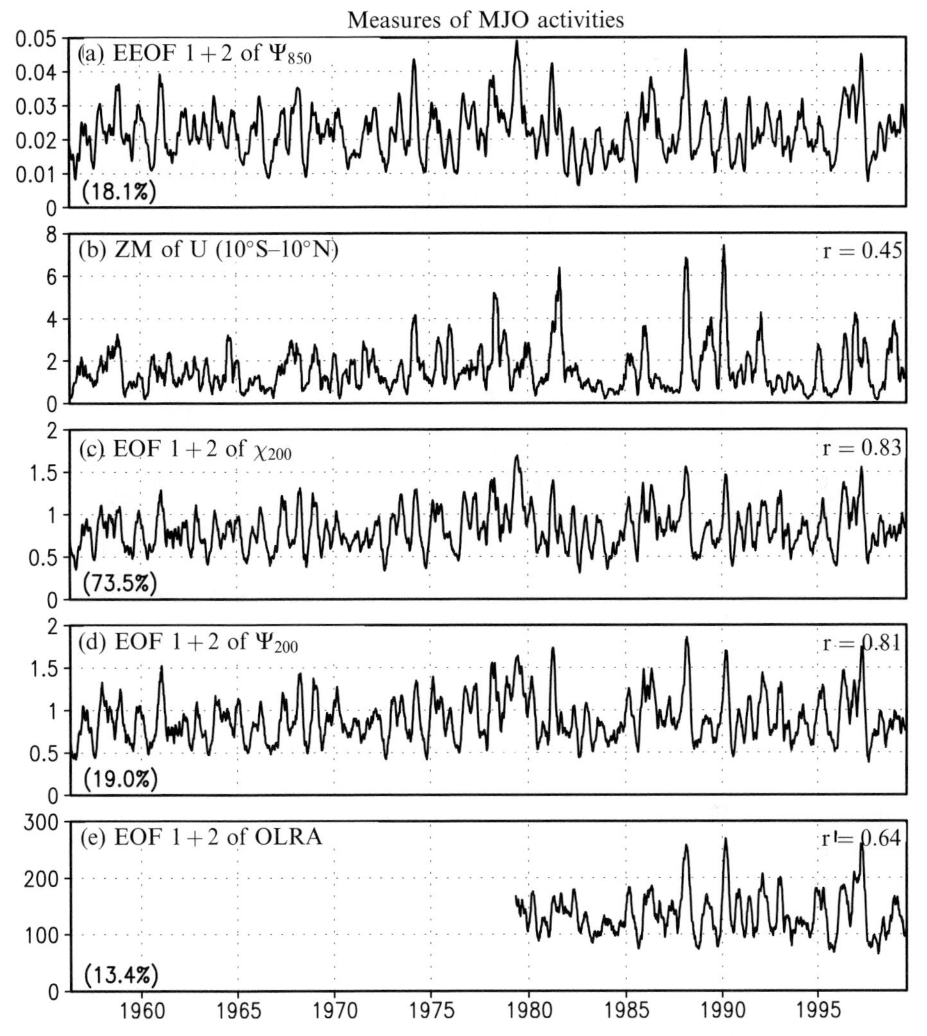

Figure 9.11. A comparison of time series of MJO indices: (a) the EPM index used in this chapter, (b) 20–70-day band-passed zonal mean 200-mb zonal wind averaged from 10°S–10°N (c), an index derived from the first two EOFs of 200-mb velocity potential, (d) 200-mb streamfunction, and (e) OLR anomalies. Numbers in brackets denotes variance explained by the first two dominant EOFs used for the index. Correlation coefficient (r) with the EPM index is shown on upper right-hand corner of each panel. Unit for zonal wind is in m s^{-1}. All other indices have non-dimensional units.

deep convection. On the other hand, the new ISO index, contains a much wider spectrum of variability including effects of rotational and divergent winds, from the tropics and extra-tropics, and is therefore a more inclusive index for describing ISO–ENSO relationships.

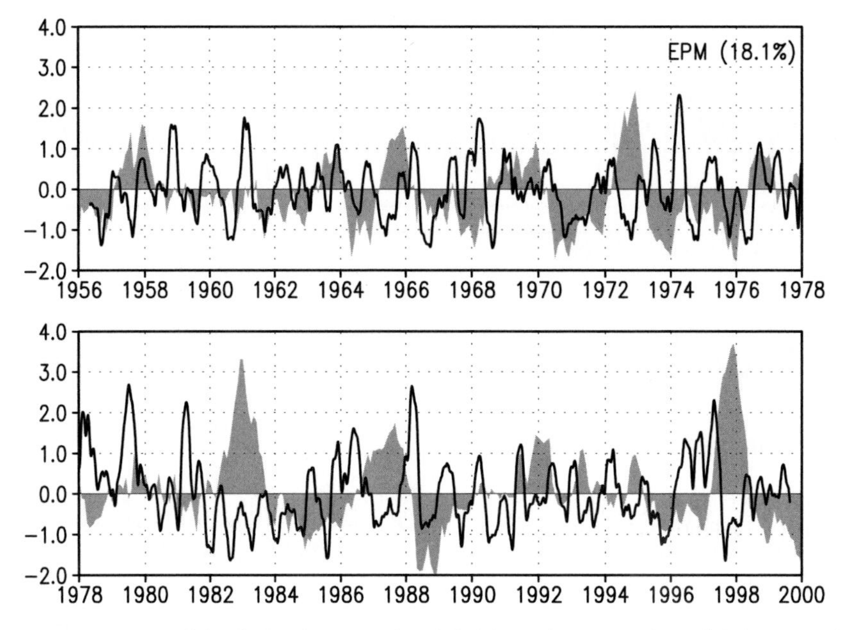

Figure 9.12. EPM activity index (see text for definition) shown as the solid line. Number in bracket shows percentage variance explained. Niño-3 SST anomaly shown as shaded plots. Units of SST in °C.

Two indices of ISO activities have been constructed based on the 90-day windowed variances of the principal components of the first two EEOFs, and of the third and the fourth EEOFs, respectively, for the EPM and QSM. Both modes show strong seasonality, with the EPM peaking in March–April, and the QSM in January–February (not shown). Figure 9.12 shows the indices of EPM activity (EEOF 1 and EEOF 2) superimposed on the monthly time series of the Niño-3 SST anomaly. For most El Niños (i.e., 1997/1998, 1987/1988, 1982/1983, 1972/1973, 1969/1970), there was an increase in EPM activity prior to the abrupt rise in Niño-3 SST, followed by a decrease in EPM activity when the Niño-3 SST was substantially above normal. For 1982/1983, the relationship was not as strong, because the increased EPM occurred more than a year before the initial warming. Notably, there were El Niño events that were not clearly preceded by enhanced EPM (i.e., 1957/1958, 1976/1977, and the series of warm episodes in 1991–1995). Conversely, there were strong EPMs that were not followed by onset of major warm events. These features are in agreement with the impacts of stochastic forcing of El Niño by the MJO (see Section 9.5.2). Figure 9.13 shows the QSM index with the Niño-3 SST. Here, with few exceptions, enhanced (reduced) QSM activity was found when the Niño-3 SST was substantially above (below) normal. This feature can be seen clearly in all the major El Niño–La Niña of 1997–2000, 1982–1985, and 1972–1974.

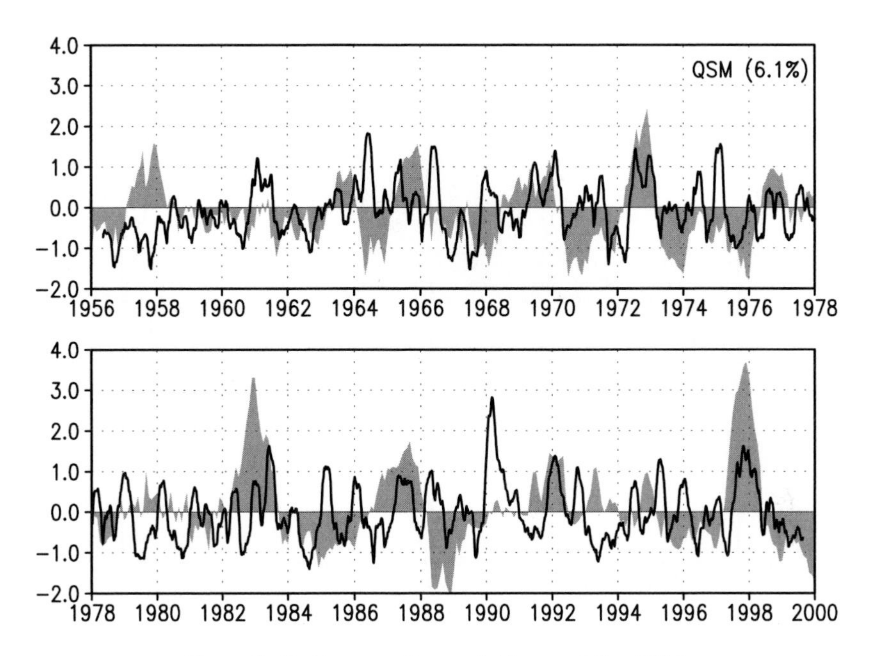

Figure 9.13. Same as Figure 9.12 except for QSM.

9.6.2 Composite events

Figure 9.14 shows the composite of EPM and QSM activities and Niño-3 SST for all past major El Niño events, as a function of the calendar month centered on January during the peak of the warm event. Increased EPM activity from the previous fall through the antecedent spring is evident prior to the onset of a warm event, which typically occurs in April–June. The most pronounced pulse of increased EPM occurs in April–May. As the warm event grows in June–July, the EPM abruptly drops off to a very low level in August–September, and remains suppressed throughout the warm phase, only to recover during the development of the cold phase in the following spring. In contrast to EPM, QSM is suppressed prior to and during the initial warming phase from the previous winter to late fall. The QSM activity notably increases after the peak of Niño-3 SST is reached in December, and remains enhanced for the rest of the winter through late spring, with a second peak in April. The QSM becomes suppressed during the La Niña phase. Overall, there is a 3–6-month lag between the peak of the Niño-3 SST and maximum activity in QSM.

The ISO–ENSO relationships implied by Figure 9.14 are coherent across the Indo-Pacific, as is evident in Figures 9.15 and 9.16 which show respectively the time–longitude sections of the lagged covariance of the 850-mb zonal wind and SST anomalies with reference to EPM and QSM. The EPM maximum activity occurs at time equal to zero ($T(0)$) when the eastern Pacific is colder than normal (Figure 9.15(b)), with anomalous easterlies over the central Pacific, and anomalous westerlies

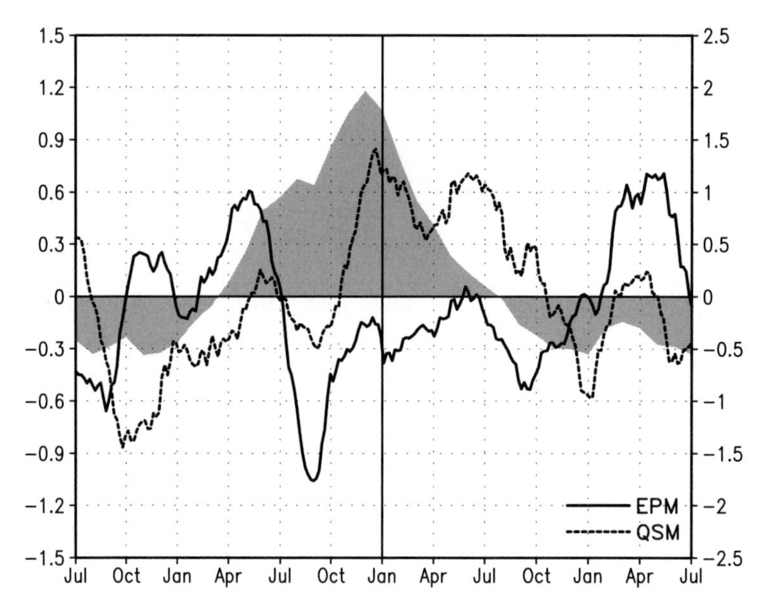

Figure 9.14. Composite of Niño-3 SST (shaded), superimposed on windowed variance of EPM (solid) and QSM (dashed), normalized by standard deviation.

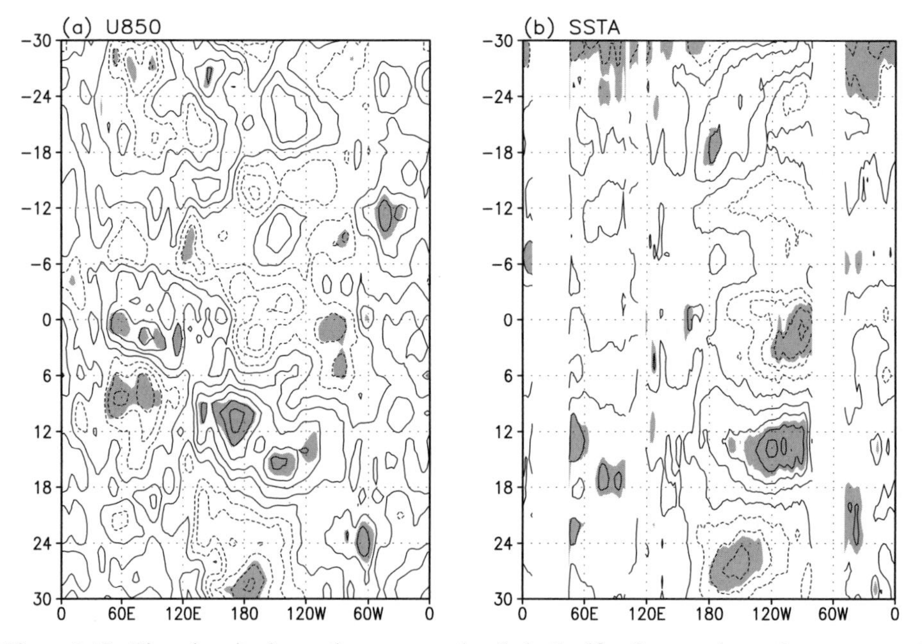

Figure 9.15. Time–longitude section across the Indo-Pacific Ocean along the equator of lagged covariance with reference to the EPM activity: (a) 850-mb zonal wind anomalies and (b) SST anomalies. Values exceeding the 95% significance are highlighted.

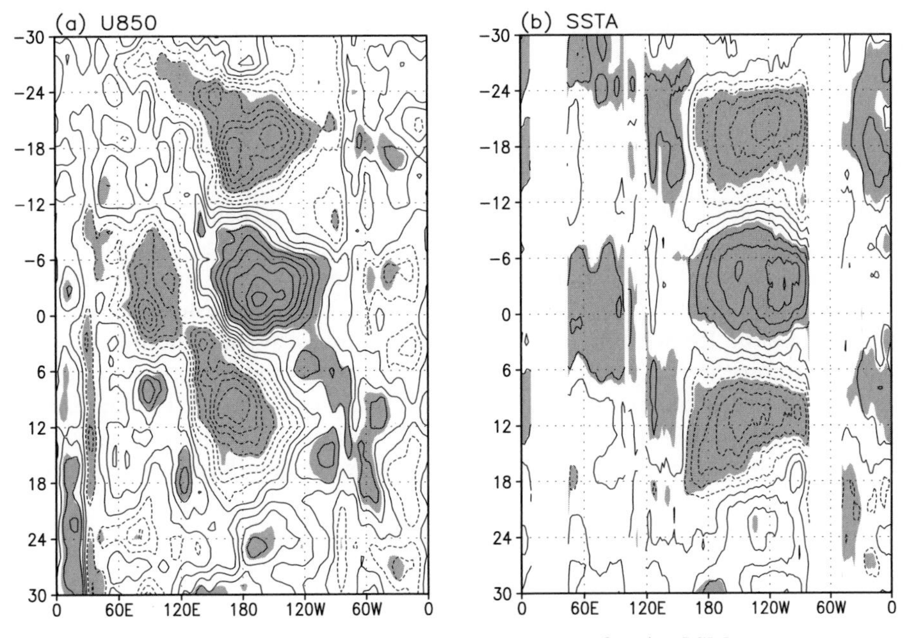

Figure 9.16. Same as Figure 9.15 except for the QSM.

over the Indian Ocean (Figure 9.15(a)). The wind anomalies propagate eastward to the central Pacific about 12 months later and lead to a warm event in the eastern Pacific (Figure 9.15(b)). Following the maximum EPM at $T(0)$, the SST and wind structures exhibit well-defined alternating warm (cold) events coupled to 850-mb westerlies (easterlies) in the central Pacific, occurring at approximately 12–14-month intervals. Before $T(-12)$, the wind and SST structures are not very well organized. The temporal asymmetry suggests that enhanced EPM activities may be responsible for the excitation of the warm and cold SSTAs, but not necessarily the converse. The timescale of the SST oscillation shown in Figure 9.16(b) is quasi-biennial (i.e., 24–26 months) and not the dominant ENSO cycle timescale of 36–48 months. As evident in the relatively small regions with 95% statistical signifiance (indicated by shaded area in Figure 9.16), the coherence in winds and SST variations with EPM on interannual timescales is not very high. This is consistent with the stochastic nature of the MJO–ENSO interaction, occurring only episodically, during a preferred phase of the seasonal and ENSO cycles. Figure 9.16 shows that QSM variability is strongly linked to the variation of the ENSO cycle. The maximum QSM signal is preceded, approximately 2–3 months, by the development of a warm event manifested in an east–west wind dipole with surface westerlies over the central Pacific, easterlies over the Indian Ocean (Figure 9.16(a)), and above normal SST in the eastern Pacific (Figure 9.16(b)). Each warm (cold) episode in the equatorial eastern Pacific is coupled to westerly (easterly) wind anomalies over the central Pacific, which can be traced to eastward wind propagation originating from the Indian Ocean. Clearly, the QSM component is associated with teleconnection signals transmitted from the tropics to the extra-tropics and is an integral part of the ENSO cycling process.

9.6.3 An ISV–ENSO biennial rhythm

Based on the new observational evidence, we present a new scenario of ISV–ENSO biennial cycle interaction (Figure 9.17). In this scenario, the tropical ocean–atmosphere system is considered to possess two climate states; one cold and one warm. In the absence of MJO/WWE forcing, transition to warm and cold states occur at the basic ENSO time intervals of four to five years, as determined by the time required for the charge and discharge of heat content in the tropical ocean with each state lasting for approximately two years (Wang *et al.*, 2003). Enhanced EPM activity denoted by EPM(+) in Figure 9.17, and associated westerly wind forcings (WWE(+)) in boreal spring may induce oceanic downwelling Kelvin waves (K(+)),

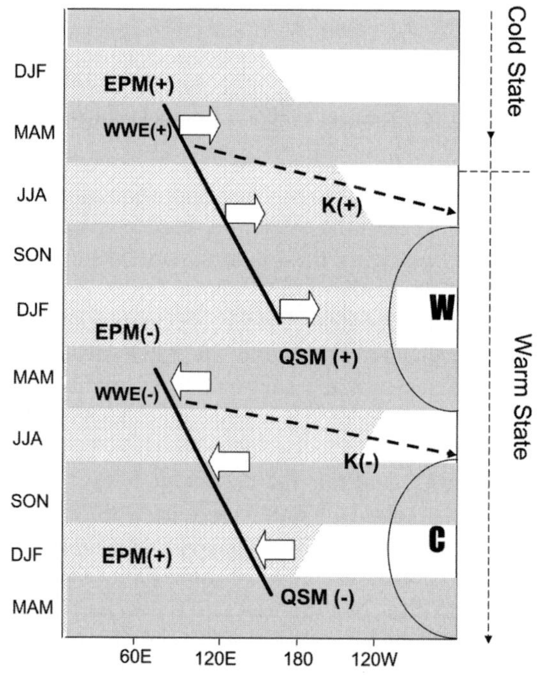

Figure 9.17. A schematic time–longitude section showing the interaction of EPM, QSM, and WWEs in setting up a binnial oscillation in the tropical ocean–atmosphere system along the equator. The phases of the EPM and QSM are indicated by the positive and negative signs. The slanting thick solid lines indicate the eastward migration of the center of ENSO–MJO activities, labeled as WWE(+) and WWE(−). The thick dashed lines denote the propagation of the downwelling oceanic Kelvin waves, K(+) initiating the warm state (**W**) and the upwelling oceanic Kelvin waves (K(−)) initiating the cold state (**C**). The region of anomalous westerly winds are indicated by the stippled background.

initiating rapid warming of the equatorial wave guide and the coastal region of South America. If EPM(+) occurs when the ocean is pre-conditioned for a warm event (i.e., heat content is already accumulating in the tropical Pacific ocean), as indicated by the transition from a cold to warm state in Figure 9.17, the EPM wind will accelerate the coupled air–sea interaction, advance the timing and amplify the warming process, and increase the likeihood of a full-blown warm event (**W**). As the warm event (**W**) develops, the region of surface westerlies expands eastward with enhanced convection over the equatorial eastern Pacific and suppressed convection over the western Pacific (see also Figure 9.5). During this phase, the excess oceanic heat content accumulated during the previous 9–12 months will be transported from tropics to extra-tropics, as manifested in part by the increased QSM(+) activity following the peak of the warm event. In the Indo-Pacific monsoon region (60°E–120°E), the subsidence associated with suppressed convection will induce over the eastern Indian Ocean anomalous surface easterly events denoted by EPM(−), which propagate eastward to the western and central Pacific as a coupled system. The ISV easterly wind anomalies (WWE(−)) embedded within the EPM(−) will drive upwelling Kelvin waves (K(−)) along the equatorial wave guide, arrest the eastern Pacific warming, and initiate a cold event (**C**). Subsequently, similar interactions, but with opposite signs to the warm event, take place. Hence, a consequence of the MJO–ENSO interaction is to produce a pronounced biennial tendency to the ENSO cycle. However, it is unlikely that the biennial rhythm will continue more than two half-cycles, because if in the first half-cycle (**W**) the biennial tendency reinforces the quasi-4-year basic ENSO cycle, the second half-cycle (**C**) will oppose it. Thus the MJO-ENSO interacting with the seasonal cycle may sow the seeds of its own destruction, through suppression and enhancement of westerly wind forcing in the western Indo-Pacific monsoon region. The scenario proposed here is consistent with results of recent model studies which stressed the importance of the western Pacific wind forcing in generating the biennial tendency in ENSO cycle (Kim and Lau, 2001; Wisebert and Wang, 1997).

9.7 TISV AND PREDICTABILITY

This chapter would not be complete without a discussion of the relationship between TISV and ENSO predictability. While MJO/WWE signals undoubtedly provides useful information, which can be harnessed to improve ENSO prediction, it is clear from previous discussions in this chapter that TISV is also a source of climate noise, limiting the potential predictability of the climate system on inter-annual and longer timescales. It has been suggested that a useful analogy of TISV–ENSO coupling is that of a slightly damped oscillator subject to both external periodic forcing (the annual cycle), internal dynamics (ENSO mode), and random noise forcings (TISV) (Lau, 1985a; Fedorov *et al.*, 2003). The noise forcing renders the oscillation irregular. In the absence of such random forcings, ENSO would be perfectly predictable. However, for ENSO, the response of the system to the noise forcing is sensitive to the phase of the annual cycle and the ENSO cycle itself. During the boreal spring, the MJO and related WWEs are confined to the proximity of the equator and are therefore more effective, compared to other seasons, in driving Kelvin waves along the equatorial wave guide, triggering the onset of El Niño.

The impact of stochastic forcing in limiting the potential predictability of ENSO has been investigated by many authors. Eckert and Latif (1997) found that the effectiveness of ISV noise forcing typically led to a loss of predictability before a cycle period has elapsed. Moore and Kleeman (1998) showed that ISV forcing associated with statistical optimals in a coupled system can be used to mimic the presence of initial condition errors and high-frequency stochastic noise to provide a measure of the reliability and skill in ensemble forecasts. Blanke *et al.* (1997) found that periodic states in a seasonally varying coupled ocean–atmosphere model can be modified by WWE forcings to produce irregular behavior as well as strong phase locking to sub-harmonics of the annual cycle, in particular the biennial tendency. The broadening of the model ENSO spectral peaks by noise forcing provides a much richer spatial and temporal pattern consistent with observations.

More recently, Federov *et al.* (2003) provided an insightful analysis of predictability of ENSO and ISV forcings, which is re-captured in the following. Figure 9.18 shows the possible responses of coupled ocean–atmosphere states to an initial burst of a WWE in the coupled system of Neelin (1990), depending on whether the model ENSO state is: (a) damped; (b) has periodic oscillations, that are affected by ISV; or (c) exhibits chaotic behavior. These situations correspond to Hypothesis III, II, and I discussed in the previous subsection. It is argued that if the ENSO state is damped (Hypothesis III), then ENSO can only be generated by WWE stochastic forcings. But, Hypothesis III does not explain why some El Niños are generated by WWEs and others are not, and why there is a quasi-periodic 4–5-year ENSO cycle. On the other hand, if the system is chaotic (Hypothesis I), like that shown in Figure 9.18(c), the role of ISV forcing is irrelevant, because it is just like any other noise component, including those generated by chaotic behavior inherent in the coupled system. Under this hypothesis, predictability is limited. ISV plays no unique role. Figure 9.18(b) may well describe a system which is quasi-periodic, but also responds to MJO/WWE forcings in ways that are dependent on the phase of the annual cycle, as well as the phase of the ENSO cycle itself (Hypothesis II). This scenario is the most consistent with observations so far. In fact, the variation of SST here is similar to those portrayed in Figure 9.16, following strong activity in MJO/WWE events. It is likely that predictability of ENSO depends on the interplay of two sets of phenomena: a low-frequency signal that governs the timescale of the ENSO cycles (years) and a high-frequency component that readjusts the actual time of the onset and the amplitude of the event in a statistical sense. The readjustment could sometimes be amplified to substantially accelerate the onset time, or to abort an ongoing warming event. For this reason, dynamical ENSO forecasts should be probabilistic, with large ensembles, so that the shift in probability distribution functions, with respect to wind and convection fluctuations in the Indo-Pacific region, can be detected. There is now increasing observational evidence that ENSO predictions can benefit from inclusion of realistic MJO/ISV precursory signals in the Indo-Pacific region, in addition to those that describe the slowly

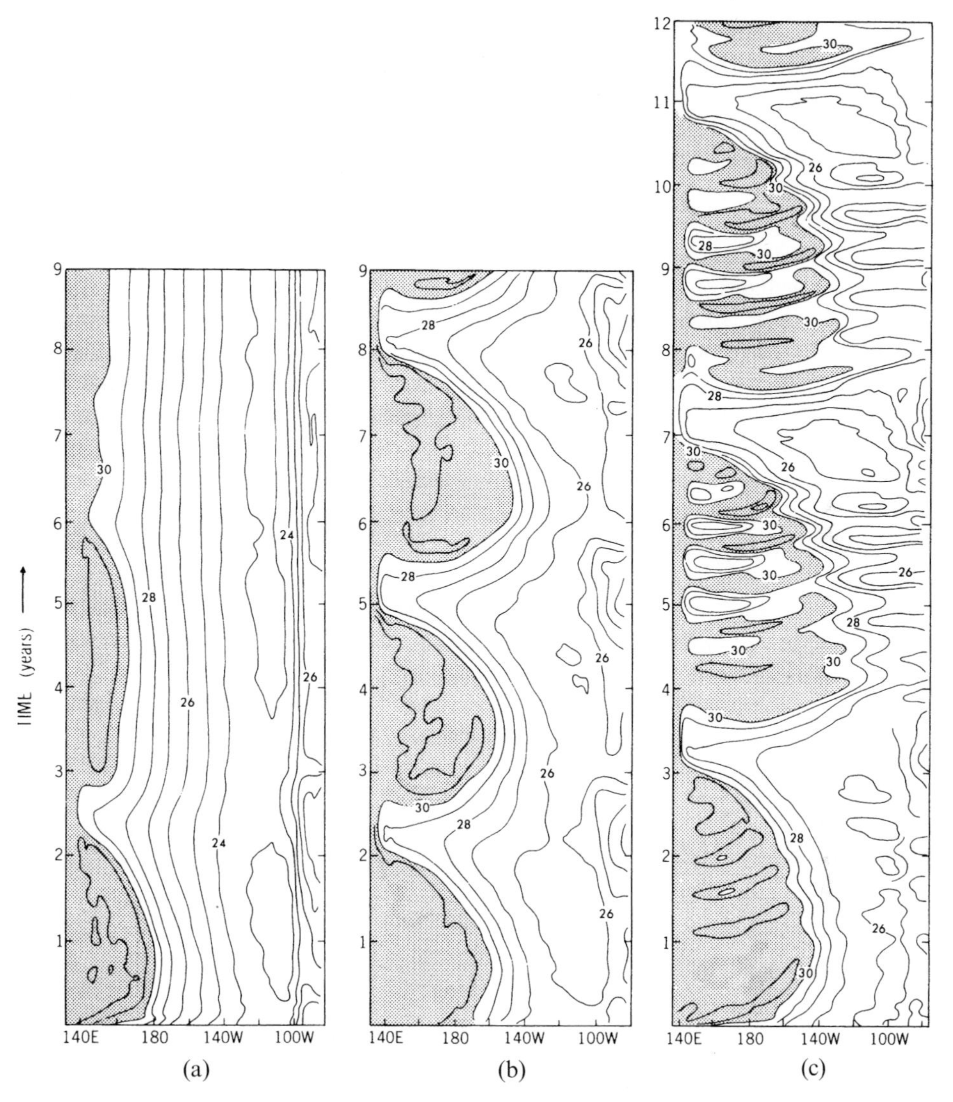

Figure 9.18. Time–longitude section along the equator, showing the evolution of SST (°C) in response to a westerly wind burst imposed at $T = 0$ in a coupled model, which (a) is damped, (b) has periodic oscillations, and (c) exhibits chaotic behavior.
From Federov *et al.* (2003)

evolving ocean–atmosphere states such as sea level, SST, and heat content (Kessler and Kleeman, 2000).

Recently, Curtis *et al.* (2002) found enhanced precipitation and wind gradients in the east equatorial Indian Ocean associated with MJO fluctuation, 3–9 months preceding the five strongest El Niño events since 1979 (1982/1983, 1986/1987, 1991/1992, 1994/1995, and 1997/1998). They successfully predicted the development of the

2001–2002 El Niño using a thresholding technique using an index based on MJO signals. Clarke and Gorder (2003) have shown that updating the most current space–time information on Indo-Pacific winds, and equatorial Pacific upper-ocean heat content associated with the MJO/WWE can lead to improved empirical forecasts of El Niño. Zhang and Gottschalck (2002) found a lead signal of 6–12 months before the onset of major El Niño in a wind stress index in the equatorial Pacific related to oceanic Kelvin wave forcing associated with the MJO. For improving dynamical prediction, the challenging tasks are to harness these observational precursory signals by including realistic representation of the ocean–atmosphere coupling across the whole spectrum of spatial and temporal scales from intraseaonal to interannual, and by including accurate initial conditions for the ocean–atmosphere states associated with the MJO/WWEs to carry out probabilistic forecasts using ensemble methods. These tasks are daunting, but should be achievable within the present decade.

9.8 ACKNOWLEDGMENT

This work was supported by the NASA Global Modeling and Analysis Program, the TRMM Project of the NASA Earth Science Enterprise. Dr. K. M. Kim provided valuable literature search and programming support.

9.9 REFERENCES

Battisti, D. S. and A. C. Hirst (1989) Interannual variability in a tropical atmosphere–ocean model: Influences of the basic state, ocean geometry and nonlinearity. *J. Atmos. Sci.*, **46**, 1687–1712.

Belamari, S., J. Redelsperger, and M. Pontaud (2003) Dynamic role of a westerly wind burst in triggering equatorial Pacific warm event. *J. Climate*, **16**, 1869–1890.

Bergman, J. W., H. H. Hendon, and K. M. Weickmann (2001) Intraseasonal air–sea interaction at the onset of El Niño. *J. Climate*, **14**, 1702–1719.

Blanke, B., J. D. Neelin, and D. Gutzler (1997) Estimating the effect of stochastic wind stress forcing on ENSO irregularity. *J. Climate*, **10**, 1473–1486.

Cane, M. A. and S. E. Zebiak (1985) A theory of El Niño and the Southern Oscillation. *Science*, **228**, 1085–1087.

Cane, M. (1983) Oceanographic events during El Niño. *Science*, **222**, 1189–1195.

Chang, C. P and K. M. Lau (1980) Northeasterly cold surges and near-equatorial disturbances over the winter MONEX area during December. Part II: Large-scale aspects. *Mon. Wea. Rev.*, **108**, 298–312.

Chang, C. P. and H. Lim (1988) Kelvin wave CISK: A plausible mechanism for the 30–50 day oscillation. *J. Atmos. Sci.*, **45**, 1709–1720.

Clarke, A. J. and S. Van Gorder (2003) Improving El Niño prediction using a space–time integration of Indo-Pacific winds and equatorial Pacific upper ocean heat content. *Geophys. Res. Letters.*, **30**, 7, 1399, doi:10.1029/2002GL-16673.

Curtis, S., G. J. Huffman, and R. F. Adler (2002) Precipitation anomalies in the tropical Indian Ocean and their relationship to initiation of El Niño. *Geophys. Res. Lett.*, **29**(10) 1441.

Eckert, C. and M. Latif (1997) Predictability of a stochastically forced hybrid coupled model of El Niño. *J. Climate*, **10**, 1488–1504.

Emanuel, K. (1988) An air–sea interaction model of intraseasonal oscillations in the tropics. *J. Atmos. Sci.*, **44**, 2324–2340.

Federov, A. V., S. L. Harper, S. G. Philander, B. Winter, and A. Wittenberg (2003) How predictable is El Niño? *Bull. Am. Meteor. Soc.*, **84**, 911–919.

Fink, A. and P. Speth (1997) Some potential forcing mechanisms of the year-to-year variability of the tropical convection and its intraseasonal (25–70 day) variability. *Int. J. Climatol.*, **17**, 1513–1534.

Flatau M., P. J. Flatau, P. Phoebus, and P. P. Niiler (1997) The feedback between equatorial convection and local radiative and evaporative processses: The implications for intraseasonal oscillations. *J. Atmos. Sci.*, **54**, 2373–2386.

Fu, X., B. Wang, T. Li, and J. McCreary (2003) Coupling between northward propagating boreal summer ISO and Indian Ocean SST revealed in an atmosphere–ocean coupled model. *J. Atmos. Sci.*, **60**, 1733–1753.

Giese, B. S. and D. E. Harrison (1990) Aspects of Kelvin wave response to episodic wind forcing. *J. Geophys. Res.*, **95**(C5), 7289–7312.

Gill, A. and E. M. Rasmusson (1983) The 1982/83 climate anomaly in the equatorial Pacific. *Nature*, **306**, 229–234.

Gill, A. E. (1984) Elements of coupled ocean–atmosphere models for the tropics. In: J. C. J. Nihoul (ed.) *Coupled Ocean Atmosphere Models* (Elsevier Oceanography Series 40). Elsevier, Amsterdam, 767 pp.

Gruber, A. and A. F. Krueger (1984) The status of the NOAA outgoing longwave radiation data set. *Bull. Am. Meteor. Soc.*, **65**, 958–962.

Gualdi, S., A. Navarra, and G. Tinarelli (1999) The interannual variability of the Madden–Julian Oscillation in an ensemble of GCM simulations. *Clim. Dyn.*, **15**, 643–658.

Harrison, D. E. (1984) On the appearance of sustained equatorial westerlies during the 1982 Pacific warm event. *Science*, **225**, 1099–1102.

Harrison, D. E. and B. Geise (1988) Remote westerly wind forcing of the eastern equatorial Pacific: Some model results. *Geophys. Res. Letters*, **15**, 804–807.

Harrison, D. E. and P. Schopf (1984) Kelvin-wave induced anomalous advection and the onset of surface warming in El Niño events. *Mon. Wea. Rev.*, **112**, 923–933.

Harrison, D. E. and G. A. Vecchi (1997) Westerly wind events in the tropical Pacific: 1986–95. *J. Climate*, **10**, 3131–3156.

Hayashi, Y. Y. and A. Sumi (1986) The 30–60 day oscillations simulated in an aqua-planet model. *J. Meteor. Soc. Jap.*, **64**, 451–467.

Hendon, H. H. (1988) A simple model of the 40–50 day oscillation. *J. Atmos. Sci.*, **45**, 569–584.

Hendon, H. H., C. Zhang, and J. D. Glick (1999) Interannual variation of the Madden–Julian Oscillation during austral summer. *J. Climate*, **12**, 2538–2550.

Hendon, H. H., B. Liebmann, and J. D. Glick (1998) Oceanic Kelvin waves and the Madden–Julian Oscillation. *J. Atmos. Sci.*, **55**, 88–101.

Hirst, A. C. (1986) Unstable and damped equatorial modes in simple coupled ocean–atmosphere models. *J. Atmos. Sci.*, **43**, 606–630.

Hirst, A. and K. M. Lau (1990) Intraseasonal and interannual variability in coupled ocean–atmosphere models. *J. Climate*, **3**, 713–725.

Hsu, H. H. (1996) A global view of intraseasonal oscillation during northern winter. *J. Climate*, **9**, 2396–2406.

Ji, M. A., A. Leetmaa, and J. Derber (1995) An ocean analysis system for seasonal to interannual climate study. *Mon. Wea. Rev.*, **123**, 460–481.

Kalnay, E., M. Halem, and W. E. Baker (1981) The FGGE observing system during SOP-2. *Bull. Amer. Meteor. Soc.*, **62**, 897–898.

Keen, R. A. (1982) The role of cross-equatorial cyclone pairs in the Southern Oscillation. *Mon. Wea. Rev.*, **110**, 1405–1416.

Kessler, W. S. (2001) EOF representations of the Madden–Julian Oscillation and its connection with ENSO. *J. Climate*, **14**, 3055–3061.

Kessler, W. S. and R. Kleeman (2000) Rectification of the Madden–Julian Oscillation into the ENSO cycle. *J. Climate*, **13**, 3560–3575.

Kessler, W. S., M. J. McPhaden, and K. M. Weickmann (1996) Forcing of intraseasonal Kelvin waves in the equatorial Pacific. *J. Geophys. Res.*, **100**, 10613–10631.

Kim, K. M. and K. M. Lau (2001) Monsoon induced biennial variability in ENSO. *Geophys. Res. Lett.*, **28**, 315–318.

Kindle, J. C. and P. A. Phoebus (1995) The ocean response to operational wind bursts during the 1991–1992 El Niño. *J. Geophys. Res.*, **100**(C3), 4803–4920.

Kleeman, R. and A. M. Moore (1997) A theory for the limitations of ENSO predictability due to stochastic atmospheric transients. *J. Atmos. Sci.*, **54**, 753–767.

Knutson, T. R. and K. M. Weickmann (1987) 30–60 day atmospheric oscillation: Composite cycles of convection and circulation anomalies. *Mon. Wea. Rev.*, **115**, 1407–1436.

Krishnamurti, T. N., P. K. Jayakumar, J. Shen, N. Surgi, and A. Kumar (1985) Divergent circulation on the 30–50 day time scale. *J. Atmos. Sci.*, **42**, 364–374.

Krishnamurti, T. N. and D. Subrahmanyam (1983) The 30–50 day mode at 850 mb during MONEX. *J. Atmos. Sci.*, **39**, 2088–2095.

Krishnamurti, T. N., D. K. Oosterhof, and A. V Mehta (1988) Air–sea interaction on the time scale of 30–60 days. *J. Atmos. Sci.*, **45**, 1304–1322.

Kutsuwada, K. and M. McPhaden (2002) Intraseasonal variations in the upper equatorial Pacific Ocean prior to and during the 1997–98 El Niño. *J. Phys. Oceanogr.*, **32**, 1133–1149.

Latif, M., T. P. Barnett, M. A. Cane, M. Flugel, N. E. Graham, H. Von Storch, J. S. Xu, and S. E. Zebiak (1994) A review of ENSO prediction studies. *Climate Dynamics*, **9**, 167–179.

Lau, K. M. (1981) Oscillations in a simple equatorial climate system. *J. Atmos. Sci.*, **38**, 248–261.

Lau, K. M. and P. H. Chan (1983a) Short-term climate variability and atmospheric teleconnection as inferred from satellite derived outgoing longwave radiation. Part I: Simultaneous correlations. *J. Atmos. Sci.*, **40**, 2735–2750.

Lau, K. M. and P. H. Chan (1983b) Short-term climate variability and atmospheric teleconnection from satellite derived outgoing longwave radiation. Part II: Lagged correlations. *J. Atmos. Sci.*, **40**, 2752–2767.

Lau, K. M., C. P. Chang, and P. H. Chan (1983) Short-term planetary scale interactions over the tropics and midlatitudes. Part II. Winter-MONEX period. *Mon. Wea. Rev.*, **111**, 1372–1388.

Lau, K. M. (1985a) Elements of a stochastic-dynamical theory of the long-term variability of the El Niño/Southern Oscillation. *J. Atmos. Sci.*, **42**, 1552–1558.

Lau, K. M. (1985b) Subseasonal scale oscillations, bimodal climate state and the El Niño/ Southern Oscillation. In: J. C. J. Nihoul (ed.), *Coupled Atmosphere-Ocean Models* (Elsevier Oceanographic Series 40). Elsevier, Amsterdam, 767 pp.
Lau, K. M. and P. H. Chan (1985) Aspects of the 40–50 day oscillation during northern winter as inferred from outgoing longwave radiation. *Mon. Wea. Rev.*, **113**, 1889–1909.
Lau, K. M. and P. H. Chan (1986a) The 40–50 day oscillation and the El Niño/Southern Oscillation: A new perspective. *Bull. Amer. Meteor. Soc.*, 533–534.
Lau, K. M. and P. H. Chan (1986b) Aspects of the 40–50 day oscillation during northern summer as inferred from outgoing longwave radiation. *Mon. Wea. Rev.*, **114**, 1354–1367.
Lau, K. M. and L. Peng (1987) Origin of low frequency (intraseasonal) oscillation in the tropical atmosphere. Part I: Basic theory. *J. Atmos. Sci.*, **44**, 950–972.
Lau, K. M. and P. H. Chan (1988) Intraseasonal and interannual variability of tropical convection: A possible link between the 40–50 day oscillation and ENSO. *J. Atmos. Sci.*, **45**, 506–521.
Lau, K. M., T. Nakazawa, and C. H. Sui (1991) Observations of cloud cluster hierarchies over the tropical western Pacific. *J. Geophys. Res.*, **96**, 3197–3208.
Lau, K. M. and S. Shen (1988) Dynamics of Intraseasonal Oscillations and ENSO. *J. Atmos. Sci.*, **45**, 1781–1797.
Lau, K. M., L. Peng, C. H. Sui, and T. Nakazawa (1989) Super cloud clusters, westerly wind burst, 30–60 day oscillations and ENSO: A unified view. *J. Meteor. Soc. Jap.*, **67**, 205–219.
Lau, K-M. and C.-H. Sui (1997) Mechanisms of short-term sea surface temperature regulation: Observations during TOGA–COARE. *J. Climate*, **9**, 465–472.
Lorenc, A. C. (1984) The evolution of planetary scale 200 mb divergences during the FGGE year. *Quart. J. Roy. Meteor. Soc.*, **110**, 427–442.
Luther, D. S. and D. E. Harrison (1984) Observing long-period fluctuations of surface winds in the tropical Pacific: Initial results from island data. *Mon. Wea. Rev.*, **112**, 285–302.
Luther, R. S., D. E. Harrison, and R. A. Knox (1983) Zonal winds in the central equatorial Pacific and El Niño. *Science*, **222**, 327–330.
Madden, R. and P. Julian (1971) Detection of a 40–50 day oscillation in station pressure and zonal winds in the tropical Pacific. *J. Atmos. Sci.*, **28**, 702–708.
Madden, R. and P. Julian (1972) Description of global scale circulation cells in the tropics with a 40–50 day period. *J. Atmos. Sci.*, **29**, 1109–1123.
Madden, R. (1986) Seasonal variations of the 40–50 day oscillation in the tropics. *J. Atmos. Sci.*, **43**, 3138–3158.
Maloney, E. D. and D. L. Hartmann (2000) The sensitivity in intraseasonal variability in the NCAR CCM3 to changes in convective parameterization. *J. Climate*, **14**, 2015–2034.
McGovern, W. and R. Fleming (1977) FGGE and its implications for numerical weather prediction. *Bull. Amer. Meteor. Soc.*, **58**, 113.
McCreary, J. P. and D. L. T. Anderson (1984) A simple model of El Niño and the Southern Oscillation. *Mon. Wea. Rev.*, **112**, 934–946.
McPhaden, M. J. (1999) Genesis and evolution of the 1997–98 El Niño. *Science*, **283**, 950–954.
McPhaden, M. J. and X. Yu (1999) Equatorial waves and the 1997–98 El Niño. *Geophys. Res. Lett.*, **26**, 2961–2964.
McPhaden, M. J., H. P. Freitag, S. P. Hayes, B. A. Taft, Z. Chen, and K. Wyrtki (1988) The response of the equatorial Pacific Ocean to a westerly wind burst in May 1986. *J. Geophys. Res.*, **93**(C9), 10589–10603.
Meyer, G., J. R. Donguy, and R. K. Reed (1986) Evaporative cooling of the western equatorial Pacific ocean by anomalous winds. *Nature*, **323**, 523–526.

Moore, A. M. and R. Kleeman (1998) Skill assessment of ENSO using ensemble prediction. *Quart. J. Royal Meteor. Soc.*, **124**, 557–584.

Moore, A. M. and R. Kleeman (1999) Stochastic forcing of ENSO by the intraseasonal oscillation. *J. Climate*, **12**, 1199–1220.

Murakami, T., T. Nakazawa, and J. He (1984) On the 40–50 day oscillations during the 1979 northern hemisphere summer. Part I: Phase propagation. *J. Meteor. Soc. Jap.*, **62**, 440–468.

Murakami, T. and T. Nakazawa (1985) Tropical 45-day oscillations during the 1979 northern hemisphere summer. *J. Atmos. Sci.*, **42**, 1107–1122.

Nakazawa, T. (1988) Tropical super clusters within intraseasonal variations over the western Pacific. *J. Meteor. Soc. Jap.*, **66**, 823–839.

Nakazawa, T. (2000) MJO and tropical cyclone activity during 1997/98 ENSO. *Adv. Space Res.*, **25**, 953–958.

National Research Council Report (1996) *Learning to Predict Climate Variations Associated with El Niño and the Southern Oscillation: Accomplishments and Legacies of the TOGA Program.* National Academy Press, Washington, D.C., 235 pp.

Neelin, J. D. (1990) A hybrid coupled general circulation model for El Niño studies. *J. Atmos. Sci.*, **47**, 674–693.

Neelin, J. D., D. S. Battisti, A. C. Hirst, F. F. Jin, Y. Wakata, T. Yamagata, and S. E. Zebiak (1998) ENSO theory. *J. Geophys. Res.*, **103**, 14261–14290.

Penland, C. and P. D. Sardeshmukh (1995) The optimal growth of tropical sea surface temperature anomalies. *J. Climate*, **8**, 1999–2024.

Penland, C. (1996) A stochastic model of Indo-Pacific sea surface temperature anomalies. *Physica D*, **98**, 534–558.

Philander, S. G. H., T. Yamagata, and C. Pacanowski (1984) Unstable air–sea interactions in the tropics. *J. Atmos. Sci.*, **41**, 604–613.

Roulston, M. S. and J. D. Neelin (2000) The response of an ENSO model to climate noise, weather noise and intraseasonal forcing. *Geophys. Res. Lett.*, **27**, 3723–3726.

Schrage, J. M., D. G. Vincent, and A. H. Fink (1999) Modulation of intraseasonal (25–70 day) processes by the superimposed ENSO cycle across the Pacific basin. *Meteor. Atmos. Phys.*, **70**, 15–27.

Slingo, J. M., D. P. Rowell, and K. R. Sperber (1999) On the predictability of the interannual behavior of the Madden–Julian Oscillation and its relationship with El Niño. *Q. J. Roy. Meteor. Soc.*, **125**, 583–609.

Slingo, J. M. and AMIP collaborators (1997) Intraseasonal oscillations in 15 atmospheric general circulation mdoels: Results from an AMIP diagnostic subproject. *Climate Dynamics*, **12**, 325–357.

Sperber, K. R., J. M. Slingo, P. M. Inness, and K. M. Lau (1997) On the maintenance and initiation of intraseasonal oscillation in the NCEP/NCAR reanalysis and the GLA and UKMO AMIP simulations. *Climate Dynamics*, **13**, 769–795.

Suarez, M. and P. Schopf (1988) A delayed action oscillator for ENSO. *J. Atmos. Sci.*, **45**, 3283–3287.

Takayabu, Y. N., T. Iguchi, M. Kachi, A. Shibata, and H. Kanzawa (1999) Abrupt termination of the 1997–98 El Niño in response to a Madden–Julian Oscillation. *Nature*, **402**, 279–282.

Vautard, R. and M. Ghil (1989) Singular spectrum analysis in nonlinear dynamics with applications to paleoclimatic time series. *Physica*, **35D**, 395–424.

Vecchi, G. A. and D. E. Harrison (2000) Tropical Pacific sea surface temperature anomalies, El Niño and equatorial westerly wind events. *J. Climate*, **13**, 1814–1830.

Waliser, D., Z. Zhang, K. M. Lau, and J. H. Kim (2001) Interannual sea surface temperature variability and the predictability of tropical intraseasonal variability. *J. Atmos. Sci.*, **58**, 2595–2615.

Waliser, D. E., K. M. Lau, and J. H. Kim (1999) The influence of coupled sea surface temperatures on the Madden–Julian Oscillation: A model perturbation experiment. *J. Atmos. Sci.*, **56**, 333–358.

Waliser, D. E., K. Jin, I. S. Kang, W. F. Stern, S. D. Schubert, M. L. Wu, K. M. Lau, M. I. Lee, J. Shukla, V. Krishnamurthy, *et al.* (2003) AGCM simulations of intraseasonal variability associated with the Asian summer monsoon. *Clim. Dyn.*, **21**, 423–446.

Waliser, D. E., N. E. Graham, and C. Gautier (1993) Comparison of the highly reflective cloud and outgoing longwave radiation datasets for use in estimating tropical deep convection. *J. Climate*, **6**, 331–353.

Wang, B. and H. Rui (1990) Dynamics of the coupled moist Kelvin–Rossby wave on an equatorial β-plane. *J. Atmos. Sci.*, **47**, 397–413.

Wang, B. and X. Xie (1998) Coupled modes of the warm pool climate system. Part I: The role of air–sea interaction in maintaining the Madden–Julian Oscillation. *J. Climate*, **11**, 2116–2135.

Wang, X. C., F. F. Jin, and Y. Q. Wang (2003) A tropical ocean recharge mechanism for climate variability. Part I: Equatorial heat content changes induced by off-equatorial wind. *J. Climate*, **16**, 3585–3598.

Weickmann, K. M., G. R. Lussky, and J. E. Kutzbach (1985) Intraseasonal (30–60 day) fluctuations of outgoing longwave radiation and 250 mb streamfunction during northern winter. *Mon. Wea. Rev.*, **113**, 941–961.

Weickmann, K. M. (1983) Intraseasonal circulation and outgoing longwave radiation modes during northern hemisphere winter. *Mon. Wea. Rev.*, **111**, 1838–1858.

Weickmann, K. M. (1991) El Niño/Southern Oscillation and Madden and Julian (30–60 day) oscillations during 1981–82. *J. Geophys. Res. Oceans*, **96**, 3187–3195.

Weisberg, R. H. and C. Wang (1977) A western Pacific oscillator paradigm for the El Niño/Southern Oscillation. *Geophys. Res. Lett.*, **24**, 770–782.

Wyrtki, K. (1975) El Niño – The dynamic response of the equatorial Pacific Ocean to atmospheric forcing. *J. Phys. Oceanogr.*, **5**, 572–584.

Wyrtki, K. (1979) Sea level variations: Monitoring the breath of the Pacific. *EOS Trans. AGU.*, **60**, 25–27.

Yasunari, T. (1982) Structure of the Indian summer monsoon system with a period around 40 days. *J. Meteor. Soc. Jap.*, **60**, 336–354.

Yu, L. and M. M. Rienecker (1998) Evidence of an extratropical atmospheric influence during the onset of the 1997–98 El Niño. *Geophys. Res. Lett.*, **25**, 3537–3540.

Zebiak, S. E. (1982) A simple atmospheric model of relevance to El Niño. *J. Atmos. Sci.*, **39**, 2017–2027.

Zebiak, S. and M. Cane (1987) A model El Niño Southern Oscillation. *Mon. Wea. Rev.*, **115**, 2262–2278.

Zebiak, S. (1989) On the 30–60 day oscillation and the prediction of ENSO. *J. Climate*, **2**, 1381–1387.

Zhang, C. and H. Hendon (1997) Propagating and standing components of the intraseasonal oscillation in tropical convection. *J. Atmos. Sci.*, **54**, 741–752.

Zhang, C., H. H. Hendon, W. S. Kessler, and A. Rosati (2001) A workshop on the MJO and ENSO. *Bull. Amer. Meteor. Soc.*, **82**, 971–976.

Zhang, C. and J. Gottschalck (2002) SST anomalies of ENSO and the MJO in the equatorial Pacific. *J. Climate*, **15**, 2429–2445.

10

Theory

Bin Wang

10.1 INTRODUCTION

In the last two decades, many studies have been devoted to developing a theoretical understanding of the tropical intraseasonal oscillation (TISO) in order to improve the models and their predictions of these disturbances. Progress in modeling and predicting the ISO will only happen if the mechanisms underlying its complex interactions and fundamental dynamics are more fully understood. Significant progress in theoretical understanding has been achieved, although some aspects of the theories remain disputable and incomplete.

One of the purposes of this chapter is to review the basic ideas and limitations of these existing theories and hypotheses. These invoke a broad range of processes and mechanisms to account for the growth and maintenance of the low-frequency disturbances. The major mechanisms that have been proposed are categorized based on their key processes and feedback mechanisms. They are discussed in Section 10.3.2.

A significant effort has been made in this chapter to identify the essential physics of the ISO and to formulate a general theoretical framework. It is hoped that this general framework can coherently tie together various hypotheses and mechanisms that have been put forward to account for the ISO. The proposed theoretical model is simplistic but integrates several key mechanisms, providing a prototype model for describing basic dynamics relevant to both the Madden–Julian Oscillation (MJO) and its boreal summer counterpart – referred to here as the monsoon ISO. The essential physical processes involved in the model and their mathematical description are presented in Section 10.3.3.

The ultimate goal of a meteorological theory is to explain an observed phenomenon and to predict it. A large body of this chapter (Sections 10.3.4, 10.3.5, and 10.3.6) is devoted to this goal. What are the fundamental characteristics of the ISO that a theory must explain? Given the complexity of the phenomenon and limitations in the current observations, determining these features is not easy. Nevertheless, such

W. K. M. Lau and D. E. Waliser (eds), *Intraseasonal Variability in the Atmosphere–Ocean Climate System*.

a list is necessary and useful for validating models and defining the theoretical targets. The following set of statistical features is proposed which defines the wavelength, propagation (direction and phase speed), and spatial structure of the ISO: (1) planetary-scale circulation coupled with a large-scale convective complex (Madden and Julian, 1972); (2) a horizontal circulation comprising equatorial Kelvin and Rossby waves (Rui and Wang, 1990) and a baroclinic circulation with boundary layer convergence preceding the major convection region (Hendon and Salby, 1994); (3) slow eastward propagation (about $5\,\mathrm{m\,s^{-1}}$) in the eastern hemisphere (Knutson and Weickmann, 1986) and longitudinal dependence of the amplification (Wang and Rui, 1990a); and (4) prominent northward propagation (Yasunari, 1979, 1980) and off-equatorial westward propagation (Murakami, 1980) during boreal summer in the Asian monsoon region. More stringent validation characteristics might include the multi-scale structure of the convective complex (Nakazawa, 1988) and the associated sea surface temperature (SST) variability (Krishnamurti *et al.*, 1988, and many others). The above characteristics provide focal points for theoretical analyses and discussions.

An account is given of a series of previous and current theoretical studies, aiming to *promote* our understanding of the fundamental dynamics behind the MJO (Section 10.3.4) and monsoon ISO (Section 10.3.5). Section 10.3.6 is devoted to enlighten on the role of air–sea coupling in enhancing the MJO. The physical explanation presented in these sections primarily reflects the author's personal advocacy. The effort to explain the phenomenon is perhaps more significant than the explanation itself, because the author's purpose is to stimulate better and deeper insights into the physics of the phenomenon.

In spite of all the work, some aspects of the ISO are not yet adequately understood and present the scientific community with a challenge. The last section in this chapter discusses the success and limitations of the presented theory and outstanding issues for improving our theories and models.

10.2 REVIEW OF ISO THEORIES

Madden and Julian's (1971, 1972) pioneering work not only discovered the 40–50-day oscillation, but also went beyond the pure descriptive to propose a hypothesis for the origin of the oscillation, which depends on an equatorial eastward propagating planetary-scale circulation anomaly that couples with a convection anomaly of a few thousand kilometers (see Chapter 1). Thus, their studies laid down a foundation for our understanding of the MJO dynamics.

The vertical structure and eastward propagation of the MJO visualized by Madden and Julian stimulated an earlier explanation in terms of Kelvin waves. But the gravest vertical mode ("free") Kelvin wave propagates 5 times too fast to account for the slow timescale of the MJO. To account for the slow propagation, Chang (1977) proposed that the MJO might be convectively driven Kelvin waves that are subjected to a damping arising from Newtonian cooling or cumulus friction. In the early 1980s, the horizontal structure of the MJO was further documented in detail (e.g., Weickmann, 1983; Lau and Chan, 1985; Knutson and Weickmann, 1986). Perhaps motivated by Gill's (1980) theory, many scientists interpreted the MJO as an atmospheric response to a localized heat source that pulsates with a 40–50-day period (e.g., Yamagata and Hayashi, 1984; Yamagata, 1987; Anderson, 1987; Hayashi and Miyahara, 1987), or to a mobile heat source with a given speed (Chao, 1987). These "response" hypotheses, however, could not explain what selects the oscillation frequency or what causes the movement of the heat source.

In the mid-1980s, a number of general circulation models (GCMs) had shown capability to realistically simulate some features of the observed MJO, in particular the eastward propagation of the large-scale circulation and convective anomalies (e.g., Hayashi and Sumi, 1986; Lau and Lau, 1986). It was soon recognized that a key to the explanation of the MJO is the interaction between large-scale motion and moist processes. Three major lines of thinking emerged. The first invokes an instability arising from interaction between convective heating and large-scale wave motion (i.e., wave–CISK (Conditional Instability of the Second Kind), Lau and Peng, 1987). The second mechanism pinpoints the importance of wind-induced surface heat exchange (WISHE), also known as wind-evaporation feedback (Emanuel, 1987; Neelin *et al.*, 1987). The third mechanism is an instability arising from friction-induced moisture-convergence feedback (Wang, 1988a). These ideas are centered on the interaction of convective heating and large-scale dynamics. Later, the roles of thermodynamic feedback processes were recognized, which include water vapor accumulation (Blade and Hartmann, 1993) and convective–radiative feedback (Hu and Randall, 1994, 1995). To explain the seasonal behavior of the ISO, the impacts of the seasonal mean circulation and moist static energy distribution were considered to be essential (Wang and Xie, 1997). After the Tropical Ocean Global Atmosphere–Coupled Ocean-Atmosphere Response Experiment (TOGA–COARE), the thermodynamic feedback between atmosphere and oceanic mixed layer was also recognized to play a role in development of the MJO (Flatau *et al.*, 1997; Wang and Xie, 1998; Waliser *et al.*, 1999).

Based on the key processes, the MJO theories/hypotheses are summarized below.

10.2.1 Wave–CISK

Use of the term CISK dates back to Ooyama (1964) and Charney and Eliassen (1964) who explain hurricane formation by a cooperative feedback between the collective effect of small-scale convection and large-scale low-level convergence due to Ekman pumping. This idea was consequently applied to the possible cooperative interaction between tropical wave convergence and organized convection (i.e., wave-CISK) (Yamasaki, 1969; Hayashi, 1970; Lindzen, 1974).

Lau and Peng (1987) proposed a mobile wave-CISK mechanism for the MJO. They suggested that the observed eastward propagation of the MJO arises from an intrinsic mode of interaction between convection and Kelvin waves. In their 5-level model with positive-only condensational heating (a.k.a. conditional heating or non-linear heating), Kelvin waves are selectively amplified. The periodicity of the oscillation is determined by the time taken by the moist Kelvin wave to complete one circuit around the globe. Their linear analysis indicates that both growth rate and phase speed depends on the vertical structure of the heating profile and the static stability of the basic state. Using a global model whose cumulus parameterization was based on wave–CISK, Lau *et al.* (1989) further showed that the slow eastward propagation of the "supercloud cluster" (SSC; a model MJO system) is driven by the formation of new convection centers in the convergent regions of a Kelvin wave, while the westward motion of individual cloud clusters are a manifestation of Rossby waves.

Many wave–CISK models produce faster than observed eastward propagation. Chang and Lim (1988), based on their multi-level model results, suggested that slow phase propagation speed of the wave–CISK relies on reduction in effective static stability and the coupling of two internal modes that are locked in phase vertically. The slow propagation was also attributed to a specified heating maximized in the lower troposphere in multi-layer models (e.g., Takahashi, 1987; Sui and Lau, 1989); but the question of what vertical heating profile favors the MJO-like mode remains controversial. For example, Cho and Pendlebury (1997) showed that an unstable large-scale mode emerges only when the heating profile is sufficiently top-heavy; the model results of Mapes (2000) supported this point of view (i.e., unstable mode occurs when the specified heating contains a sufficient amount of the second vertical mode of the troposphere).

Most wave–CISK models used Kuo (1974)-type cumulus parameterization; some used the Arakawa-Schubert (1974) scheme. When other schemes are used, the unstable wave–CISK mode might not be found. Neelin and Yu (1994) have examined the interaction between the cumulus convection and large-scale wave dynamics using the Betts–Miller moist convective adjustment parameterization (Betts, 1986; Betts and Miller, 1986). They found that under the Betts–Miller scheme a sufficiently long Kelvin wave becomes a slowly decaying mode under reasonable conditions. Introduction of a finite convective adjustment timescale has the property of selectively damping the smallest scales while certain vertical modes at planetary scales decay only slowly.

A caveat of the wave–CISK theory is that the *linear* theory does not explain how the disturbance can maintain the planetary-scale structure against the explosive growth of the short-wave components (Chao, 1995). Lim *et al.* (1990) claim that positive-only heating may remedy this catastrophe in a finite resolution numerical model. However, in the linear dynamic framework, analytical solutions with the positive-only heating were obtained (Dunkerton and Crum, 1991; Wang and Xue, 1992), which show that the positive-only heating may result in planetary-scale descending regions on both sides of a concentrated convective region, but the convective region of the most unstable mode has an infinitesimal width. Hendon (1988), in his 2-level model study, found that the effects of non-linear advection make the growing CISK modes rapidly stabilize the atmosphere. The stability increases greatest to the west of the CISK heating. It appears that the non-linear advection supported the eastward propagating disturbances with finite wavelengths.

10.2.2 Wind–evaporation feedback or WISHE

Emanuel (1987) and Neelin *et al.* (1987) proposed that the waves producing the MJO arise from an instability driven by a wind-induced surface latent heat flux. This mechanism was called wind–evaporation feedback in Neelin *et al.* (1987) and WISHE in Yano and Emanuel (1991). In Emanuel's conceptual model (Emanuel, 1987), no cumulus convective scheme was involved. The entropy is redistributed through the troposphere by local convection because the atmosphere is assumed to be neutrally stratified. A basic assumption is that the mean surface winds are easterlies. Thus, the enhancement of the mean easterlies ahead of an eastward propagating trough leads to enhanced transfer of entropy from the ocean; the associated atmospheric warming lowers pressure and moves the trough eastward. Similar to the wave–CISK, the linear WISHE mechanism favors short-wave growth. Emanuel (1993) introduced a time lag between the large-scale forcing of convection and its response, which damps short waves. But the time lag in the vertical transport of water vapor was shown to make the phase speed too large by a factor of 4–5 (Brown and Bretherton, 1995).

The WISHE instability has been examined in models with different cumulus parameterization schemes. Brown and Bretherton (1995) investigated WISHE with a simple 2-D, non-rotational model on the equatorial plane. They found that the evaporationally driven unsaturated downdrafts play a major role in damping short waves and only the long-wavelength WISHE mode is unstable, although the phase speed is too large. On the other hand, Lin *et al.* (2000) investigated ISO with their intermediate tropical circulation model in which non-linearity, radiative–convective feedback, and a Betts–Miller scheme are included. Their experiments with specified SST show that the wind-evaporation feedback partially organizes the model intra-seasonal variability (ISV) by reducing damping, but it is not by itself sufficient to sustain the oscillation for the most realistic parameters.

Without invoking wave dynamics, the WISHE mechanism alone has difficulty to explain the slow eastward propagation. The eastward propagation of the WISHE mode in the Emanuel's (1987) model is driven by the asymmetry in surface heat fluxes: enhanced surface evaporation in the perturbation easterly phase and suppressed heat fluxes in the westerly phase. Such an asymmetric heat flux pattern relies on the existence of mean surface easterlies. Yet, the observed mean winds are equatorial westerlies over the Indian and western Pacific where the MJO is most active (Wang, 1988b). TOGA–COARE observations have revealed that the surface evaporation associated with the MJO has an opposite zonal asymmetry (i.e., a large anomalous latent heat flux occurs in the westerly phase behind the convection, e.g., Jones and Weare, 1996; Lin and Johnson, 1996; Zhang and Anderson, 2003). Using the Hadley Center Unified Model, Matthews *et al.* (1999) found when fast Kelvin waves propagate with a speed of $55\,\mathrm{m\,s^{-1}}$ through the eastern Pacific, the anomalous surface easterlies associated with this fast Kelvin wave enhanced the climatological mean easterlies and led to positive convective anomalies over the eastern Pacific consistent with the WISHE mechanism; however, WISHE alone was not able to account for the eastward development of the convective anomalies over the Indian Ocean and western Pacific region.

In spite of the aforementioned weakness, the WISHE theory accounts for an important energy source for the ISO. Neelin *et al.* (1987) found that the evaporation is important to ISO in a GCM with a zonally symmetric climate, although the existence of the energy peak on the intraseasonal timescale does not depend on it. Furthermore, when equatorial wave dynamics are included in simple models with conditional heating parameterization, the WISHE mechanism is found to lead to instability and to enhance eastward propagation, which does not necessarily require the existence of mean easterlies (Xie *et al.*, 1993; Wang and Li, 1994).

10.2.3 Frictional convergence feedback

Wang (1988a) proposed that the MJO is driven by an instability arising from boundary layer friction-induced moisture convergence. The frictional moisture convergence couples the equatorial Kelvin and Rossby waves through organizing convective heating, and selects a slowly eastward moving, planetary-scale unstable mode (Wang and Rui, 1990a). The unstable mode resulting from frictional moisture convergence is characterized by a boundary layer convergence ahead of the free tropospheric wave convergence and major precipitation anomaly. Different from wave–CISK, the frictional paradigm emphasizes that the wave-induced moisture convergence itself does not result in instability (Wang, 1988a; Xie and Kubakawa, 1990).

The frictional moisture convergence mechanism has been further examined in more complex models with different convective parameterization schemes. Salby *et al.* (1994) used a column-integrated moisture flux convergence to represent convection. They showed that friction-induced convergence renders the gravest zonal dimensions most unstable. In an equatorial β-plane very-high-resolution model of SCCs and Kelvin waves, Ohuchi and Yamasaki (1997) showed that the convergence in the boundary layer exhibits a phase shift slightly eastward relative to the convergence aloft, which is the key that includes an effective feedback between convection and the wave motion. They found that the resultant unstable wave and SSCs are characterized by a slow phase speed of less than $10\,\mathrm{m\,s^{-1}}$ and by a weak dependence of its growth rate on wavelength. Moskowitz and Bretherton (2000) used a Betts–Miller-like convective parameterization. Friction is found to be modestly destabilizing for the moist Kelvin mode and the gravest moist Rossby mode. Frictionally forced boundary layer convergence promotes wave amplification by enhancing convective heating along the equator in the warm sector of the wave. With a radiation upper boundary condition, the longest wave has the largest growth rate. The effect of frictional convergence feedback was reported to be fairly insensitive to the convective parameterization used (Moskowitz and Bretherton, 2000). In simulation of SSCs with cloud resolving models, gravity wave features may contribute to set up the convergence pattern ahead of the convective center, and leading to eastward and westward propagation of convective elements away from the center of the MJO convection (Peng *et al.*, 2001; Moncrieff and Liu, 1999).

The key feature that the frictional moisture convergence is ahead of precipitation in the MJO has been observed and documented by using a variety of data sets (Hendon and Salby, 1994; Jones and Weare, 1996; Maloney and Hartmann, 1998; Matthews, 2000; Sperber, 2003; Lin *et al.*, 2004). Maloney and Hartmann (1998) analyzed column integrated water vapor from the surface to 300 hPa that is associated with the MJO life cycle and found that a significant correlation exists between surface convergence and column water vapor anomalies in the western Pacific and Indian Oceans. They showed that frictional moisture convergence fosters growth of positive water vapor anomalies to the east of convection. The frictional convergence in front of convection helps to slowly moisten the atmosphere to a state that is favorable for convection. A number of diagnostic studies of the model simulations have also confirmed the importance of the boundary layer frictional convergence in various GCMs with different cumulus parameterization schemes (Lau and Lau, 1986; Lau *et al.*, 1988; Kuma, 1994; Maloney, 2002; Lee *et al.*, 2003) and in a coupled GCM (Waliser *et al.*, 1999).

One issue concerning the frictional moisture feedback is the value of the Rayleigh friction coefficient E in the boundary layer. Wang (1988) took $E = 3 \times 10^{-5}\,\mathrm{s}^{-1}$. Moskowitz and Bretherton (2000) have suggested that this value is an order of magnitude too large, although their use of a small value yielded qualitatively similar results. Of note is that for low-frequency motions the Rayleigh friction coefficient represents not only friction but also other damping effects such as high-frequency transient Reynolds stress. Calculations of the tropical winds from sea-level pressure fields indicate that the adequate value for E is O ($10^{-5}\,\mathrm{s}^{-1}$) (Murphree and van den Dool, 1988; Murakami *et al.*, 1992). If E is on an O ($10^{-6}\,\mathrm{s}^{-1}$), the observed surface pressure would yield an unrealistically large boundary layer wind field for tropical low-frequency variability. Diagnosis of the surface momentum balance over the tropical Pacific Ocean suggests that the Rayleigh friction coefficient in the meridional direction should be three times larger or on an O ($3 \times 10^{-5}\,\mathrm{s}^{-1}$) (Deser, 1993; Li and Wang, 1994).

10.2.4 Cloud–radiation feedback

Hu and Randall (1994, 1995) suggested that the non-linear interactions among radiation, convection, and surface flux of moisture and sensible heat might result in a non-propagating ISO rather than a steady state. They constructed a highly simplified column radiative–convective model in which the free tropospheric thermal and moisture structure is maintained by Newtonian cooling and moist convection. The convection was parameterized by moist convective adjustment (Manabe, 1965) in Part I of their series of papers and by a single-cloud-type scheme Arakawa-Schubert (1974) in Part II. Their model, however, excluded large-scale dynamics and fixed the surface wind speed and SST in calculating surface heat flux (thus the WISHE mechanism is also excluded). The radiative cooling and surface moisture flux tend to destabilize the model atmosphere, while convection tends to maintain a convectively neutral state by reducing the boundary layer moisture and lapse rate. The oscillations are favored by a warm sea surface and weak surface wind speed. This mechanism was speculated to play a role in explaining the stationary component of the MJO over the western Pacific and Indian Ocean warm pool (e.g., Lau and Chan, 1985; Zhu and Wang, 1993), although the results of Zhang and Hendon (1997) suggest there was not much of a stationary component.

Raymond (2001) proposed a new model of radiative–convective instability. The large-scale convective overturning given by a combination of latent heat release and radiative heating anomalies induced by cloud–radiation interactions are thought to be the primary driving mechanism of the model ISO. A pre-existing precipitation anomaly has a region of mid-to-upper-level stratiform cloudiness associated with it. The suppression of outgoing long-wave radiation (OLR) results in a heating anomaly relative to its surroundings, which causes lifting and further precipitation. The effects of surface heat flux variability also substantially modify the behavior of the unstable mode. The cumulus parameterization used in the model allows a lag of several days to exist between the strongest surface heat flux into the column and the subsequent development of heavy precipitation in that column. One could argue that this lag mimics the time to moisten the free atmosphere, but it should be verified by observations because the model oscillation depends on this lag.

Cloud–radiation feedback has been studied with atmospheric GCMs. The high clouds, with a thick optical depth, play an important role in driving large-scale diabatic circulations in the tropics (Slingo and Slingo, 1988; Randall *et al.*, 1989). In addition to the condensational heating, the radiation process, which is affected by cloud properties, also contributes to the vertical distribution of diabatic heating and static stability, thus affecting convective instability and vertical distribution of latent heating. Slingo and Madden (1991) found that long-wave cloud radiative forcing is not crucial in simulating the MJO. They suggested that the intensity of the simulated MJO depends on cloud–radiation interaction but the period is not significantly affected. In later studies, the role of long-wave radiation became controversial. For instance, Mehta and Smith (1997) suggested the importance of long-wave cooling in maintaining the MJO. On the contrary, Lee *et al.* (2001) found that the inclusion of cloud–radiation weakens the model MJO. Since the magnitude of the long-wave cooling is greater than the short-wave in the MJO-related tropospheric radiative heating (Lee *et al.*, 2001), the roles of the long-wave interaction and the processes by which the cloud–radiation feedback influences the MJO are a subject that calls for further studies.

10.2.5 Thermal relaxation and water vapor feedback

Blade and Hartmann (1993) proposed a "discharge–recharge" hypothesis whereby the 40-day recurrence period in their model is set by the growth and duration times of a convective episode together with the recharge time for the instability. Kemball-Cook and Weare (2001), based on analysis of radiosonde data, suggested that prior to the onset of the MJO convection, the atmosphere is destabilized through a combination of low-level moist static energy build-up and concurrent drying of the middle atmosphere by subsidence in the wake of the previous cycle of MJO convection. In model studies, Goswami and Mathew (1994) suggested that the large-scale flow is not in an exact quasi-equilibrium with the precipitation heating; and a time-dependent moisture equation is necessary. In his numerical experiments with a model that treats the moisture budget explicitly, Itoh (1989) found that in order to obtain a SCC and associated wave number 1 circulation, the cumulus convection scheme has to restrain the occurrence of deep convection so that dry regions occur over a wide area of the tropics where weak moisture convergence exists. In these dry regions, the accumulation of moisture is presumably preconditioning MJO deep convection.

Woolnough *et al.* (2000) have suggested that the feedback between convection and water vapour appears to be important in regulating the strength and propagation speed of the MJO. Tompkins (2001) proposed that the water vapour–convection feedback can cause self-aggregation (i.e., the occurrence of convection makes future convection more likely to occur in the same location through an organized positive feedback between convection and water vapor feedback). Using a global model that applies a cloud-resolving convection parameterization, Grabowski (2003) obtained large-scale organization of convection in the equatorial wave guide. If large-scale fluctuations of convectively generated free-atmospheric moisture are removed on a timescale of a few hours, the coherent structure of the MJO-like disturbance does not develop.

The "discharge–recharge" and the moisture–convection feedback processes may play a significant role in the development of the ISO convective anomalies. However, these moist processes alone are not sufficient to explain the wavelength selection and slow eastward propagation of the MJO.

10.2.6 Northward and westward propagation mechanism in monsoon regions

During boreal summer, the eastward propagating MJO mode substantially weakens (Madden, 1986; Wang and Rui, 1990b), whereas northward propagation becomes a prominent feature of ISO in the Indian summer monsoon region (Chapter 2; Yasunari, 1979, 1980; Sikka and Gadgil, 1980; Krishnamurti and Subrahmanyam, 1982). Several mechanisms have been suggested to account for the northward propagation over the Indian Ocean sector.

Webster (1983) proposed that the northward propagation could result from feedback between the hydrological cycle and dynamics over India. The land surface heat fluxes into the boundary layer can destabilize the atmosphere ahead of the ascending zone, causing a northward shift of the convection zone. Goswami and Shukla (1984) suggested that the low-frequency oscillation simulated in their axially symmetric atmospheric model results from a convective–thermal relaxation feedback. The convective activity increases atmospheric stability, which would depress convection itself; meanwhile, dynamic and radiative relaxation brings the atmosphere to a new convectively unstable state. Anderson and Stevens (1987) found, in a zonally symmetric primitive equation model, that inclusion of a divergent basic state Hadley cell leads to the formation of a slowly oscillating mode. Lau and Peng (1990), based on their numerical experiments, suggested that the interaction of the equatorial moist Kelvin waves with large-scale monsoon flows can generate unstable quasi-geostrophic baroclinic waves over the Indian monsoon region along 15°N–20°N; meanwhile the equatorial disturbance weakens. They infer that the fast northward advance of the ISO in the Indian monsoon region may be related to this process.

Based on their model results, Wang and Xie (1997) have proposed that the northward propagating rain band associated ISO is a manifestation of the continuous north-westward emanation of Rossby waves from the equatorial Kelvin–Rossby wave packet when the latter passes through the maritime continent. More prominent emanation occurs in the western Pacific when the equatorial coupled Kelvin–Rossby wave packet rapidly decays in the central Pacific. This theory explains the observed northward propagation component over both the North Indian Ocean and the western North Pacific, but why the Rossby wave propagation has a northward component was not addressed. In the recent works of Drbohlav and Wang (2004) and Jiang *et al.* (2004), they showed how the easterly vertical shears, boundary layer moisture advection, and air–sea interaction could contribute to the northward propagation.

The westward propagation generated over the monsoon region may be a manifestation of unstable baroclinic waves (Lau and Peng, 1990) or equatorial Rossby waves destabilized by easterly vertical shear and interactive convective heating (Xie and Wang, 1996). These disturbances may be emanated from and modulated by an equatorial eastward propagating ISO (Wang and Xie, 1997; Kemball-Cook and Wang, 2001) or formed by merging of an equatorial eastward moving convective system with a westward propagating low-level convergence anomaly located in the subtropics (Hsu and Weng, 2001).

10.2.7 Atmosphere–ocean interaction

The TOGA–COARE has provided firm evidence that active air–sea interaction occurred during two MJO events in late 1992 and early 1993. The coupled structure of the MJO and oceanic mixed layer variability were documented in the mid-1990s (see Section 10.6 and Chapter 7). Stimulated by these observations, there has been a surge of theoretical and numerical model studies of the nature of the air–sea interaction in the Indo-Pacific warm oceans and its role in the development of ISO.

In theoretical model studies, Wang and Xie (1998) introduced a simple coupled atmosphere–ocean model suitable for study of the coupled instability of the warm pool climate system. This model emphasizes oceanic mixed layer physics and its thermodynamic coupling (through surface heat exchanges) with transient atmospheric motion. The fastest growing coupled mode in the model has a planetary zonal scale and an intraseasonal timescale, as well as a realistic SST–convection relationship. The wind–evaporation/entrainment feedback plays a primary role in generating the coupled instability, while the contribution of cloud–SST coupling becomes significant only when the wind effect is weak (see Section 10.6). Recently, Sobel and Gildor (2003) introduced a simple model for the evolution of SST in a localized region of a warm ocean. The model consists of a 0-D atmosphere coupled to an ocean mixed layer. For plausible parameter values, the steady state of the system can oscillate with periods ranging from intraseasonal to sub-annual. The basic mechanism for the instability and oscillation comes from cloud–radiative and wind–evaporation feedback, which agree qualitatively with the model results of Wang and Xie (1998). In their model, however, these two processes play the same roles. This latter conclusion may be due to the neglect of atmospheric dynamics. In the presence of atmospheric dynamics, the regions of active interaction associated with these two feedback processes would have a spatial phase shift with their roles being different as discussed by Wang and Zhang (2002).

In numerical modeling studies, Flatau *et al.* (1997) used an atmospheric global circulation model with a parameterized, empirical relationship between wind speed and SST tendency to examine the effect of SST feedback on the equatorial convection on an aqua-planet. The model MJO-like fluctuations were slowed down and became more organized compared to those with fixed SST distribution.

Waliser *et al.* (1999), using a GCM coupled to a slab ocean mixed layer, showed that air–sea coupling improves the simulation of the MJO. The improvement includes increased MJO variability, a closer match of the timescale of oscillation with observations, reduced eastward phase speed in the eastern hemisphere, and an increased seasonal signature in the MJO with more events occurring in the December–May period. The subsequent numerical model studies have generally confirmed the positive contributions of the air–sea interaction in enhancing eastward propagating MJOs and the northward propagation of boreal summer ISOs. For instance, analysis of the European Centre for Medium-range Weather Forecast–Hamburg atmospheric model (ECHAM4) and its coupled version with the University of Hawaii 2.5-layer tropical ocean model, Kemball-Cook *et al.* (2002) found that upon coupling, pronounced northward propagation of convection and circulation anomalies appear in the May–June Asian monsoon season.

However, it has also been recognized that in order for air–sea interaction to enhance ISO, realistic simulation of the mean state in the coupled model appears to be necessary (Gualdi *et al.*, 1999; Hendon, 2000). Kemball-Cook *et al.* (2002) found that their coupled model failed simulating the August–October ISO in the western North Pacific, because the mean SST in the coupled model was too cold and the monsoon vertical easterly shears were absent. Inness and Slingo (2003) also found the air–sea coupling improves eastward propagation of the convection across the Indian Ocean; but there was no eastward propagation in the western Pacific, because the errors in the mean low-level zonal wind component in the west Pacific prevented the MJO from propagating into this region (Inness *et al.*, 2003).

In the coupled model of Waliser *et al.* (1999), the SST variation is primarily due to changes in latent heat flux and to a lesser degree, changes in the surface shortwave flux. However, the slab ocean model might be too simple to address the question of what causes intraseasonal SST variations. The cause of the SST variability has been further studied using a coupled atmosphere–ocean general circulation model by Cubucku and Krishnamurti (2002), who found that the intraseasonal SST oscillations in the warm pool are primarily caused by the tendency of solar radiation, while evaporative cooling was of secondary importance.

How the SST anomalies (SSTAs) feedback to the MJO is a key, and a more complex, issue. Waliser *et al.* (1999) showed that, in their coupled model, the enhanced SST to the east of the convection reinforces meridional convergence associated with the frictional moisture convergence. The resulting increase in moist static energy helped destabilize the disturbance and maintain it against dissipation more effectively relative to the case without coupling. Kemball-Cook *et al.* (2002) concurred with this idea and they attributed the improved northward propagation of monsoon ISO to the increased low-level convergence into the regions where a positive SST anomaly exists (i.e., ahead of the convective anomaly). Fu and Wang (2004) pointed out that the positive SSTAs ahead (north) of the convection organize convection through destabilization of the moist Rossby waves and local adjustment of the atmospheric convection to the SSTAs, thus enhancing the northward propagating ISO.

The observed positive SSTAs tend to lead convective anomalies by about one-quarter of a wavelength. What creates this phase lag is controversial. Woolnough *et al.* (2001) investigated the response of convection to an idealized imposed mobile intraseasonal SSTA in an atmospheric GCM on an aqua-planet. The convection was found to organize on the spatial and temporal scales of the imposed SSTAs and the location of the maximum in precipitation relative to the SSTAs is in good agreement with observations. They suggested that the free-tropospheric humidity plays a critical role for determining the location and magnitude of the precipitation response. On the other hand, Wu *et al.* (2002) compared an observed strong case of the MJO and its counterparts, simulated by 10 different atmospheric GCMs forced with the same observed weekly SST. In the observations, the positive SSTAs develop upstream of the main convection center while in the simulations, the forced component is in phase with the SST.

The coupled modeling study by Fu and Wang (2004) demonstrates that the air–sea interaction significantly enhances the northward propagation of ISO compared to the forced run in which the same atmospheric model is forced by the daily SST that is produced by the coupled model. They pointed out that the coupled and forced solutions are fundamentally different. Without coupling the SST and convection anomalies are nearly in phase, but in the coupled run the SST–convection has a structure similar to the observed. Neglect of atmospheric feedback makes the forced solution depart from the coupled solution in the presence of initial noises or tiny errors in the lower boundary.

10.3 A GENERAL THEORETICAL FRAMEWORK

10.3.1 Fundamental physical processes

What is the energy source for the MJO disturbances? Given the fact that the vertical structure of the MJO is dominated by the gravest baroclinic mode, it is rational to assume that the MJO is stimulated and sustained by the diabatic heating in the middle troposphere. Wang (1988a) presented a detailed scale analysis for the

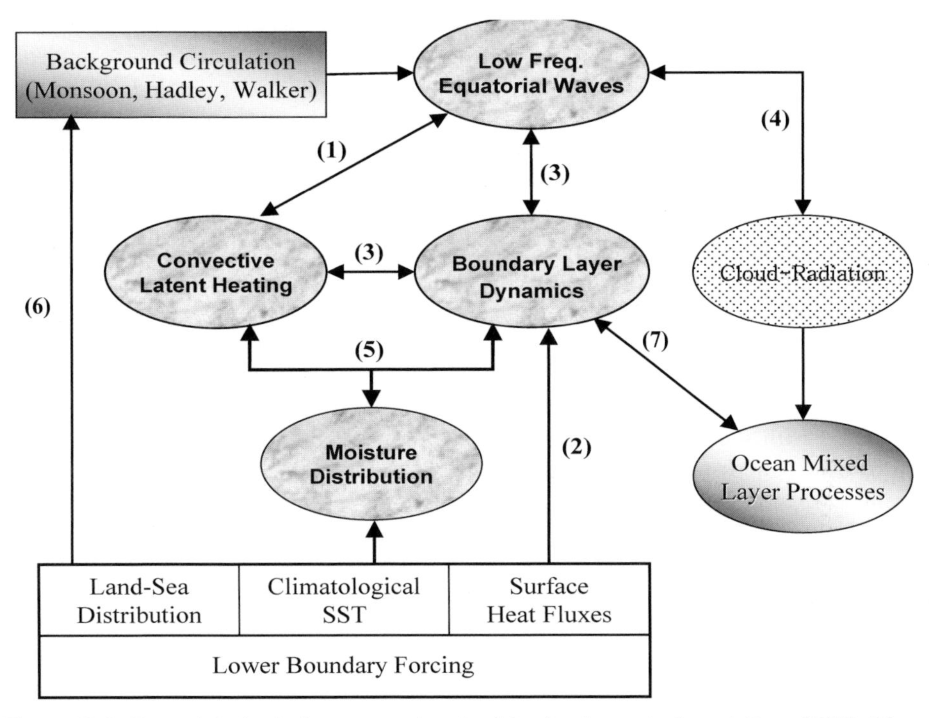

Figure 10.1. Essential physical processes involved in the theoretical modeling of ISO. These processes are numbered in this figure in an order consistent with the reviews in Section 10.2.

MJO. He demonstrated that if the observed baroclinic structure of the MJO is maintained, the required divergence is on an order of ($5 \times 10^{-7}\,s^{-1}$), which has to be sustained by a diabatic heating rate on an order of 2–3 mm day^{-1} because of the relatively small magnitude of the other energy sources in the tropics. This scaling argument is consistent with observations (Krishnamurti *et al.*, 1985) and model results such as the Geophysical Fluid Dynamics Laboratory (GFDL) model (Lau *et al.*, 1988). In this chapter it is assumed that the diabatic heating, especially the latent heat released in precipitation, drives the ISO, albeit some studies have emphasized the roles of extra-tropical forcing (e.g., Hsu *et al.*, 1990; Blade and Hartmann 1993; Lin *et al.*, 2000) in exciting and maintenance of the MJO.

Figure 10.1 presents a schematic summary of the fundamental processes relevant to ISO. Of central importance is the convective interaction with dynamics (CID), following Neelin and Yu (1994). The CID is marked by the ovals with marble shading in Figure 10.1. The CID comprises feedbacks among collective effects of convective heating, low-frequency equatorial Kelvin and Rossby waves, and planetary boundary layer processes including surface heat and momentum exchanges with the lower boundary and moisture feedback with convection and dynamics. These feedback processes integrate the following mechanisms reviewed in Section 10.2: convection–wave convergence feedback, frictional moisture convergence feedback, evaporation–wind feedback, and moisture feedback. The CID is considered essential for understanding the basic internal dynamics of the MJO. The planetary boundary layer processes are of central importance in CID. Dynamically, it provides a large-scale control of the moisture that fuels convection; it contributes to accumulation of moist static energy, and to triggering shallow and deep convections. Thermodynamically, the wind-induced heat flux exchange is key in changing the moist static energy in the boundary layer, thus providing an energy source for the MJO (Emanuel, 1993). Besides the potential role in determining the location and strength of the precipitation heating, the boundary layer friction is also an efficient energy sink for the low-frequency motion.

Other processes include the cloud–radiation feedback, impacts of seasonal mean flows, and air–sea interaction (Figure 10.1). The tropospheric radiative heating associated with the MJO is dominated by long-wave radiative cooling (Lee *et al.*, 2001), which in turn is determined by the properties of clouds. In the following formulation, the simplistic Newtonian cooling is adopted to represent the net radiative heating effect. Keep in mind that, as suggested by Lin and Mapes (2004), in the cloud region, the net radiative heating may be slightly positive, which could enhance latent heating to the first order. In addition, the radiative heating/cooling by short-wave and long-wave radiation at upper-level clouds associated with the MJO convection might be important in destabilizing the atmosphere and leading to deep convection, thus the Newtonian cooling may be an oversimplified parameterization of radiation processes.

The seasonality of the ISO suggests a possible regulation of the moist wave dynamics by seasonal varying background flows. The amplitude of the MJO circulation variations is typically small compared to that of the seasonal variations of the tropical circulations, such as the monsoon circulation. Also, the transient momentum and heat fluxes tend to play a negligible role in determining the tropical mean circulation (Ting, 1994; Hoskins and Rodwell, 1995). Thus, to the first approximation the ISO is treated as a perturbation motion in this chapter. The ISO disturbances are influenced by the seasonal mean circulation and climatological distribution of moisture through SST (Figure 10.1). The impact of the background circulation is essential in explaining the seasonal behavior of the MJO (Section 10.5).

While the ISO is, to a large extent, determined by the atmospheric internal dynamics, its coupling to the oceanic mixed layer may have a considerable impact on its behavior. The coupled formulation will be presented in Section 10.6.

10.3.2 Governing equations

The basic equations governing hydrostatic *perturbation* motion consist of conservations of momentum, mass, thermodynamic energy, and water vapor. These can be written in pressure coordinates on an equatorial β-plane:

$$\frac{\partial u}{\partial t} + \beta y v = -\frac{\partial \phi}{\partial x} + F_x + M(u) \tag{10.1a}$$

$$\frac{\partial v}{\partial t} - \beta y u = -\frac{\partial \phi}{\partial y} + F_y + M(v) \tag{10.1b}$$

$$\frac{\partial u}{\partial x} + \frac{\partial v}{\partial y} + \frac{\partial \omega}{\partial p} = 0 \tag{10.1c}$$

$$\frac{\partial}{\partial t}\frac{\partial \phi}{\partial p} + S(p)\omega = -\mu\frac{\partial \phi}{\partial p} - \frac{R}{C_p p} Q_c(p) + M\left(\frac{\partial \phi}{\partial p}\right) \tag{10.1d}$$

$$\frac{\partial}{\partial t} M_c + \frac{1}{g}\int_{p_u}^{p_s} \nabla \cdot (\bar{q}\vec{V})\, dp = E_v - P_r + M(q) \tag{10.1e}$$

where $M(u)$, $M(v)$, $M(\partial\phi/\partial p)$, and $M(q)$ represent the following mean flow terms in which the quantities with an over bar denote basic state quantities:

$$M(u) = \left(-\bar{u}\frac{\partial u}{\partial x} - u\frac{\partial \bar{u}}{\partial x} - \bar{v}\frac{\partial u}{\partial y} - v\frac{\partial \bar{u}}{\partial y} - \bar{\omega}\frac{\partial u}{\partial p} - \omega\frac{\partial \bar{u}}{\partial p}\right) \tag{10.2a}$$

$$M(v) = \left(-\bar{u}\frac{\partial v}{\partial x} - u\frac{\partial \bar{v}}{\partial x} - \bar{v}\frac{\partial v}{\partial y} - v\frac{\partial \bar{v}}{\partial y} - \bar{\omega}\frac{\partial v}{\partial p} - \omega\frac{\partial \bar{v}}{\partial p}\right) \tag{10.2b}$$

$$M\left(\frac{\partial \phi}{\partial p}\right) = -\bar{u}\frac{\partial^2 \phi}{\partial x \partial p} - u\frac{\partial^2 \bar{\phi}}{\partial x \partial p} - \bar{v}\frac{\partial^2 \phi}{\partial y \partial p} - v\frac{\partial^2 \bar{\phi}}{\partial y \partial p} \tag{10.2c}$$

$$M(q) = -\int_{p_u}^{p_s}\left(\bar{u}\frac{\partial q}{\partial x} + \bar{v}\frac{\partial q}{\partial y}\right)\frac{dp}{g} \tag{10.2d}$$

In (10.1) the dependent variables, u, v, ω, and ϕ denote zonal and meridional wind, vertical pressure velocity, and geopotential height, respectively; β is the meridional variation of the Coriolis parameter; $S(p)$ is the static stability parameter; and F_x and F_y denote frictions. In thermodynamic equation (10.1d), two diabatic heating terms are included: the condensational latent heat and long-wave radiation, in which μ denotes a constant coefficient for Newtonian cooling; and Q_c which expresses condensational heating rate per unit mass, with R and C_p the gas constant of the air and the specific heat at constant pressure, respectively.

The conservation of the column-integrated water vapor M_c in (10.1e) requires the perturbation precipitation rate P_r being balanced by the sum of the perturbation surface evaporation rate E_v, the column integrated moisture convergence, and the local rate of change of M_c. In the moisture convergence term, $\vec{V}$ represents horizontal wind and p_u and p_s are the pressures at the upper boundary and the surface, respectively. The moisture convergence depends on the basic state specific humidity $\bar{q}$, which provides latent energy for the perturbation motion and it can be expressed as a function of pressure and mean SST over the ocean. Assume the absolute humidity of the basic state atmosphere to fall off with height exponentially with a water scale height $H_1 = 2.2\,\text{km}$. The mean specific humidity in an arbitrary vertical layer between pressure p_1 and p_2 where $p_2 > p_1$ is (Wang, 1988a):

$$\bar{q}(p_1, p_2) = q_0 \frac{(p_2^m - p_1^m)}{m(p_2 - p_1)} \tag{10.3a}$$

where $m = H/H_1$ is the ratio of the density scale height H to the water vapor scale height H_1, and q_0 is the air specific humidity at the surface. Over ocean and on the timescale of a month or so, q_0 is well correlated with SST and may be approximated by the following empirical formula (Li and Wang, 1994):

$$q_0 = q_0(\text{SST}) = (0.94 \times \text{SST}(^\circ\text{C}) - 7.64) \times 10^{-3} \tag{10.3b}$$

The condensational heating rate Q_c must be constrained by precipitation rate (i.e., the column integrated condensational heating rate is linked with the precipitation rate):

$$\delta L_c P_r = \frac{1}{g} \int_{p_u}^{p_s} Q_c(p)\, dp \tag{10.3c}$$

where L_c is latent heat of condensation and δ represents a switch-on tracer for nonlinear heating in the absence of basic state rainfall: δ equals unity in regions of positive precipitation and zero otherwise. The heating is linear when $\delta \equiv 1$.

10.3.3 The boundary layer dynamics near the equator

In view of the importance of the frictional moisture convergence in the CID, in this subsection we derive the barotropic boundary layer dynamics in detail. The perturbation equations for a slab mixed layer flow have the form:

$$\frac{\partial u}{\partial t} + \beta y v = -\frac{\partial \phi_e}{\partial x} + g \frac{\partial \tau_{xp}}{\partial p} \tag{10.4a}$$

$$\frac{\partial v}{\partial t} - \beta y u = -\frac{\partial \phi_e}{\partial y} + g \frac{\partial \tau_{yp}}{\partial p} \tag{10.4b}$$

$$\frac{\partial \omega}{\partial p} = -\left(\frac{\partial u}{\partial x} + \frac{\partial v}{\partial y}\right) \tag{10.4c}$$

where ϕ_e is the geopotential at p_e. Matching conditions require that the vertical velocity and turbulent Reynolds stresses (τ_{xp}, τ_{yp}) are continuous at p_e, the latter implies:

$$(\tau_{xp}, \tau_{yp}) = 0 \qquad \text{at } p = p_e \tag{10.5a}$$

At the lower boundary, $p = p_s$, the Reynolds stress is related to the surface wind by a simple linearized stress relationship, i.e.:

$$(\tau_{xp}, \tau_{xp}) = -\rho_e K_D(u_s, v_s) \qquad \text{at } p = p_s \tag{10.5b}$$

where ρ_e is air density in the boundary layer, K_D is a measure of the surface drag, and $u_s = u(p_s)$, $v_s = v(p_s)$. To determine K_D, we subdivide the boundary layer into a surface layer and an outer boundary layer (Blackadar and Tenneker, 1968). In the

Table 10.1. The model parameter values used in Sections 10.4 and 10.5.

P_e	Pressure at the top of the boundary layer	900 hPa
Δ_P	Half-pressure depth of the free troposphere	400 hPa
SST	Sea surface temperature	29°C
C_0	Dry gravity wave speed of the baroclinic mode	$50\,\mathrm{m\,s^{-1}}$
r	Horizontal momentum diffusion coefficient	$10^6\,\mathrm{m^2\,s^{-1}}$
b	Precipitation efficiency coefficient	0.9
A_z	Vertical turbulent viscosity in the boundary layer	$10\,\mathrm{m^2\,s^{-1}}$
h_0	Depth of the atmospheric surface layer	40 m
z_0	Surface roughness depth	0.01 m
C_E	Heat exchange coefficient	1.5×10^{-3}
I	Heating coefficient due to wave convergence	0.84
B	Heating coefficient due to friction convergence	1.73
F	Heating coefficient due to evaporation	0.59
E	Ekman number in the boundary layer	3×10^{-5}
d	Non-dimensional boundary-layer depth	0.25
h	Half-depth of the free troposphere	3.6 km

outer boundary layer, (10.4) applies while in the surface layer the Reynolds stress is nearly constant and the wind profile is logarithmic. Matching the outer boundary layer with the surface layer at their interface leads to $K_D = Az/[h_0 \ln(h_0/Z_0)]$, where Az is turbulent viscosity, h_0 the depth of the surface layer, and Z_0 the surface roughness length. Integrating (10.4a–c) with respect to p from p_e to p_s, dividing the resulting equations by $(p_s - p_e)$, and using conditions (10.5a, b), we obtain:

$$\frac{\partial u_b}{\partial t} + \beta y v_b = -\frac{\partial \phi_e}{\partial x} - E u_b \tag{10.6a}$$

$$\frac{\partial v_b}{\partial t} - \beta y u_b = -\frac{\partial \phi_e}{\partial y} - E v_b \tag{10.6b}$$

where u_b and v_b are the vertical averaged boundary-layer winds, and E is the friction coefficient in the boundary layer, where $E = \rho_e g K_D/(p_s - p_e)$. For typical parameter values, E is O $(10^{-5}\,\mathrm{s^{-1}})$ (Table 10.1).

Due to change of sign of the Coriolis force at the equator, equatorial boundary layer dynamics is distinct from the quasi-geostrophic Ekman theory. In the subtropics or extra-tropics, the vertical velocity induced by boundary layer friction (the Ekman pumping) is determined by the vorticity at the top of the boundary layer (Eliassen, 1971). Near the equator, however, the solution of (10.6) for steady motions yields:

$$D = -\frac{E}{E^2 + \beta^2 y^2}\left(\nabla^2 \phi_e + \beta u_b + \frac{\beta^2}{E} y v_b\right) \tag{10.7}$$

where D is the boundary layer divergence. Thus, the frictional convergence in the deep tropics is determined by the Laplacian of the pressure at the top of the boundary layer and the strengths of the eastward and poleward surface winds.

This explains why the boundary layer convergence zone often occurs on the equatorward side of an off-equatorial monsoon trough. To the equatorward side of a monsoon trough the winds have an eastward and poleward component. The β effect acting on both components causes convergence according to the second and third terms on the right-hand side of (10.7).

10.3.4 The $1\frac{1}{2}$-layer model for the MJO

In this subsection, a simplistic model for ISO is formulated with the effects of the boundary layer included. This model will be used in the modeling of the basic dynamics of the MJO in Section 10.4.

To simplify the vertical resolution of the model, we begin with analysis of the structure of the vertical normal modes in the governing equation (10.1). For this purpose, let us consider the linear, frictionless, adiabatic motion of the dry atmosphere in a quiescent environment, namely, the friction terms, mean flow terms, all heating terms, and the moisture equations being neglected in (10.1). The static stability parameter $S(p)$ is a function of pressure only. With these approximations, the vertical modes that satisfy the rigid boundary condition at the surface and at the top of the atmosphere (for instance, $\omega = 0$ at $p = p_u$ and p_s) are separable; each satisfies the so-called "shallow-water equation" with differing equivalent depth (or gravity wave speed) (e.g., Gill, 1980). The vertical structures of these modes are solely determined by the basic-state stratification.

For an idealized yet realistic dry atmosphere (e.g., an isothermal atmosphere), the static stability parameter $S(p)$ is proportional to the inverse of the pressure square. For such a stratified atmosphere, the vertical structures and phase speeds of the vertical modes were derived analytically (Wang and Chen, 1989). The phase speed for the lowest four ($m = 1$, 2, 3, and 4) modes are approximately 50, 26, 18, and 13 $\mathrm{m\,s^{-1}}$, respectively. Higher vertical modes have smaller equivalent depth and slower phase speed. The vertical velocity profiles for the lowest four vertical modes are shown in Figure 10.2. The gravest baroclinic mode has maximum vertical velocity in the middle of the atmosphere. The higher baroclinic modes have more nodes and shorter wavelengths. Since the vertical structure of the MJO is dominated by the gravest baroclinic mode, the simplest model of the MJO should consist of two layers in the free troposphere (Figure 10.3).

In the absence of basic flows, all terms in (10.2) vanish. One can obtain a 2-level free-atmosphere system by writing the horizontal momentum and continuity equations (10.1a–c) at p_1 and p_3 and thermodynamic equation (10.1d) at mid-level p_2 of the model free atmosphere. The motion in the 2-level system can be alternatively represented by a baroclinic and a barotropic mode. In the presence of boundary layer friction, these two vertical modes are coupled through frictional convergence induced vertical motion at p_e. To save space, the equations for this 2-level system are not given here. The interested readers can find them in Wang and Rui (1990a). Note that only the baroclinic mode is subjected to diabatic heating. The

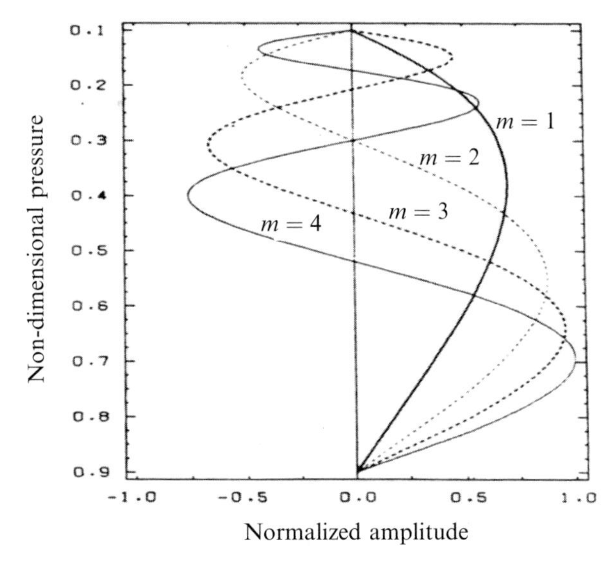

Figure 10.2. The vertical structures of the vertical pressure velocity for the first four internal modes computed for an isothermal atmosphere in which the static stability parameter is proportional to the inverse of the pressure square. The vertical pressure velocity vanishes at the upper ($p = 0.1$) and lower ($p = 0.9$) boundary.
Adapted from Wang and Chen (1989).

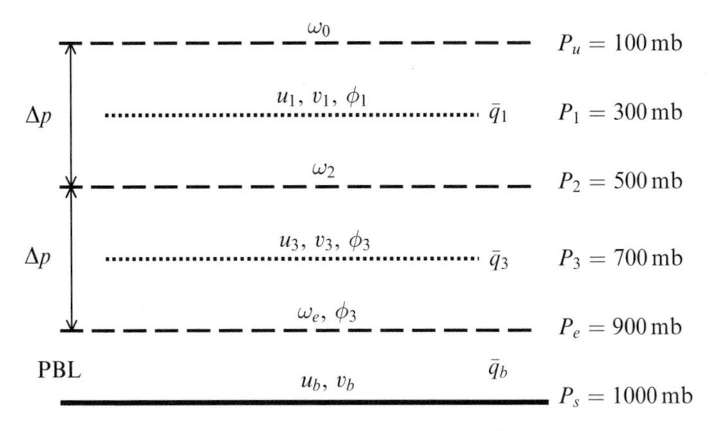

Figure 10.3. Schematic vertical structure of the $2\frac{1}{2}$-layer model of ISO.

condensational heating is linked to the precipitation rate (10.3c). With the limited vertical resolution of the 2-level system, the precipitation rate is expressed by:

$$\delta P_r' = \delta b\{[-\omega_2 \bar{q}_3 - \omega_e(\bar{q}_e - \bar{q}_3)]/g + \rho_s C_E |V_b|(q_s - q_0)\} \tag{10.8}$$

where ω_e and ω_2 represent, respectively, vertical pressure velocities at the top of the boundary layer (p_e) and the mid-troposphere (p_2); g, ρ_s, and C_E are gravity, surface air density, and heat exchange coefficients, respectively; V_b the wind speed at surface $p = p_s$ that will be approximated by the model boundary-layer wind; q_s the saturation specific humidity at the SST, which can be calculated from the Clausius–Clapeyron equation. Equation (10.8) enables the governing equations to be a closed system.

It has been demonstrated that *in the absence of the basic flows* the magnitude of the barotropic mode is an order of magnitude smaller than that of the baroclinic mode (Wang and Rui, 1990a; Wang and Li, 1993). Thus, a simplification can be made to neglect the barotropic mode by assuming a vanishing column integral of divergence in the free troposphere. The baroclinic mode in the free troposphere is governed by the following equations on an equatorial β-plane, after ω_2 and ω_e are eliminated by using the continuity equation:

$$\frac{\partial u}{\partial t} + \beta yv = -\frac{\partial \phi}{\partial x} \tag{10.9a}$$

$$\frac{\partial v}{\partial t} - \beta yu = -\frac{\partial \phi}{\partial y} \tag{10.9b}$$

$$C_0^{-2}\frac{\partial \phi}{\partial t} + (1 - \delta I)\nabla \cdot \vec{V} = d(\delta B - 1)\nabla \cdot \vec{V}_b - \delta F C_E |\vec{V}_b|/h \tag{10.9c}$$

where u, v, and ϕ represent the lower-troposphere zonal and meridional wind and geopotential height, respectively (the upper-tropospheric zonal and meridional wind are $-u$ and $-v$, respectively); $\vec{V}_b$ denotes the boundary layer barotropic wind whose components (u_b, v_b) satisfy (10.6a, b); $C_0 = 50\,\mathrm{m\,s^{-1}}$ and denotes the dry gravity wave speed of the free-troposphere baroclinic mode (corresponding to the gravest baroclinic mode in a vertically continuous model); $d = (p_s - p_e)/\Delta p$ is the dimensionless depth of the boundary layer; and $h = \Delta p/\rho_e g$, where Δp is the one-half pressure depth of the free troposphere. In the thermodynamic equation (10.9c) there are three non-dimensional heating parameters, which are defined by:

$I = \bar{q}_3/q_c$	Heating coefficient due to wave convergence,	(10.10a)
$B = \bar{q}_e/q_c$	Heating coefficient due to frictional convergence,	(10.10b)
$F = (q_s - q_0)/q_c$	Heating coefficient associated with evaporation,	(10.10c)

where $q_c = 2C_p p_s C_0^2/(bR\Delta p L_c)$ stands for a vertical mean specific humidity in the lower-tropospheric layer, which corresponds to a vanishing effective static stability in the presence of convective heating. The standard values of model parameters used in this chapter are given in Table 10.1.

The equations (10.9a–c) and (10.6a, b) (with the assumption $\phi_e = \phi$) consist of a close set of equations, which describes moist dynamics of a single free-troposphere baroclinic mode that is coupled with the boundary-layer motion. This model is an extension of the Matsuno (1966) model by including diabatic heating and effects of the boundary layer. Such a model is referred to as a $1\frac{1}{2}$-layer model.

Note that, in a two-level free-atmospheric model, the heating is released in the middle of the troposphere; the closure assumption for condensational heating is provided solely by conservation laws of the moisture and thermal energy through the linkage between vertical integrated condensational heating rate and the precipitation rate in the same column (10.3c). Any type of cumulus parameterization, when boiled down to a 2-level approximation, must obey the same physical principles. Therefore, use of (10.8) should not be considered a version of Kuo or any other specific parameterization schemes. The only approximation made in (10.8) is the neglect of the local change of moisture and the moisture in the upper-tropospheric layer. An adjustable parameter b is introduced to compensate the omission of the moisture storage in the atmosphere. The parameter b represents the condensation efficiency measuring the fraction of total moisture convergence that condenses out as precipitation. This simplification facilitates eigenvalue analysis. A 2-level version of the time-dependent moisture equation (10.1e) and a transient boundary layer (rather than a steady boundary layer) had also been used; the results are not qualitatively different from those derived with these simplifications.

10.3.5 The $2\frac{1}{2}$-layer model including effects of basic flows

As shown by Wang and Xie (1996), the presence of mean flow directly couples the baroclinic and barotropic modes with the barotropic mode having a significant magnitude that is no longer negligible. Thus in the presence of the mean flows, a full 2-level free troposphere is required. Similar to the formulation of the $1\frac{1}{2}$-layer model, one can obtain a 2-level free-atmosphere system by writing the horizontal momentum and continuity equations (10.1a–c) at p_1 and p_3 and thermodynamic equation (10.1d) at mid-level p_2 of the model free atmosphere. In this case, the mean flow terms in (10.2) are included. The motion in this 2-level system can be alternatively represented by a baroclinic and a barotropic mode. To save space, these equations for barotropic and baroclinic components are not shown here. Interested readers may refer to Wang and Xie (1996).

The ω_e in the free-tropospheric equations is provided as a lower boundary condition and it is determined from the boundary layer equations (10.6a, b). For a steady boundary layer, it can be shown that:

$$\omega_e = D_1 \frac{\partial^2 \phi_e}{\partial x^2} + D_2 \frac{\partial \phi_e}{\partial x} + D_3 \frac{\partial^2 \phi_e}{\partial y^2} + D_4 \frac{\partial \phi_e}{\partial y} \tag{10.11}$$

where coefficients D_1 through D_4 are functions of latitude and model parameters (for details refer to Wang and Xie (1997)). Here ω_e is related to the free atmospheric convergence by mass conservation in a vertical column:

$$\omega_e = -\Delta p \sum_{k=1}^{2} \left(\frac{\partial u_k}{\partial x} + \frac{\partial v_k}{\partial y} \right) \tag{10.12}$$

By assuming $\phi_e = \phi_3$, (10.11) and (10.12), along with the 2-level finite difference versions of (10.1) and (10.2), constitute a close set of equations. Since the barotropic boundary layer is included in this 2-level system, this set of equations will be referred to as a $2\frac{1}{2}$-layer system. For numerical details of solving the system readers are referred to Wang and Xie (1997). This $2\frac{1}{2}$-layer model will be used in Section 10.5 for study of the seasonal behavior of the ISO.

10.4 DYNAMICS OF THE MJO

The elementary dynamics of the low-frequency disturbances producing the MJO may be elucidated by examining the behavior of the convectively interactive low-frequency motion in a quiescent atmosphere with underlying uniform SST. The simplest $1\frac{1}{2}$-layer model described in Section 10.3.4 is used. The simplicity of the model allows us to focus on basic atmospheric internal dynamics, such as the frictional CID. The model is solved for both the boundary value and initial value problems. The behavior of the moist low-frequency motion will be compared against observed features of the MJO.

10.4.1 Low-frequency equatorial waves and the associated Ekman pumping

Let us begin with analysis of the basic wave motions relevant to MJO disturbances. For clarity, let us neglect the diabatic heating for the time being in (10.9a–c) (i.e., $\delta = 0$). The resulting equations describing the adiabatic baroclinic motion of the free troposphere becomes a shallow-water equation (e.g., Matsuno, 1966).

Because of the observed anisotropic length scales (zonal scale is an order of magnitude larger than the meridional scale) of the MJO, the geostrophic approximation can be applied to the v-momentum equation. This approximation is known as long-wave approximation (Gill, 1980), ruling out the high-frequency inertio-gravity waves. Thus, only low-frequency Kelvin and Rossby waves are relevant wave motions and communicators in the MJO.

Figure 10.4(a) illustrates the horizontal structure of the pressure fields for Kelvin waves and the most equatorially trapped Rossby waves. These waves are slightly damped due to the presence of boundary layer friction (10.6a–b). The eastward propagating Kelvin waves are strongly trapped to the equator and owe their existence to the vanishing Coriolis parameter there. Away from the equator, the geostrophic balance between pressure gradient force and Coriolis force dominates the frictionless atmospheric motion, which is a characteristic of Rossby waves. The meridional variation of the Coriolis force strongly constrains the speed of the westward propagation of the Rossby waves. Overall the structures are similar to their corresponding inviscid counterparts (Matsuno, 1966). But a notable modification on Rossby waves is the slant of the troughs and ridges equatorward and eastward.

Figure 10.4(b) shows the vertical motions at the top of the boundary layer associated with the Kelvin wave and the most trapped symmetric Rossby wave shown in Figure 10.4(a). The calculations were based on a steady version of (10.6). For these waves, the Laplacian of pressure terms generally dominates the frictional convergence (10.7). Thus, for Kelvin waves the friction-induced upward motion is located in its low-pressure or easterly phase, while for the most trapped

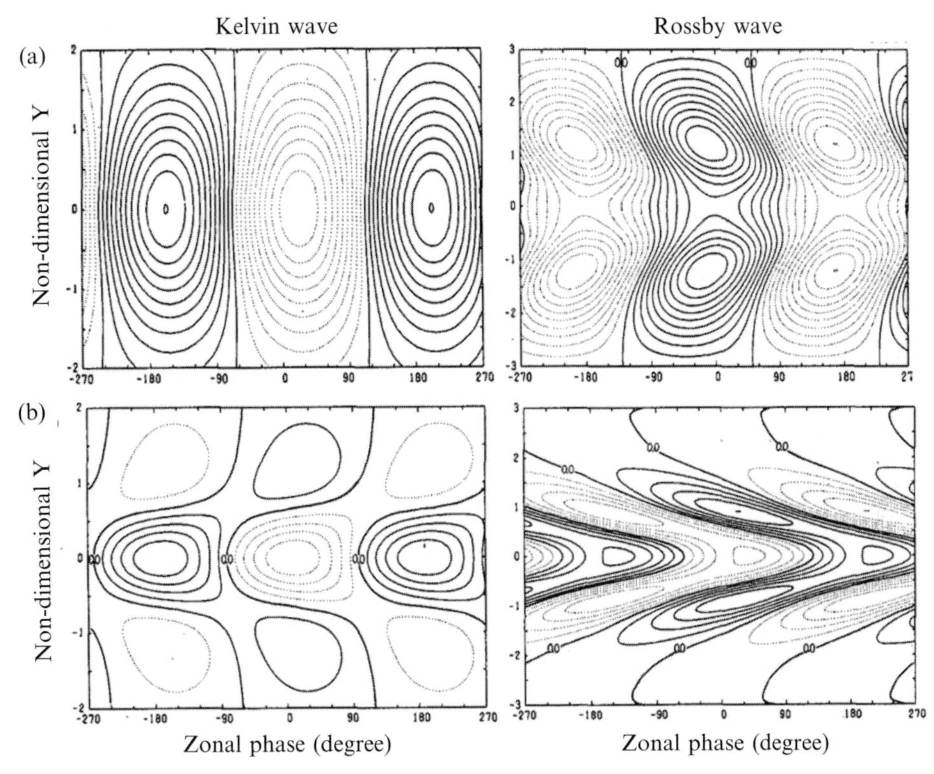

Figure 10.4. Horizontal structures of the equatorial Kelvin wave (*left*) and the most trapped equatorial Rossby wave (*right*) in the presence of boundary layer damping: (a) geopotential height (*upper panels*) and (b) vertical pressure velocity at the top of the boundary layer (*lower panels*). The meridional scale is the Rossby radius of deformation, whose unit corresponds to about 1,500 km.

equatorial Rossby waves, the ascent occurs in both the off-equatorial low pressures and the equatorial trough between the two off-equatorial anticyclones. As a result, along the equator the maximum ascent (descent) leads the corresponding westerly (easterly) by about one-eighth of a wavelength. Overall, the upward motion in the Rossby wave is generally shifted eastward compared to the minimum pressure and strongest equatorial westerlies. This has important ramifications for the selection of the eastward propagating unstable mode in the presence of heating.

10.4.2 The unstable mode due to frictional moisture convergence

Here we further consider the effect of convective heating (the terms with the tracer δ) in the $1\frac{1}{2}$-layer model (Section 10.3.4) In the presence of interactive convective heating, the latent heat that drives the equatorial waves is associated, respectively, with free-tropospheric wave convergence, the boundary layer frictional convergence, and the surface evaporation (10.10). The wave-induced heating measured by the parameter I can be estimated from equations (10.10a, 10.3a, and 10.3b), which is a function of SST: $I = 0.72$ for SST $= 26°C$ and $I = 0.88$ for SST $= 30°C$ for typical parameter values of the tropical atmosphere listed in the Table 10.1. The condition $I < 1$ means that the latent heating rate due to free-tropospheric wave convergence is smaller than the adiabatic cooling rate arising from the ascending motion in the mid-level of the model. Thus, the wave-induced convergence feedback in the model does not produce instability *per se*. Note that if there were no boundary layer, the same set of parameters would yield a parameter $I > 1$ when SST exceeds 29°C. Thus, this stable regime is not due to artificial parameter tuning, rather it reflects the fact that the free-tropospheric wave convergence can only control a portion of the moisture convergence, which in reality cannot produce direct overturning instability.

In a dynamic regime stable to wave–CISK, the growth or maintenance of the low-frequency waves has to rely on destabilization by other mechanisms such as the frictional convergence feedback (B), surface wind–evaporation feedback (F), or cloud–radiative enhancement. When the boundary-layer moisture concentration is sufficiently high (or underlying SST exceeds a critical value), the positive contribution of boundary-layer frictional moisture convergence to wave growth would exceed its dissipative effect. The frictional moisture convergence thus acts to generate instability. The unstable mode generated by frictional moisture convergence feedback was originally termed as frictional wave–CISK mode (Wang, 1988a). Since this unstable mode occurs in a dynamic regime stable to wave–CISK and since the instability fits in the general concept of "convective interaction with dynamics (CID)", it is more meaningful to call this type of unstable mode the *frictional CID mode*.

To investigate the nature of the frictional CID mode, we first examine the behavior of the normal modes of the $1\frac{1}{2}$-layer model with a linear heating ($\delta = 1$). The parameters used in the analysis are listed in Table 10.1. It can be shown that an unstable eastward propagating frictional CID mode exists as long as the basic state SST exceeds a critical value (Figure 10.5(a)). The growth rate of the unstable mode increases with increasing background SST. In contrast to the short-wave blow up of wave–CISK, the longest wave is most unstable until the equatorial SST exceeds 29.5°C, above which the wavelength of the fastest growing wave shifts to 30,000 km. The propagation speed decreases with increasing SST and wavelength (Figure 10.5(b)). The phase speeds of the fastest growing waves are slow, about 5–10 m s^{-1}.

The growing mode exhibits equatorial symmetric and trapped geopotential and zonal wind fields, which resemble Kelvin waves. But it also has a significant meridional wind component, which is antisymmetric about the equator and resembles that of the most equatorially trapped Rossby waves. Thus the friction-induced ascending motion is a mixed Kelvin and Rossby wave; along the equator it is located to the east of the free-tropospheric precipitation and rising motion (Figure 10.5(c)). This horizontal and vertical structure compares favorably to the observations (Figure 10.5(d)), so do the spatial scales and slow propagation and amplification.

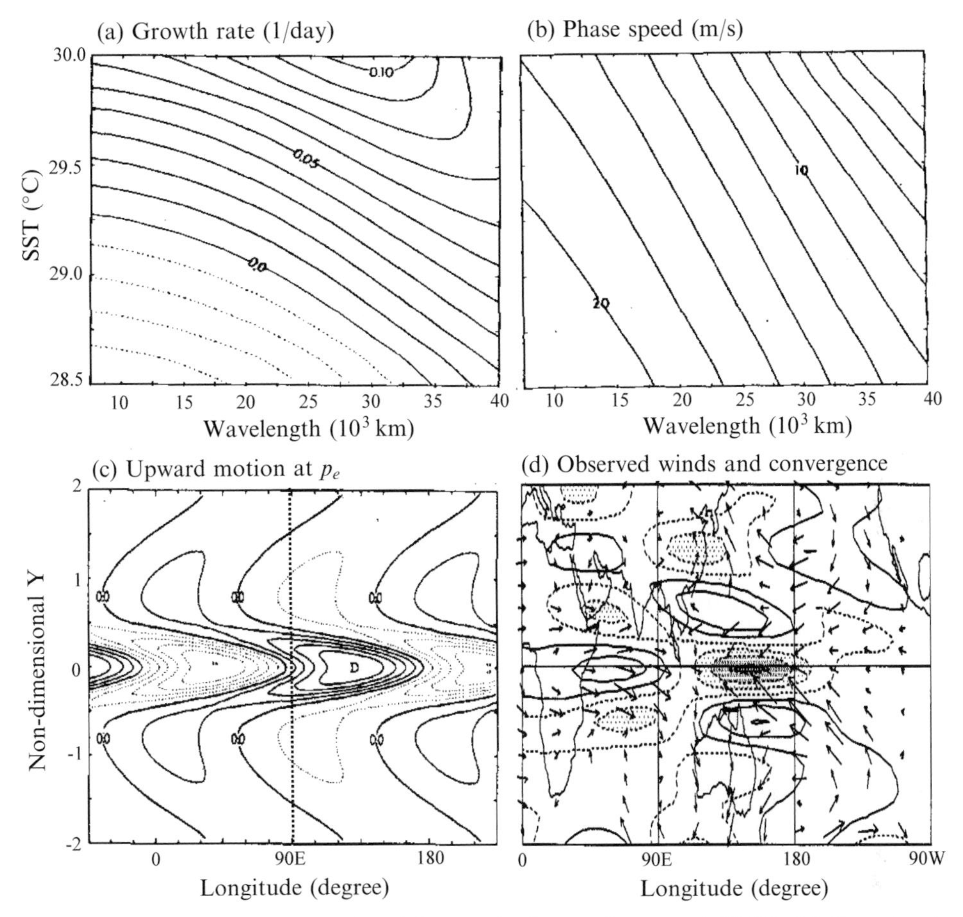

Figure 10.5. Behavior of the frictional CID mode that produces model MJO. (a) Growth rate and (b) zonal phase speed as functions of wavelength and the maximum SST at the equator. (c) Normalized upward motion at the top of the boundary layer computed for the growing frictional CID mode under SST = 29.5°C. The panel (d) shows observed surface winds and convergence (contours). The meridional scale in (c) is the Rossby radius of deformation (about 1,500 km).

(a)–(c) Adapted from Wang and Rui (1990a). (d) Adapted from figure 3 of Hendon and Salby (1994).

10.4.3 Dynamics of the frictional CID mode

Why does the unstable mode in the $1\frac{1}{2}$-layer model have a low-frequency growth rate and favor planetary scales? A fundamental reason is that the frictional convergence that supplies a large amount of moisture is not in phase with the wave-induced moisture convergence. This effectively reduces the strength of the interaction between wave–induced heating and circulation, prohibiting unstable wave-CISK. The energy source driving the instability comes from generation of the eddy available potential energy, which is proportional to the covariance between the warming and heating. The frictional convergence to the east of the major convection induces condensational heating, which overlaps a positive temperature anomaly, thereby generating eddy available potential energy for the growth of the unstable mode. Wang (1988a) has shown that the rate of generation of eddy energy by frictional moisture convergence increases with increasing zonal scale so that the planetary-scale mode is preferred.

In the model, a region of an organized condensational heating may generate both Kelvin and Rossby waves. The convectively interactive Kelvin and Rossby waves would soon decouple and propagate in opposite directions (e.g., Li and Cho, 1997). Then what mechanism can hold them together and make eastward propagation? Why does it have a rearward tilt of rising motion against the direction of propagation? Again, the frictional convergence paradigm is instrumental in addressing these questions. As shown in Figure 10.4, the Rossby wave-induced boundary layer convergence favors in part the development of moist Kelvin waves by producing equatorial convergence at the easterly phase, but the Kelvin wave-induced frictional convergence favors its own growth. Therefore, the frictional organization of convective heating couples the Kelvin and Rossby waves together but selects eastward propagation. As such, the frictional coupling creates a realistic mixed Kelvin and Rossby wave structure. In addition, the boundary layer convergence coincides with the low pressure (easterly) of the Kelvin wave response to the east of the precipitation heating, thus the boundary layer convergence leads to eastward propagating precipitation anomalies.

What gives rise to the slow propagation speed so that the oscillation has an intraseasonal timescale? A primary cause is that the wave convergence induced heating, as measured by parameter I (10.10a), acts to reduce the effective static stability by a factor of $(1 - I)^{1/2}$ (about 0.35 at 30°C of SST), hence reducing the propagation speed of the equatorial waves by a factor of 3. Results in Figure 10.5(b) indicate that the speed of the unstable mode is much slower than the pure moist Kelvin wave speed which is $C_0(1 - I)^{1/2}$ (about $17\,\mathrm{m\,s^{-1}}$), suggesting that the frictional coupling of Kelvin and Rossby waves operates as a brake on the eastward movement. The reason is that the coupling-induced off-equatorial twin cyclonic cells resist eastward movement because the meridional transport of planetary vorticity constantly generates a westward moving tendency for Rossby waves.

10.4.4 Kelvin–Rossby wave packet under positive-only heating

In Section 10.4.2, the normal mode behavior under linear heating was examined using the $1\frac{1}{2}$-layer model. In this subsection, we further investigate the behavior of time evolution of the low-frequency motion in the same model by solving the initial value problem. In the time integration of (10.9a–c) and (10.6a–b) a positive-only and SST-dependent non-linear heating is used. This non-linear SST-dependent heating is not only controlled by positive-only precipitation but also controlled by underlying SST. This SST-dependent heating was motivated by the observed relationship between SST and deep convection (Wang and Li, 1993). Physically, this formulation reflects the impact of the underlying SST on deep convection through changing the convective instability of the atmosphere. The SST-dependent heating assumes that when SST is below 26°C, no convective heating occurs; when SST increases from 26°C to 28°C, the heating coefficient increases linearly from 0 to 1; and when SST exceeds 28°C, the heating coefficient equals 1. In order to eliminate small-scale numerical noise, two momentum diffusion terms that are proportional to the Laplacian of u and v, are added, respectively, to (10.9a–b) with a horizontal momentum diffusion coefficient r being $10^6\,\mathrm{m^2\,s^{-1}}$ (Table 10.1). All other parameter values used in the computation are given in Table 10.1. The integration is initiated with a pure Kelvin wave perturbation in the free troposphere. While the results confirm the major conclusions derived from the linear analysis, some new features are notable.

As is shown in Figure 10.6, the initial dry disturbance rapidly evolves into a multi-scale wave packet: a global scale circulation coupled with a large-scale (several thousand kilometer) convective complex, which consist of a few synoptic-scale precipitation cells. Thus non-linear heating renders the model low-frequency waves to have a planetary circulation scale with a concentrated precipitation, a feature resembling the observed MJO structure. Why does the circulation have a planetary wave number 1 structure, while precipitation is confined? The positive-only heating creates a precipitation core and widespread dry descending regions away from the core. The precipitation core moves slowly due to the reduced effective static stability, but in the descending regions the dry Kelvin wave moves eastward with a speed of $C_0 = 50\,\mathrm{m\,s^{-1}}$ and dry Rossby wave moves westward with a speed of $17\,\mathrm{m\,s^{-1}}$ (about one-third of the Kelvin wave speed). These dry waves expand the dry regions until it is constrained by Earth's finite geometry. Thus, the spreading of energy by fast propagating dry Kelvin and Rossby waves away from the precipitation complex forms the planetary circulation scale. The fast Kelvin waves have been seen in data (Milliff and Madden, 1996) and in AGCM experiments (Matthews *et al.*, 1999).

The heating released in the precipitation complex couples the equatorial waves, forming a *dispersive* wave packet in which the energy propagation is slower than that of the individual cells within the complex. This offers a slowing-down mechanism for the MJO in addition to the reduction of effective static stability and coupling of Kelvin and Rossby waves. In the boundary layer, notable westerlies are located beneath the major precipitation cells and convergence occurs to the east of the precipitation complex (figure not shown, see Wang and Li, 1994). Without the boundary layer friction, the multi-scale structure would disappear.

Wang and Li (1994) compared the growth rates that are induced by wave–CISK (I), frictional moisture convergence feedback (B), and the evaporation–wind feedback (F), using the same $1\frac{1}{2}$-layer model with the SST-dependent, positive-only heating and the same parameter values given in Table 10.1. It was found that both the wind–evaporation feedback and wave–CISK favors a synoptic-scale growth rate of O ($10^{-5}\,\mathrm{s^{-1}}$) in the absence of boundary layer friction, while the instability generated by frictional feedback is of low frequency with a typical growth rate of O ($10^{-6}\,\mathrm{s^{-1}}$). The observed development of the MJO over the Indian Ocean for instance takes a week or so to double the amplitude, which is much slower than

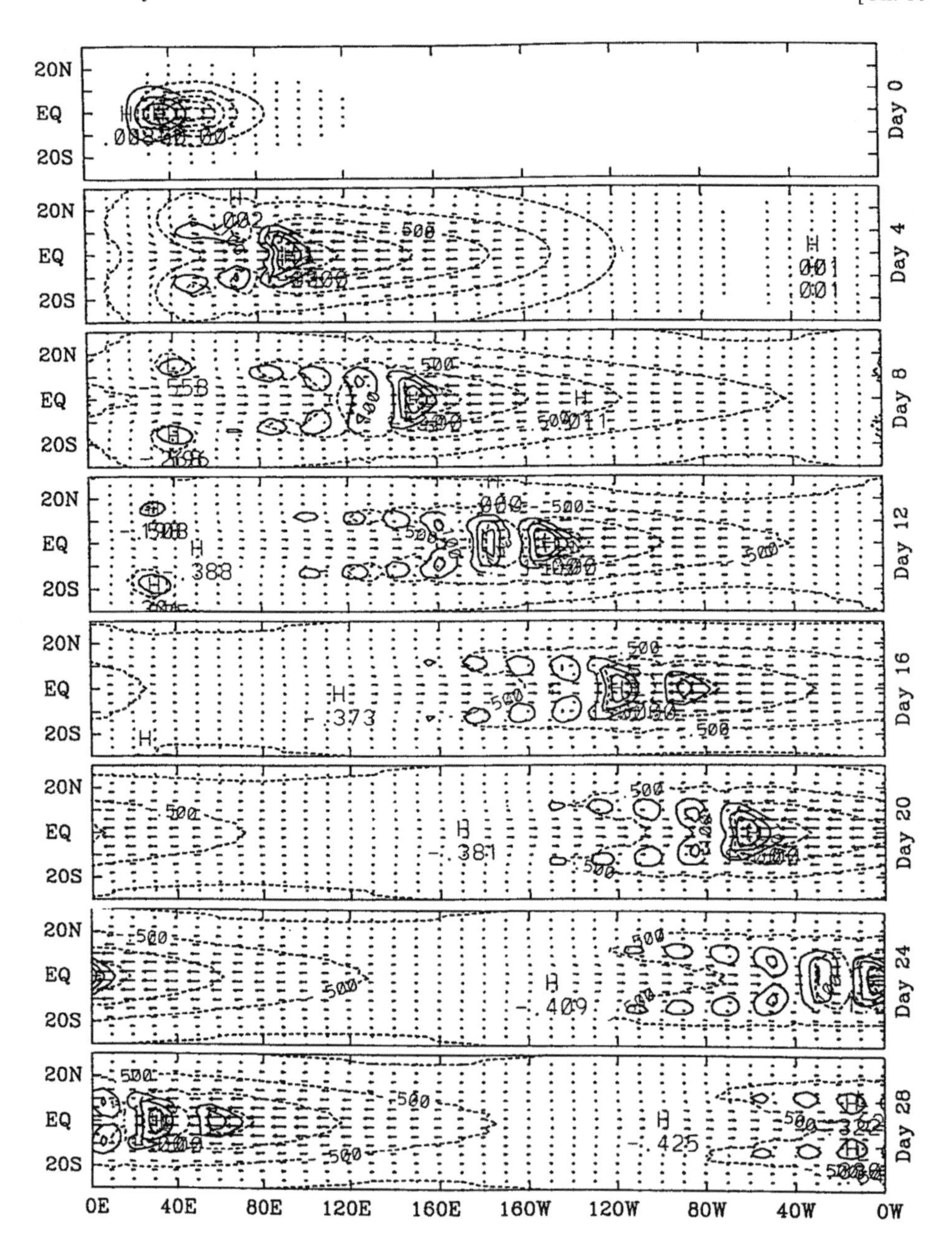

Figure 10.6. Sequential maps of the precipitation rate (solid contours), lower troposphere geopotential perturbation (dashed contours), and winds (arrows) for the Kelvin–Rossby wave packet induced by frictional convergence under non-linear (positive only and SST-dependent) heating. All three fields are normalized by their respective maxima at each panel. The contour starts from 0.1 and the interval is 0.2.
Adapted from Wang and Li (1994).

synoptic-scale growth. When both the frictional moisture convergence and wind–evaporation feedback are included, the resulting growth rate and other properties are very close to those of the frictional CID mode that is without WISHE. For more details the interested reader is referred to Wang and Li (1994).

10.4.5 Longitudinal variations of MJO disturbances

The aqua-planet model cannot explain the pronounced longitudinal variability of the propagation speed (Knutson and Weickmann, 1986) and development and decay (Wang and Rui, 1990b). These longitudinal variations are due to underlying SST variations. Climatological SSTs determine the atmospheric convective instability and availability of the moist energy. The longitudinal variation of SST has a major impact over the cold sectors of the tropics where the atmosphere is sufficiently stable, and the MJO disturbance propagates in the form of a damped moisture-modified Kelvin wave. In general, the MJO perturbation could travel around the globe and periodically regenerate and amplify over the warm ocean pools in response to a local build up of instability as shown by Salby *et al.* (1994).

During northern winter and spring, the MJO shows most coherent eastward propagation along the equator. The reason is that SST distribution is largely symmetric about the equator and the background flow effect is not critical except for the modulation of the Intertropical Convergence Zone (ITCZ) on the MJO. It has been shown that the greatest amplification of the equatorial Kelvin wave and associated subtropical Rossby gyres occurs when the maximum SST is located at the equator (Wang and Rui, 1990a) and when the atmospheric heating is strongest at the equator (Salby *et al.*, 1994).

10.5 DYNAMICS OF BOREAL SUMMER ISO

During boreal summer, the MJO disturbances weaken significantly and major centers of ISO variability in convection and precipitation move to the northern hemisphere Asian–Pacific summer monsoon region. Prominent northward propagation takes place in the Indian monsoon region. In the off-equatorial regions of the western North Pacific (WNP), westward and north-westward propagation prevails (Murakami, 1980; Lau and Chan, 1986; Chen and Murakami, 1988; Wang and Xu, 1997). In addition, there exists a stationary component, a convective see-saw between the equatorial Indian Ocean and the WNP (Zhu and Wang, 1993; Zhang and Hendon, 1997). Therefore, the boreal summer ISO behaves in a much more complicated manner than the MJO (see Chapters 2 and 3). This section aims to explain the complex behavior of the boreal summer ISO.

Recent observations have established two fundamental features of the boreal summer ISO. First, the dominant mode of ISO exhibits an eastward moving precipitation band that is tilted north-westward from the equator, tailing the main center of the equatorial convection associated with the MJO (Ferranti *et al.*, 1997; Annamalai and Slingo, 2001; Waliser *et al.*, 2003b). Second, the equatorial eastward propagating MJO tends to bifurcate poleward near Sumatra (Maloney and Hartmann, 1998; Kemball-Cook and Wang, 2001; Lawrence and Webster, 2002). It is important to achieve an understanding of what is responsible for these observed features. This is the aim of the current analysis.

10.5.1 Effects of mean flows on ISO

Wang and Xie (1997) proposed that the complexity of ISO during boreal summer could be understood as consequences of the impact of the seasonal mean circulations and SST (or surface specific humidity). Based on this premise, they constructed a prototype model for explaining the seasonal behavior of the ISO, which was described in Section 10.3.5. In their model, they prescribed the climatological July mean flows and the surface specific humidity (or equivalently the SST) as the typical boreal summer basic state. The model upper and lower layer basic flows and the surface specific humidity are shown in Figure 10.7. An initial perturbation is a Kelvin wave-like zonal wind perturbation with a circular shape and a diameter of 4,000 km being centered at 40°E on the equator. The wind variations follow a cosine function in both the zonal and meridional directions. The geopotential and temperature fields are determined by semi-geostrophic and hydrostatic balance, respectively.

Figure 10.8 shows snapshot views of the lower-tropospheric wind and precipitation rate every 4 days. Bear in mind that in this model the upper-tropospheric perturbation winds are nearly 180° out of phase with the lower-tropospheric winds. The initial disturbance moves eastward along the equator, as its major component is an equatorial Kelvin wave (Figure 10.8(a)). Because the boundary layer friction generates meridional flow that feeds back to convection, by day 4 the perturbation develops into a precipitation complex consisting of an equatorial cell and 2 off-equatorial cells, indicating that the perturbation evolves into a Kelvin–Rossby wave packet coupled by convective heating which is similar to that shown in Figure 10.6. When the wave packet approaches the maritime continent, it weakens because of the reduction of the basic flow specific humidity (Figure 10.7(c)), meanwhile Rossby wave cells having a typical zonal scale of 2,000–4,000 km start to emit out of the packet and move north-westward. By day 6 the emanated Rossby wave cells produce a north-west–south-east tilted rain band from India to Borneo (Figure 10.8(b)). When the equatorial packet arrives at the western Pacific at day 10, it starts to emanate Rossby cells again (Figure 10.8(c)). By day 14 the equatorial disturbance weakens and stalls east of the dateline (Figure 10.8(d)), meanwhile sending fast eastward propagating Kelvin waves crossing the eastern Pacific (Figure 10.8(e)) and South America (Figure 10.8(f)) and dissipating in the Atlantic (Figure 10.8(g)). On the other hand, the emitted moist Rossby cells over the Philippine Sea at day 10 continuously migrate north-westward through the South China Sea and back to India. When the northern cell decays in the Arabian Sea due to the "blocking" of the sinking dry air mass over north Africa, the southern cell re-initiates an equatorial perturbation (Figure 10.8(f)) and starts the next cycle. The whole life cycle spans about 4 weeks. One should note that the intraseasonal disturbances in this model experience development and decay locally due to the

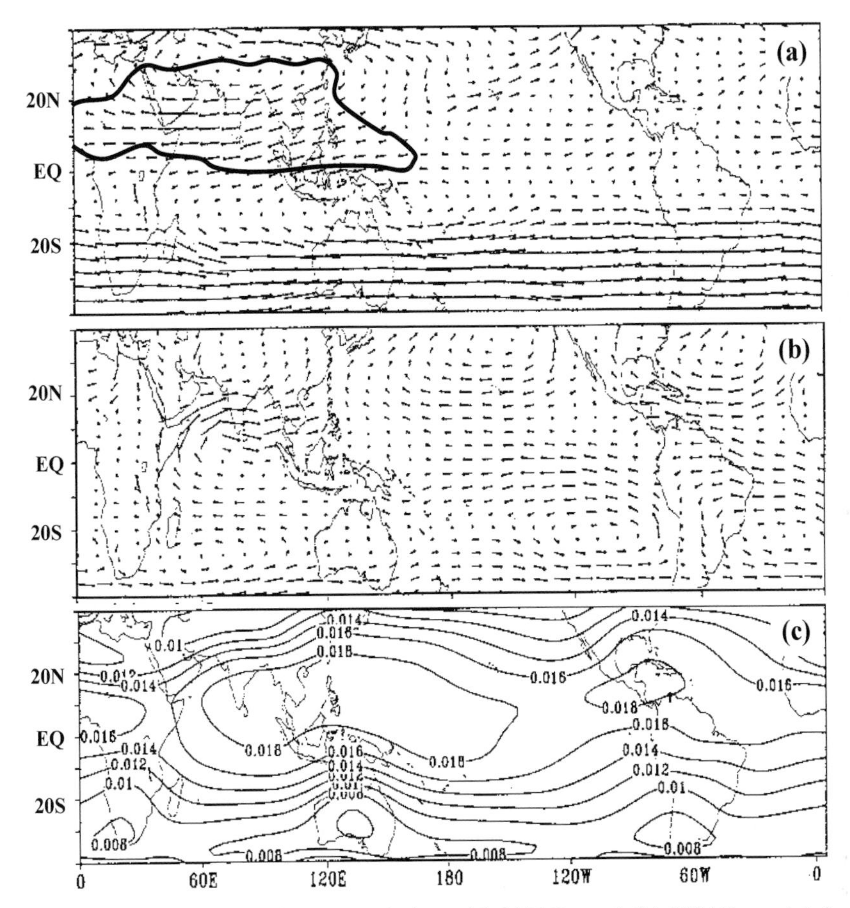

Figure 10.7. Climatological July mean winds at (a) 200 hPa and (b) 850 hPa and July mean specific humidity at (c) 1,000 hPa. The thick contour in panel (a) represents the contour of $-4\,\mathrm{m\,s^{-1}}$ of July mean U_{200}–U_{850}, which outlines the regions of significant easterly vertical shears. The data used were derived from the European Centre for Medium-range Weather Forecasts (ECMWF) reanalysis for the period 1979–1992.

variations of the basic state moisture distribution (as reflected by the surface specific humidity) and the influences of the basic state circulation. While this idealized life cycle exaggerates the strength of the off-equatorial westward propagation, there are some notable features that might provide hints for understanding observed boreal summer ISO.

First, the model low-frequency disturbances not only invokes the equatorial trapped Kelvin–Rossby wave packet but also invokes off-equatorial Rossby wave activity. The result here suggests that the boreal summer mean circulation and the spatial variations of moist static energy of the mean flows can trap moist Kelvin and Rossby waves within the northern summer monsoon domain that is defined by the region of easterly vertical shear (Figure 10.7(a)) and the surface specific humidity

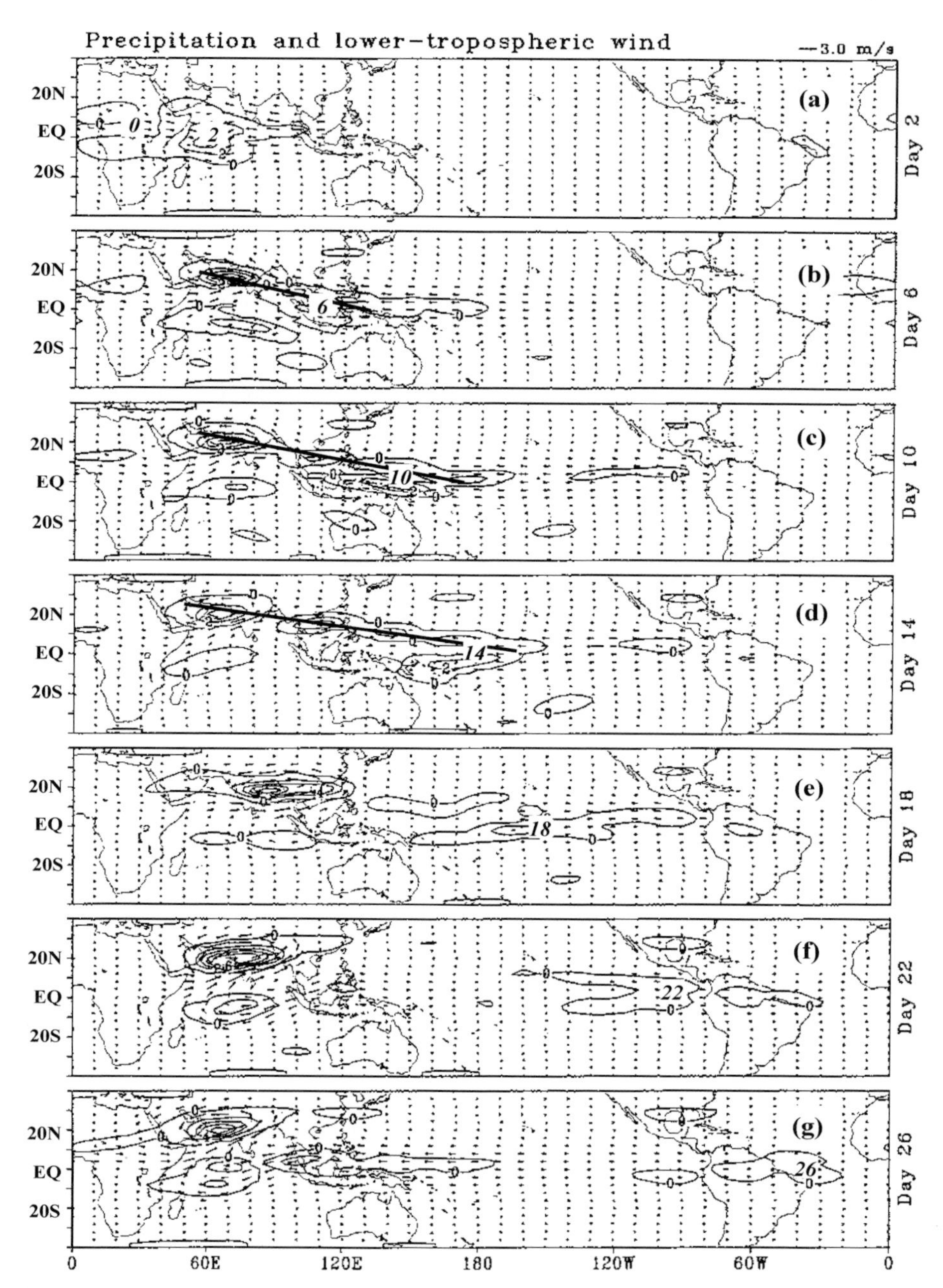

Figure 10.8. Sequential maps of the lower-tropospheric winds and precipitation rate (contour interval 2 mm day^{-1}) for the Kelvin–Rossby wave packet induced by frictional convergence under non-linear heating and in July mean basic state. The straight line indicates the tilted precipitation bands. The numbers denote the day of model integration, which traces the locations of the major precipitation centers.

Modified from Wang and Xie (1997).

exceeding 18 g kg^{-1} (Figure 10.7(c)). The westward propagating perturbations in the model are readily identified as the gravest meridional mode of moist equatorial Rossby waves that are destabilized and modified by monsoon easterly vertical shears. Wang and Xie (1996) and Xie and Wang (1996) have shown that an easterly vertical shear and convective heating can destabilize equatorial Rossby waves; the resulting most unstable wavelength is about 4,000 km; furthermore, when the easterly shear is confined to the northern hemisphere as in the case of the northern summer monsoon, the structure of the Rossby waves can become remarkably asymmetric about the equator with the southern cell severely suppressed and close to the equator. Krishnan *et al.* (2000) suggested that the rapid north-westward propagating Rossby waves from the central Bay of Bengal toward north-west India and the decoupling of the eastward propagating equatorial anomaly determine the transition from a wet phase to a break phase of the Indian summer monsoon.

Second, the model is able to simulate the north-west–south-eastward tilted rain band (Figure 10.8; see also Figure 2.11). The model results suggest that this tilted rain band consists of emanated moist Rossby cells moving west-north-westward. The radiation of moist Rossby waves is longitudinally phase locked with weakening or disintegration of the equatorial wave packets over Indonesia and over the equatorial central Pacific due to the decrease in SST or the mean state latent energy. The decay of equatorial eastward propagating disturbances over these two longitudes is an observed feature. The decay near the dateline is due mainly to the sharp decrease in the SST. However, the die out over Indonesia is not solely due to the reduction of mean state latent energy, it might involve multi-factors that are not included in the model. The topographic blocking of the Sumatra Island (a mountain range higher than 2 km) could be destructive to the boundary layer organization of the MJO convection. The strong diurnal cycle over the Indonesian region is also an unfavorable condition for the MJO because it constantly releases convective energy and destroys the energy accumulation needed on MJO timescales. Another possibility is the destructive land effects on air–sea interaction that is a positive contributor to the MJO.

Third, the model result is instrumental for understanding the nature of the northward propagation over the northern Indian Ocean. The longitude–time diagrams along 90°E and 110°E show slow northward migration of the precipitation (Figure 10.9). The model northward propagation over the eastern Indian Ocean and north of Indonesia is an integrated part of the movement of the north-west–south-east tilted rain band. The north-westward propagation of moist Rossby waves that is modulated by the equatorial eastward MJO mode is also relevant to explaining the observed north-westward propagation of the low-frequency cloud and vorticity anomalies on the bi-weekly and 30-day period observed over the WNP (Nitta, 1987; Lau and Lau, 1990).

Fourth, the life cycle shown in Figure 10.8 tends to repeat itself in the 90-day model integration (figure not shown, see Wang and Xie, 1997). This recurrence of the ISO cycle suggests a self-sustaining mechanism of the monsoon ISO. The results obtained from a suite of reduced physics experiments in Wang and Xie (1997)

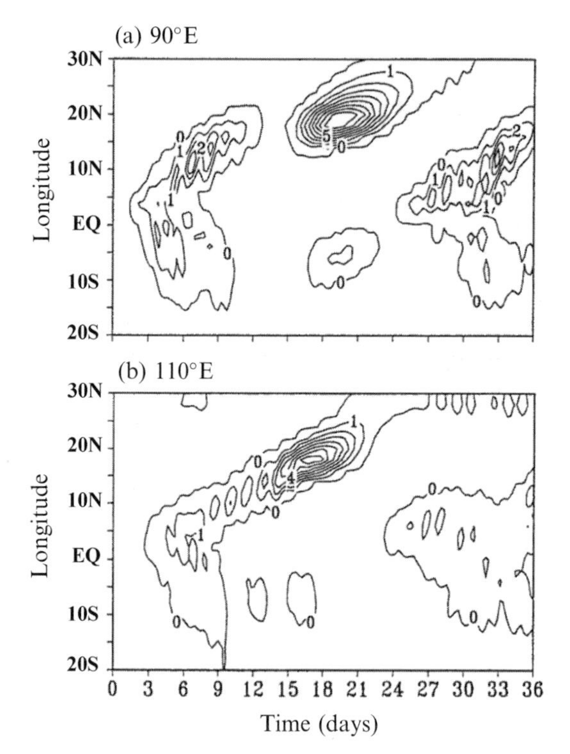

Figure 10.9. (a) Time–longitude cross sections of precipitation rate along (a) 90°E and (b) 110°E for the experimental results shown in Figure 10.8. Contour interval is 1 mm day^{-1}. Modified from Wang and Xie (1997).

suggest that the basic state meridional and Walker circulations play a critical role in regeneration of the disturbances over the equatorial Indian Ocean. When these mean circulations are removed from the model basic flows, the southern cell of the Rossby waves are suppressed and re-initiation of the equatorial perturbation is so weak that the ISO can not be sustained.

10.5.2 Mechanism of northward propagation

A critical question in explaining the north-west–south-east tilt of the precipitation band and associated northward propagation of ISO is what makes the Rossby waves have a northward propagation component. Obviously, without mean flows, the emanated Rossby waves move only westward as shown in Figure 10.6. It is the presence of the basic flow that induces a northward propagation component. However, what specific factors in the summer mean circulation are responsible for the northward propagation? This remains an issue. The works of Drbohlav and Wang (2004) and Jiang *et al.* (2004) identified that the effect of the easterly vertical shear is an important internal dynamic factor.

To illustrate this mechanism, let us consider a simplified 2-D version of the model of Wang and Xie (1997) in which the zonal variations of the basic state and dependent variables are neglected. The vorticity equations for the barotropic component (denoted by the subscript "+") is (refer to Wang and Xie for derivation):

$$\frac{\partial \zeta_+}{\partial t} = -\beta v_+ - U_T \left(\frac{\partial \omega}{\partial y} \right) \tag{10.13}$$

where U_T denotes the constant vertical shear of the basic zonal flow. Equation (10.13) indicates that in the presence of vertical easterly shear $U_T < 0$, a northward decrease in the perturbation upward motion can generate positive barotropic vorticity to the north of the convection. This process is illustrated in Figure 10.10. A mean flow with easterly vertical shear has horizontal relative vorticity with an equatorward component, which the perturbation motion can tap. Rossby-wave-induced heating generates a perturbation vertical motion that decreases north of the convection. This vertical motion field twists the mean flow horizontal vorticity and generates a vorticity with a positive vertical component north of the convection region. The positive vorticity in turn induces convergence in the boundary layer, which would destabilize the atmosphere and trigger new convection to the north of the convection. For the same reasons, negative vorticity and divergence in the boundary layer develop and suppress convection south of the convection region. Thus, the twisting of the mean flow horizontal vorticity by the vertical motion field associated with the Rossby waves creates conditions that favor northward movement of the enhanced rainfall. This conclusion is supported by observations made by Jiang *et al.* (2004) who showed that the barotropic vorticity in the free troposphere is located about 4° to the north of the northward propagating convection anomalies. Jiang *et al.* (2004) also argued that the advection of the mean state specific humidity by meridional winds of ISO in the boundary layer favors the northward propagation. The third factor that may enhance the northward propagation is the intraseasonal variation of SST, which is shown to guide convection in the northward propagating ISO (Kemball-Cook and Wang, 2001; Fu *et al.*, 2003). The SST feedback to convection is essentially the same mechanism as that for the MJO that is discussed in the next section.

10.6 ROLES OF ATMOSPHERIC–OCEAN INTERACTION

Figure 10.11(a) presents a schematic summary of the observed structure of the MJO and associated oceanic mixed layer based on the TOGA–COARE observations. The "wet" region of the MJO in the equatorial zonal plane features a large-scale convective envelope whose core consists of SCCs. This convective region is accompanied by large-scale rising motion and planetary-scale equatorial upper-level easterly and low-level westerly anomalies (Lin and Johnson, 1996). The core of the low-level and surface westerly anomalies, however, lag behind enhanced convection by slightly less than one-quarter of a wavelength (e.g., Chen *et al.*, 1996; Chou *et al.*, 1995; Fasullo and Webster, 1995). The westerly wind bursts associated with the convective phase of the MJO cause SST to drop off more than 1°C and profoundly change the mixed

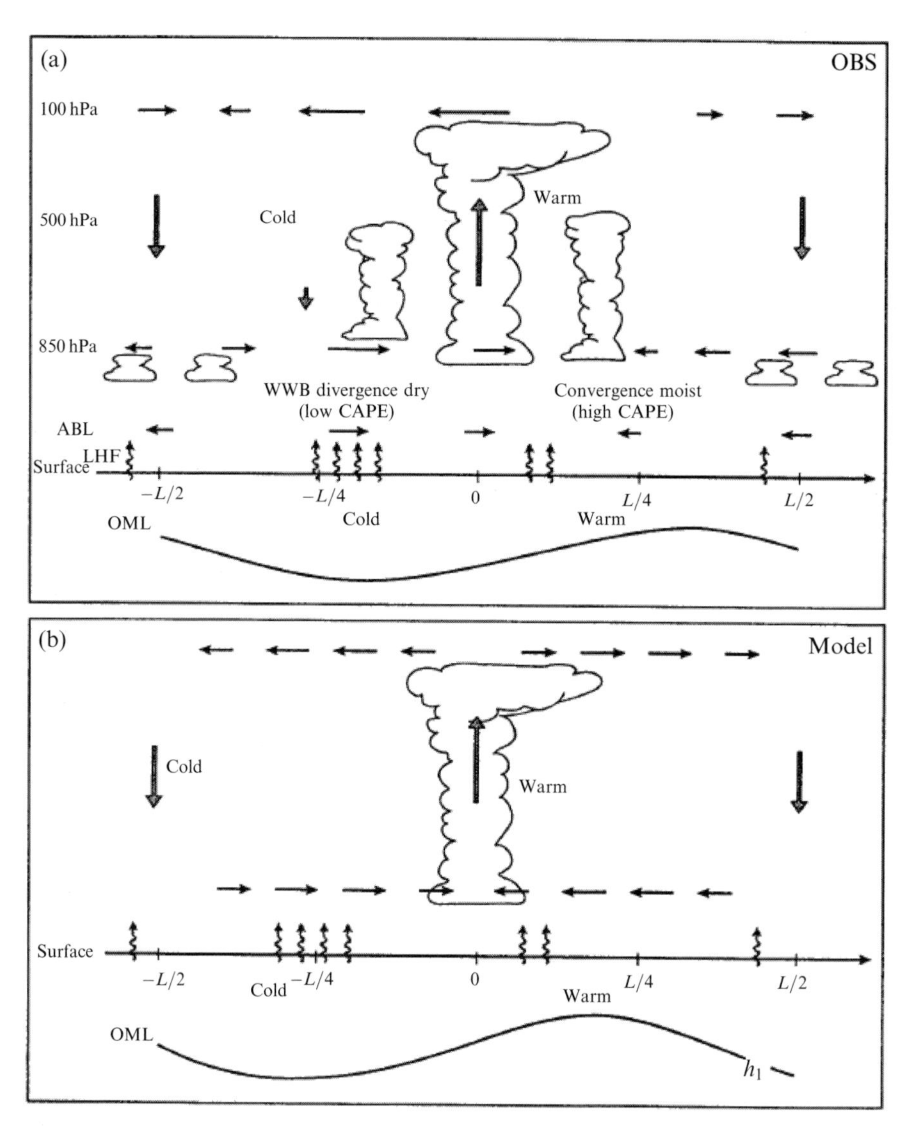

Figure 10.11. Schematic diagram illustrating (a) the equatorial vertical structure of the MJO observed in TOGA–COARE and (b) the most unstable coupled mode in the theoretical model. The squiggly lines denote surface latent heat flux. The symbols ABL, OML, LHF, WWB, CAPE, and L represent, respectively, atmospheric boundary layer, ocean mixed layer, latent heat flux, westerly wind burst, convective available potential energy, and wavelength.
Adapted from Wang and Xie (1998).

layer structure and currents (Webster, 1994; Weller and Anderson, 1996; Lau and Sui, 1997; Jones *et al.*, 1998; Shinoda *et al.*, 1998; Shinoda and Hendon, 1998; Zhang and Anderson, 2003). The SST decreases under and to the west of the enhanced convection due to an enhanced evaporation and latent heat flux that nearly coincides with low-level westerly anomalies. To the east of the convection in the MJO low-level easterly phase the SST rises due to the reduced wind speed, shallow mixed layer, and the increased insolation in the suppressed convection region (Jones and Weare, 1996; Zhang, 1996; Lin and Johnson, 1996). Thus, positive SSTAs lead enhanced convection by about one-quarter of a wavelength.

Over the Indian Ocean, in boreal summer, the coherent negative/positive SSTAs generated by surface heat fluxes that move northward were found to follow regions of active and suppressed convection anomaly (Kemball-Cook and Wang, 2001; Sengupta *et al.*, 2001). In the WNP, SSTAs associated with the north-westward propagation of ISO and their possible feedbacks were also noted (Kemball-Cook and Wang, 2001; Hsu and Weng, 2001).

As reviewed in Section 10.2.7, these observational analyses have stimulated numerous theoretical and numerical modeling studies. In this section, an attempt is made to elucidate the nature and impacts of air–sea interaction on the MJO in terms of a simple theoretical coupled model.

As shown by Hirst and Lau (1990), the unstable coupled modes on an intraseasonal timescale results from atmospheric waves, which contrasts the coupled ENSO modes that arise from oceanic wave adjustment (Philander *et al.*, 1984) or slow SST variation (Neelin, 1990). In Hirst and Lau's model formulation, however, the models and the atmosphere–ocean interaction are essentially the same as those used for ENSO studies except for inclusion of atmospheric transient waves. The ENSO-type model is not suitable for study of the warm pool of the tropical Indian and western Pacific Oceans, because both the climatological mean state and the processes of atmosphere–ocean coupling in the warm pool differ fundamentally from those in the cold tongue of SST in the eastern tropical Pacific.

The theoretical coupled model of Wang and Xie (1998) consists of a single vertical mode atmospheric model coupled to a linearized ocean mixed layer model. The ocean component differs from those used in the previous coupled stability analysis for the eastern Pacific (e.g., Hirst and Lau, 1990), which describes ocean mixed layer physics and thermodynamic coupling of the atmosphere and ocean through surface heat exchanges. The dynamical coupling that plays an essential role in the eastern Pacific is neglected, because it is unimportant in changing SST in the warm pool oceans. The ocean mixed layer is described by the following linearized equations:

$$\frac{\partial h_1}{\partial t} = \varepsilon \bar{U} U + \bar{w}_e \left(\frac{3U}{\bar{U}} - \frac{h_1}{H_1} \right) \tag{10.14a}$$

$$\frac{\partial T}{\partial t} = D_{rad} \left(\frac{\partial U}{\partial x} + \frac{\partial V}{\partial y} \right) - (3D_{ent} + D_{eva}) \frac{U}{\bar{U}} + D_{ent} \frac{h_1}{H_1} - dT \tag{10.14b}$$

where h_1 and T denote mixed layer depth and temperature, respectively. $H_1 = 50\,\text{m}$ is the mean depth of the mixed layer, $\bar{U}$ the mean surface wind, and $\bar{w}_e$ the mean entrainment rate; d is the thermal damping coefficient; and the coefficients D_{rad}, D_{ent}, and D_{eva} measure the heating rate associated with, respectively, the downward short-wave radiation, entrainment, and evaporation processes. Here, the downward solar radiation flux is assumed to decrease with increasing atmospheric moisture convergence (hence atmospheric cloudiness); the surface evaporation is assumed to enhance over regions of anomalous westerlies as the mean surface winds are from the west.

The atmospheric component of the coupled model describes linear motion of the lowest baroclinic mode, which is similar to Davey and Gill (1987) except that the local rate of changes of momentum and temperature are added. The equations take the shallow-water form:

$$\frac{\partial u}{\partial t} + \varepsilon_a U - \beta y V = -\frac{\partial \phi}{\partial x} \tag{10.15a}$$

$$\frac{\partial V}{\partial t} + \varepsilon_a V + \beta y U = -\frac{\partial \phi}{\partial y} \tag{10.15b}$$

$$\frac{\partial \phi}{\partial t} + \mu_a \phi + C_a^2(1 - I)\left(\frac{\partial U}{\partial x} + \frac{\partial V}{\partial y}\right) = -\frac{Rg}{2C_p p_2}\alpha T \tag{10.15c}$$

where U, V, and ϕ are the lower-tropospheric zonal and meridional wind and geopotential, respectively; ε_a and μ_a are, respectively, coefficients for Rayleigh friction and Newtonian cooling; C_a is a dry atmospheric Kelvin wave speed; α is a latent heating coefficient; T is the SST or ocean mixed layer temperature anomaly; as in (10.10), the moist atmospheric Kelvin wave speed is $C_a(1 - I)^{1/2}$. Expressions for all coefficients and the derivation of (10.14a, b) are given in Wang and Xie (1998).

Figure 10.12 shows the results derived from the instability analysis of the coupled system (10.14) and (10.15): how the growth rate and phase speed of the fastest growing coupled mode vary with the two coupling coefficients, the cloud–SST coupling coefficient D_{rad} and wind–SST coupling coefficient $D_{wind} = D_{eva} + 3D_{ent}$. The fastest growing coupled mode has a planetary zonal scale (Figure 10.12(a)). Obviously, the wind–SST coupling plays a primary role in generating the coupled instability (Figure 10.12(b)). The cloud–SST coupling can significantly contribute to the growth only when the wind effect is relatively weak. The coupled modes have an eastward phase speed less than $10\,\mathrm{m\,s^{-1}}$ (Figure 10.12(c)). The unstable coupled mode is originated from the atmospheric moist Kelvin waves. These results indicate that the warm pool basic state is conducive to the coupled unstable mode on intraseasonal timescales.

The structure of the unstable coupled mode is illustrated in Figure 10.11(b). This schematic diagram was strictly based on the model results (figure not shown, see Wang and Xie, 1998). The coupled mode in the model has a realistic SST–convection relationship: positive SSTAs are located to the east of the convection anomalies by about one-sixth of a wavelength, while lagging behind the surface easterly wind anomalies by about one-twelfth of a wavelength. Note, however, the phase relationship between equatorial zonal wind (lower and upper-level) and convection anomalies is not correctly captured. This appears to be a common weakness of the current theoretical models of the MJO.

In the model the SSTA-induced heating tends to increase atmospheric temperature (or thickness) locally so that the positive covariance between heating and warming generates perturbation available potential energy for the growing

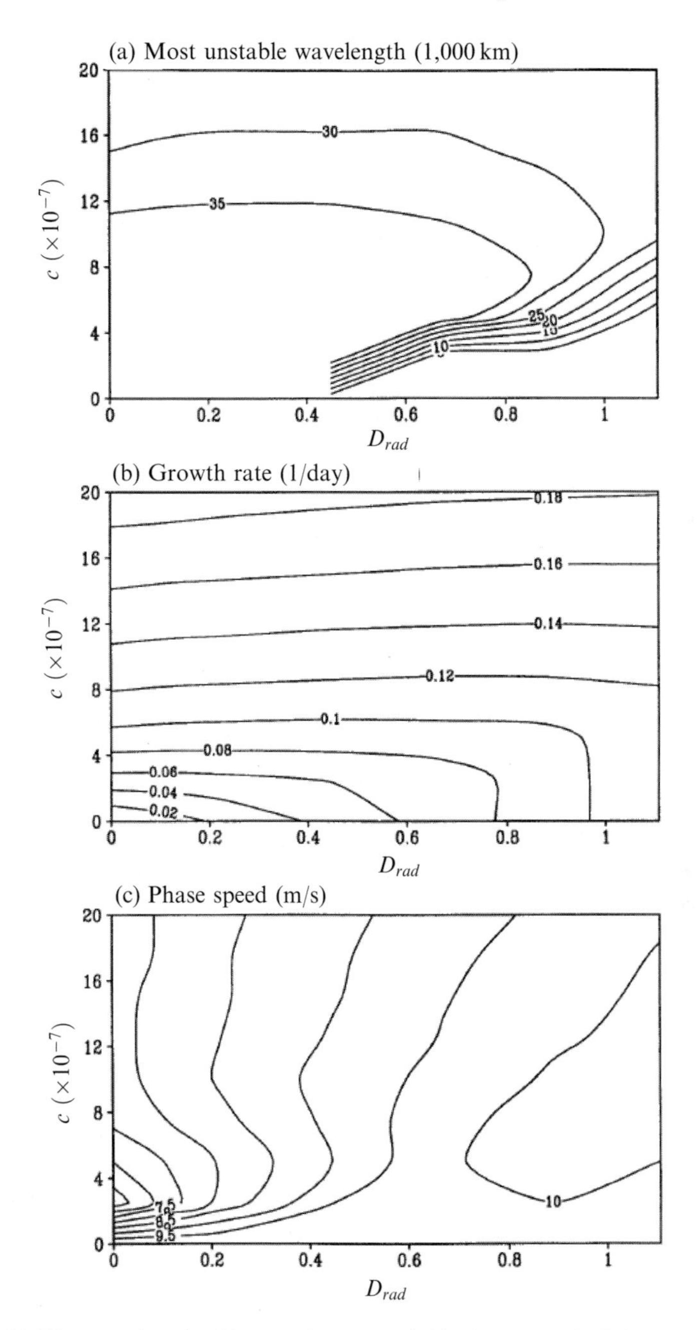

Figure 10.12. (a) The wavelength, (b) growth rate, and (c) phase speed of the most unstable coupled mode as functions of the cloud–SST coupling coefficient D_{rad} (K) and wind–SST coupling coefficient D_{wind} (K s^{-1}) in a coupled atmosphere–ocean model for the warm pool climate system.

Adapted from Wang and Xie (1998).

coupled mode. It is shown that the coupling effects are more effective for planetary-scale perturbations because the planetary-scale disturbance can have a sufficiently long time to change SST; also, the larger the change in SST the stronger the feedback from the SST.

The coupled model results suggest that while the atmospheric internal dynamics are primary causes for the MJO, the ocean mixed layer thermodynamic processes interacting with the atmosphere may play a significant part in sustaining the MJO by adding instability to atmospheric moist low-frequency perturbations and by providing a mechanism for long-wave selection and slow eastward propagation.

10.7 SUMMARY AND DISCUSSION

10.7.1 Understanding gained from the frictional CID theory

Based on a review of the existing theories, the essential physics of the ISO is illustrated in Figure 10.1 and discussed in Section 10.3.1. Based on this conceptual framework a theoretical model is put forward in Section 10.3. This simplistic model is a time-dependent primitive equation model on an equatorial β-plane that has a two-level free troposphere and a well-mixed planetary boundary layer (Figure 10.3). As shown in Figure 10.1, at the center of the model are the non-linear interactions including condensational heating, low-frequency equatorial (Kelvin and Rossby) waves, boundary layer dynamics, moisture feedback, and wind-induced heat exchanges at the surface. These non-linear interactions are called, for short, "convective interaction with dynamics" (CID). The model also includes the impacts of 3-D background circulations and specific humidity (or SST) as well as the effect of the interactive SST feedback. Thus, the model physics integrate, to varying degrees, the mechanisms listed in the review (Section 10.2) except that a simple Newtonian cooling was used for representing the net cloud–radiative heating. Regardless of its simplicity, the model is able to reproduce atmospheric disturbances that closely resemble features of the observed MJO and the boreal summer monsoon ISO, thus providing a unifying framework for the tropical ISO.

The theory based on this general framework is termed, for short, "frictional CID" theory. Here the boundary layer frictional moisture convergence feedback is emphasized because the low-frequency equatorial Kelvin–Rossby waves that characterize the MJO are essentially coupled through boundary layer frictional moisture convergence-induced condensational heating. The disturbance originating from the fastest growing frictional CID mode has the following characteristics resembling the MJO (Figure 10.13): an east–west circulation that spans the globe and is coupled to a large-scale complex of convective cells; a baroclinic structure, with winds converging in the boundary layer in front of the main precipitation; a horizontal circulation consisting of both equatorial Kelvin and Rossby waves, and a slow eastward (about 5–10 $\mathrm{m\,s^{-1}}$) movement that gives rise to an intraseasonal timescale (30–60 days) (Figure 10.5(a, b)). Given the seasonally varying basic circulation and SST distribution, the simulated ISO shows appropriate seasonality, having prominent northward propagation and off-equatorial westward propagation of disturbances in the Asian monsoon region during the boreal summer (Figure 10.8).

Perhaps the most important understanding from analyzing the theoretical model is the role played by friction-induced moisture convergence. The deep convective heating released in the major convective region excites large-scale eastward moving Kelvin waves and westward moving equatorial Rossby waves. These waves are dissipated away from the heating region, leaving an equatorial low-pressure trough of Kelvin wave response (K-L) east of the main convection and 2 off-equatorial lows of the Rossby wave (R-L) response (Figure 10.13, see color section). A central question to be addressed for a theory is why eastward propagation associated with the Kelvin waves was selected over the westward propagation of the Rossby waves. As shown in Figure 10.4, not only the Kelvin wave but also the most trapped equatorial symmetric Rossby waves induce a maximum boundary layer convergence on the equatorial trough region, thereby both types of waves creating a unified boundary layer convergence to support the equatorial convective heating. The Kelvin wave-induced frictional convergence favors itself while the Rossby wave-induced boundary layer convergence also partially favors the Kelvin waves. Thus, the frictional moisture convergence-induced heating couples the equatorial Kelvin and Rossby waves and selects an eastward moving, unstable mode. The condensational heating induced by free tropospheric waves, while alone does not produce instability, acts to reduce the effective static stability and contributes to a slow eastward phase speed. The slow eastward propagation is also attributed to frictional coupling of the moist Kelvin and Rossby waves, because the eastward movement of the Kelvin wave is slowed by the westward tendency of the Rossby wave that results from the β-effect. The coupled Rossby–Kelvin wave takes on the approximate eastward speed of the MJO. The upper-level returning flow of the eastward moving Kelvin wave sinks and warms the region east of the major convection. On the other hand, in the same region the friction-induced moist converging air rises and condenses. It is the latent heat released in this frictional convergence region that is in phase with atmospheric warming and provides eddy available potential energy to support the growth of the perturbation. Being ahead of the main convective region, this energy source for instability is not converted to kinetic energy for CISK as in the development of tropical cyclones. A portion of the energy is carried away from the convective regions by fast moving, dry Kelvin and Rossby waves, which spread perturbed circulation around the globe. This energy dispersion accounts for the planetary circulation scale of the MJO.

The model results suggest that the northward propagation of ISO in the Asian–Pacific summer monsoon region is due to north-westward propagation of moist Rossby waves that emanated from the equatorial disturbances when the latter decay over Indonesia and near the dateline. What induces the northward propagation component for the emitted moist Rossby waves? The model experiments suggest that the monsoon easterly vertical shear provides such an atmospheric internal dynamic mechanism. This mechanism, as illustrated in Figure 10.10 (see color section), is essentially due to the twisting of the mean flow horizontal vorticity by the vertical motion field associated with the Rossby waves. This twisting process generates positive vorticity north of the convection in the troposphere, thus creating boundary layer moisture convergence that favors northward movement of the enhanced rainfall. The interactive SST and surface heat fluxes also contribute to the northward propagation.

The coupled model results suggest that while the atmospheric internal dynamics are essential in generating the MJO, the interaction between the atmosphere and ocean mixed-layer may further enhance and better organize the eastward propagating MJO through additional coupled instability amplifying moist atmospheric low-frequency perturbations. The basic state of the warm pool is conducive to the occurrence of the coupled unstable mode on intraseasonal timescales. The wind–evaporation–SST feedback is central to the coupled instability (Figure 10.12).

10.7.2 Model limitations

The major drawback of the present theoretical model is that its simple representation of diabatic heating cannot deal with the interactions among scales and that it presumes direct coupling between the MJO disturbance and convections. The MJO is a multi-scale tropical atmospheric motion. The planetary-scale MJO does not directly organize convection and the convective latent heat release is largely consumed directly by mesoscale and synoptic-scale disturbances. The model also neglects the role of the diurnal cycle in the MJO, which might be important for sustaining the MJO (Slingo *et al.*, 2003).

The crude vertical structure confines the description of the heating-induced baroclinic motion to the gravest baroclinic mode. In reality, the higher vertical modes are necessary for explaining realistic vertical structure, namely, the magnitude of wind anomalies at the upper level is considerably larger than its low-level counterpart and a rearward tilt of the large-scale vertical velocity against its direction of propagation. The higher modes have a slower phase speed and may be important in explaining the slow eastward propagation of the MJO (Mapes, 2000).

The crude vertical resolution of the model can only describe vertical integrated condensational heating that is constrained by water vapor conservation. Therefore, the heating representation does not represent any specific type of cumulus parameterization scheme. This representation does not allow partitioning of the condensational heating vertically, especially between deep and shallow convection. In nature, the frictional convergence occurring ahead of the major deep convection is capped by subsidence and it cannot trigger immediate development of deep convection. It is likely that this large-scale frictional forcing acts to collect boundary layer water vapor and to foster convective instability (Maloney and Hartmann, 1998), which would first lead to development of shallow convection (Kikuchi and Takayabu, 2004). Wu (2003) has suggested that warm rain (low to middle level condensational processes) may play a significant role in explaining the build up of the MJO convection. This idea is consistent with Johnson *et al.* (1999) on the trimodal distribution of convection in the western Pacific and Tropical Rain Measuring Mission (TRMM) observations (Short and Nakamura, 2000; Lau and Wu, 2003).

Increased vertical resolution of the numerical model may certainly better resolve the melting layer and cumulus congestus, thus providing more realistic simulation of the MJO (Slingo *et al.*, 2003).

The model's simplicity does not allow description of the complex cloud–radiation feedback process, which may play an important role in sustaining oscillations on intraseasonal timescales (Hu and Randall, 1994; Raymond, 2001). The simple ocean mixed-layer model used in Section 10.6 has neglected the effect of the salinity barrier layer, which can potentially provide much stronger local coupling between the atmosphere and ocean.

10.7.3 Outstanding issues

Accurate modeling and predicting of the ISO may improve seasonal-to-interannual climate prediction and bridge the gap between weather forecast and seasonal prediction (Waliser *et al.*, 2003a; see also Chapter 12). Unfortunately, current global circulation models still have great difficulty in simulating the properties of the tropical oscillation correctly (Slingo *et al.*, 1996; Wu *et al.*, 2002; Waliser *et al.*, 2003b). One might ask if the frictional CID theory is the key to the ISO, why would some AGCMs that have well-constrained formulations of a similar processes give such poor simulations?

In simple model such as the one presented in Section 10.3, a direct linkage between the large-scale low-frequency wave motion and collective effect of convective heating, was established through moisture and heat energy conservation without going to details in resolving the vertical heating distribution or scale interactions. Such models have no difficulty in producing MJO-like low-frequency oscillations because these models avoid the complex interactions among many physical processes that take place on different scales of time and motion. Modeling of the ISO in complex models, however, must entail a series of interacting parameterizations including moisture transport, cloud and convection, and radiation transfer. Uncertainties in mathematical descriptions of these interactive parameterizations could jeopardize the model's capability in simulating the ISO.

The theoretical model results here suggest that in order to simulate the MJO realistically, the cumulus parameterization scheme in complex models has to allow the large-scale low-frequency waves (and associated boundary layer motion) to be affected by the parameterized convective heating and to allow these low-frequency waves to have some effects on the parameterized heating either directly (through grid-scale precipitation for example) or indirectly (through correct description of the multi-scale interactions). If all convective heating were consumed by high-frequency small-scale disturbances, and if there were no appropriate description of upscale transport of energy, how could the model maintain the low-frequency MJO?

In complex GCMs, one does not know what the correct heating partitioning is between the convective and stable precipitation and between the small-scale high-frequency and large-scale low-frequency disturbances. Recent TRMM precipitation radar measurements show that stratiform precipitation contributes more to the intraseasonal rainfall variations than it does to seasonal mean rainfall (Lin *et al.*,

2004). The author suspects that some AGCMs might have underestimated the portion of condensational heating released by stable precipitation. In summary, the inadequate treatment of cumulus parameterization and the multi-scale interaction processes could be the major hurdles for realistic simulation of the MJO.

The sensitivity of ISO simulations to various cumulus parameterization schemes has been evaluated with a single model by differing cumulus schemes. Both Chao and Deng (1998) and Lee *et al.* (2003) compared three different schemes, the moist convective adjustment (MCA) (Manabe *et al.*, 1965), the Kuo (1974) scheme, and the modified Arakawa–Schubert (1974) (AS) scheme. Both studies found that MCA produces the strongest ISO variability while the AS scheme the smallest. What causes this sensitivity deserves further investigation. Wang and Schlesinger (1999) used the University of Illinois AGCM with the above three types of cumulus parameterization schemes to simulate the MJO. For each parameterization a relative humidity criterion (RHc) for convection or convective heating to occur was used. They found that as the RHc increases, the simulated ISO becomes stronger for all three parameterizations. They suggest that when large values of RHc were used the triggering convection required the moist static energy in the lower troposphere to be accumulated to a certain amount through moisture convergence; this elevated RHc weakened the interaction between the circulation and heating for small-scale perturbations and allowed the ISO to occur at low frequencies. On the other hand, Maloney and Hartmann (2001) found that the ISO in National Center for Atmospheric Research (NCAR) community climate model (CCM3) with a relaxed AS scheme is not improved by increasing the RHc. They reported that the ISO is highly sensitive to the parameterization of the convective precipitation evaporation in unsaturated environmental air and saturated downdrafts. Concerning improvement of the cumulus scheme, three aspects of notable interests have emerged, the vertical profile of diabatic heating, closure assumptions used in the parameterization, and the role of shallow vs. deep cumulus clouds.

Learning how upscale energy transfer and the interactions among the various space and time scales sustain the MJO presents a major challenge, but it might shed light on why many current GCMs fail to simulate the MJO accurately. Krishnamurti *et al.* (2003) have shown that about 30–50% of the total surface heat flux on the MJO timescale comes from the triad interaction of the MJO with two other synoptic timescales. Developing the systematic multi-scale model for the MJO (Majda and Klein, 2003) and investigation of upscale transfer of kinetic and thermal energy generated by organized mesoscale and synoptic-scale circulations in maintaining the ISO (Houze *et al.*, 2000; Majda and Biello, 2004) are among the important steps toward a deeper understanding of MJO dynamics.

Validating theories and developing new ideas rely on improved observations. Different from weather systems, the MJO or the intraseasonal variation in general is a "broad frequency band" phenomenon (Madden and Julian, 1994). Current observational analyses of the MJO have been often focussed on its statistical behavior or averaged features of many events. TOGA–COARE provided invaluable observations on two MJO events and analyses of TOGA–COARE observations have greatly advanced our knowledge on the structure of the MJO and its associated surface heat flux exchanges and air–sea interaction. The information gained from these analyses has been extremely useful for validating theories and furthering our theoretical understanding. Yet, we still do not have sufficient information on the differences among individual events, which may be as important as their common features. Most observed features have been derived using temporal or spatially filtered data, which tends to artificially separate high frequency and intraseasonal variations in a linear fashion that might undermine the inherent non-linearity. In addition, due to a void of accurate observation over tropical oceans, we do not have sufficient information about the spatial structure and the nature of the clouds and diabatic heating that drives the MJO, which is critical for improving theories and numerical simulations.

The current theoretical model results are useful in the sense that they provide clues for understanding basic mechanisms that may be in action in nature. However, to improve representation of the MJO in GCMs, numerical experiments are necessary with full physical representations that establish the sensitivity of the MJO to various processes and that validate these processes with observations. At present, a complete understanding of the complex interactive processes involved in the initiation and maintenance of the ISO and a faithful simulation of it by GCMs remain elusive.

10.8 ACKNOWLEDGMENTS

The author thanks anonymous reviewers and Drs. R. Madden, B. Mapes, D. Waliser, K.-M. Lau, and P. V. Joseph for their critical comments on an earlier version of the manuscript which have contributed to a significant improvement of the paper. This work is supported by Climate Dynamics Program National Science Foundation award ATM03-29531.

10.9 REFERENCES

Anderson, J. R. (1987) Response of the tropical atmosphere to low-frequency thermal forcing. *J. Atmos. Sci.*, **44**, 676–686.

Anderson, J. R. and D. E. Stevens (1987) Presence of linear wavelike modes in a zonally symmetric model of the tropical atmosphere. *J. Atmos. Sci.*, **44**, 2115–2117.

Annamalai, H. and J. M. Slingo (2001) Active/break cycles: Diagnosis of the intraseasonal variability of the Asian Summer Monsoon. *Clim. Dyn.*, **18**, 85–102.

Arakawa, A. and W. H. Schubert (1974) Interaction of a cumulus cloud ensemble with the large-scale environment. Part I: *J. Atmos. Sci.*, **31**, 674–701.

Betts, A. K. (1986) New convective adjustment scheme. Part 1: Observational and theoretical basis. *Quart. J. Roy. Meteor. Soc.*, **112**, 677–691.

Betts, A. K. and M. J. Miller (1986) New convective adjustment scheme. Part 2: Single column tests using GATE wave, BOMEX, ATEX, and Arctic air-mass data sets. *Quart. J. Roy. Meteor. Soc.*, **112**, 693–709.

Blackadar, A. K. and H. Tenneker (1968) Asymptotic similarity in neutral barotropic planetary boundary layer. *J. Atmos. Sci.*, **225**, 1015–1020.

Blade, I. and D. L. Hartmann (1993) Tropical intraseasonal oscillations in a simple nonlinear model. *J. Atmos. Sci.*, **50**, 2922–2939.

Brown, R. G. and C. S. Bretherton (1995) Tropical wave instabilities: convective interaction with dynamics using the Emanuel convective parameterization. *J. Atmos. Sci.*, **52**, 67–82.

Chang, C.-P. (1977) Some theoretical problems of the planetary-scale monsoons. *Pure and Appl. Geophys.*, **115**, 1089–1109.

Chang, C.-P. and H. Lim (1988) Kelvin wave-CISK: A possible mechanism for the 30–50 day oscillations. *J. Atmos. Sci.*, **45**, 1709–1720.

Chao, W. C. (1987) On the origin of the tropical intraseasonal oscillation. *J. Atmos. Sci.*, **44**, 1940–1949.

Chao, W. C. (1995) A critique of wave-CISK as an explanation for the 40–50 day tropical intraseasonal oscillation. *J. Meteor. Soc. Jap.*, **73**, 677–684.

Chao, W. C. and L. Deng (1998) Tropical intraseasonal oscillation, super cloud clusters, and cumulus convection schemes. Part II: 3D aquaplanet simulations. *J. Atmos. Sci.*, **55**, 690–709.

Charney, J. G. and A. Eliassen (1964) On the Growth of the Hurricane Depression. *J. Atmos. Sci.*, **21**, 68–75.

Chen, T.-C, and M. Murakami (1988) The 30–50 day variation of convective activity over the western Pacific Ocean with the emphasis on the northwestern region. *Mon. Wea. Rev.*, **116**, 892–906.

Chen, S. S., R. A. Houze Jr., and B. E. Mapes (1996) Multiscale variability of deep convection in relation to large-scale circulation during TOGA COARE. *J. Atmos. Sci.*, **53**, 1380–1409.

Cho, H.-R. and D. Pendlebury (1997) Wave CISK of equatorial waves and the vertical distribution of cumulus heating. *J. Atmos. Sci.*, **54**, 2429–2440.

Chou, S.-H., C.-L. Shie, R. M. Atlas, and J. Ardizzone (1995) The December 1992 westerly wind burst and its impact on evaporation determined from SSMI data. *Proc. Int. Scientific Conf. on the Tropical Ocean Global Atmosphere Program, Melbourne, Australia, World Meteor. Org., 489–493.*

Cubukcu, N. and T. N. Krishnamurti (2002) Low-frequency controls on the thresholds of sea surface temperature over the western tropical Pacific. *J. Climate*, **15**, 1626–1642.

Davey, M. K. and A. E. Gill (1987) Experiments on tropical circulation with a simple moist model. *Quart. J. Roy. Meteor. Soc.*, **113**, 1237–1269.

Deser, C. (1993) Diagnosis of the surface momentum balance over the tropical Pacific Ocean. *J. Climate*, **6**, 64–74.

Drbohlav, H.-K. L. and B. Wang (2004) Mechanism of the northward propagating intraseasonal oscillation in the south Asian monsoon region: Results from a zonally averaged model. *J. Climate*, in press.

Dunkerton, T. J. and F. X. Crum (1991) Scale selection and propagation of wave-CISK with conditional heating. *J. Meteor. Soc. Jap.*, **69**, 449–458.

Eliassen, A. (1971) On the Ekman layer in a circular vortex. *J. Meteor. Soc. Jap.*, **49**(special issue), 784–789.

Emanuel, K. A. (1987) Air–sea interaction model of intraseasonal oscillations in the Tropics. *J. Atmos. Sci.*, **44**, 2324–2340.

Emanuel, K. A. (1993) The effect of convective response time on WISHE modes. *J. Atmos. Sci.*, **50**, 1763–1776.

Fasullo, J., and P. J. Webster (1995) Aspects of ocean/atmosphere interaction during westerly wind bursts. *Proc. Int. Scientific Conf. on the Tropical Ocean Global Atmosphere Program, Melbourne, Australia, World Meteor. Org., 39–43.*

Ferranti, L., J. M. Slingo, T. N. Palmer, and B. J. Hoskins (1997) Relations between interannual and intraseasonal monsoon variability as diagnosed from AMIP integrations. *Quart. J. Roy. Meteor. Soc.*, **123**, 1323–1357.

Flatau, M., P. J. Flatau, P. Phoebus, and P. P. Niiler (1997) The feedback between equatorial convection and local radiative and evaporative processes: The implication for intraseasonal oscillations. *J. Atmos. Sci.*, **54**, 2373–2386.

Fu, X. and B. Wang (2004) Differences of boreal summer intraseasonal oscillations simulated in an atmosphere–ocean coupled model and an atmosphere-only model. *J. Climate*, **17**, 1263–1271.

Fu, X., B. Wang, T. Li, and J. P. McCreary (2003) Coupling between northward propagating, intraseasonal oscillations and sea-surface temperature in the Indian Ocean. *J. Atmos. Sci.*, **60**, 1733–1753.

Gill, A. E. (1980) Some simple solutions for heat-induced tropical circulation. *Quart. J. Roy. Meteor. Soc.*, **106**, 447–462.

Goswami, B. N. and J. Shukla (1984) Quasi-periodic oscillations in a symmetric general circulation model. *J. Atmos. Sci.*, **41**, 20–37.

Goswami, P. and V. Mathew (1994) A mechanism of scale selection in tropical circulation at observed intraseasonal frequencies. *J. Atmos. Sci.*, **51**, 3155–3166.

Grabowski, W. W. (2003) MJO-like coherent structures: Sensitivity simulations using the Cloud-Resolving Convection Parameterization (CRCP). *J. Atmos. Sci.*, **60**, 847–864.

Gualdi, S., A. Navarra, and M. Ficher (1999) The tropical intraseasonal oscillation in a coupled ocean–atmosphere general circulation model. *Geophys. Res. Lett.*, **26**, 2973–2976.

Hayashi, Y. (1970) A theory of large scale equatorial waves generated by condensation heat and accelerating the zonal wind. *J. Meteor. Soc. Jap.*, **48**, 140–160.

Hayashi, Y. and S. Miyahara (1987) Three-dimensional linear response model of the tropical intraseasonal oscillation. *J. Meteor. Soc. Jap.*, **65**, 843–852.

Hayashi, Y. Y. and A. Sumi (1986) 30–40-day oscillations simulated in an "aqua planet" model. *J. Meteor. Soc. Jap.*, **64**, 451–467.

Hendon, H. H. (1988) Simple model of the 40–50 day oscillation. *J. Atmos. Sci.*, **45**, 569–584.

Hendon, H. H. (2000) Impact of air-sea coupling on the Madden–Julian Oscillation in a general circulation model. *J. Atmos. Sci.*, **57**, 3939–3952.

Hendon, H. H. and M. L. Salby (1994) The life cycle of the Madden–Julian Oscillation. *J. Atmos. Sci.*, **51**, 2225–2237.

Hirst, A. C. and K.-M. Lau (1990) Intraseasonal and interannual oscillations in coupled ocean–atmosphere models. *J. Climate*, **3**, 713–725.

Hoskins, B. J. and M. J. Rodwell (1995) A model of the Asian summer monsoon. Part I: The global scale. *J. Atmos. Sci.*, **52**, 1329–1340.

Houze, R. A., S. S. Chen, D. K. Kingsmill, Y. Serra, and S. E. Yuter (2000) Convection over the Pacific warm pool in relation to the atmospheric Kelvin–Rossby wave. *J. Atmos. Sci.*, **57**, 3058–3089.

Hsu, H.-H. and C. H. Weng (2001) Northwestward propagation of the intraseasonal oscillation in the western north Pacific during the Boreal Summer: Structure and mechanism. *J. Climate*, **14**, 3834–3850.

Hsu, H.-H., B. J. Hoskins, and F.-F. Jin (1990) The 1985/86 intraseasonal oscillation and the role of the extratropics. *J. Atmos. Sci.*, **47**, 823–839.

Hu, Q. and D. A. Randall (1994) Low-frequency oscillations in radiative–convective systems. *J. Atmos. Sci.*, **51**, 1089–1099.

Hu, Q. and D. A. Randall (1995) Low-frequency oscillations in radiative–convective systems. Part II: An idealized model. *J. Atmos. Sci.*, **52**, 478–490.

Inness, P. M. and J. M. Slingo (2003) Simulation of the Madden–Julian Oscillation in a coupled general circulation model. Part I: Comparison with observations and an atmospheric only GCM. *J. Climate*, **16**, 345–364.

Inness, P. M., J. M. Slingo, E. Guilyardi, and C. Jeffrey (2003) Simulation of the Madden–Julian Oscillation in a coupled general circulation model. Part II: The role of the basic state. *J. Climate*, **16**, 365–382.

Itoh, H. (1989) The mechanism for the scale selection of tropical intraseasonal oscillations. Part I: Selection of wavenumber 1 and the three-scale structure. *J. Atmos. Sci.*, **46**, 1779–1798.

Jiang, X., T. Li, and B. Wang (2004) Structures and mechanisms of the northward propagating boreal summer intraseasonal oscillation. *J. Climate*, in press.

Johnson, R., T. M. Rickenbarch, S. A. Rutledge, P. E. Ciesielski and W. H. Schubert (1999) Trimodal characteristics of tropical convection. *J. Climate*, **12**, 2397–2417.

Jones, C. and B. C. Weare (1996) The role of low-level moisture convergence and ocean latent heat fluxes in the Madden–Julian Oscillation: An observational analysis using ISCCP data and ECMWF analyses. *J. Climate*, **9**, 3086–3104.

Jones, C., D. E. Waliser and C. Gautier (1998) The influence of the Madden–Julian Oscillation on ocean surface heat fluxes and sea surface temperature. *J. Climate*, **11**, 1057–1072.

Kemball-Cook, S. and B. Wang (2001) Equatorial waves and air–sea interaction in the Boreal summer intraseasonal oscillation. *J. Climate*, **14**, 2923–2942.

Kemball-Cook, S., B. Wang, and X. Fu (2002) Simulation of the intraseasonal oscillation in ECHAM4 Model: The impact of coupling with an ocean model. *J. Atmos. Sci.*, **59**, 1433–1453.

Kemball-Cook, S. and B. C. Weare (2001) The onset of convection in the Madden–Julian Oscillation. *J. Climate*, **14**, 780–793.

Kikuchi, K. and Y. N. Takayabu (2004) Equatorial circumnavigation of moisture signal associated with the Madden–Julian Oscillation (MJO) during the boreal winter. *J. Meteor. Soc. Jap.*, in press.

Knutson, T. R., K. M. Weickmann, and J. E. Kutzbach (1986) Global-scale intraseasonal oscillations of outgoing longwave radiation and 250 mb zonal wind during northern hemisphere summer. *Mon. Wea. Rev.*, **114**, 605–623.

Krishnamurti, T. N. and D. Subrahmanyam (1982) The 30–50 day mode at 850 mb during MONEX. *J. Atmos. Sci.*, **39**, 2088–2095.

Krishnamurti, T. N., D. K. Oosterhof, and A. V. Mehta (1988) Air–sea interaction on the time scale of 30 to 50 days. *J. Atmos. Sci.*, **45**, 1304–1322.

Krishnamurti, T. N., P. K. Jayakumar, J. Sheng, N. Surgi and A. Kumar (1985) Divergent circulations on the 30 to 50 day time scale. *J. Atmos. Sci.*, **42**, 364–375.

Krishnamurti, T. N., D. R. Chakraborty, N. Cubukcu, L. Stefanova, and T. S. V. Kumar (2003) A mechanism of the Madden–Julian Oscillation based on interactions in the frequency domain. *Quart. J. Roy. Meteor. Soc.*, **129**, 2559–2590.

Krishnan, R., C. Zhang, and M. Sugi (2000) Dynamics of breaks in the Indian summer monsoon. *J. Atmos. Sci.*, **57**, 1354–1372.

Kuma, K.-I. (1994) The Madden–Julian Oscillation and tropical disturbances in an aqua-planet version of JMA global model with T63 and T159 resolution. *J. Meteor. Soc. Jap.*, **72**, 147–172.

Kuo, H. L. (1974) Further studies of the parameterization of the influence of cumulus convection on large-scale flow. *J. Atmos. Sci.*, **31**, 1232–1240.

Lau, K. H. and N.-C. Lau (1990) Observed structure and propagation characteristics of tropical summertime synoptic scale disturbances. *Mon. Wea. Rev.*, **118**, 1888–1913.

Lau, K. M. and P. H. Chan (1985) Aspects of the 40–50 day oscillation during northern winter as inferred from OLR. *Mon. Wea. Rev.*, **113**, 1889–1909.

Lau, K. M. and P. H. Chan (1986) Aspects of the 40–50 day oscillation during northern summer as inferred from OLR. *Mon. Wea. Rev.*, **114**, 1354–1367.

Lau, K. M. and L. Peng (1987) Origin of low-frequency (intraseasonal) oscillations in the tropical atmosphere. Part I: Basic theory. *J. Atmos. Sci.*, **44**, 950–972.

Lau, K. M. and L. Peng (1990) Origin of low frequency (intraseasonal) oscillations in the tropical atmosphere. Part III: Monsoon dynamics. *J. Atmos. Sci.*, **47**, 1443–1462.

Lau, K. M., and C. H. Sui (1997) Mechanisms of short-term sea surface temperature regulation: Observations during TOGA-COARE. *J. Climate*, **10**, 465–472.

Lau, K. M., L. Peng, L. C. H. Sui., and T. Nakazawa (1989) Dynamics of super cloud clusters, westerly wind bursts, 30–60 day oscillations and ENSO: A unified view. *J. Meteor. Soc. Jap.*, **67**, 205–219.

Lau, K. M. and H.-T. Wu (2003) Warm rain processes over the tropical ocean and climate implications. *Geophys. Res. Lett.*, **30**(24), 2290, doi:10.1029/2003GL018567.

Lau, N.-C. and K. M. Lau (1986) The structure and propagation of intraseasonal oscillation appearing in a GFDL general circulation model. *J. Atmos. Sci.*, **43**, 2023–2047.

Lau, N.-C., I. M. Held and J. D. Neelin (1988) The Madden–Julian Oscillation in an idealized GCM model. *J. Atmos. Sci.*, **45**, 3810–3832.

Lawrence, D. M. and P. J. Webster (2002) The boreal summer intraseasonal oscillation: Relationship between northward and eastward movement of convection. *J. Atmos. Sci.*, **59**, 1593–1606.

Lee, M. I., I. S. Kang, and B. E. Mapes (2003) Impacts of cumulus convecton parameterization on aqua-planet AGCM simulations of tropical intraseasonal variability. *J. Meteor. Soc. Jap.*, **81**, 963–992.

Lee, M. J., I. S. Kang., J. K. Kim, and B. E. Mapes (2001) Influence of cloud-radiation interaction on simulating tropical intraseasonal oscillation with an atmospheric general circulation model. *J. Geophys. Res.*, **106**, 14219–14233.

Li, T. and B. Wang (1994) A thermodynamic equilibrium climate model for monthly mean surface winds and precipitation over the tropical Pacific. *J. Atmos. Sci.*, **51**, 1372–1385.

Li, X. and H.-R. Cho (1997) Development and propagation of equatorial waves. *Adv. Atmos. Sci. China*, **14**, 323–338.

Lim, H., T. K. Lim, and C.-P Chang (1990) Reexamination of Wave-CISK theory: Existence and properties of nonlinear Wave-CISK modes. *J. Atmos. Sci.*, **47**, 3078–3091.

Lin, J. and B. E. Mapes (2004) Radiation budget of the tropical intraseasonal oscillation. *J. Atmos. Sci.*, **61**, 2050–2062.

Lin, J., B. E. Mapes, M. Zhang, and M. Newman (2004) Stratiform precipitation, vertical heating profiles, and the Madden–Julian Oscillation. *J. Atmos. Sci.*, **61**, 296–309.

Lin, J. W.-B., J. Neelin, and N. Zeng (2000) Maintenance of tropical intraseasonal variability: Impact of evaporation-wind feedback and midlatitude storms. *J. Atmos. Sci.*, **57**, 2793–2823.

Lin, X. and R. H. Johnson (1996) Kinematic and thermodynamic characteristics of the flow over the western Pacific warm pool during TOGA-COARE. *J. Atmos. Sci.*, **53**, 695–715.

Lindzen, R. S. (1974) Wave-CISK and tropical spectra. *J. Atmos. Sci.*, **31**, 1447–1449.
Madden, R. A. (1986) Seasonal variations of the 40–50 day oscillation in the tropics. *J. Atmos. Sci.*, **43**, 3138–3158.
Madden, R. A. and P. R. Julian (1971) Detection of a 40–50 day oscillation in the zonal wind in the tropical Pacific. *J. Atmos. Sci.*, **28**, 702–708.
Madden, R. A. and P. R. Julian (1972) Description of global-scale circulation cells in the tropics with a 40–50 day period. *J. Atmos. Sci.*, **29**, 1109–1123.
Madden, R. A. and P. R. Julian (1994) Observations of the tropical 40–50 day Oscillation-review. *Mon. Wea. Rev.*, **122**, 814–837.
Majda, A. J. and R. Klein (2003) Systematic Multiscale Models for the Tropics. *J. Atmos. Sci.*, **60**, 393–408.
Majda, A. J. and J. A. Biello (2004) A multiscale model for tropical intraseasonal oscillations. *Proc. Nat. Acad. Sci.*, **101**, 4736–4741.
Maloney, E. D. (2002) An intraseasonal oscillation composite life cycle in the NCAR CCM3.6 with modified convection. *J. Climate*, **15**, 964–982.
Maloney, E. D. and D. L. Hartmann (1998) Frictional moisture convergence in a composite life cycle of the Madden–Julian Oscillation. *J. Climate*, **11**, 2387–2403.
Maloney, E. D. and D. L. Hartmann (2001) The sensitivity of intraseasonal variability in the NCAR CCM3 to changes in convective parameterization. *J. Climate*, **14**, 2015–2034.
Manabe, S., J. Smagorinsky, and R. F. Strickler (1965) Simulated climatology of a general circulation model with a hydrologic cycle. *Mon. Wea. Rev.*, **93**, 769–798.
Mapes, B. E. (2000) Convective inhibition, subgrid-scale triggering energy, and stratiform instability in a toy tropical wave model. *J. Atmos. Sci.*, **57**, 1515–1535.
Matsuno, T. (1966) Quasigeostrophic motions in the equatorial area. *J. Meteor. Soc. Jap.*, **44**, 25–43.
Matthews, A. J. (2000) Propagation mechanisms for the Madden–Julian Oscillation. *Quart. J. Roy. Meteor. Soc.*, **126**, 2637–2651.
Matthews, A. J., J. M. Slingo, B. J. Hoskins, and P. M. Inness (1999) Fast and slow Kelvin waves in the Madden–Julian Oscillation of a GCM. *Quart. J. Roy. Meteor. Soc.*, **125**, 1473–1498.
Mehta, A. V. and E. A. Smith (1997) Variability of radiative cooling during the Asian summer monsoon and its influence on intraseasonal waves. *J. Atmos. Sci.*, **54**, 941–966.
Moncrieff, M. W., and C. Liu (1999) Convective initiation by density currents: Role of convergence, shear and dynamicall organization. *Mon. Wea. Rev.*, **127**, 2455–2464.
Moskowitz, B. M. and C. S. Bretherton (2000) An analysis of frictional feedback on a moist equatorial Kelvin mode. *J. Atmos. Sci.*, **57**, 2188–2206.
Milliff, R. F. and R. A. Madden (1996) The existence and vertical structure of the fast, eastward-moving disturbances in the equatorial troposphere. *J. Atmos. Sci.*, **53**, 586–597.
Murakami, T. (1980) Empirical orthogonal function analysis of satellite observed outgoing longwave radiation during summer. *Mon. Wea. Rev.*, **108**, 205–222.
Murakami, T., B. Wang, and S. W. Lyons (1992) Summer monsoons over the Bay of Bengal and the eastern North Pacific. *J. Meteor. Soc. Jap.*, **70**, 191–210.
Murphree, T. and H. van den Dool (1988) Calculating winds from time mean sea level pressure fields. *J. Atm. Sci.*, **45**, 3269–3281.
Nakazawa, T. (1988) Tropical super clusters within intraseasonal variations over the western Pacific. *J. Meteor. Soc. Jap.*, **66**, 823–839.
Neelin, J. D. (1990) A hybrid coupled general circulation model for El Niño studies. *J. Atmos. Sci.*, **47**, 674–693.

Neelin, J. D. and J.-Y. Yu (1994) Modes of tropical variability under convective adjustment and the Madden–Julian Oscillation. Part I: Analytical theory. *J. Atmos. Sci.*, **51**, 1876–1894.

Neelin, J. D., I. M. Held, and K. H. Cook (1987) Evaporation-wind feedback and low-frequency variability in the tropical atmosphere. *J. Atmos. Sci.*, **44**, 2341–2348.

Nitta, T. (1987) Convective activities in the tropical western Pacific and their impact on the Northern Hemisphere summer monsoon. *J. Meteor. Soc. Jap.*, **65**, 373–390.

Ohuchi, K. and M. Yamasaki (1997) Kelvin wave-CISK controlled by surface friction: A possible mechanism of super cloud cluster. *J. Meteor. Soc. Jap.*, **75**, 497–511.

Ooyama, K. (1964) A dynamic model for the study of tropical cyclone development. *Geofits. Int.* (Mexico), **4**, 187–198.

Pedlosky, J. (1979) *Geophysical Fluid Dynamics*. Springer Verlag, New York, 710 pp.

Peng, L., C.-H. Sui, K.-M. Lau, and W. K. Tao (2001) Genesis and evolution of hierarchical cloud clusters in a two dimentional cumulus resolving model. *J. Atmos. Sci.*, **58**, 877–895.

Philander, S. G. H., T. Yamagata, and R. C. Pacanowski (1984) Unstable air–sea interactions in the Tropics. *J. Atmos. Sci.*, **41**, 604–613.

Randall, D. A., Harshvardhan, D. A. Dazlich, and T. G. Corsetti (1989) Interactions among radiation, convection, and large-scale dynamics in a general circulation model. *J. Atmos. Sci.*, **46**(13), 1943–1970.

Raymond, D. J. (2001) A new model of the Madden–Julian Oscillation. *J. Atmos. Sci.*, **58**, 2807–2819.

Rui, H. and B. Wang (1990) Development characteristics and dynamic structure of tropical intraseasonal convection anomalies. *J. Atmos. Sci.*, **47**, 357–379.

Salby, M. L., R. R. Garcia, and H. H. Hendon (1994) Planetary-scale circulations in the presence of climatological and wave-induced heating. *J. Atmos. Sci.*, **51**, 2344–2367.

Sengupta, D., B. N. Goswami, and R. Senan (2001) Coherent intraseasonal oscillations of ocean and atmosphere during the Asian summer monsoon. *Geophys. Res. Lett.*, **28**, 4127–4130.

Shinoda, T. and H. H. Hendon (1998) Mixed layer modeling of intraseasonal variability in the tropical western Pacific and Indian Ocean. *J. Climate*, **11**, 2668–2685.

Shinoda, T., H. H. Hendon, and J. Glick (1998) Intraseasonal variability of surface fluxes and sea surface temperature in the tropical Western Pacific and Indian Oceans. *J. Climate*, **11**, 1685–1702.

Short, D. and K. Nakamura (2000) TRMM radar observations of shallow precipitation over tropical oceans. *J. Climate*, **13**, 4107–4124.

Sikka, D. R. and S. Gadgil (1980) On the maximum cloud zone and the ITCZ over Indian longitudes during the southwest monsoon. *Mon. Wea. Rev.*, **108**, 1840–1853.

Slingo, A. and J. M. Slingo (1988) Response of a general circulation model to cloud long-wave radiative forcing. Introduction and initial experiments. *Quart. J. Roy. Meteor. Soc.*, **114**, 1027–1062.

Slingo, A., J. M. Boyle, J. S. Ceron, J.-P. Dix, M. Dugas, B. Ebisuzaki, W. Fyfe, J. Gregory, D. Gueremy, and J.-F. Hack (1996) Intraseasonal oscillations in 15 atmospheric general circulation models: Results from an AMIP diagnostic subproject. *Clim. Dyn.*, **12**, 325–357.

Slingo, A., P. Inness, R. Neale, S. Woolnough, and G.-Y. Yang (2003) Scale interaction on diurnal to seasonal timescales and their relevance to model systematic errors. *Geophys. Ann.*, **46**, 139–155.

Slingo, A., and R. A. Madden (1991) Characteristics of the tropical intraseasonal oscillation in the NCAR community climate model. *Quart. J. Roy. Meteor. Soc.*, **117**, 1129–1169.

Sobel, A. H. and H. Gildor (2003) A simple time-dependent model of SST hot spas. *J. Climate*, **16**, 3978–3992.

Sperber, K. R. (2003) Propagation and vertical structure of the Madden–Julian Oscillation. *Mon. Wea. Rev.*, **131**, 3018–3037.

Sui, C.-H. and K.-M. Lau (1989) Origin of low-frequency (intraseasonal) oscillations in the tropical atmosphere. Part 2: Structure and propagation of mobile wave-CISK modes and their modification by lower boundary forcings. *J. Atmos. Sci.*, **46**, 37–56.

Takahashi, M. (1987) Theory of the slow phase speed of the intraseasonal oscillation using the wave-CISK. *J. Meteor. Soc. Jap.*, **65**, 43–49.

Ting, M. (1994) Maintenance of northern summer stationary waves in a GCM. *J. Atmos. Sci.*, **51**, 3286–3308.

Tompkins, A. M. (2001) On the relationship between tropical convection and sea surface temperature. *J. Atmos. Sci.*, **58**, 529–545.

Waliser, D. E., K. M. Lau, and J.-H. Kim (1999) The influence of coupled sea surface temperatures on the Madden–Julian Oscillation: A model perturbation experiment. *J. Atmos. Sci.*, **56**, 333–358.

Waliser, D. E., K. M. Lau, W. Stern, and C. Jones (2003a) Potential predictability of the Madden–Julian Oscillation. *Bull. Amer. Meteor. Soc.*, **84**, 33–50.

Waliser, D. E., K. Jin, I. S. Kang, W. F. Stern, S. D. Schubert, M. L. Wu, K. M. Lau, M. I. Lee, J. Shukla, V. Krishnamurthy, *et al.* (2003b) AGCM Simulations of intraseasonal variability associated with the Asian summer monsoon. *Clim. Dyn.*, **21**, 423–446.

Wang, B. (1988a) Dynamics of tropical low-frequency waves: an analysis of the moist Kelvin wave. *J. Atmos. Sci.*, **45**, 2051–2065.

Wang, B. (1988b) Comments on "An air–sea interaction model of intraseasonal oscillation in the tropics". *J. Atmos. Sci.*, **45**, 3521–3525.

Wang, B. and J. K. Chen (1989) On the zonal-scale selection and vertical structure of equatorial intraseasonal waves. *Quart. J. Roy. Meteor. Soc.*, **115**, 1301–1323.

Wang, B. and T. Li (1993) A simple tropical atmosphere model of relevance to short-term climate variations. *J. Atmos. Sci.*, **50**, 260–284.

Wang, B. and T. Li (1994) Convective interaction with boundary-layer dynamics in the development of a tropical intraseasonal system. *J. Atmos. Sci.*, **51**, 1386–1400.

Wang, B. and H. Rui (1990a) Dynamics of the coupled moist Kelvin–Rossby wave on an equatorial beta-plane. *J. Atmos. Sci.*, **47**, 397–413.

Wang, B. and H. Rui (1990b) Synoptic climatology of transient tropical intraseasonal convection anomalies: 1975–1985. *Meteorol. Atmos. Phys.*, **44**, 43–61.

Wang, B. and X. Xie (1996) Low-frequency equatorial waves in vertically sheared zonal flow. Part I: Stable waves. *J. Atmos. Sci.*, **53**, 449–467.

Wang, B and X. Xie (1997) A model for the boreal summer intraseasonal oscillation. *J. Atmos. Sci.*, **54**, 72–86.

Wang, B. and X. Xie (1998) Coupled modes of the warm pool climate system. Part I: The role of air–sea interaction in maintaining Madden–Julian Oscillation. *J. Atmos. Sci.*, **11**, 2116–2135.

Wang, B. and X. Xu (1997) Northern Hemisphere summer monsoon singularities and climatological intraseasonal oscillation. *J. Climate*, **10**, 1071–1085.

Wang, B. and Y. Xue (1992) Behavior of a moist Kelvin wave packet with nonlinear heating. *J. Atmos. Sci.*, **49**, 549–559.

Wang, B. and Q. Zhang (2002) Pacific-East Asian teleconnection. Part II: How the Philippine Sea anticyclone established during development of El Niño. *J. Climate*, **15**, 3252–3265.

Wang, W. and M. E. Schlesinger (1999) The dependence on convective parametrization of the tropical intraseasonal oscillation simulated by the UIUC 11-layer atmospheric GCM. *J. Climate*, **12**, 1423–1457.

Webster, P. J. (1983) Mechanisms of monsoon low-frequency variability: Surface hydrological effects. *J. Atmos. Sci.*, **40**, 2110–2124.

Webster, P. J. (1994) The role of hydrological processes in ocean–atmosphere interactions. *Rev. Geophys.*, **32**, 427–476.

Weickmann, K. M. (1983) Intraseasonal circulation and outgoing longwave radiation modes during Northern Hemisphere winter. *Mon. Wea. Rev.*, **111**, 1838–1858.

Weller, R. A. and S. P. Anderson (1996) Surface meteorology and air–sea fluxes in the western equatorial Pacific warm pool during the TOGA Coupled Ocean-Atmosphere Response Experiment. *J. Climate*, **9**, 1959–1990.

Wheeler *et al.* (2000)

Woolnough, S. J., J. M. Slingo, and B. J. Hoskins (2000) The relationship between convection and sea surface temperature on intraseasonal timescales. *J. Climate*, **13**, 2086–2104.

Woolnough, S. J., J. M. Slingo, and B. J. Hoskins (2001) The organization of tropical convection by intraseasonal sea surface temperature anomalies. *Quart. J. Roy. Meteor. Soc.*, **127**, 887–907.

Wu, M. L. C., S. Schubert, I. S. Kang, and D. E. Waliser (2002) Forced and free intraseasonal variability over the south Asian monsoon region simulated by 10 AGCMs. *J. Climate*, **15**, 2862–2880.

Wu, Z. (2003) A shallow CISK, deep equilibrium mechanism for the interaction between large scale convection and large scale circulations in the tropics. *J. Atmos. Sci.*, **60**, 377–392.

Xie, S.-P. and A. Kubokawa (1990) On the wave-CISK in the presence of a frictional boundary layer. *J. Meteor. Soc. Jap.*, **68**, 651–657.

Xie, S.-P., A. Kubokawa, and K. Hanawa (1993) Evaporation-wind feedback and the organizing of tropical convection on the planetary scale. Part II: Nonlinear evolution. *J. Atmos. Sci.*, **50**, 3884–3893.

Xie, X. and B. Wang (1996) Low-frequency equatorial waves in vertically sheared zonal flows. Part II: unstable waves. *J. Atmos. Sci.*, **53**, 3589–3605.

Yamagata, T. (1987) Simple moist model relevant to the origin of intraseasonal disturbances in the Tropics. *J. Meteor. Soc. Jap.*, **65**, 153–165.

Yamagata, T. and Y. Hayashi (1984) Simple diagnostic model for the 30–50 day oscillation in the Tropics. *J. Meteor. Soc. Jap.*, **62**, 709–717.

Yamasaki, M. (1969) Large-scale disturbances in the conditionally unstable atmosphere in low latitudes. *Papers Meteor. Geophys.*, **20**, 289–336.

Yano, J.-I. and K. Emanuel (1991) An improved model of the equatorial troposphere and its coupling with the stratosphere. *J. Atmos. Sci.*, **48**, 377–389.

Yasunari, T. (1979) Cloudiness fluctuations associated with the Northern Hemisphere summer monsoon. *J. Meteor. Soc. Jap.*, **57**, 227–242.

Yasunari, T. (1980) A quasi-stationary appearance of 30–40 day period in the cloudiness fluctuations during the summer monsoon over India. *J. Meteor. Soc. Jap.*, **58**, 225–229.

Zhang, C. D. (1996) Atmospheric intraseasonal variability at the surface in the tropical western Pacific Ocean. *J. Atmos. Sci.*, **53**, 739–758.

Zhang, C. D. and S. P. Anderson (2003) Sensitivity of intraseasonal perturbations in SST to the structure of the MJO. *J. Atmos. Sci.*, **60**, 2196–2207.

Zhang, C. D. and H. H. Hendon (1997) Propagating and standing components of the intraseasonal oscillation in tropical convection. *J. Atmos. Sci.*, **54**, 741–752.
Zhu, B. and B. Wang (1993) The 30–60 day convection seesaw between the tropical Indian and western Pacific Oceans. *J. Atmos. Sci.*, **50**, 184–199.

11

Modeling

J. M. Slingo, P. M. Inness, and K. R. Sperber

11.1 INTRODUCTION

The Madden–Julian Oscillation (MJO) has long been an aspect of the global climate that has provided a challenging test for the climate modeling community. Since the 1980s there have been numerous studies of the simulation of the MJO in atmospheric general circulation models (GCMs), ranging from Hayashi and Golder (1986, 1988) and Lau and Lau (1986), through to more recent studies such as Wang and Schlesinger (1999) and Wu *et al.* (2002). Of course, attempts to reproduce the MJO in climate models have proceeded in parallel with developments in our understanding of what the MJO is and what drives it. In fact, many advances in understanding the MJO have come through modeling studies. In particular, failure of climate models to simulate various aspects of the MJO has prompted investigations into the mechanisms that are important to its initiation and maintenance, leading to improvements both in our understanding of, and ability to simulate, the MJO.

Most of the early studies concentrated on the ability of models to simulate the signal of the MJO in the upper level winds (e.g., Swinbank *et al.*, 1988), partly because these were the fields in which the MJO was originally identified in observations, and partly because the dynamical signal of the MJO has often been more reliable in GCMs than its convective signal. Many quite simple GCMs with coarse resolution were shown to produce a peak at approximately the right frequency in the spectrum of upper tropospheric wind variability, along with many of the characteristics of the observed oscillation (e.g., Slingo and Madden, 1991; Hayashi and Golder, 1993). Furthermore, these studies showed that the simulated oscillation resembled the observed structure of a Kelvin wave coupled to a forced Rossby wave, with the typical baroclinic structure in the vertical (e.g., Knutson and Weickmann, 1987; Sperber *et al.*, 1997; Matthews *et al.*, 1999). However, there remained some substantial deficiencies; in particular, the periodicity of the simulated oscillation tended to be too short, nearer 25–30 days than 40–50 days,

W. K. M. Lau and D. E. Waliser (eds), *Intraseasonal Variability in the Atmosphere–Ocean Climate System.*

and the eastward propagation of the convective anomaly across the warm pool of the Indian and West Pacific Oceans was poorly simulated.

In the 1990s, following the more limited intercomparison of Park *et al.* (1990), a comprehensive study of the ability to simulate the MJO by the then state-of-the-art atmospheric models was carried out by Slingo *et al.* (1996) as part of the first Atmospheric Model Intercomparison Project (AMIP I; Gates *et al.*, 1999). In that study, the following key questions for the simulation of the MJO were addressed:

- Can characteristics of the convective parameterization, such as the vertical profile of the heating and the closure (e.g., moisture convergence), be identified, which might influence the existence of intraseasonal variability (ISV)?
- How does the intraseasonal oscillation (ISO) depend on aspects of a model's basic climate?
- What seasonal and interannual variability (IAV) in the activity of the MJO is simulated? How does it compare with reality?

Slingo *et al.* (1996) showed that, although there were GCMs that could simulate some aspects of the MJO, all the models in their survey were deficient in some respect. In particular, the period of the oscillation was too fast in many models, and the amplitude of the MJO signal in the upper level winds was often too weak. No model was able to capture the pronounced spectral peak associated with the observed MJO. In reality, the MJO is strongest and most coherent in northern winter/spring, whereas many models showed no seasonality for the MJO. Furthermore, as the envelope of enhanced convection associated with the variations in the upper wind field develops over the Indian Ocean and propagates eastwards into the West Pacific, the propagation speed of the oscillation is observed to slow down. Many models failed to capture this geographical dependence. In an extension of the study of Slingo *et al.* (1996), Sperber *et al.* (1997) focused on the most skilful models in AMIP I, and showed that, at best, the models produced a pattern of standing oscillations, with convective anomalies developing and decaying over the Indian Ocean on intraseasonal timescales, with out-of-phase oscillations occurring over the West Pacific.

A more recent, limited intercomparison by Wu *et al.* (2002) has shown that models are still unable to reproduce the observed concentration of power at the 40–50-day timescale. However, progress in simulating the MJO is being made. At a workshop on simulation and prediction of subseasonal variability in 2003 (Waliser *et al.*, 2003d), most of the models presented were able to simulate at least some aspects of the MJO. In contrast to the study of Slingo *et al.* (1996), some of the modeling results presented at this workshop showed an MJO that was actually too strong or propagated more slowly than the observed oscillation. It is probably true to say that as our understanding of the MJO increases we are setting our GCMs more stringent tests in terms of what constitutes a "good" MJO simulation. Even so, the questions posed in 1996 by Slingo *et al.* are still very relevant.

The initial focus of this chapter will be on modeling the MJO during northern winter, when it is characterized as a predominantly eastward propagating mode and is most readily seen in observations. Aspects of the simulation of the MJO will be discussed in the context of its sensitivity to the formulation of the atmospheric model, and the increasing evidence that it may be a coupled ocean–atmosphere phenomenon. Later, we will discuss the challenges regarding the simulation of boreal summer ISV, which is more complex since it is a combination of the eastward propagating MJO and the northward propagation of the tropical convergence zone. Finally some concluding remarks on future directions in modeling the MJO and its relationship with other timescales of variability in the tropics will be made.

11.2 MODELING THE MJO IN BOREAL WINTER

11.2.1 Interannual and decadal variability of the MJO

Slingo *et al.* (1996) introduced an index of MJO activity based on the near-equatorial zonal wind at 200 hPa, to provide a preliminary measure of MJO variability in models and to describe the interannual and decadal variations in MJO activity (Figure 11.1). This index uses the fact that the MJO projects on to the zonal mean of the equatorial zonal wind component through its Kelvin and Rossby wave characteristics (Slingo *et al.*, 1999).

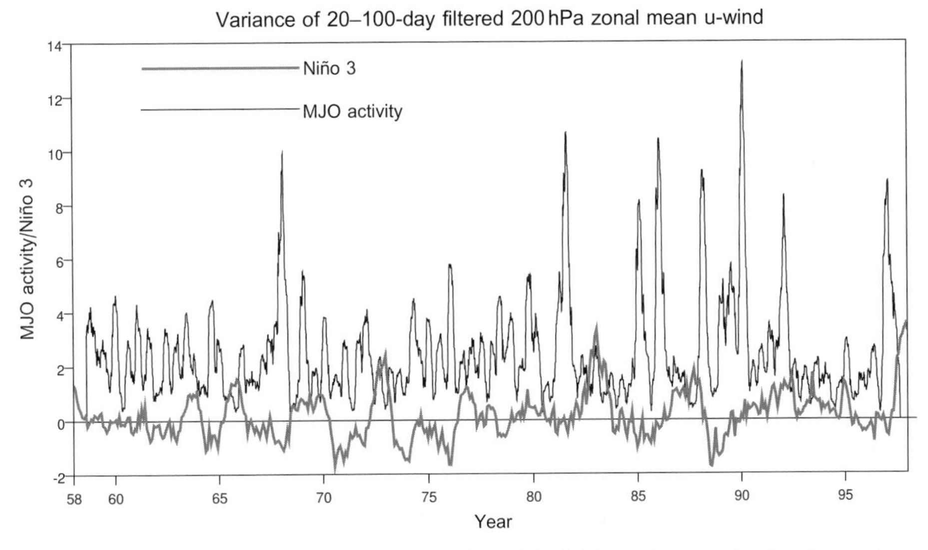

Figure 11.1. Interannual variability in the activity of the MJO as depicted by the time series of the variance ($m^2 s^{-2}$) of the 20–100-day band-pass filtered zonal mean zonal wind from the recent ECMWF Reanalysis for 1958–1997 (ERA-40). A 100-day running mean has been applied to the variance time series. The lower, grey curve is the sea surface temperature anomaly (SSTA) (K) for the Niño-3 region (5°N–5°S, 90°W–150°W). See Slingo *et al.* (1990) for more details on the calculation of the MJO index.

This index also shows that there is substantial IAV in the activity of the MJO, which Slingo *et al.* (1999) and Hendon *et al.* (1999) found was not strongly related to sea surface temperatures (SSTs) (Figure 11.1 also includes the time series of the Niño-3 region sea surface temperature anomaly (SSTA)). This lack of predictability was also seen in a 4-member ensemble of 45-year integrations with the Hadley Centre climate model (HADAM2a), forced by observed SSTs for 1949–1993, suggesting that the interannual behavior of the MJO is not controlled by the phase of El Niño and would appear to be mainly chaotic in character. In a related study, Gualdi *et al.* (1999) also showed that only with a very large ensemble was it possible to detect any predictability for the interannual behavior of the MJO. These results may have important implications for the predictability of the coupled system through the influence of the MJO on westerly wind activity and hence on the development and amplification of El Niño (e.g., McPhaden, 1999; Lengaigne *et al.*, 2004).

Also evident in Figure 11.1 is a marked decadal change in the activity of the MJO. Prior to the mid-1970s, the activity of the MJO was consistently lower than during the latter part of the record. This may be related to either inadequate data coverage, particularly over the tropical Indian Ocean prior to the introduction of satellite observations, or to the real effects of a decadal timescale warming in the tropical SSTs. However, as described by Slingo *et al.* (1999), the ensemble of integrations with the Hadley Centre model were able to reproduce the low frequency, decadal timescale variability of MJO activity seen in Figure 11.1. The activity of the MJO is consistently lower in all realizations prior to the mid-1970s, suggesting that the MJO may indeed become more active as tropical SSTs become warmer with implications for the effects of global warming on the coupled tropical atmosphere–ocean system. Zveryaev (2002) also notes that interdecadal changes in ISV during the Asian summer monsoon. Slingo *et al.* (1999) based their results on the U.S. National Centers for Environmental Protection/National Center for Atmospheric Research (NCEP–NCAR) Reanalyses. The fact that very similar results have been obtained from the more recent European Centre for Medium-range Weather Forecasts (ECMWF) 40-year Reanalysis (ERA-40) as shown in Figure 11.1 adds credence to the decadal variability identified earlier.

11.2.2 Sensitivity to the formulation of the atmospheric model

In the 1980s, the resolution of GCMs was low (typically spectral T21, R15, equivalent to a grid of $\sim 5^\circ$) by comparison with the current generation of models, and much of the early success in simulating an eastward propagating mode was achieved with models whose resolution was not sufficient to resolve tropical synoptic systems. Since the active phase of the MJO is often characterized by smaller scale organized convection associated with tropical synoptic systems, this lack of resolution was considered a possible cause for the errors in the simulation of the MJO. In the early 1990s, Slingo *et al.* (1992) analyzed the tropical variability in high-resolution (spectral T106, $\sim 1^\circ$) simulations with the ECMWF model and showed that the various aspects of tropical synoptic variability, such as easterly waves, could be captured with considerable skill. Their integrations were not long enough, however, to say anything conclusive about the MJO.

In AMIP I, the majority of models were run at resolutions capable of capturing synoptic variability (typically spectral T42, equivalent to a grid of at least 3° and above). However, the results from the study by Slingo *et al.* (1996) suggested that horizontal resolution did not play an important role in determining a model's intraseasonal activity. Even at much higher resolutions, up to as much as T576, recent evidence from ECMWF suggests no improvement in the simulation of the MJO (Jung and Tompkins, 2003). Hence, at this stage there is no clear evidence that increasing the horizontal resolution in the atmospheric model will improve the simulation of the MJO, possibly because of more fundamental errors in representing convection and its interaction with dynamics. Support for this hypothesis has come recently from the studies of Grabowski (2003) and Randall *et al.* (2003) in which the convective parameterization has been replaced by a 2-D cloud-resolving model (CRM) – the "cloud-resolving convective parameterization" or super-parameterization approach. By representing the interaction between the convective clouds and the dynamics more completely, their studies have shown dramatic improvements in the organization of convection on both synoptic and intraseasonal timescales. These are very preliminary results, and although there are some important questions regarding the 2-D nature and orientation of the CRM (the use of which is currently prohibitively expensive), they potentially provide useful insights into fundamental aspects of organized convection in the tropics and how to address sub-grid-scale processes.

Even though there is no compelling evidence to suggest that horizontal resolution is important for the simulation of the MJO, this appears not be the case for vertical resolution. Experiments with the Meteorological Office Unified Model (UM, version HadAM3) using two different vertical resolutions (19 and 30 levels) have shown significant differences in the amount of variability in the tropical upper tropospheric zonal wind component associated with the MJO (Inness *et al.*, 2001; Figure 11.2).

Most of the extra levels were placed in the middle and upper troposphere, decreasing the layer thickness in the mid-troposphere from 100 hPa to 50 hPa, and giving a much better representation of the temperature and humidity structure around the freezing level. The model results suggested a change in the temporal organization of convection which was investigated further using an aqua-planet version of the UM. These experiments, described in detail in Inness *et al.* (2001), showed that when the vertical resolution was increased in the UM, the spectrum of tropical cloud-top heights changed from a bi-modal to a tri-modal distribution, with a third peak in the mid-troposphere, near the freezing level. Associated with periods when these mid-level clouds were dominant, the detrainment from these clouds significantly moistened the mid-troposphere. In comparison, the 19-level version of the model shows no evidence of a tri-modal distribution in convection and no such moistening events.

Many conceptual models of tropical convection are based on a bi-modal cloud distribution, emphasizing shallow "trade wind" or boundary layer cumuli and deep

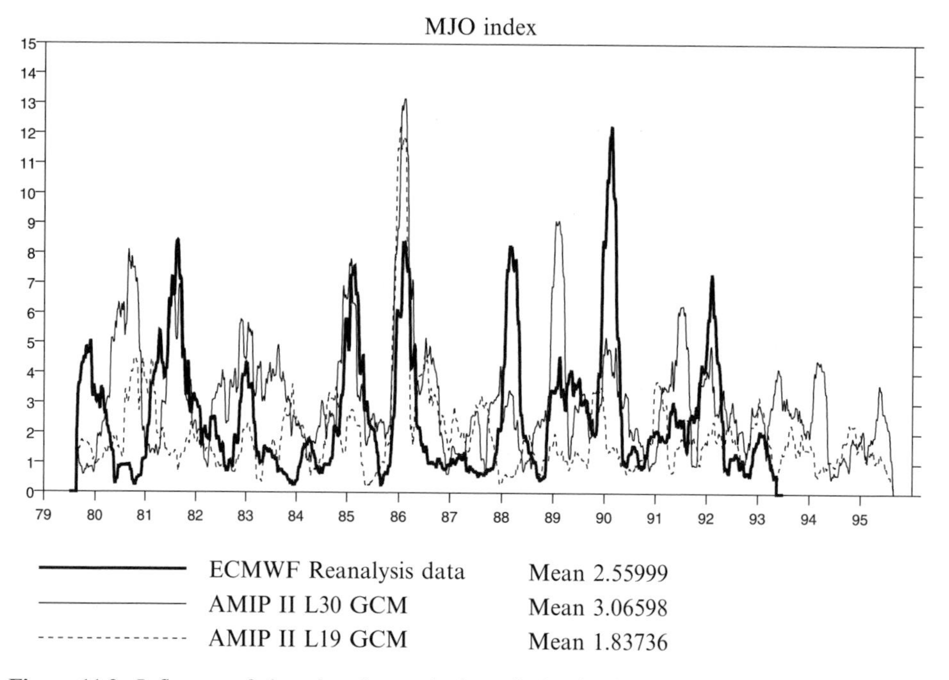

Figure 11.2. Influence of changing the vertical resolution in the Hadley Centre's atmospheric model (HadAM3) on the strength of the MJO as described by the index used in Figure 11.1. Note the increased amplitude of MJO activity in the L30 version of the model and the improved seasonality with respect to the ECMWF Reanalyses.
From Inness *et al.* (2001).

cumulonimbi. However, Tropical Ocean Global Atmosphere–Coupled Ocean–Atmosphere Response Experiment (TOGA–COARE) results have shown the dominance of cumulus congestus clouds, and point to a tri-modal cloud distribution in which the freezing level inversion is the key. Observational studies have shown that, during the suppressed phase of the MJO, tropical convection is dominated by clouds that terminate around the stable layer at the 0°C level (Johnson *et al.*, 1999), and that these clouds provide a source of moisture to the mid-troposphere (Lin and Johnson, 1996). Inness *et al.* (2001) argued that the development of a stable layer around the tropical melting level, which is frequently observed over the tropical oceans, acts to reinforce the transition from the enhanced convective phase to the suppressed phase of the MJO. Subsequently, the moistening of the mid-troposphere during the suppressed phase acts to reinforce the transition back to the active phase. This is consistent with the "recharge–discharge" theory for the MJO proposed by Bladé and Hartmann (1993) in which the MJO timescale may be set by the time it takes for the moist static energy to build up following the decay of the previous convective event. It may be that the recharging of the moist static energy is achieved in part by the injection of moisture into the mid-troposphere by the cumulus congestus clouds that dominate during the suppressed phase of the MJO.

The appearance of these congestus clouds has been postulated as the reason for the improvement in the simulation of the MJO in the 30-level version of the UM. This is shown to be partly due to improved resolution of the freezing level and of the convective processes occurring at this level. However, the results also suggest that convection and cloud microphysics schemes must be able to represent cumulus congestus clouds which, being neither shallow nor deep cumulus as well as often weakly precipitating, tend not to be explicitly represented in current schemes. In addition, this study has highlighted the importance of understanding and modeling the suppressed phase of the MJO; over the last two decades most of the attention has been given, understandably, to the active phase of the MJO, but with limited success. Further evidence of the importance of cumulus congestus in the life cycle of the MJO comes from a theoretical and simple modeling study by Wu (2003). This study presents a "shallow CISK, deep equilibrium" mechanism for the interaction of convection and large-scale circulations in the tropics, emphasizing the role of the heating by congestus clouds as a precursor to the outbreak of deep convection corresponding to the active phase of the MJO.

The results of Inness *et al.* (2001) highlighted the importance of vertical resolution, in line with the study of Tompkins and Emanuel (2000), as well as the need to properly represent the tri-modal structure of tropical convection. The importance of the cumulus congestus stage of tropical convection is being stressed here as a potentially important ingredient for the MJO. This means that vertical resolution in the free troposphere must be adequate to resolve the formation of the freezing level inversion and the cooling associated with melting precipitation.

That the MJO is intimately linked to convection is undeniable, and numerous modeling studies have demonstrated that changes to the convection scheme can produce radical changes in the simulation of the MJO. For example, Slingo *et al.* (1994) replaced the Kuo convection scheme (Kuo, 1974; closed on moisture convergence) by the convective adjustment scheme of Betts and Miller (Betts, 1986; closed on buoyancy) and showed extreme sensitivity in the representation of organized tropical convection at synoptic to intraseasonal timescales, with the Kuo scheme unable to capture realistic levels of tropical variability. This suggested that a dependence of convective activity on moisture convergence might be a factor contributing to the poor simulation of the MJO. This was further supported by Nordeng (1994), who showed that when the moisture convergence dependence of the ECMWF convection scheme was replaced by a buoyancy criterion, there was a marked improvement (i.e., increase) in transient activity in the tropics of the ECMWF model.

More recently, the closure of the convection scheme of the Australian Bureau of Meteorology Research Center's seasonal prediction GCM has been modified from moisture convergence to CAPE relaxation, whereby convective available potential energy is removed over a specified timescale, with a resulting increase in eastward-moving power at MJO frequencies (Section IIB in Waliser *et al.*, 2003d). At a broader level, Slingo *et al.* (1996) also suggested that those models in AMIP I with a reasonable level of intraseasonal activity used convection schemes that were closed on buoyancy rather than moisture supply. However, as Wang and Schlesinger (1999) demonstrated, it is possible to change the strength of the MJO substantially by modifying the particular closure used within the convection scheme, as well as the fundamental design of the convection scheme itself. But as they point out, some configurations of the convection schemes did not produce realistic mean climates, which as will be discussed later, can compromise the simulation of the MJO. Studies such as those of Maloney and Hartmann (2001) and Lee *et al.* (2003) have also demonstrated that considerable changes to the simulation of the MJO can be brought about by modifications to the convective parameterization. In this case, the imposition of a minimum entrainment rate for deep convective plumes in the Arakawa–Schubert convection scheme (Arakawa and Schubert, 1974; Tokioka *et al.*, 1988) in an aqua-planet configuration of the Seoul National University GCM resulted in a much stronger MJO-like signal.

Although there has recently been a move away from convection schemes that are closed on moisture convergence towards those based on buoyancy considerations, difficulties still remain. Many schemes use an equilibrium approach to convection, which assumes that instabilities are removed completely at each time step. Sensitivity experiments with non-equilibrium closures suggest that improvements in the intraseasonal organization of convection can be achieved, but often at the expense of the quality of the mean climate. Indeed, separating the effects of the changes to the convection scheme on the organization of convection, from the effects on the mean climate of the tropics has been notoriously difficult. For example, Inness and Gregory (1997) showed that the inclusion of the vertical transport of momentum by the convection scheme considerably weakened the upper tropospheric signal of the MJO in the UM, possibly due to changes in the basic state winds in tropical latitudes.

Although much of the focus of attention for the simulation of the MJO has been on the convective parameterization, there are other aspects of the physics that deserve attention. For example, a study by Salby *et al.* (1994) has suggested that the oscillation may be very sensitive to boundary layer friction in which the sympathetic interaction between the convection and the large-scale circulation, through the process termed "frictional wave–CISK" (see Chapter 10), can explain many aspects of the observed behavior of the MJO in the eastern hemisphere. Due to frictional effects the surface convergence is shifted some 40°–50° to the east of the heating, towards low pressure and in phase with the temperature anomaly associated with the Kelvin wave. This study also emphasized the importance of the Rossby gyres generated by the heating. In the amplifying phase of the MJO their position is such as to reinforce the moisture convergence to the east of the heating, so providing the necessary conditions for the heating to amplify and propagate eastwards. Salby *et al.* (1994) showed that their solutions were very sensitive to the boundary layer friction, suggesting that this may be an important factor in GCMs, but one that, so far, has not been pursued. However, Sperber *et al.* (1997) investigated the role of wind induced surface heat exchange (WISHE; Emanuel, 1987) and frictional wave–CISK in the MJO as simulated by the most skilful models in AMIP I, and concluded that neither mechanism was represented, at least in atmosphere-only models. On the other hand, Waliser *et al.* (1999) noted that when coupling between the atmosphere and ocean was introduced (see Section 11.2.3), then frictional wave–CISK was enhanced and became an important factor in the improved simulation of the MJO.

With the low-level moisture convergence leading the convection, as suggested by Salby *et al.* (1994), there is a pronounced westward vertical tilt in the divergence, vertical velocity, zonal wind, and specific humidity, as demonstrated by Sperber (2003) and Seo and Kim (2003) using the NCEP–NCAR Reanalysis. The strongest zonal inflow into the convective region occurs in the free troposphere between 600–700 hPa. The conditions to the east of the center of convection promote the eastward propagation of the MJO, while to the west they erode the convection. Thus, free-tropospheric interactions are also an essential component of the MJO that models need to represent. The ability of the models to represent these features will be sensitive to the simulated diabatic heating profile, and thus to the aforementioned sensitivities to the convection scheme and vertical resolution. Unfortunately, such detailed analyses of models are not the norm due to the extensive archive of data required. However, further progress in understanding a models ability to capture the MJO will necessitate more comprehensive model output to become routine.

In a recent paper, Raymond (2001) suggested that cloud–radiation interaction might be important for the simulation of the MJO. Slingo and Madden (1991), in their study of the MJO simulated by the NCAR Community Climate Model, investigated the role of atmospheric cloud long-wave forcing in the behavior of the MJO. They showed that cloud–radiation interaction had little effect on the periodicity of the MJO and its basic characteristics. Without cloud–radiation interaction, the simulated MJO was slightly more regular. However, this issue probably deserves revisiting with the current models that have a more sophisticated representation of cloud microphysics. In fact, this area is indeed being investigated more fully in the context of the "cloud-resolving convective parametrization" approach discussed earlier in this chapter (e.g., Grabowski and Moncrieff, 2002). In this approach, the convective parametrization is replaced by a CRM in each grid column and so the representation of cloud microphysics is far more detailed than in a conventional GCM. Initial results do indicate that the interaction of the clouds and radiation does indeed have a part to play in the large-scale organization of convection.

11.2.3 Modeling the MJO as a coupled ocean–atmosphere phenomenon

One of the biggest advances in modeling the MJO during the last few years has been in the recognition that it almost certainly involves coupling with the ocean, as discussed in Chapter 7 and references cited therein. There is now convincing evidence from observations that the MJO interacts with the upper ocean in such a way for it to be a coupled phenomenon, and which may therefore require an interactive ocean system for its proper simulation.

In a comprehensive analysis of observational and reanalysis data, Woolnough *et al.* (2000) showed that, for the Indian Ocean and West Pacific, a coherent relationship exists between MJO convection, surface fluxes, and SST, in which the SSTs are warmer than normal about 10 days prior to, and east of, the maximum in convective activity (Figure 11.3). As shown in Figure 11.3, this warming is associated with increased solar radiation, reduced surface evaporation, and light winds, which

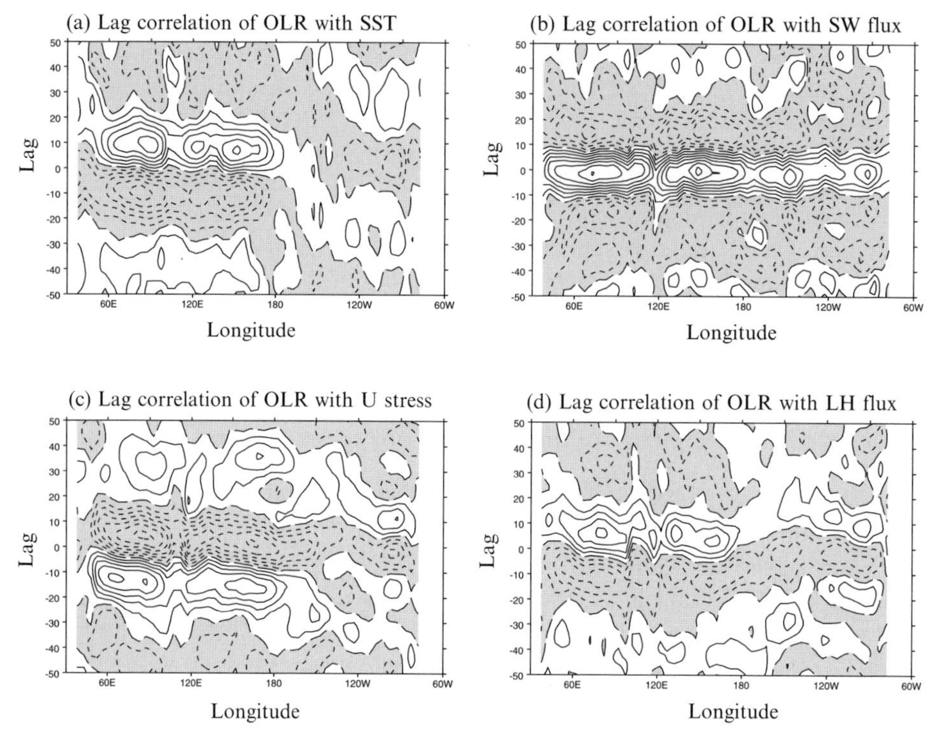

Figure 11.3. Lag correlations between observed outgoing long-wave radiation (convection) and surface fields: (a) SST, (b) short-wave radiation, (c) zonal wind stress, and (d) latent heat flux. Negative lags indicate that the convection lags the surface field, positive lags indicate that the convection leads the surface fields. The sign convention is such that positive correlations indicate that enhanced convection (a negative OLR anomaly) is correlated with a negative SSTA, reduced short-wave radiation at the surface, enhanced evaporation, or an easterly wind stress anomaly.
From Woolnough *et al.* (2001).

reduces vertical mixing. To the west of the convective maximum, the SSTs cool due to reduced solar radiation and enhanced evaporation associated with stronger winds. A key requirement for the observed temporal and spatial phase relationship between the latent heat flux, winds, and convection is the presence of a surface westerly basic state, an issue that emerges later as being crucial for the improved simulation of the MJO in coupled models. In addition to the SSTA pattern, Figure 11.3 also shows the phasing of the surface flux and wind stress anomalies relative to the convective maximum.

Having established that the surface fluxes and winds associated with the MJO can force intraseasonal variations in the SSTs, which can typically reach 1 K in individual events, it then needs to be confirmed that the atmosphere can respond to these SST variations. In a related study, Woolnough *et al.* (2001) therefore used the observed SST perturbations associated with the MJO to form the basis of a series of experiments with the aqua-planet version of the UM to investigate firstly the organization of tropical convection by these intraseasonal anomalies, and secondly, how this organization depends on the temporal behavior of these SSTAs. The study showed that the boundary layer humidity adjusts rapidly to the presence of the SSTA. However, the free atmosphere takes longer to adjust. Initial convective plumes triggered by the presence of warm SSTs are rapidly eroded by entrainment of dry air in the free troposphere and so terminate relatively low down in the troposphere. However, the detrainment of the terminating plumes moistens the atmosphere allowing subsequent convective plumes to penetrate further before decaying. Eventually the atmosphere is moist enough to support deep convection through most of the depth of the troposphere. This type of pre-conditioning behavior means that the most intense convection occurs, not directly over the warm SSTA, but to the west over the maximum gradient in SST between the warm and cold anomalies, as observed in the MJO. The timescale of about 5 days for the pre-conditioning of the tropical atmosphere for deep convection has recently been confirmed in a detailed study of reanalysis data by Sperber (2003). Associated with this adjustment timescale, the experiments of Woolnough *et al.* (2001) also showed that intraseasonal SSTAs could potentially organize convection in a manner that favors the longer timescales ($\sim$60 days), typical of the observed MJO, and which produces a phase relationship between the convection and SST, consistent with the observed structure over the Indian and West Pacific Oceans.

Sperber *et al.* (1997) had already suggested that a possible reason for the lack of realistic propagation of convective anomalies in atmospheric models used in AMIP I was that the MJO may be, at least in part, a coupled mode. The results of Woolnough *et al.* (2000, 2001) appeared to support this hypothesis. Flatau *et al.* (1997) also proposed that the eastward propagation of MJO convection might involve a coupled mechanism, and performed a simple numerical experiment to test their hypothesis. Using a low-resolution (spectral R15) GCM, configured as an aqua-planet model, they modeled the dependence of SST on surface fluxes empirically by relating SST fluctuations to changes in the strength of the low-level winds, based on observed SST changes and wind speeds from drifter buoys in the tropical Pacific. Their results showed that oscillations in the low-level winds on intraseasonal timescales became more organized when the variations of SST with wind speed were included, producing a coherent, eastward propagating signal which resembled the MJO in some respects.

A similar modeling study was carried out by Waliser *et al.* (1999), but using a more complex GCM and a more realistic parameterization of SSTAs in the tropics, based on a slab ocean model of fixed depth in which SSTAs developed in association with changes in net surface heat flux according to the formula:

$$\frac{dT'}{dt} = \frac{F'}{(\rho C_p H)} - \gamma T'$$

Here T' is the SSTA, F' is the surface flux anomaly, H is the depth of the mixed layer (fixed at 50 m), and γ is a damping factor, set to (50 days)$^{-1}$. Changes in SST due to this formula were small, however, being of the order of 0.1–0.15°C and were due largely to changes in the latent heat flux ahead of and behind the convective region, and to changes in the short-wave flux associated with the variations in convective cloudiness. It is worth noting that in their study the use of a fixed mixed layer depth underestimated the SST variability associated with the MJO since the warming during the suppressed phase is, in reality, strongly amplified by the shoaling of the mixed layer during light wind conditions (e.g., Weller and Anderson, 1996). Nevertheless, their results showed that the MJO simulation was improved in a number of respects. The period of the oscillation slowed down to be closer to the observed period, the variability of upper level winds and convective activity on intraseasonal timescales became stronger, the number of MJO events occurring during northern hemisphere winter and spring increased significantly, and the phase speed of the oscillation slowed in the eastern hemisphere in association with more organized convection.

The results of Waliser *et al.* (1999) were very encouraging and suggested that a more comprehensive and realistic approach to simulating the coupled aspects of the MJO might be fruitful. However, there have only been a limited number of studies of the MJO in coupled GCMs in the literature. There are several reasons for this. First, until quite recently the cost of running coupled GCMs has been prohibitively high for many research centers and so their use had been limited to a few institutes. Second, the development of coupled GCMs has historically been motivated by the requirements of long-term climate prediction and, more recently, seasonal prediction, so the ability of models to capture variability on timescales of less than a season has not been a primary consideration to the groups involved. Third, it has been only recently that coupled GCMs have been developed without the need for flux-adjustment to maintain a stable mean climate (e.g., Gordon *et al.*, 2000), and there had been concerns that the flux adjustment might compromise the ISV of the coupled system.

Initial studies by Gualdi *et al.* (1999) and Hendon (2000) of the MJO in fully coupled models concluded that an interactive ocean did not improve the simulation. Instead they found that accompanying changes in the mean climate of the model and deficiencies in the representation of surface flux anomalies were the main factors affecting the behavior of the MJO. However, more recently Kemball-Cook *et al.* (2002), Inness and Slingo (2003), and Inness *et al.* (2003) demonstrated that the coupling improves the organization and propagation characteristics of the MJO in comparison with the results from the atmosphere-only models, at least for the boreal winter (Figure 11.4). Whereas the atmosphere-only model had a predominantly standing oscillation in the convection (Figure 11.4(b)), the coupled model produced a more realistic eastward propagating signal (Figure 11.4(a)). This was associated with coherent variations in SST (Figure 11.4(c)), which showed a similar phase relationship with convection as in observations (Figure 11.3(a)), with warmer SSTs preceding the maximum in convection by between 5 and 10 days.

Due to the increased number of degrees of freedom in a fully coupled, un-flux

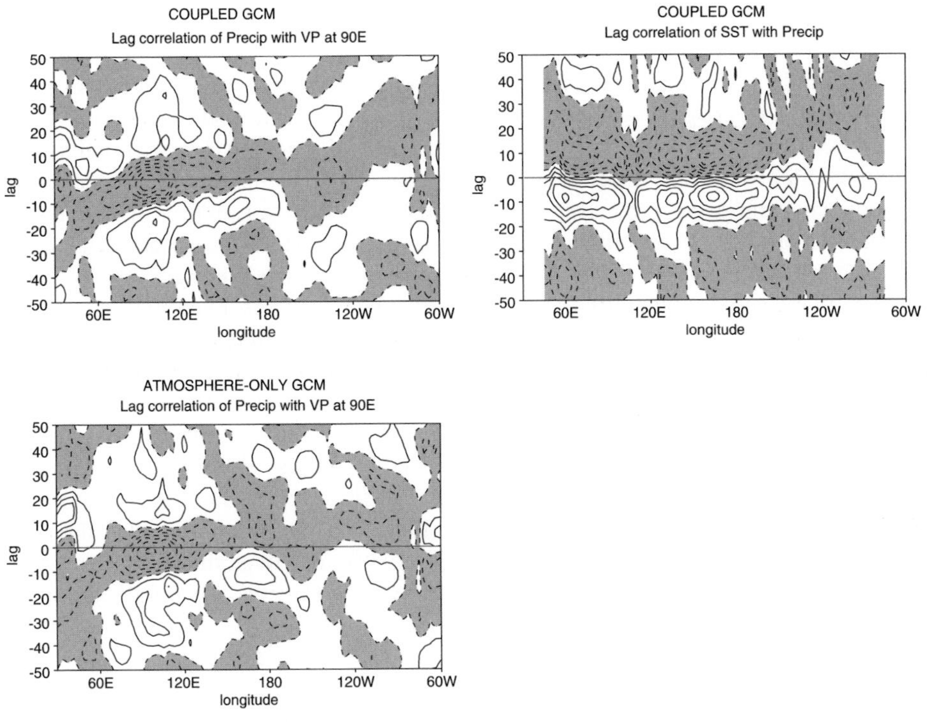

Figure 11.4. Lag correlations between precipitation at every longitude and an index of MJO activity at 90°E, based on the 20–100-day filtered 200-hPa velocity potential, from (a) a version of the coupled ocean–atmosphere model, HadCM3, and (b) the equivalent atmosphere-only model, HadAM3. (c) Shows the simulated lag correlations between the precipitation and SST at every longitude (as in Figure 11.3(a)) from HadCM3.
From Slingo *et al.* (2003).

corrected GCM, it is much more likely that there will be errors in the basic state than in an atmosphere-only GCM constrained by realistically prescribed SSTs. This has emerged as a crucial factor in the simulation of the MJO in coupled models. In particular the low-level climatological westerlies across the Indo-Pacific warm pool, associated with the Austral monsoon, are critical for the air–sea interaction mechanism of the MJO. It is only when these winds are westerly that the wind perturbations associated with the MJO can give enhanced latent heat fluxes (i.e., cooling of the ocean) to the west of the convection and reduced fluxes to the east (i.e., warming of the ocean). Inness *et al.* (2003) showed conclusively that the easterly bias over the West Pacific, typical of the majority of coupled models, acts to restrict the eastward propagation of the MJO by disabling the air–sea interaction mechanism. Consequently, improving the mean simulation in coupled models is a major issue facing future improvements in modeling the MJO.

11.3 MODELING BOREAL SUMMER ISV

As noted in the introduction, the MJO during boreal summer is much more complex, and the eastward propagation is often accompanied by northward propagation over the Indian Ocean sector. A brief discussion of boreal summer intraseasonal variability (BSISV) follows in order to characterize the basic challenges to the modeling community. A more comprehensive discussion of observed variability is presented in Chapters 2 and 3. The BSISV is important because it is intimately related to the active/break cycles of the Asian summer monsoon (Webster *et al.*, 1998; Krishnamurti and Bhalme, 1976; Sikka, 1980; Gadgil and Asha, 1992). Observed years of below-normal Indian monsoon rainfall tend to be associated with prolonged breaks in the monsoon, and conversely, fewer breaks of shorter duration tend to occur during years of normal or above-normal monsoon rainfall. During northern summer, the MJO is modified substantially by the off-equatorial heating associated with the Asian summer monsoon. It has a mixed character of both northward and eastward propagation. Northward propagation of the tropical convergence zone on timescales of 30–50 days over the Indian longitudes was initially identified by Yasunari (1979, 1980) and Sikka and Gadgil (1980), and over the West Pacific by Murakami *et al.* (1984), and Lau and Chan (1986). Wang and Rui (1990) classified intraseasonal propagating events over the monsoon domain, including isolating northward propagation that occurred independent of eastward propagation. Later, Lawrence and Webster (2002) found that 78% of northward propagating intraseasonal events were accompanied by eastward propagation, and it is mainly on these events that we concentrate. Figure 11.5(a) (see color section) shows the composite rainfall from observations corresponding to active convection over India, extending to the south-east into the western Pacific. As this tilted rain band propagates to the east, rainfall occurs further north at a given longitude.

Lau and Peng (1990) proposed that the northward propagation is due to coupled Kelvin–Rossby wave interactions. The theory of tropical ISOs is discussed in Chapter 10. The intermediate complexity model of Wang and Xie (1997) replicated the north-west–south-east tilt of the rain band due to Kelvin–Rossby wave interactions. Observational evidence that the tilt is due to the emanation of Rossby waves has been found by Annamalai and Slingo (2001), Kemball-Cook and Wang (2001), and Lawrence and Webster (2002). Annamalai and Sperber (2004) used a linear barotropic model forced with heating proportional to the rainfall rate for different phases of the BSISV life cycle. They were able to reproduce the observed low-level circulation, and showed that the development of the forced Rossby waves could only occur in the presence of easterly zonal shear, as suggested by Lau and Peng (1990) and Wang and Xie (1997). Additionally, they concluded that the ISV over the Indian Ocean and the West Pacific are mutually dependent systems. That is, the convection over the West Pacific helps initiate the monsoon break over India, while the Indian Ocean convection can modulate the active and break phase over the West Pacific.

Additionally, low-level moisture convergence is important for maintaining the eastward propagation as it destabilizes the atmosphere ahead of the main center of convection. In the boreal summer, the northward propagation also exhibits the tendency for low-level moisture convergence to lead the convection (Kemball-Cook and Wang, 2001). Thus, the mechanisms involved in BSISV are akin to those during the boreal winter MJO. Additionally, over the western North Pacific it has been suggested that subtropical westward propagating low-level convergence anomalies contribute to the north-westward propagation of the rain band (Hsu and Weng, 2001). Thus, the complex nature of the BSISV makes it especially challenging to simulate.

11.3.1 Atmospheric model simulations

Modeling studies of BSISV have been relatively limited, partly due to the difficulties in simulating both the mean monsoon and its variability (Sperber and Palmer, 1996; Sperber *et al.*, 2000). Given the complex orography over the summer monsoon domain, deficiencies in simulating rainfall were noted by Hahn and Manabe (1975) and Gilchrist (1977). Subsequently, numerous studies have evaluated the monsoon sensitivity to horizontal resolution, though most studies concentrated on the time-mean behavior (e.g., Tibaldi *et al.*, 1990). Typical results indicated a better representation of the rainfall along the western Ghats and their downwind rain-shadow effect, as well as improvement in the foothills of the Himalayas.

The ability of atmospheric models to simulate the dominant intraseasonal rainfall pattern has remained problematic, as shown in Figure 11.5 (see color section). These results, from a comparison study by Waliser *et al.* (2003a), demonstrate that models have difficulty in representing the tilted rain band and its propagation characteristics. When the full life cycle of the dominant mode is considered, only half of the models in the study exhibited any north-eastward propagation, and none of the models exhibited any systematic intraseasonal rainfall variability over the Indian Ocean.

As with the boreal winter MJO, studies of the sensitivity of BSISV to horizontal resolution have been inconclusive. Using the Geophysical Fluid Dynamics Laboratory GCM, Hayashi and Golder (1986) found that R30 ($\sim 3^{\circ}$) better represented the space–time spectra of rainfall compared to the R15 ($\sim 5^{\circ}$) model version. Of special note was the ability of the model to simulate the poleward propagation of rainfall over the monsoon domain, including the observed asymmetry, with the northern hemisphere propagation being stronger than that in the southern hemisphere. Using a T21 ($\sim 5^{\circ}$) model from ECMWF, Gadgil and Srinivasan (1990) found that this model produced northward propagation of the rain belt over the Bay of Bengal. However, using a later version of the ECMWF model, Sperber *et al.* (1994) found that a resolution of T106 ($\sim 1^{\circ}$) was needed to represent the northward propagation of the tropical convergence zone and the sudden jump of the Mei-yu front over China, although later work has suggested coarser resolution models may have similar capabilities (Lau and Yang, 1996; Martin, 1999). In fact the differences among models are mainly associated with the combinations of, improvement in, and the addition of, physical parameterizations.

11.3.2 Air–sea interaction and BSISV

Modeling of BSISV has also benefited from an understanding of the important role that air–sea interaction has played in representing the boreal winter MJO. Kemball-Cook *et al.* (2002), using the ECHAM4 model in coupled and uncoupled configurations, showed that with air–sea feedback, space–time spectra of OLR displayed a more realistic partitioning of variance between eastward and westward propagation near the equator. They also found that "coupling is helping to destabilize the northward moving mode by enhancing low-level convergence into the positive SST anomaly." However, unlike the reanalysis, the short-wave surface heat flux was more important than the latent heat flux for forcing the SSTAs that are in quadrature with the convection. In addition, the model also overestimated the strength of the low-level convergence. Thus, the model appears to compensate for the weak latent heat flux anomalies, suggesting that the BSISV is arising from the wrong combination of interactions. Despite this, the indication is that the net surface heat flux is important for generating realistic SSTAs, which in turn are important for modulating the propagation of the BSISV.

Kemball-Cook *et al.* (2002) also found that the failure to generate easterly wind shear in the late summer precluded the emanation of Rossby waves and prohibited the north-westward propagating mode. As in the boreal winter case, this attests to the importance of simulating a realistic basic state to properly capture ISV. In cases where there is an eastward propagating equatorial convective component, Kelvin–Rossby wave interactions and air–sea interaction both promote the northward propagation of precipitation resulting in the tilted rain band.

Using the Meteorological Research Institute CGCM2, Rajendran *et al.* (2004) presented additional evidence that air–sea interaction results in a more realistic BSISV. Compared to the uncoupled integration, the coupled model had 50% more northward propagating events, and exhibited surface flux, convection, and SST feedbacks that resulted in a more realistic life cycle of the BSISV. Thus it appears that air–sea interaction gives rise to a more accurate simulation of ISV, provided the model has a realistic mean state. An important question for the future is: What are the relative contributions of the Kelvin–Rossby wave interactions vs. the air–sea interaction for promoting the northward propagation? Fu *et al.* (2003) suggest that air–sea interaction is the most important process. They note cases of northward propagation that occur independently of an eastward equatorial propagating convective component, hence having no contribution due to the Kelvin–Rossby wave interactions, so that the northward propagation occurs solely due to air–sea interaction. Conversely, numerous GCM studies have shown some ability to generate northward propagation using prescribed SST, suggesting that processes other than air–sea interaction are also important. What is needed is a better understanding of the hierarchy of subseasonal modes of monsoon variability (e.g., Wang and Rui, 1990; Sperber *et al.*, 2000), and the mechanisms that control them.

11.3.3 Modeling studies of the links between boreal summer intraseasonal and interannual variability

In the mid-1990s, modeling studies of BSISV and its possible link to interannual variations outpaced our ability to firmly establish such a link in observations. Fennessy and Shukla (1994) used the Center for Ocean Land Atmosphere's (COLA) atmospheric general circulation model to simulate the weak (strong) Indian monsoon of 1987 (1988). They found that the spatial pattern of interannual rainfall difference was nearly identical to the difference due to break and active phases of the monsoon. Ferranti *et al.* (1997) found a similar result with the ECMWF model in AMIP simulations forced with observed SST for 1979–1988. Using canonical correlation analysis (CCA), they found the 850-hPa relative vorticity exhibited a common mode of variability on interannual and intraseasonal timescales, being characterized by an alternation of the tropical convergence zone between the tropical Indian Ocean and over the continental landmass, centered at about 15°N. However, the oceanic and continental locations of the tropical convergence zone were regime transitions that were not associated with northward propagating intraseasonal events.

With the advent of reanalysis, it became possible to investigate the link between intraseasonal and interannual variability based on a dynamically consistent representation of the atmosphere using a uniform model and data assimilation system (Gibson *et al.*, 1996, 1997; Kalnay *et al.*, 1996). From reanalysis, winds and vorticity are more reliable than rainfall or OLR (Kalnay *et al.*, 1996), they provide a longer record compared with satellite derived OLR, and are more spatially complete compared with observed rainfall. Using 850-hPa relative vorticity, Annamalai *et al.* (1999) showed that both the ECMWF and NCEP–NCAR Reanalyses had nearly identical dominant modes of ISV, characterized by a north-west–south-east tilt and northward propagation (Figure 11.6). Additionally, these modes were linked to the active and break monsoon over India. Compared with these results of Annamalai *et al.* (1999), the aforementioned model results of Ferranti *et al.* (1997) and Martin (1999) exhibited intraseasonal patterns that were too zonal, with the transition from ocean to the continent being more regime-like rather than propagating. Furthermore, the first mode in the models explained far more of the sub-seasonal variance than in the observations.

Observational evidence for a common mode of intraseasonal and interannual variability was found by Sperber *et al.* (2000) and Goswami and Ajaya Mohan (2001). This mode, shown in Figure 11.6(c), is characterized by cyclonic flow at 850 hPa over India and an anticyclone to the south over the Indian Ocean. It shows a strong link to all-India rainfall manifested as a systematic shift in the mean of the frequency distribution of the principal component time series when stratified between years of above-normal and below-normal all-India rainfall (Sperber *et al.*, 2000). Unfortunately, a direct link of this mode to slowly varying boundary conditions, which could be the source of predictability, has remained elusive. Other modes in the 850-hPa wind are associated with the northward propagation of the tropical convergence zone, associated with the onset of northward

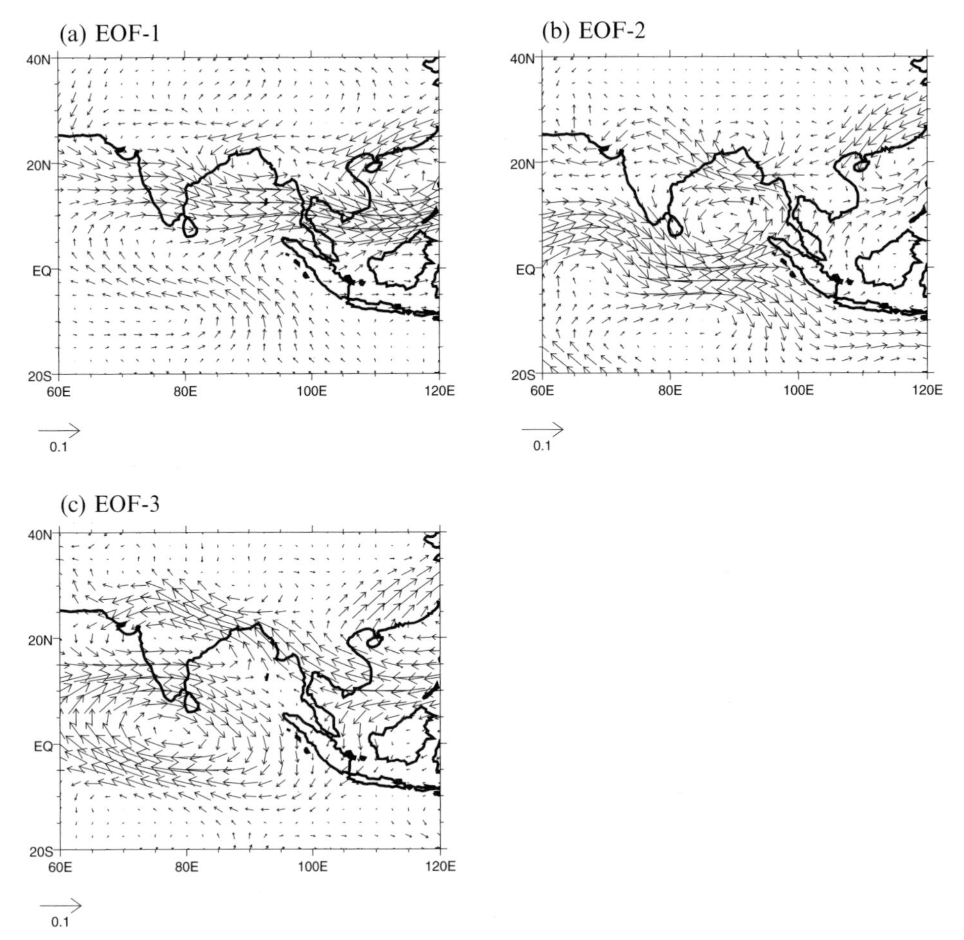

Figure 11.6. The dominant modes of BSISV in the 850-hPa winds from the NCEP–NCAR Reanalysis.
After Sperber *et al.* (2001).

propagation being linked to the phase of the El Niño Southern Oscillation (ENSO) (Sperber *et al.*, 2000). While encouraging from the viewpoint of predictability, this is not the dominant mode of intraseasonal variability, and thus the chaotic nature of the other components of the BSISV can obscure a boundary forced signal.

The ability of atmospheric general circulation models to simulate the dominant modes of the BSISV in the 850-hPa winds using hindcast experiments run with observed SST was evaluated by Sperber *et al.* (2001). While the models were largely successful at representing the observed patterns (Figure 11.6), they over-emphasized the role of EOF-1, and unlike the observations, most models linked this mode to the boundary forcing. As a result the models were predisposed to incorrectly project the sub-seasonal variability onto the seasonal rainfall, thus poorly representing the IAV. Similar to Ferranti *et al.* (1997), Molteni *et al.* (2003) found zonally oriented anomalies to be common between interannual and intraseasonal timescales using a more comprehensive suite of hindcast experiments with a later version of the ECMWF model. Though the principal component (PC) of the dominant mode was not correlated with ENSO, it did exhibit "multiple-regime behavior" related to the strength of zonal asymmetry in equatorial Pacific SST, a characteristic yet to be seen in observations. As in Sperber *et al.* (2001) they noted "significant discrepancies from observations in the partition of variance between modes with different regional characteristics."

Overall, models show some ability to represent the observed spatial patterns of the 850-hPa intraseasonal wind field, and poorer ability to represent the northward and eastward propagating rain band associated with the 30–50-day BSISV. Numerous factors complicate dynamical seasonal predictability of the summer monsoon. These include, but are not limited to, (i) the inability of models to realistically partition the relative importance of the dominant modes, (ii) the failure of models to link these modes to the boundary forcing as observed, and (iii) the fact that the ENSO forced mode is not the dominant mode of variability.

11.3.4 Predictability studies of BSISV

The aforementioned results indicate that forecasting the statistics of BSISV and its impact on the seasonal mean is a challenge that is limited by chaotic variability as well as model shortcomings. Despite this, there is another form of predictability that can be exploited because of the long timescale over which the BSISV evolves. The questions to be addressed include:

- If BSISV is present, how far into the future can we predict its influence?
- Is the degree of predictability dependent upon the strength and/or phase in the life cycle from which the forecast is made?
- What aspects of the BSISV are most predictable? These are important questions to address since they can potentially influence crop selection and planting time, as well as water resource management.

Such an investigation was undertaken by Waliser *et al.* (2003b) using the NASA Goddard Laboratory for Atmosphere's atmospheric GCM, which has been used in a similar capacity to investigate potential predictability of the boreal winter MJO. Predictability was assessed based on the ratio of the deterministic intraseasonal signal from the control run to the mean-squared error from perturbed initial condition integrations. Overall, 200-hPa velocity potential (rainfall) was predictable out to 25 (15) days, with the subset of strong intraseasonal cases having predictability extended by 10 (5) days compared with weak intraseasonal events. The suppressed phase exhibited a better signal-to-noise ratio at longer lead times compared with the convective phase. The results of this study are purely model generated, and thus are limited by the ability of the model to represent the BSISV. Other limitations include the use of climatological SSTs as the surface boundary condition, while as discussed earlier, air–sea interaction is important in the life cycle of the MJO. Plans for experimental sub-seasonal forecasts have been outlined in a workshop summary (Waliser *et al.*, 2003d), which also includes an assessment of model performance, indicating that coupled models tend to more realistically represent the MJO than their uncoupled counterparts, as discussed earlier in Section 11.2.3.

11.4 CONCLUDING REMARKS

It is certainly true that the simulation of the MJO by general circulation models is improving, along with our understanding of what are the key processes for its initiation and maintenance. However, it is still not the case that a good representation of all aspects of the MJO is inherent in the majority of the current generation of GCMs. Recent research has pointed to possible avenues that might lead to improvements in the simulation of the MJO in the coming years. First, greater emphasis is being placed on understanding the suppressed phase of the MJO and the processes that recharge the tropical troposphere for the next period of active convection. Steps are being taken to represent cumulus congestus clouds in convection schemes, including warm rain processes, which are key to the life cycle of these clouds.

Second, there is good evidence that the MJO in both boreal winter and summer manifestations is, at least to some extent, a coupled ocean–atmosphere mode. Whilst coupled models are capable of producing the correct relationship between convection and SST on intraseasonal timescales, these models still underestimate the activity of the MJO (e.g., Inness *et al.*, 2003) and the magnitude of the SST perturbations is smaller than observed. This is despite variations in the surface fluxes that are similar to those observed. This suggests that the representation of the upper layers of the ocean may not be responding realistically to sub-seasonal variations in winds and fluxes.

Most coupled climate models have a relatively coarse vertical resolution in the upper ocean, typically of the order of 10 m, but observations by tethered buoys, such as the Woods Hole IMET buoy during TOGA–COARE (e.g., Anderson *et al.*, 1996), have shown that the upper ocean has a very complex structure, which undergoes dramatic changes during the life cycle of the MJO. A particularly noteworthy aspect of these buoy observations is the diurnal variation in SST that only occurs during suppressed phases of the MJO, when the winds are light, the net heat flux into the ocean is large, and the mixed layer is very shallow. In studies with a very high vertical resolution mixed layer model, Shinoda and Hendon (1998) and Bernie *et al.* (2004) have shown that the rectification of these diurnal variations on to intraseasonal timescales is significant and accounts for a large proportion of the intraseasonal warming of the ocean during the suppressed phase of the MJO. Clearly, the coarse resolution of the upper ocean in current coupled models and the lack of resolution of the diurnal cycle in the coupling frequency means that these diurnal variations in SST and their rectification on to intraseasonal timescales are not represented. Bernie *et al.* (2004) concluded that a resolution of 1 m for the

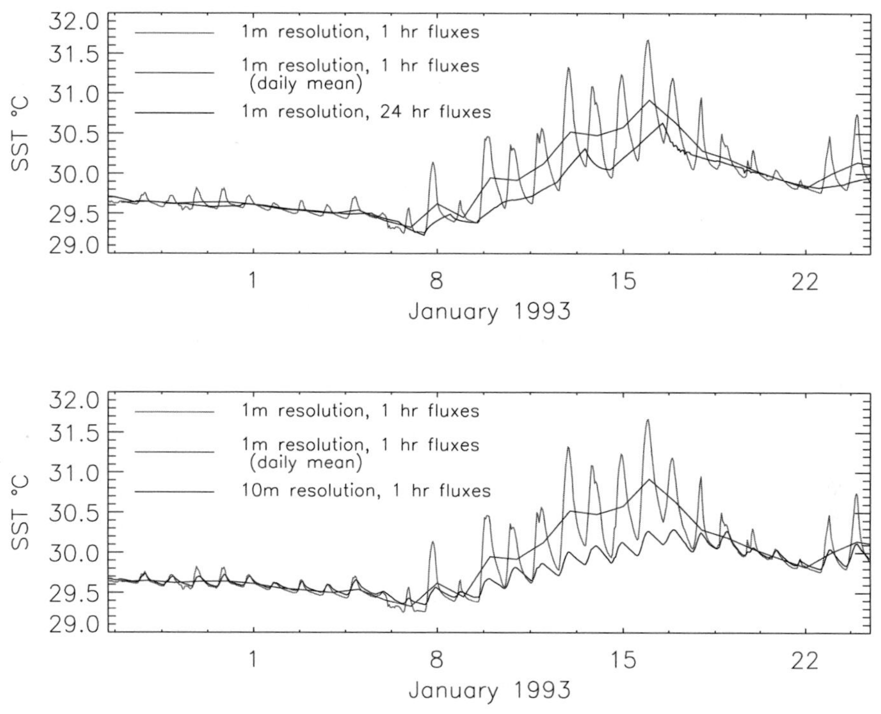

Figure 11.7. Impact of coupling frequency (*upper panel*) and resolution of uppermost ocean (*lower panel*) on simulations of the diurnal and intraseasonal variations in SST from TOGA–COARE with a mixed layer ocean model. The observed SSTs are very close to the 1-m resolution 1-hr fluxes curves.
From Bernie *et al.* (2004).

skin layer of the ocean and a coupling frequency of at least every 3 hours are needed to adequately capture diurnal and intraseasonal SST variability. As Figure 11.7 shows, only simulations with high-frequency coupling and a shallow top layer are capable of reproducing the observed signal.

The diurnal SST variations may also be important for the MJO in other ways. Johnson *et al.* (1999) showed that cumulus congestus clouds are most prevalent during light wind conditions in the presence of a strong diurnal cycle in SST. These clouds occur most frequently in the late afternoon, with a behavior that more closely resembles the diurnal cycle in land convection, suggesting that they may be triggered by the diurnal cycle in SST. The fact that these clouds appear to be key players during the suppressed phase of the MJO adds further weight to the need for taking a complete atmosphere–upper ocean approach to simulating the MJO.

Although the focus of this chapter has been on modeling intraseasonal variability, other aspects of sub-seasonal tropical variability need to be considered. Interactions between multiple timescales of variability in the tropics have been the subject of several papers (e.g., Nakazawa, 1988; Lau *et al.*, 1991), suggesting that the synoptic scale, higher frequency modes of convective activity are modulated by the MJO. How much the synoptic and mesoscale activity embedded within the MJO is responsible for the evolution of the oscillation itself remains an open question (e.g., Hendon and Liebmann, 1994). The importance of equatorial wave modes for organizing tropical convection in general has been highlighted by Wheeler and Kiladis (1999) and Yang *et al.* (2003). In fact, the results of Yang *et al.* (2003) suggest that the majority of tropical convection is associated with equatorial Kelvin, Rossby, and mixed Rossby–gravity waves, which undergo Doppler shifting and changes in vertical structure depending on the basic state wind and vertical shear. Yang *et al.* (2003) also showed that the structure of the waves is substantially modified over the Indo-Pacific warm pool by equatorial convection induced through wind–evaporation feedbacks. However, a preliminary analysis of these waves in the AMIP II models (M. Wheeler, pers. commun.) and in the Hadley Centre's climate model (G. Yang, pers. comm.) has shown major deficiencies in their structure and their coupling with convection. Since these waves are the building blocks of the tropical climate and are fundamental to the simulation of the MJO, future efforts to model the MJO must also address the more general issue of convectively coupled equatorial waves.

The measures used to determine the quality of the MJO simulation are very important. Early GCM studies of the MJO tended to concentrate on the signal in upper tropospheric tropical winds or velocity potential. It could be that *in situ* intraseasonal modulation of the main convective region over the Indo-Pacific warm pool produces an equatorially trapped Kelvin wave response, which resembles the MJO signal in the upper level winds, without actually being accompanied by an eastward propagation of the main convective region through the Indian Ocean and into the West Pacific. The need to use a reasonable range of diagnostics to determine the quality of the MJO simulation is clearly important, in which the signal of the MJO in the upper tropospheric winds should be regarded as a bare minimum indication of the presence of the MJO. The evolution of convection through the cycle of the MJO, with particular emphasis on the eastward propagation, and in boreal summer the northward propagation, must be examined. Recent research has emphasized the complex 3-D structure of the MJO, in particular the vertical distribution of the humidity field, and these should provide stringent tests for the model simulations. Finally, the ISV of surface fluxes and their impact on SST should be diagnosed, ensuring that the coupled nature of the simulated MJO is properly represented.

Compared with the MJO, modeling studies of the BSISV have been less extensive and arguably less successful, partly because the simulation of the mean climate of the Asian summer monsoon continues to prove a challenge. Furthermore our basic understanding of what drives the BSISV and its northward vs. eastward propagation is not so advanced, and we do not fully understand the role that land surface processes and the Tibetan Plateau may play in the evolution of the BSISV. Yet the social and economic benefits from even extended range prediction of BSISV could be huge making this a major challenge for the modeling community in the coming years.

11.5 ACKNOWLEDGMENTS

Julia Slingo and Peter Inness acknowledge support through the NERC Centres for Atmospheric Science. K. R. Sperber was supported under the auspices of the University of California Lawrence Livermore National Laboratory under contract W-7405-ENG-48. We thank Professor Duane Waliser for kindly providing Figure 11.4. KRS would like to thank Drs H. Annamalai and X. Fu for helpful discussions.

11.6 REFERENCES

Anderson, S. P., R. A. Weller, and R. B. Lukas (1996) Surface buoyancy forcing and the mixed layer of the Western Pacific warm pool: Observations and 1D model results. *J. Clim.*, **9**, 3056–3085.

Annamalai, H., J. M. Slingo, K. R. Sperber, and K. Hodges (1999) The mean evolution and variability of the Asian summer monsoon: Comparison between ECMWF and NCEP-NCAR Reanalyses. *Mon. Wea. Rev.*, **127**, 1157–1186.

Annamalai, H. and J. M. Slingo (2001) Active/break cycles: Diagnosis of the intraseasonal variability of the Asian summer monsoon. *Clim. Dynam.*, **18**, 85–102.

Annamalai, H. and K. R. Sperber (2004) Regional heat sources and the active and break phases of boreal summer intraseasonal variability. *J. Atmos. Sci.*, submitted.

Arakawa, A. and W. H. Schubert (1974) Interaction of a cumulus cloud ensemble with the large-scale environment, Part I. *J. Atmos. Sci.*, **31**, 674–701.

Bernie, D. J., S. J. Woolnough, J. M. Slingo, and E. Guilyardi (2004) Modelling diurnal and intraseasonal variability of the ocean mixed layer. *J. Clim.*, in press.

Betts, A. K. (1986) A new convective adjustment scheme. Part I: Observational and theoretical basis. *Q. J. R. Meteorol. Soc.*, **112**, 677–691.

Bladé, I. and D. L. Hartmann (1993) Tropical intraseasonal oscillations in a simple nonlinear model. *J. Atmos. Sci.*, **50**, 2922–2939.

Emanuel, K. A. (1987) An air–sea interaction model of intraseasonal oscillations in the tropics. *J. Atmos. Sci.*, **44**, 2324–2340.

Fennessy, M. J. and J. Shukla (1994) GCM simulations of active and break periods. *Proceedings of the International Conference on Monsoon Variability and Prediction, Trieste, Italy* (WCRP-84, WMO/TD-No. 619, Vol. 2, 576–585).

Ferranti, L., J. M. Slingo, T. N. Palmer, and B. J. Hoskins (1997) Relations between interannual and intraseasonal variability as diagnosed from AMIP integrations. *Q. J. R. Meteorol. Soc.*, **123**, 1323–1357.

Flatau, M., P. J. Flatau, P. Phoebus, and P. P. Niiler (1997) The feedback between equatorial convection and local radiative and evaporative processes: The implications for intraseasonal oscillations. *J. Atmos. Sci.*, **54**, 2373–2386.

Fu, X., B. Wang, T. Li, and J. P. McCreary (2003) Coupling between northward propagation, intraseasonal oscillations and sea surface temperature in the Indian Ocean. *J. Atmos. Sci.*, **60**, 1733–1753.

Gadgil, S. and J. Srinivasan (1990) Low frequency variation of tropical convergence zones. *Meteorol. Atmos. Phys.*, **44**, 119–132.

Gadgil, S. and G. Asha (1992) Intraseasonal variation of the summer monsoon. Part I: Observational aspects. *J. Meteorol. Soc. Jap.*, **70**, 517–527.

Gates, W. L. (1992) AMIP: The atmospheric model intercomparison project. *Bull. Amer. Met. Soc.*, **73**, 1962–1970.

Gates, W. L., J. S. Boyle, C. Covey, C. G. Dease, C. M. Doutriaux, R. S. Drach, M. Fiorino, P. Gleckler, J. J. Huilo, S. M. Marlais, *et al.* (1999) An overview of the results of the Atmospheric Model Intercomparison Project (AMIP 1). *Bull. Amer. Meteorol. Soc.*, **80**, 29–56.

Gibson, J. K., P. Kallberg, and S. Uppala (1996) The ECMWF ReAnalysis (ERA) project. *ECMWF Newsletter*, **73**, 7–17.

Gibson, J. K., P. Kallberg, S. Uppala, A. Hernandez, A. Nomura, and E. Serrano (1997) *ECMWF ReAnalysis Project Report, Series 1. ECMWF*, Reading, U.K., 77 pp.

Gilchrist, A. (1977) The simulation of the Asian summer monsoon by general circulation models. *Pageopf*, **115**, 1431–1448.

Gordon, C., C. Cooper, C. A. Senior, H. Banks, J. M. Gregory, T. C. Johns, J. F. B. Mitchell, and R. A. Wood (2000) The simulation of SST, sea ice extents and ocean heat transports in a version of the Hadley Centre coupled model without flux adjustments. *Clim. Dyn.*, **16**, 147–168.

Goswami, B. N. and R. S. Ajaya Mohan (2001) Intraseasonal oscillations in interannual variability of the Indian summer monsoon. *J. Clim.*, **14**, 1180–1198.

Grabowski, W. W. (2003) MJO-like coherent structures: Sensitivity simulations using the Cloud-Resolving Convection Parameterization (CRCP). *J. Atmos. Sci.*, **60**, 847–864.

Grabowski, W. W. and M. W. Moncrieff (2002) Large-scale organization of tropical convection in two-dimensional explicit numerical simulations: Effects of interactive radiation. *Q. J. R. Meteorol. Soc.*, **128**, 2349–2375.

Gualdi, S., A. Navarra, and M. Fischer (1999) The tropical intraseasonal oscillation in a coupled ocean–atmosphere general circulation model. *Geophys. Res. Lett.*, **26**, 2973–2976.

Gualdi, S., A. Navarra, and G. Tinarelli (1999) The interannula variability of the Madden–Julian Oscillation in an ensemble of GCM simulations. *Clim. Dyn.*, **15**, 643–658.

Hahn, D. G. and S. Manabe (1975) The role of mountains in the south Asian monsoon circulation. *J. Atmos. Sci.*, **32**, 1515–1541.

Hayashi, Y.-Y. and D. G. Golder (1986) Tropical intraseasonal oscillations appearing in a GFDL general circulation model and FGGE data. Part I: Phase propagation. *J. Atmos. Sci.*, **43**, 3058–3067.

Hayashi, Y.-Y. and D. G. Golder (1988) Tropical intraseasonal oscillations appearing in a GFDL general circulation model and FGGE data. Part II: Structure. *J. Atmos. Sci.*, **45**, 3017–3033.

Hayashi, Y. and D. G. Golder (1993) Tropical 40–50 and 25–30 day oscillations appearing in realistic and idealized GFDL climate models and ECMWF dataset. *J. Atmos. Sci.*, **50**, 464–494.

Hendon, H. H. (2000) Impact of air–sea coupling on the Madden–Julian Oscillation in a general circulation model. *J. Atmos. Sci.*, **57**, 3939–3952.

Hendon, H. H. and B. Liebmann (1994) Organization of convection within the Madden–Julian Oscillation. *J. Geophys. Res.*, **99**, 8073–8083.

Hendon, H. H., C. D. Zhang, and J. D. Glick (1999) Interannual variation of the Madden–Julian Oscillation during austral summer. *J. Clim.*, **12**, 2538–2550.

Hsu, H.-H. and C.-H. Weng (2001) Northwestward propagation of the intraseasonal oscillation in the western north Pacific during the boreal summer: structure and mechanism. *J. Clim.*, **14**, 3834–3850.

Inness, P. M. and D. Gregory (1997) Aspects of the intraseasonal oscillation simulated by the Hadley Centre Atmosphere Model. *Clim. Dyn.*, **13**, 441–458.

Inness, P. M., J. M. Slingo, S. J. Woolnough, R. B. Neale, and V. D. Pope (2001) Organization of tropical convection in a GCM with varying vertical resolution: Implications for the simulation of the Madden–Julian Oscillation. *Climate Dynamics*, **17**, 777–793.

Inness, P. M. and J. M. Slingo (2003) Simulation of the MJO in a coupled GCM. Part I: Comparison with observations and an atmosphere-only GCM. *J. Clim.*, **16**, 345–364.

Inness, P. M., J. M. Slingo, E. Guilyardi, and J. Cole (2003) Simulation of the MJO in a coupled GCM. Part II: The role of the basic state. *J. Clim.*, **16**, 365–382.

Johnson, R. H., T. M. Rickenbach, S. A. Rutledge, P. E. Ciesielski, and W. H. Schubert (1999) Trimodal characteristics of tropical convection. *J. Clim.*, **12**, 2397–2418.

Jung, T. and A. Tompkins (2003) *Systematic Errors in the ECMWF Forecasting System* (ECMWF Technical Memorandum No. 422). ECMWF, Reading, U.K.

Kalnay, E., M. Kanamitsu, R. Kistler, W. Collins, D. Deven, L. Gandin, M. Iredell, S. Saha, G. White, J. Woollen, *et al.* (1996) The NCEP/NCAR 40-year reanalysis project. *Bull. Amer. Meteorol. Soc.*, **77**, 437–471.

Kemball-Cook, S. and B. Wang (2001) Equatorial waves and air–sea interactions in the boreal summer intraseasonal oscillation. *J. Clim.*, **14**, 2923–2942.

Kemball-Cook, S., B. Wang, and X. Fu (2002) Simulation of the intraseasonal oscillation in the ECHAM-4 Model: The impact of coupling with an ocean model. *J. Atmos. Sci.*, **59**, 1433–1453.

Krishnamurti, T. N. and H. N. Bhalme (1976) Oscillations of the monsoon system. Part 1. Observational aspects. *J Atmos. Sci.*, **33**, 1937–1954.

Knutson, T. R. and K. M. Weickmann (1987) 30–60 day atmospheric oscillations: Composite life-cycles of convection and circulation anomalies. *Mon. Wea. Rev.*, **115**, 1407–1436.

Kuo, H. L. (1974) Further studies of the parameterization of the influence of cumulus convection on large-scale flow. *J. Atmos. Sci.*, **31**, 1232–1240.

Lau, K. M. and P. H. Chan (1986) Aspects of the 40–50 day oscillation during the northern summer as inferred from outgoing longwave radiation. *Mon. Wea. Rev.*, **114**, 1354–1367.

Lau, N. C. and K. M. Lau (1986) Structure and propagation of intraseasonal oscillations appearing in a GFDL GCM. *J. Atmos. Sci.*, **43**, 2023–2047.

Lau, K. M. and S. Yang (1996) Seasonal variation, abrupt transition, and intraseasonal variability associated with the Asian summer monsoon in the GLA GCM. *J. Clim.*, **9**, 965–985.

Lau, K. M. and L. Peng (1990) Origin of low frequency (intraseasonal) oscillations in the tropical Atmosphere. Part III: Monsoon dynamics. *J. Atmos. Sci.*, **47**, 1443–1462.

Lau, K.-M., T. Nakazawa, and C. H. Sui (1991) Observations of cloud cluster hierarchies over the tropical western Pacific. *J. Geophys. Res.*, **96**, 3197–3208.

Lawrence, D. M. and P. J. Webster (2002) The boreal summer intraseasonal oscillation: Relationship between northward and eastward movement of convection. *J. Atmos. Sci.*, **59**, 1593–1606.

Lee, M. I., I. S. Kang, and B. E. Mapes (2003) Impacts of convection parametrization on aqua-planet AGCM simulations of tropical intraseasonal variability. *J. Meteorol. Soc. Jap.*, **81**. 963–992.

Lengaigne, M., E. Guilyardi, J.-P. Boulanger, C. Menkes, P. Delecluse, P. Inness, J. Cole, and J. M. Slingo (2004) Triggering of El Niño by westerly wind events in a coupled general circulation model. *Climate Dynamics*, **23** [doi:10.1007/500382-004-0457-2].

Lin, X. and R. H. Johnson (1996) Heating, moistening and rainfall over the western Pacific warm pool during TOGA COARE. *J. Atmos. Sci.*, **53**, 3367–3383.

Maloney, E. D. and D. L. Hartmann (2001) The sensitivity of intraseasonal variability in the NCAR CCM3 to changes in convective parametrization. *J. Clim.*, **14**, 2015–2034.

Martin, G. (1999) The simulation of the Asian summer monsoon, and its sensitivity to horizontal resolution, in the UK Meteorological Office Unified Model. *Q. J. R. Meteorol. Soc.*, **125**, 1499–1525.

Matthews, A. J., J. M. Slingo, B. J. Hoskins, and P. M. Inness (1999) Fast and slow Kelvin waves in the Madden–Julian Oscillation of a GCM. *Q. J. R. Meteorol. Soc.*, **125**, 1473–1498.

McPhaden, M. J. (1999) Genesis and evolution of the 1997–1998 El Niño. *Science*, **283**, 950–954.

Molteni, F., S. Corti, L. Ferranti, and J. M. Slingo (2003) Predictability experiments for the Asian summer monsoon: Impact of SST anomalies on interannual and intraseasonal variability. *J. Clim.*, **16**, 4001–4021.

Murakami, T., T. Nakazawa, and J. He (1984) On the 40–50 day oscillations during the 1979 northern hemisphere summer. Part I: Phase propagation. *J. Meteorol. Soc. Jap.*, **62**, 440–468.

Nordeng, T. E. (1994) *Extended Versions of the Convective Parametrization Scheme at ECMWF and their Impact on the Mean and Transient Activity of the Model in the Tropics* (ECMWF Technical Memorandum No. 206). ECMWF, Reading, U.K.

Nakazawa, T. (1988) Tropical superclusters within intraseasonal variations over the western Pacific. *J. Meteorol. Soc. Jap.*, **66**, 823–839.

Park, C. K., D. M. Straus, and K. M. Lau (1990) An evaluation of the structure of tropical intraseasonal oscillations in 3 general circulation models. *J. Meteorol. Soc. Jap.*, **68**, 403–417.

Rajendran, K., A. Kitoh, and O. Arakawa (2004) Monsoon low-frequency intraseasonal oscillation and ocean–atmosphere coupling over the Indian Ocean. *Geophys. Res. Lett.*, **31**, doi:10.1029/2003GL019031.

Randall, D., M. Khairoutdinov, A. Arakawa Akio, and W. Grabowski (2003) Breaking the cloud parameterization deadlock. *Bull. Amer. Met. Soc.*, **84**, 1547–1564.

Raymond, D. J. (2001) A new model of the Madden–Julian Oscillation. *J. Atmos. Sci.*, **58**, 2807–2819.

Salby, M. M., H. H. Hendon, and R. R. Garcia (1994) Planetary-scale circulations in the presence of climatological and wave-induced heating. *J. Atmos. Sci.*, **51**, 2344–2367.

Seo, K. H. and K. Y. Kim (2003) Propagation and initiation mechanisms of the Madden–Julian Oscillation. *J. Geophys. Res.*, **108**, doi:10.1029/2002JD002876.

Shinoda, T. and H. H. Hendon (1998) Mixed layer modeling of intraseasonal variability in the tropical Western Pacific and Indian Oceans. *J. Clim.*, **11**, 2668–2685.

Sikka, D. R. (1980) Some aspects of the large-scale fluctuations of summer monsoon rainfall over India in relation to fluctuations in planetary and regional scale circulation parameters. *Proc. Indian Acad. Sci. (Earth Planet. Sci.)*, **89**, 179–195.

Sikka, D. R. and S. Gadgil (1980) On the maximum cloud zone and the ITCZ over Indian longitudes during the southwest monsoon. *Mon. Wea. Rev.*, **108**, 1840–1853.

Slingo, J. M. and R. A. Madden (1991) Characteristics of the tropical intraseasonal oscillation in the NCAR community climate model. *Q. J. R. Meteorol. Soc.*, **117**, 1129–1169.

Slingo, J. M., K. R. Sperber, J.-J. Morcrette, and G. L. Potter (1992) Analysis of the temporal behavior of convection in the tropics of the ECMWF model. *J. Geophys. Res.*, **97**, 18119–18135.

Slingo, J. M., K. R. Sperber, J. S. Boyle, J.-P. Ceron, M. Dix, B. Dugas, W. Ebisuzaki, J. Fyfe, D. Gregory, J.-F. Gueremy, *et al.* (1996) Intraseasonal oscillations in 15 atmospheric general circulation models: Results from an AMIP diagnostic subproject. *Climate Dynamics*, **12**, 325–357.

Slingo, J. M., D. P. Rowell, K. R. Sperber, and F. Nortley (1999) On the predictability of the interannual behaviour of the Madden–Julian Oscillation and its relationship with El Niño. *Q. J. R. Meteorol. Soc.*, **125**, 583–609.

Slingo, J. M., M. Blackburn, A. Betts, R. Brugge, K. Hodges, B. Hoskins, M. Miller, L. Steenman-Clark, and J. Thuburn (1994) Mean climate and transience in the tropics of the UGAMP GCM: Sensitivity to convective parameterization. *Q. J. R. Meteorol. Soc.*, **120**, 881–922.

Slingo, J. M., P. M. Inness, R. B. Neale, S. J. Woolnough, and G.-Y. Yang (2003) Scale interactions on diurnal to seasonal timescales and their relevance to model systematic errors. *Ann. Geophys.*, **46**, 139–155.

Sperber, K. R., S. Hameed, G. L. Potter, and J. S. Boyle (1994) Simulation of the northern summer monsoon in the ECMWF model: Sensitivity to horizontal resolution. *Mon. Wea. Rev.*, **122**, 2461–2481.

Sperber, K. R. and T. N. Palmer (1996) Interannual tropical rainfall variability in general circulation model simulations associated with the atmospheric model intercomparison project. *J. Clim.*, **9**, 2727–2750.

Sperber, K. R., J. M. Slingo, P. M. Inness, and W. K.-M. Lau (1997) On the maintenance and initiation of the intraseasonal oscillation in the NCEP/NCAR Reanalysis and the GLA and UKMO AMIP simulations. *Climate Dynamics*, **13**, 769–795.

Sperber, K. R., J. M. Slingo, and H. Annamalai (2000) Predictability and the relationship between subseasonal and interannual variability during the Asian Summer Monsoon. *Q. J. R. Meteorol. Soc.*, **126**, 2545–2574.

Sperber, K. R., C. Brankovic, M. Deque, C. S. Frederiksen, R. Graham, A. Kitoh, C. Kobayashi, T. Palmer, K. Puri, W. Tennant, and E. Volodin (2001) Dynamical seasonal prediction of the Asian summer monsoon. *Mon. Wea. Rev.*, **129**, 2226–2248.

Sperber, K. R. (2003) Propagation and the vertical structure of the Madden–Julian Oscillation. *Mon. Wea. Rev.*, **131**, 3018–3037.

Swinbank, R., T. N. Palmer, and M. K. Davey (1988) Numerical simulations of the Madden–Julian Oscillation. *J. Atmos. Sci.*, **45**, 774–788.

Tibaldi, S., T. N. Palmer, C. Brankovic, and U. Cubasch (1990) Extended-range predictions with ECMWF models: Influence of horizontal resolution on systematic model error and forecast skill. *Q. J. R. Meteorol. Soc.*, **116**, 835–866.

Tokioka, T., K. Yamazaki, A. Kitoh, and T. Ose (1988) The equatorial 30–60 day oscillation and the Arakawa-Schubert penetrative cumulus parametrization. *J. Meteorol. Soc. Jap.*, **66**, 883–901.

Tompkins, A. M. and K. A. Emanuel (2000) The vertical resolution sensitivity of simulated equilibrium tropical temperature and water vapour profiles. *Q. J. R. Meteorol. Soc.*, **126**, 1219–1238.

Waliser, D. E., K. M. Lau, and J.-H. Kim (1999) The influence of coupled sea surface temperatures on the Madden–Julian Oscillation: A model perturbation experiment. *J. Atmos. Sci.*, **56**, 333–358.

Waliser, D. E., K. Jin, I.-S. Kang, W. F. Stern, S. D. Schubert, M. L. C. Wu, K.-M. Lau, M.-I. Lee, V. Krishnamurthy, A. Kitoh, *et al.* (2003a) AGCM simulations of intraseasonal variability associated with the Asian summer monsoon. *Clim. Dynam.*, **21**, 423–446.

Waliser, D. E., W. Stern, S. Schubert, and K. M. Lau (2003b) Dynamic predictability of intraseasonal variability associated with the Asian summer monsoon. *Q. J. R. Meteorol. Soc.*, **129**, 2897–2925.

Waliser, D. E., K. M. Lau, W. Stern, and C. Jones (2003c) Potential predictability of the Madden–Julian Oscillation. *Bull. Amer. Meteorol. Soc.*, **84**, 33–50.

Waliser, D. E., S. Schubert, A. Kumar, K. Weickmann, and R. Dole (2003d) Modeling, simulation, and forecasting of subseasonal variability. Technical Report Series on Global Modeling and Data Assimilation, NASA/CP-2003-104606, Vol. 25, 66 pp.

Wang, B. and H. Rui (1990) Synoptic climatology of transient tropical intraseasonal convection anomalies: 1975–1985. *Meteorol. Atmos. Phys.*, **44**, 43–61.

Wang, B. and X. Xie (1997) A model for the boreal summer intraseasonal oscillation. *J. Atmos. Sci.*, **54**, 72–86.

Wang, W. Q. and M. E. Schlesinger (1999) The dependence on convective parameterization of the tropical intraseasonal oscillation simulated by the UIUC 11-layer atmospheric GCM. *J. Clim.*, **12**, 1423–1457.

Webster, P. J., V. O. Magana, T. N. Palmer, J. Shukla, R. A. Tomas, M. Yanai, and T. Yasunari (1998) Monsoons: Processes, predictability, and the prospects for prediction. *Journal of Geophysical Research*, **103**(C7), 14451–14510.

Weller, R. A. and S. P. Anderson (1996) Surface meteorology and air–sea fluxes in the western equatorial Pacific warm pool during the TOGA coupled ocean–atmosphere experiment. *J. Clim.*, **9**, 1959–1992.

Wheeler, M. and G. N. Kiladis (1999) Convectively coupled equatorial waves: Analysis of clouds and temperature in the wavenumber–frequency domain. *J. Atmos. Sci.*, **56**, 374–399.

Woolnough, S. J., J. M. Slingo, and B. J. Hoskins (2000) The relationship between convection and sea surface temperature on intraseasonal timescales. *J. Clim.*, **13**, 2086–2104.

Woolnough, S. J., J. M. Slingo, and B. J. Hoskins (2001) The organization of tropical convection by intraseasonal sea surface temperature anomalies. *Q. J. R. Meteorol. Soc.*, **127**, 887–907.

Wu, M. L. C., S. Schubert, I. S. Kang, and D. E. Waliser (2002) Forced and free intraseasonal variability over the South Asian Monsoon region simulated by 10 AGCMs. *J. Clim.*, **15**, 2862–2880.

Wu, Z. (2003) A shallow CISK, deep equilibrium mechanism for the interaction between large-scale convection and large-scale circulations in the tropics. *J. Atmos. Sci.*, **60**, 377–392.

Yang, G.-Y., B. J. Hoskins, and J. M. Slingo (2003) Convectively coupled equatorial waves: A new methodology for identifying wave structures in observational data. *J. Atmos. Sci.*, **60**, 1637–1654.

Yasunari, T. (1979) Cloudiness fluctuations associated with the northern hemisphere summer monsoon. *J. Met. Soc. Jap.*, **57**, 227–242.

Yasunari, T. (1980) A quasi-stationary appearance of 30–40 day period in cloudiness fluctuations during the summer monsoon over India. *J. Met. Soc. Jap.*, **58**, 225–229.

Zveryaev, I. (2002) Interdecadal changes in the zonal wind and the intensity of intraseasonal oscillations during boreal summer Asian monsoon. *Tellus*, **54**, 288–298.

12

Predictability and forecasting

Duane Waliser

12.1 INTRODUCTION

In April of 2002, a workshop was held that brought together participants with a wide range of geophysical expertise to focus on the problem of sub-seasonal predictability (Schubert *et al.*, 2002). This workshop marked a relatively important milestone in the development of our predictive capability of the atmosphere, ocean, and land systems. The fact that it lured scientists with expertise in modeling, theory, and observations, as well as operational forecasters and funding agency administrators indicated that we had reached the point where sub-seasonal variability presented itself as more than a theoretical concern or vaguely observed set of phenomena. In fact, the need for such a workshop was based on the recognition that a number of sub-seasonal features could likely provide near-term opportunities for improving long-lead forecast skill. One of the keynote speakers, H. van den Dool, brought to the participants' attention the early foresight that John von Neumann (1955) had of the expected progress to be made in the area of "long-range" forecasting. In terms of present-day terminology, von Neumann recognized (see Appendix I for excerpt) that the first gains to be made in the area of [atmospheric] prediction were likely to be made at the short range where the initial conditions are expected to play an important role (i.e., 1950s–1970s). Following progress in this area, substantial gains would next be likely made at the very long range, meaning climate prediction, where surface boundary conditions (e.g., large-scale sea surface temperature (SST)) are expected to play the most important role (i.e., 1980s–1990s). Then, only after considerable understanding was obtained in each of these two extreme regimes could progress be made at the sub-seasonal timescale (e.g., 2 weeks to 2 months) where both the initial conditions and boundary conditions are expected to be important. The occurrence of this workshop and its follow-on activities (Waliser *et al.*, 2003a) indicate that by virtue of our progress with both "weather" and "climate" prediction

W. K. M. Lau and D. E. Waliser (eds), *Intraseasonal Variability in the Atmosphere–Ocean Climate System*.

problems, we had reached a point where it was feasible to consider the intermediary problem of the sub-seasonal timescale.

While the workshop mentioned above included presentations and discussion of a number of sub-seasonal phenomena, including the Pacific North America pattern, North Atlantic Oscillation, Arctic Oscillation, and blocking, it was clear that the Madden–Julian Oscillation (MJO) was one of the most underexploited in terms of the likely potential for near-term gains in the area of sub-seasonal prediction, or maybe more importantly, accounting for its effects on medium-to-extended range weather prediction. This was not only due to the characteristics of the phenomena itself (e.g., see Chapter 10) and the direct impact it has on a broad region of the tropics but because of the role it plays, via tropical diabatic heating variability, on the evolution of the extra-tropics (e.g., see Chapters 2–5). In order to fully exploit the possible benefits from MJO/sub-seasonal prediction, it is obvious from the discussion in Chapter 11 that the biggest hurdle to overcome at present is the development of forecasting models that properly represent the phenomena itself. Once this is achieved, it is bound to be an important step to making further progress in weather and climate prediction. For weather, the sub-seasonal timescale offers the hope for extending (at least occasionally) the range of useful forecasts of weather and/or weather statistics, while for the seasonal and longer term climate prediction problem proper representation of the sub-seasonal timescale is a key component of the atmospheric "noise" that is an influential factor in the climate prediction problem as well as a limitation on expected skill.

The previous chapters in this book illustrate the significant influence that the MJO has on our weather and climate. As influential as the MJO is, a fundamental question yet to be adequately addressed concerns its theoretical limit of predictability. For example, it is well known that useful skill associated with deterministic prediction of most "weather" phenomena is limited to about 6–10 days (e.g., Thompson, 1957; Lorenz, 1965; 1982; Palmer, 1993; van den Dool, 1994). Similarly, it has been found that the likely limit of predictability for the El Niño Southern Oscillation (ENSO) is on the order of 12–18 months (e.g., Cane *et al.*, 1986; Graham and Barnett, 1995; Kirtman *et al.*, 1997; Barnston *et al.*, 1999). However, it is still yet to be determined what the corresponding metric is for the MJO/intraseasonal oscillation (ISO) phenomenon. The somewhat well behaved nature of the MJO/ISO (e.g., equatorially-trapped; preference for warm SSTs, seasonality) along with its intraseasonal timescale suggests that useful predictive skill might exist out to at least 15–25 days and maybe longer. Support for this suggestion comes from statistical predictive models of the MJO/ISO, which indicate useful skill out to at least 15–20 days lead time. However, as with any statistical model, these models are sorely limited in the totality of the weather/climate system they can predict, their ability to adapt to arbitrary conditions, and their ability to take advantage of known physical constraints. Additional support for this expectation comes from a few dynamic predictability studies using the twin-experiment methodology that indicates useful predictability may extend to 25–30 days or more. However, our dynamic models still have weaknesses relative to their representations of the MJO so these values also have to be considered with caution.

This chapter will review the progress that has been made regarding our capabilities of predicting the MJO via empirical and dynamical means and our understanding of its predictability characteristics. Note that there are a number of studies that indicate an influence of the MJO on the prediction and predictability of remote (extra-tropics) and/or secondary circulations (e.g., hurricanes). These will not be discussed directly but will be alluded to, and in some cases cited, in Section 12.6. In the following section, a review of empirical methods for forecasting the MJO will be presented. In Section 12.3, an analogous discussion will be presented for forecasts based on dynamical (i.e., numerical weather prediction) models. In Section 12.4, issues regarding the inherent predictability of the MJO will be discussed. In Section 12.5, present-day efforts of real-time MJO forecasting will be described. Section 12.6 concludes with a discussion of the outstanding issues and questions regarding future progress in this area.

12.2 EMPIRICAL MODELS

By the late 1980s, many characteristics of the MJO were fairly well documented and it was clear that it was a somewhat well defined phenomenon with a number of reproducible features from one event to another as well as in events from one year to the next. Given this, and the degree that research had shown a number of important interactions of the MJO with other features of our weather and climate system, it was an obvious step to begin to consider MJO forecasting in more earnest. Since numerical weather and climate models typically had a relatively poor representation of the MJO at the time, a natural avenue to consider was the development of empirical models. Along with likely providing more skillful forecasts than numerical methods available at the time, this avenue also provided a means to establish an initial estimate of the predictability limit for the MJO – at least that which could be ascertained from the observations alone.

The first study along these lines was by von Storch and Xu (1990) who examined Principal Oscillating Patterns (POPs) of equatorial 200-mb velocity potential anomalies from a 2-year subinterval of a 5-year data set. Upon verifying against the data set as a whole (as well as against the remaining three years of data), they found that forecasts based on the first pair of POPs – which tended to emphasize the variability in austral summer (e.g., Figures 4.5 and 5.8) – produced forecasts that were better than persistence and appeared to have useful skill out to at about 15 days (Figure 12.1). While this was a somewhat encouraging result – at least relative to "weather", the limited length of data used combined with the non-stationary characteristics of the MJO over interannual timescales (e.g., Salby and Hendon, 1994; Hendon *et al.*, 1999) necessitated some caution in over-interpretation. Moreover, given the smoothly varying nature of the 200-hPa velocity potential, and the fact that it is only loosely related to near-surface meteorological variables (e.g., precipitation), also suggested caution in generalizing this result to other years, variables, and/or different techniques. In this regard, one might hope that given the roughly 50-day timescale of the MJO that it might be possible to have useful skill out to one-half of a

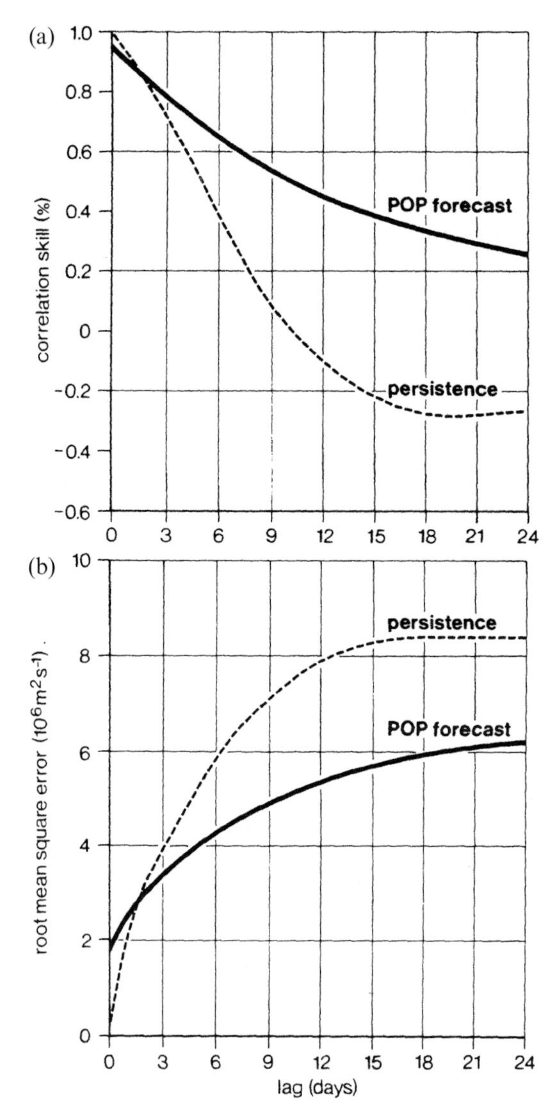

Figure 12.1. Measures of (a) correlation and (b) root-mean-square error forecast skill for persistence and the POP-based forecasting scheme developed by von Storch and Xu (1990). The skills have been derived from daily forecast experiments for the period May 1984 to April 1989. Note the model itself was developed from data between May 1986 to April 1988.

period (van den Dool and Saha, 2002) – particularly for upper level flow (e.g., 200-hPa velocity potential). Subsequent to the above, Kousky and Kayano (1993) suggested that real-time monitoring of the MJO could be achieved by projecting anomalies of a number of fields, outgoing long-wave radiation (OLR) (200-hPa velocity potential, surface pressure, etc.) onto their leading combined extended

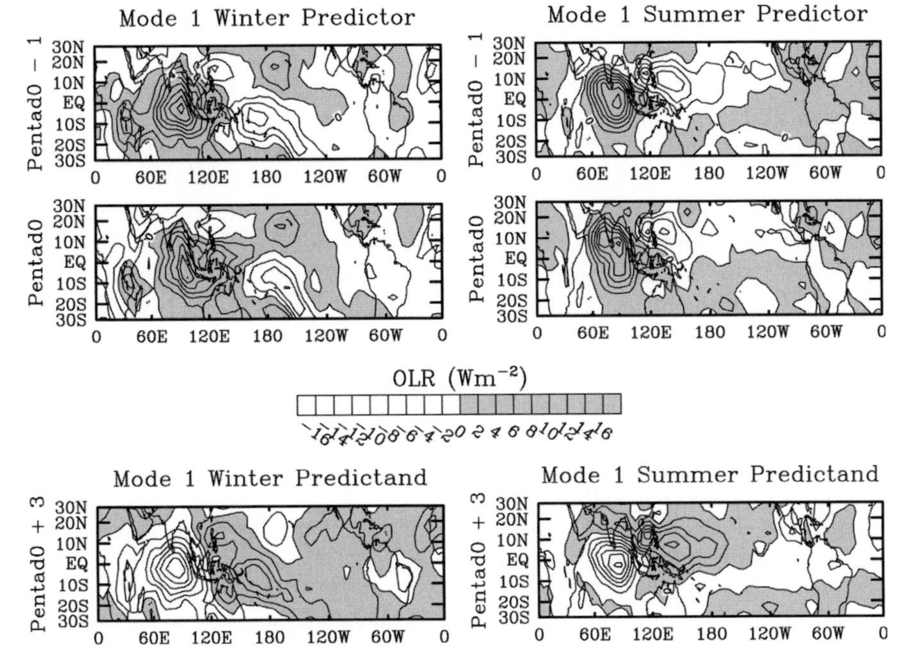

Figure 12.2. (*left column*) Mode 1 from the Singular Value Decomposition (SVD)-based MJO forecasting scheme developed by Waliser *et al.* (1999b) for the northern hemisphere winter and a 3-pented lead forecast. The top panel shows the predictor patterns for "Pentad0" (the current pentad) and "Pentad0 − 1" (the previous pentad). The bottom panel shows the associated predicted patterns for "Pentad0 + 3" (i.e., 3 pentads in the future). (*right column*) The same, except for northern hemisphere summer. Here winter (summer) is defined as November 17 to May 15 (May 16 to November 16). Note that mode 2 for each season looks similar to mode 1 but tends to be spatially in quadrature.

empirical orthogonal function patterns which would indicate the present phase and strength of the MJO in the tropical atmosphere and its likely evolution. It turns out that a number of later developments in the area of empirical MJO prediction tended to follow this suggestion in one form or another.

After a relatively long hiatus in this area, Waliser *et al.* (1999b) developed an empirical MJO forecasting method in order to use the skill results as a benchmark by which to judge the predictive skill of numerical long-range forecasts and to begin exploring the feasibility of employing such a model to augment operational long-range forecasting procedures. The model was based on a field-to-field Singular Value Decomposition that used previous and present pentads of OLR to predict future pentads of OLR (Figure 12.2). Separate models were developed for austral and boreal summer conditions (e.g., Figures 4.5 and 4.10, respectively) using 30–70-day filtered OLR data from 1979 to 1989 and validated on data from 1990 to 1996. For the validation period, the model exhibited temporal correlations to filtered observations of about 0.5–0.9 over a significant region of the eastern hemisphere at lead times from 15 to 20 days, after which the correlation dropped rapidly with increasing lead time. Correlations against observed total anomalies were of the order of 0.3 to 0.5 over a smaller region of the eastern hemisphere. While this was an equally, if not more, encouraging result than that of von Storch and Xu (1990), discussed above, the fact that the model utilized filtered data limited its real-time applicability and in this case warranted caution in considering the result too optimistic. In concluding their study, the authors provided a number of avenues for addressing this filtering problem (i.e., being able to isolate the MJO signal from both the "weather" and the interannual climate variations). For example, it was suggested that the low-frequency variations (i.e., ENSO variability) might be removed using projections on low-order empirical orthogonal functions from coarser (e.g., monthly) data, and high-frequency signals could be removed by using longer time averages, that could even overlap to retain some aspect of the high temporal resolution (e.g., overlapping 10-day averages every 5 days). In addition, it was noted that once the low-frequency variability was removed, low-pass spatial filtering might serve as a useful mechanism for low-pass temporal filtering given that the MJO variability tends to be isolated to wave numbers 1–3 and periods of about 40–60 days.

Following the above study, there were a number of empirical MJO forecasting efforts that each produced a unique and useful approach to the problem. Lo and Hendon (2000) developed a lag regression model that uses as predictors the first two and first three principal components (PCs) of spatially filtered OLR and 200-hPa streamfunction (Ψ), respectively, to predict the evolution of the OLR and 200-hPa streamfunction anomalies associated with the austral summer MJO. In order to address the filtering problem discussed above in regards to real-time application, the data had the annual cycle, interannual and high-frequency (i.e., < 30 days) components removed separately. The annual cycle was removed by subtracting out the first three annual harmonics pointwise. The interannual (e.g., ENSO) variability was removed by developing regression equations between the OLR (and Ψ) anomalies and the PC time series from the first two EOFs of tropical SST anomalies (SSTAs). Based on these regressions and the daily SST values (interpolated from weekly data), the low-frequency components of OLR (and Ψ) that could plausibly be attributed to ENSO were removed. Subsequent to this, the high-frequency temporal components of the data were removed by subjecting the data to a T12 spectral truncation – utilizing the notion that the high-frequency temporal and high wave number spatial variations tended to occur concomitantly. The resulting intraseasonally filtered OLR and Ψ anomalies were subjected to an EOF decomposition and then time-lagged regression equations were developed for predicting, at a given lead, the PC values for the EOF modes that define the MJO (e.g., mode 1 and 2). When tested on independent data, the model exhibited useful skill (correlation ~ 0.5) for predictions of these PCs out to about 15 days (Figure 12.3), with greater skill during active vs. quiescent MJO periods. In comparisons to the filtered observed OLR data, the model exhibited correlation values around 0.3–0.4 in a fairly broad region of the equatorial Indian Ocean and maritime continent.

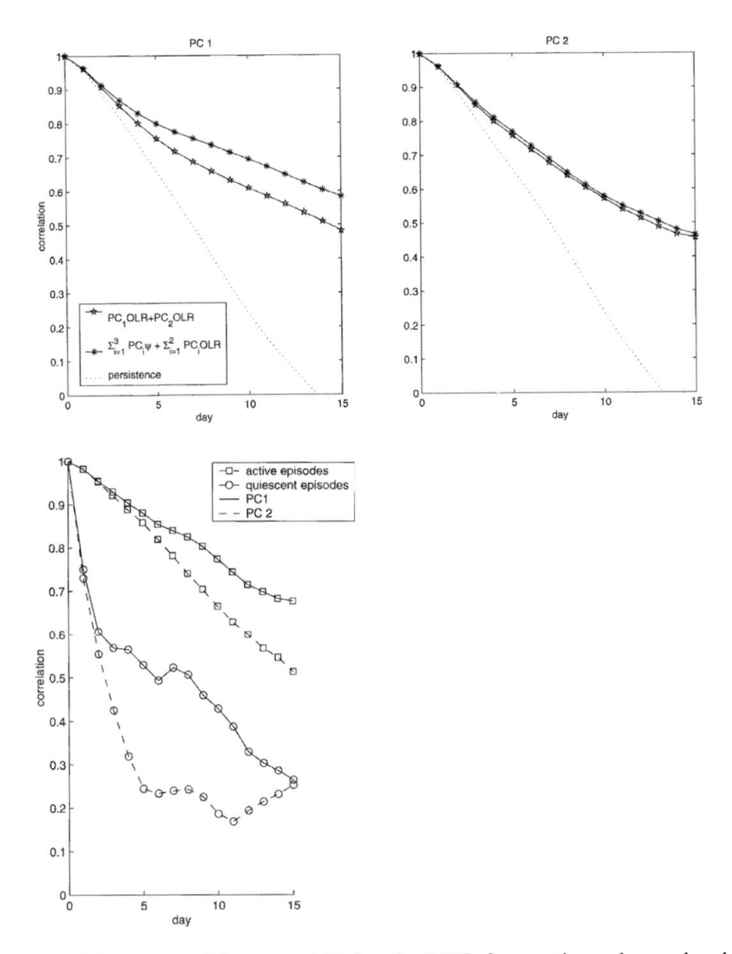

Figure 12.3. (*top*) Measure of forecast skill for the MJO forecasting scheme developed by Lo and Hendon (2000) in terms of correlations between predicted values and verifying values of the PC values associated with mode 1 (PC 1; *left*) and PC values associated with mode 2 (PC 2; *right*). Predictions were made using PC 1 and PC 2 of OLR anomalies (stars), the leading three PCs of the 200-hPa streamfunction anomalies, and leading two PCs of OLR anomalies (asterisks). Persistence is shown as the dotted curve. (*bottom*) Correlations between predicted values and verifying values of PC 1 (solid lines) and PC 2 (dashed lines) of OLR anomalies for times when the MJO was active (squares) and quiescent (circles) at the initial condition during the five winters of dependent data. Predictions were made using the leading three PCs of the 200-hPa streamfunction anomalies and leading two PCs of OLR anomalies. In all figures, correlations are shown as a function of forecast lead time and the verification is against the five winters (1984/1985–1989/1990) of independent data.

A somewhat different approach was taken by Mo (2001) who utilized empirical basis functions in time for the forecasting procedure. This was done by using a combination of singular spectrum analysis (SSA) (Vautard and Ghil, 1989) for the filtering and identification of the principal modes of variability and the maximum entropy method (MEM) (Keppenne and Ghil, 1992) for the forecasting component. The procedure was applied to monitor and forecast outgoing longwave radiation anomalies (OLRAs) in the intraseasonal band over both the Indian–Pacific sector as well as the Pan-American region. This included variability such as the MJO, higher frequency intraseasonal modes associated with the Asian monsoon (see Chapters 2 and 3), and variability related to both of these that occurs over the US west coast (see Chapter 4). For example, in the Pacific and the Pan-American region, there were three leading modes (T-EOFs) identified with periods near 40, 22, and 18 days. In this method, the leading SSA modes (T-EOFs) are determined from a training period. The OLRA time series are then projected onto T-EOFs to obtain the principal components (T-PCs). To obtain fluctuations in a given frequency band of interest (i.e., perform filtering), a subset of the T-EOFs and the related T-PCs associated with that band are summed. This filtering procedure, based on the SSA modes, is data adaptive and there is no loss of end points. This aspect makes it particularly well suited for real-time monitoring. To perform forecasts, the MEM is used to determine the autoregressive coefficients from the training period. These coefficients are used to forecast the T-PCs at future leads. The summation of the T-EOFs and T-PCs related to three retained modes used in the filtering process gives the predicted OLRAs. When tested on 8 years of independent December–February and June–August OLRA data, the averaged correlation over the tropics between the predicted and the observed anomalies was 0.65 (range 0.48–0.78) at the lead times of four pentads (20 days). An example of the forecast skill for the equatorial region for the 1992/1993 winter is given in Figure 12.4. The main activity in this record occurs prior to February 1993 and extends between the Indian and central Pacific Oceans. Although the model amplitudes are weaker that the observations, a feature not uncommon amongst the empirical models, the spatial–temporal structure is well captured out to the pentad-4 forecast.

In a quite different approach, Wheeler and Weickmann (2001) utilized tropical wave theory (Matsuno, 1966) as the basis for their filtering and forecasting technique. Essentially, a space–time Fourier analysis is performed on daily OLR data for a given time–longitude section of interest in the tropics. In a previous study, Wheeler and Kiladis (1999) showed that the spectrum from such an analysis exhibits variability that is associated with the modes that one would expect from theoretical considerations (e.g., Kelvin waves and mixed Rossby–gravity waves), as well as the expected peak of variability around wave numbers 1–3 and 40–60 days associated with the MJO. In order to monitor and predict the evolution of a given mode of interest, the specific zonal wave numbers and frequencies associated with the mode(s) of interest are retained and then the modified spectrum is inverse-Fourier analyzed. Figure 12.5 shows how the filtered values obtained for times before the end of the data set can be used for monitoring the activity of a given mode, while the filtered fields obtained for times after the end point may be used as a forecast. This idea is akin to an ocean tidal forecast, which in that case is based on harmonic analysis in time only and of course much sharper frequencies of variability. For prediction, the method exhibits useful skill for the MJO out to about 15–20 days. An advantage is that the method readily provides

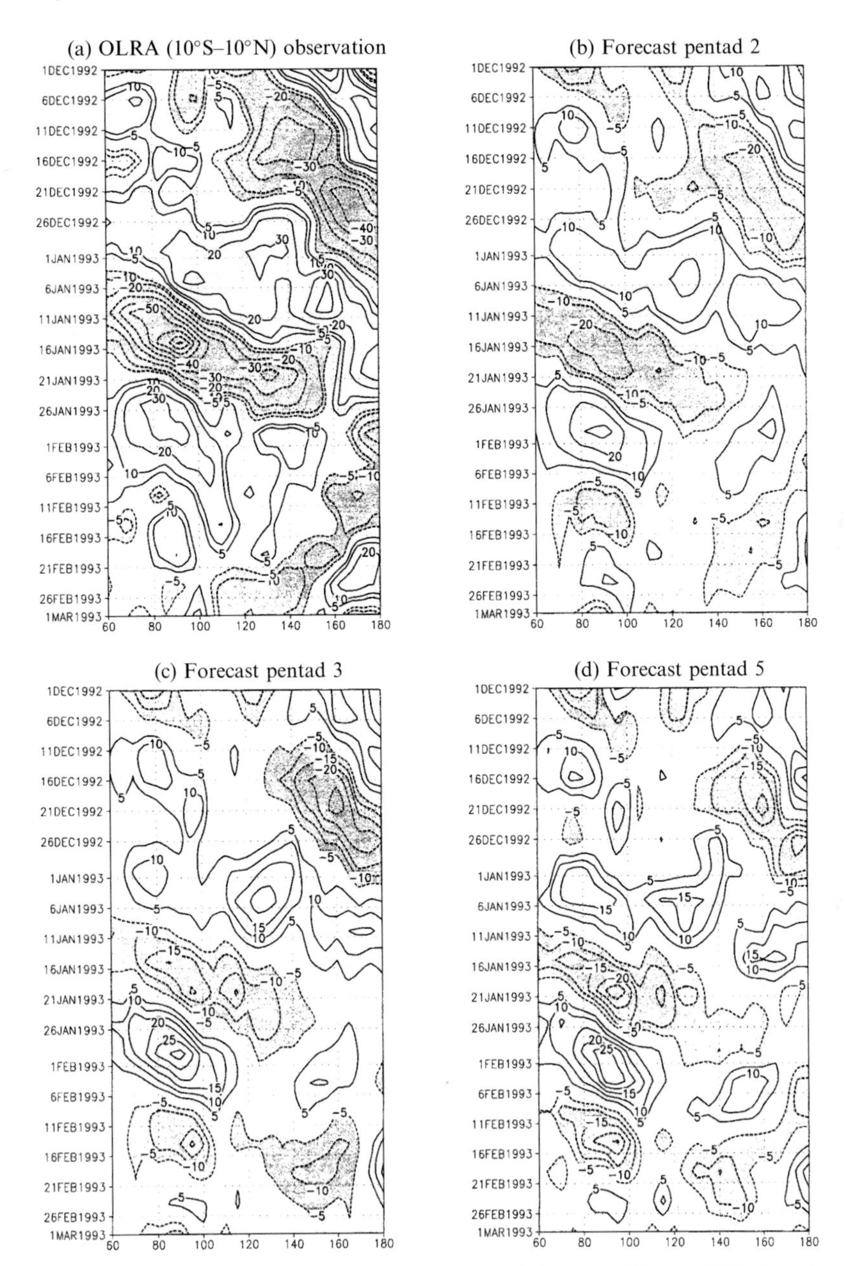

Figure 12.4. (a) 10–90-day filtered OLRAs averaged from 10°S to 10°N based on the minimum bias window for the 1992/1993 winter from observations. Contour interval is $10\,\mathrm{W\,m^{-2}}$. Contours 25 and $5\,\mathrm{W\,m^{-2}}$ are added. Negative values are shaded and zero contours are omitted. (b) Same as (a), but for 10-day (pendad 2) forecasts based on the empirical model developed by Mo (2001) verified on that day. (c) Same as (b), but for 15-day forecasts (pentad 3). Contour interval is $5\,\mathrm{W\,m^{-3}}$. (d) Same as (c), but for 20-day (pentad 4) forecasts.

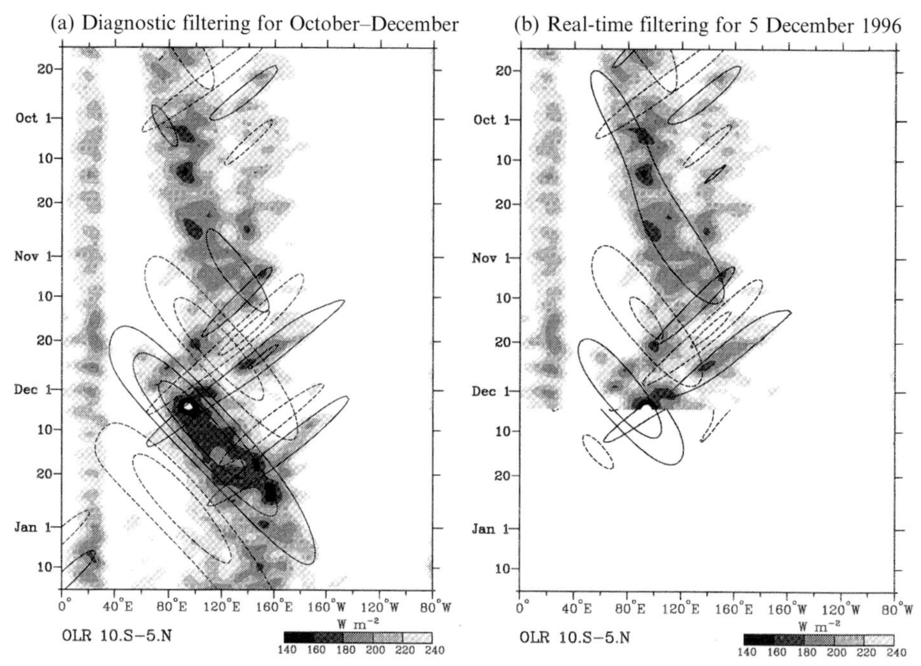

Figure 12.5. (a) Time–longitude plot of the total OLR (with R21 spatial truncation, and a 1–2–1 filter applied in time) and filtered OLRAs averaged between 10°S and 5°N during late 1996 to early 1997. Shading is for the total OLR, and contours are for the diagnostically filtered anomalies of the MJO and $n = 1$ equatorial Rossby (ER) wave. Solid contours represent negative OLRAs, while dashed contours are for positive anomalies, with the contour interval for both wave filtered bands being $10\,\mathrm{W\,m^{-2}}$, and the zero contour omitted. (b) Same as in (a) except that the filtering was performed with the last day of data being that from 5 December 1996. After 5 December, when the real-time filtered anomalies are continued into the future as a forecast, the contour interval is halved.
From Wheeler and Weickman (2001).

predictions of other well-defined, typically higher frequency, modes of large-scale tropical variability.

In an effort that focused on active and break conditions of the Indian summer monsoon, Goswami and Xavier (2003), noted that all active (break) conditions go over to break (active) phases after about 15–20 days (see Figure 2.3). The events would be highly predictable if the transitions from active to break (or vice versa) were all identical. However, the rate of transition, the magnitude of the next minimum (or maximum), and the timing of achieving the minimum (maximum) of the next phase vary from event to event. Using the rainfall-based index illustrated in Figure 2.3 and their definition of active and break conditions (see Chapter 2), Goswami and Xavier calculated the typical (i.e., ensemble average) transition from active to break (and break to active) conditions as a function of lead time. The typical size of these transitions – referred to as the "signal", and their associated intra-ensemble variance – referred to as the (ensemble) "spread", are shown in

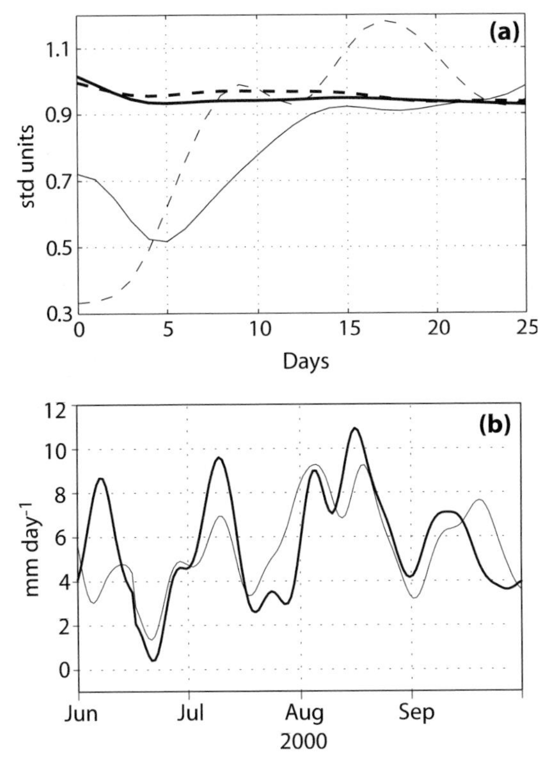

Figure 12.6. (a) The thick dashed (solid) line is the monsoon ISO "signal" starting from troughs (peaks) of the index (see Figure 2.3). The thin dashed (solid) line is the variance (or spread) of ensemble members as a function of days from the initial date corresponding to all troughs (peaks) of the index representing transitions from break to active (active to break). (b) Time series of 18-day predictions (thin line) and observations (thick line) of the rainfall (mm day^{-1}) averaged over the monsoon trough region for June–September 2000.
From Goswami and Xavier (2003).

Figure 12.6(a). While the variability among break to active transitions become as large as the associated signal in less than 10 days, it takes more than 20 days for the variance among active to break transitions to become as large as the signal. Circulation parameters such as the 850-hPa relative vorticity over the monsoon trough region also lead to the same conclusion (not shown). These results indicate that monsoon breaks are intrinsically more predictable than active monsoon conditions. Similar results were found by Waliser *et al.* (2003c) using an ensemble of twin-predictability GCM experiments (see Section 12.4). To explore the practical consequences of these results, Goswami and Xavier constructed an empirical multiple regression model for the first four PCs of 10–90-day filtered CMAP using the first four PCs of filtered rainfall data and the first two PCs of filtered surface pressure as predictors, and showed that useful prediction of monsoon breaks, up to 18 days in advance, could be made while useful forecast of active conditions could be made with lead times of only about 10 days. Eighteen-day forecasts of filtered precipitation averaged over central India and the northern Bay of Bengal (70°E–85°E, 10°N–22°N) for June to September 2000 are shown in the lower panel of Figure 12.6.

The above discussion gives a flavor of the types of empirical MJO modeling that have been developed to date and their associated levels of forecast skill. However, there is a number of additional empirical modeling studies worth describing that are presently associated with real-time efforts and these are discussed in Section 12.5. In regard to the studies above, it is useful to emphasize at this point that the skill associated with the techniques above in almost all cases has yet to be demonstrated by numerical forecast techniques (e.g., Waliser *et al.*, 1999b; Jones *et al.*, 2000; Lo and Hendon, 2000; Wheeler and Weickmann, 2001). Moreover, it is worth highlighting that in no cases are these schemes physical in nature or based on very complex techniques, and they are all based on linear methods. Thus, it is likely that we may not have yet developed and demonstrated models that have saturated the skill potential for empirical forecasting methods. In addition, it is also worth highlighting that MJO events are at best quasi-periodic in nature, meaning here that the atmosphere can be relatively quiescent in regards to MJO variability with an event suddenly developing. Each of the models above would tend to perform relatively poorly at forecasting this initial development, as they all tend to rely on the periodic nature of the MJO to forecast its evolution. For these scenarios, as well as for dealing with the heterogeneity of MJO events, it will be vital to improve our dynamic models, as they are likely to be the best means to deal with these sorts of issues. In any case, the above sorts of studies provide a useful benchmark in forecast skill for our dynamical models and suggest, based on the observations alone, that the MJO should be predictable with lead times of at least 2–3 weeks.

12.3 DYNAMICAL FORECAST MODELS

To date, there have only been a handful of studies that have examined forecast skill (i.e., verified against observations) from dynamical models. This has probably stemmed from what amounted to: (a) considerably less overall interest in forecasting the intraseasonal timescale relative to weather and ENSO, (b) the difficulty and resources required to produce an adequate sample of very long-range weather forecasts (at least 30 days), (c) the pessimism and known challenges associated with tropical weather forecasting in general, and (d) the indications that neither our forecast nor climate simulation models were very adept at simulating the MJO (see Chapter 11). In any case, as part of a more generalized forecast skill study of the planetary-scale divergent circulation, Chen and Alpert (1990) examined the MJO forecast skill from one year (June 1987–May 1988) of daily 10-day forecasts from the US National Meteorological Center's [NMC; now National Center for Environmental Prediction (NCEP)] medium-range forecast (MRF) model (based partly on MRF86 and MRF87) in terms of the 200-hPa velocity potential. In their analysis, the MRF's forecast skill, measured in terms of spatial correlations of 200-hPa

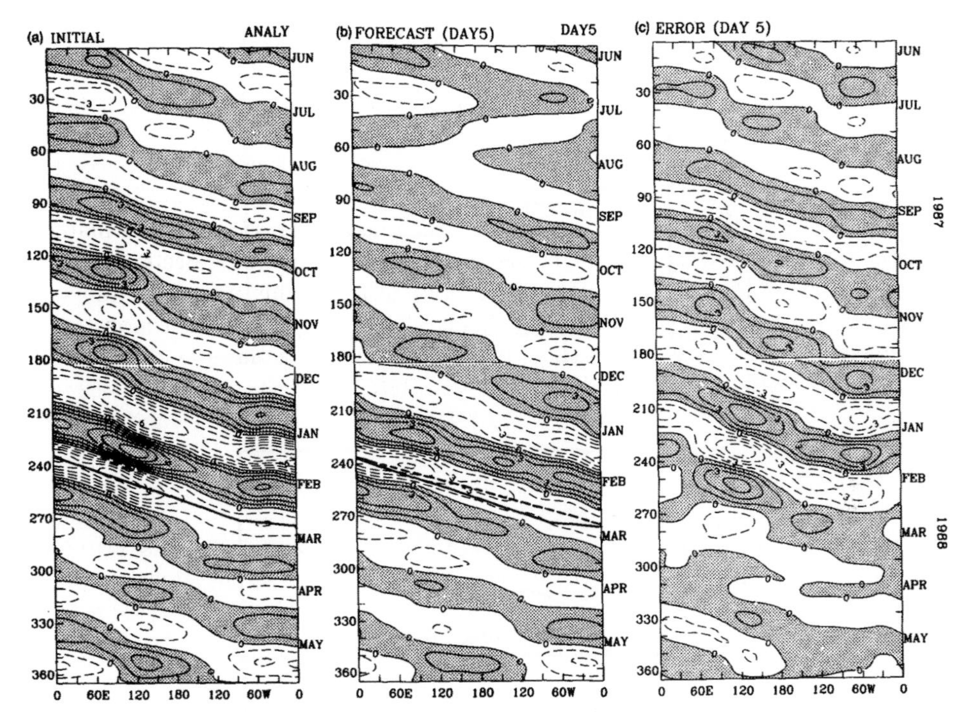

Figure 12.7. Equatorial time–longitude diagram of 200-hPa velocity potential anomalies constructed from the first two EOF modes of the data. In this case, the EOF was done separately for each lead time. Values are shown for: (a) initial conditions, (b) day 5 forecasts, and (c) day 5 forecast errors. The solid line connects maximum initial condition 200-hPa velocity potential anomalies for one-half of a life cycle of the ISO during winter in (a). The dashed line connects maximum 200-hPa velocity potential anomalies of the day 5 forecast in (b). The solid line of (a) is also shown in (b) for comparison. The contour interval is $1.0 \times 10^6\,\mathrm{m^2\,s^{-1}}$.
From Chen and Alpert (1990).

velocity potential between 50°N–50°S, declined to about 0.6 by forecast day 6 and 0.4 by forecast day 9. This relatively poor skill was attributed to: (1) the inability of the model to maintain MJO variability during a forecast (thus the model probably did not intrinsically exhibit or support an MJO of its own) and (2) the model's tendency to propagate MJO anomalies too fast. This latter aspect is illustrated in Figure 12.7. The left panel illustrates the observed 200-hPa velocity potential with a line overlaid to indicate the observed phase speed of one of the stronger events. The middle panel illustrates the model forecast values at a lead time of 5 days with the same line overlaid from the left panel which shows quite clearly that the model is propagating the anomalies too fast. Lau and Chang (1992) analyzed one season (14 December 1986–31 March 1987) of 30-day global forecasts derived from a set of Dynamical Extended Range Forecasts (DERFs) from a research version of the MRF86 model mentioned above. Their results, depicted in Figure 12.8, showed

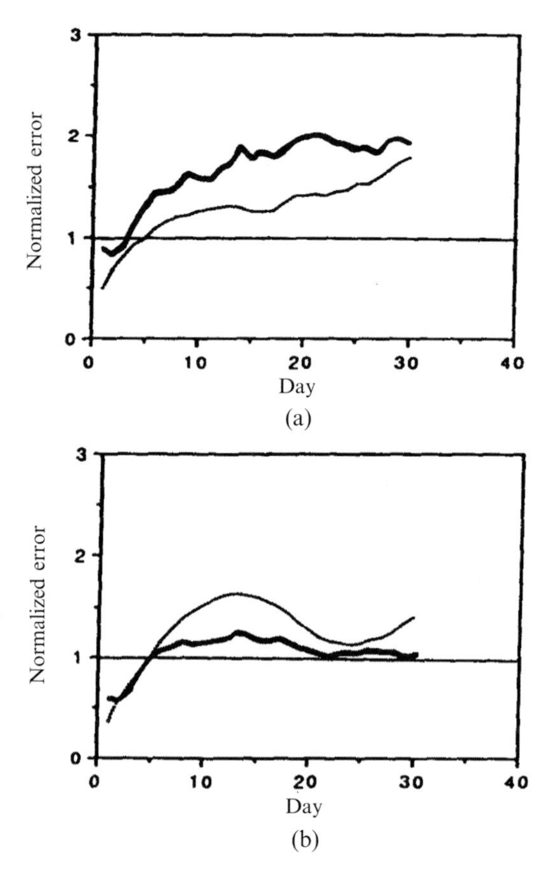

Figure 12.8. Forecast (thick) and persistence (thin) error curves for the first EOF of the global velocity potential (referred to as the tropical mode) from the NMC (now NCEP) Phase II DERF experiment. This data set consisted of 108 30-day forecasts for the period 14 December 1986–31 March 1987. The top (bottom) shows the errors for a period of weak (strong) MJO activity. The horizontal lines indicate the level of natural variability for the corresponding periods.
From Lau and Chang (1992).

that the forecast model had significant skill in predicting the global pattern of ISV in 200-hPa velocity potential and streamfunction for up to 10 days lead time, with the error growth of tropical and extra-tropical low-frequency modes less than persistence when the amplitude of the MJO was large and vice versa when the amplitude was small.

Both Hendon *et al.* (2000) and Jones *et al.* (2000) analyzed a more recent DERF experiment which used the reanalysis version (Kalnay *et al.*, 1996) of the NCEP MRF model (Schemm *et al.*, 1996). This experiment included 50-day forecasts made once a day for the period January 1985–February 1990. In both studies, the focus was on the northern hemisphere winter season. Using different analysis and filtering techniques for identifying the MJO within the forecasts, and thus for assessing forecast skill against observations, both studies concluded that this version of the NCEP MRF model also exhibited a rather poor MJO forecast skill. Specifically, the upper panel of Figure 12.9 from Jones *et al.* (2000) shows that the anomaly correlations of intraseasonally filtered values of 200-hPa zonal wind, zonally-averaged along the equator, declined from about 0.6 on day 3 to 0.2 by day 10. The lower panel shows that the model exhibited a forecast skill that was slightly better (worse) when the MJO was particularly active (weak; i.e., null case) and that the model skill might have some dependence on the phase of the MJO (see caption for details). The rather poor forecast skill was attributed to the development of systematic errors in the forecast upper-level winds, particularly over the eastern Pacific and, as above, due to the inability of this model to maintain/simulate a robust MJO phenomena itself. For example, a diagnosis of the model's representation of the MJO from a 10-year simulation using the same model showed an MJO-like phenomenon but one that was significantly less intense and propagated considerably faster than the observed phenomenon (Jones *et al.*, 2000).

The rather detailed analysis by Hendon *et al.* (2000) showed that the forecasts initialized during very active episodes of the MJO did not reproduce the observed eastward propagation of the tropical convection and circulation anomalies, rather the anomalies would typically weaken in place and even retrograde in some cases. Typically it was found that the convective anomalies would decay almost completely by day 7 of the forecast, and in nearly the same time systematic errors in the extra-tropical 200-hPa streamfunction became fully developed. They argued that the errors in the latter developed due to the collapse of the tropical heating anomalies and thus the development of an error in the Rossby wave source emanating from the tropics. Due to the types of errors, which are greatest for the largest MJO anomalies, and likely due to their categorization of active events, their analysis showed that forecast skill in the tropics and northern hemisphere extra-tropics was actually worse during the active MJO events vs. periods that exhibited very little MJO activity (Figure 12.10 – see caption for details). The above studies point to the need for the forecast models to not only have a proper representation of MJO anomalies but also to produce an unbiased mean state so initialization errors and their subsequent evolution/adjustment do not contaminate the forecast over these relatively long lead times.

In a totally independent line of research, T. N. Krishnamurti produced a number of studies in the early 1990s that examined MJO forecast skill as it relates to active and break periods of the Asian summer monsoon. Underlying these studies is the development and application of a unique and potentially promising avenue for forecasting "low-frequency modes" (as they are referred to in these studies). In the first study, Krishnamurti *et al.* (1990) laid the groundwork for the method which argues that part of the loss of forecast skill associated with low-frequency modes such as the MJO during a forecast comes about from the errors and evolution of high wave number/frequency variability. If the forecast objective is primarily the prediction of low-frequency variability (e.g., active and break periods of the monsoon), then it is plausible to filter the initial state in order to remove all but

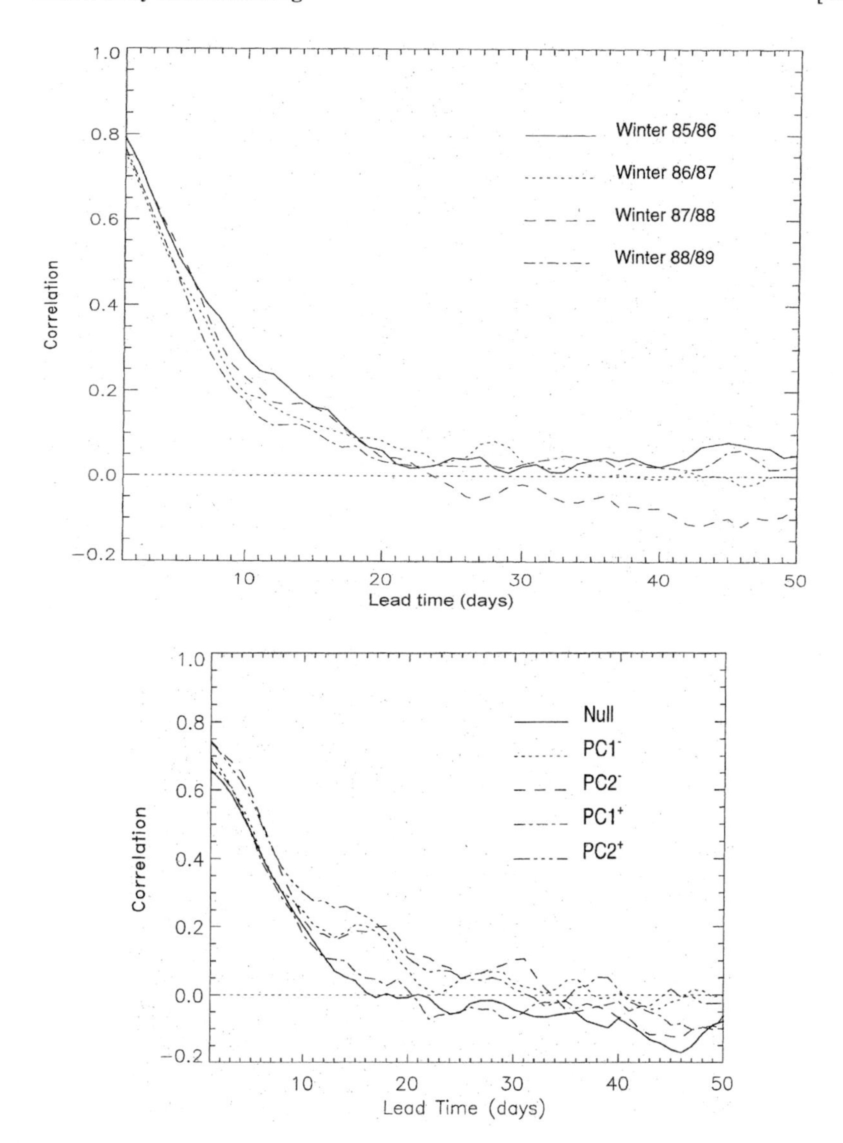

Figure 12.9. Zonal average along the equator of the anomaly correlation between the forecast and verification of 200-hPa zonal velocity as a function of lead time. Forecast data are based on the DERF experiment (Schemm *et al.*, 1996) which used the reanalysis version (Kalnay *et al.*, 1996) of the NCEP MRF model. This experiment included 50-day forecasts made once a day for the period January 1985–February 1990. Correlations are shown for (*top*) each winter season separately and for (*bottom*) the forecast separated into four different phases of the MJO, and quiescent (i.e., null) cases, using the PC values of the first two EOFs of intraseasonally filtered OLR. Values associated with PC 1+, PC 1−, PC 2+, and PC 2− are associated with the convective phase being located at 90°E, 120°E, 150°E, and 170°W, respectively, at the start of the forecast.
From Jones *et al.* (2000).

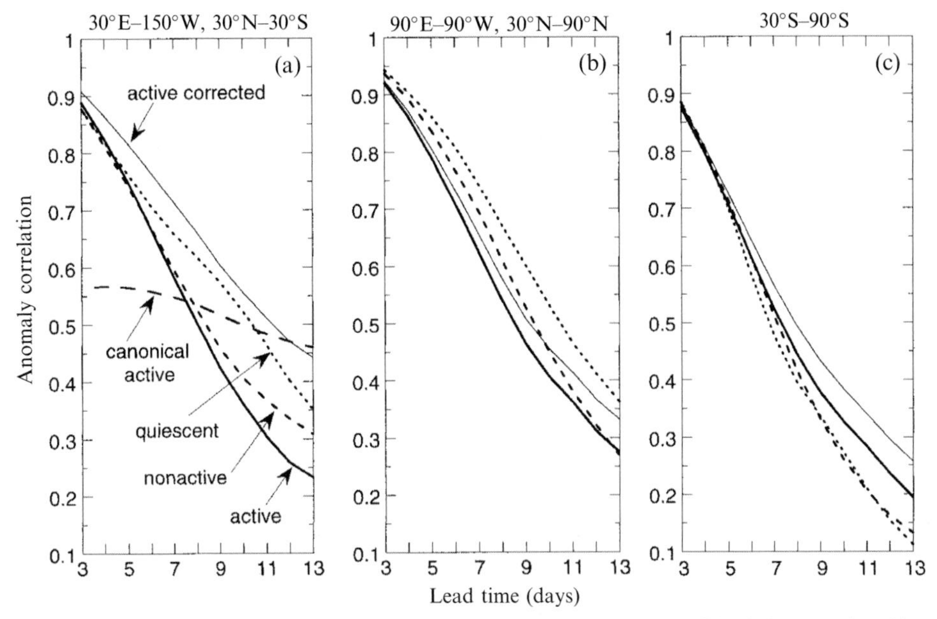

Figure 12.10. Anomaly correlations between forecasts, as functions of lead time, and verification of the 200-hPa streamfunction for: (a) the tropical region 30°N–30°S, 30°E–150°W; (b) the northern hemisphere extra-tropical region 30°–90°N, 90°E–90°W; and (c) the southern hemisphere extra-tropics 30°S–90°S. Forecast data are based on the DERF experiment (Schemm *et al.*, 1996) which used the reanalysis version (Kalnay *et al.*, 1996) of the NCEP MRF model. This experiment included 50-day forecasts made once a day for the period January 1985–February 1990. Correlations are shown for all forecasts initialized when the MJO was active and quiescent and for times when the MJO was inactive (i.e., neither active nor quiescent). Active and quiescent MJO periods were selected using the PC values of the first two EOFs of intraseasonally filtered OLR. Empirically corrected (lead-dependent systematic MJO error estimated and removed) anomaly correlations are also shown for the forecasts initialized when the MJO was active (labeled "active corrected"). Also shown in (a) is the anomaly correlation for the canonical MJO (i.e., a simple empirical model), which is formed by lag regression of the verifying analyses onto the leading two PCs of OLR at the initial forecast time.

the relevant/recent "mean" state (e.g., 45-day average conditions prior to forecast) and the low-frequency modes of interest (in this case, obtained via time filtering). Krishnamurti *et al.* (1990) argue that this will delay the "contamination of the low-frequency modes as a result of the energy exchanges from the higher frequency modes."

The above idea was tested using a T21 version of the Florida State University (FSU) global spectral model (Krishnamurti *et al.*, 1990). Observed SSTAs, filtered to include only 30–50-day variability (Krishnamurti *et al.*, 1988) and multiplied by a factor of 3.5 to help account for the coarse vertical/boundary-layer structure (i.e., 8 total layers), were specified in addition to the mean annual cycle of SST. While the

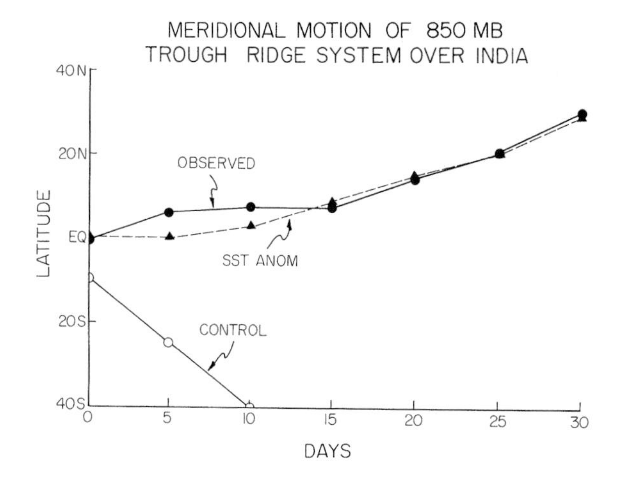

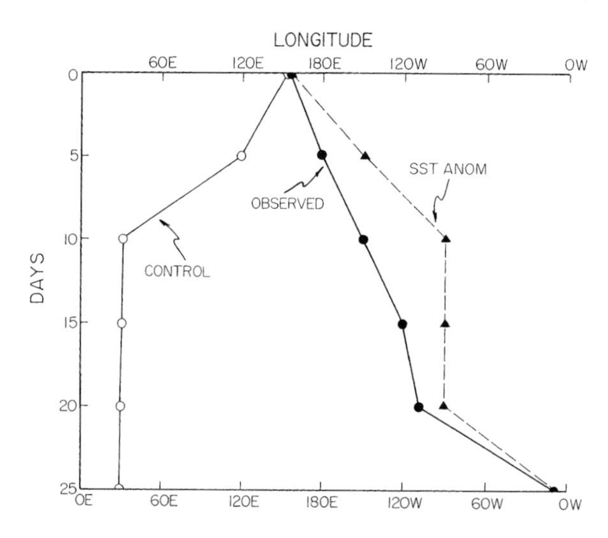

Figure 12.11. (*top*) A y–t diagram of the monsoonal low-frequency ridge line at 850 hPa. The results are shown for the observations, control experiment (no SSTA and complete initialized state), and the SSTA experiment (SSTA specified and initialized state includes time-mean and low-frequency mode only. (*bottom*) Same as (*top*), except an x–t diagram of the position of the 200-hPa divergent center. See text for details of experimental set-up and dates of forecasts. From Krishnamurti *et al.* (1990).

latter specification certainly provides the hindcast with information that a true forecast would not have, the results from the 270-day forecast case study performed from 31 July, 1979, were still encouraging given that intraseasonal SSTAs are not a dominant control at this timescale (Chapters 7, 10, and 11). Figure 12.11 shows that the model forecast exhibited considerable skill at predicting the meridional motion of the 850-hPa trough–ridge system over India and the eastward propagation of the 200-hPa divergence anomaly out to about 4 weeks. A control experiment that included all frequencies and wave numbers associated with the initial conditions (that the model would accommodate) and that did not include the SSTAs actually performed quite poorly in the first few days. In addition, it was found that if only the mean annual cycle of SST is specified, the amplitude of the low-frequency wave motion degrades considerably, thus indicating the importance of such SSTAs in such an experiment. In Krishnamurti *et al.* (1992, 1995), analogous experiments using two select case studies were performed for low-frequency "wet and dry spells" over China and Australia for each of their associated summer monsoons with essentially the same results as those indicated. As in the first study, the SSTAs were found to be vital to retaining the forecast skill. In both of these studies, simple empirical prediction of the SSTAs was incorporated and found to provide much of the necessary SST information to retain most of the long-lead forecast skill found in this suite of experiments.

12.4 PREDICTABILITY

The previous two sections provide some indication of what the inherent predictability limit might be for the MJO. From the empirical model studies, this limit might be ascertained to be at least 20–30 days. However, as with any empirical model, these models are limited in the totality of the weather/climate system they can predict, their ability to adapt to arbitrary conditions, and their ability to take advantage of known physical constraints. Thus one might conclude that if dynamical models had a realistic representation of the MJO, this limit might be extended somewhat. However, the information that can be ascertained from the above mentioned dynamical studies regarding the intrinsic predictability of the MJO is limited, due to the fact that they were either based on models with a relatively poor representation of the MJO (e.g., weak amplitude and relative fast phase propagation) or they were based on only a few select cases. Moreover, since all the dynamical studies discussed above were verified against observations, their degradation in skill with lead time includes the component associated with the natural limit of predictability of the MJO phenomenon as well as a model's systematic bias associated with the MJO.

A complimentary avenue of research for ascertaining the inherent limits of prediction for the MJO could be derived from so-called "twin-predictability" experiments in which the model employed is presumed to be "perfect" and forecast experiments are verified against others that only differ in the initial conditions (e.g., Lorenz, 1965; Shukla, 1985). This approach was taken in two recent studies by Waliser *et al.* (2003b, c). The important consideration for a study such as this is that the model provides a relatively realistic representation of the phenomenon of interest. In this case, the experiments were performed with the NASA Goddard Laboratory for Atmosphere's (GLA) GCM (Kalnay *et al.*, 1983; Sud and Walker, 1992). In a number of studies, this model has been shown to exhibit a relatively realistic MJO (Slingo *et al.*, 1996; Sperber *et al.*, 1997; Waliser *et al.*, 2003d) with reasonable amplitude, propagation speed, surface flux properties, seasonal modulation, and interannual variability (IAV) (Waliser *et al.*, 2001). One of its principal deficiencies is its relatively weak variability in the equatorial Indian Ocean, a problem quite common in AGCMs (Waliser *et al.*, 2003d).

For these studies, a 10-year control simulation using specified annual cycle SSTs was performed in order to provide initial conditions from which to perform an ensemble of twin-predictability experiments. Note that this analysis was performed separately on northern hemisphere winter MJO activity (i.e., that which typically travels eastward along the equator and South Pacific Convergence Zone (SPCZ); e.g., Figure 4.5) and northern hemisphere summer MJO activity (i.e., that which typically travels north-eastward into Indian/South East Asia; e.g., Figure 4.10). The following discussion describes the northern hemisphere winter study (Waliser *et al.*, 2003b) but the methods and results are quite similar for the northern hemisphere summer analog (Waliser *et al.*, 2003c). Initial conditions were taken from periods of strong MJO activity identified via extended EOF analysis of 30–90-day band-passed tropical rainfall during the Oct–Apr season. From the above analysis, 15 cases were chosen when the MJO convection was located over the Indian Ocean, maritime continent, western Pacific Ocean, and central Pacific Ocean, respectively, making 60 cases in total. In addition, 15 cases were selected which exhibited very little to no MJO activity. Two different sets of small random perturbations, determined in a rather *ad hoc* and simplistic manner, were added to these 75 initial states. Simulations were then performed for 90 days from each of these 150 perturbed initial conditions.

A measure of potential predictability was constructed based on a ratio of the signal associated with the MJO, in terms of band-passed (30–90-day filter) rainfall or 200-hPa velocity potential (VP200), and the mean square difference between sets of twin (band-passed) forecasts. Predictability was considered useful if this ratio was greater than one, and thus if the mean square error was less than the signal associated with the MJO. The results indicate that useful predictability for this model's MJO extends out to about 20 to 30 days for VP200 and to about 10 to 15 days for rainfall (Figure 12.12). This is in contrast to the timescales of useful predictability for the model's weather, or for cases in which the MJO is absent, which is about 12 days for VP200 and 7 days for rainfall. Note that these latter two regimes are related, in that when the MJO is quiescent, the model lacks a low-frequency component that might help it retain predictability over long timescales and is in a regime where the processes and timescales of weather are the only phenomena left to give predictability. In addition to the above, the predictability measure exhibited modest dependence on the phase of the MJO, with greater predictability for the convective phase at short ($< \sim 5$ days) lead times and for the suppressed phase at longer ($> \sim 15$ days) lead times.

While the results from these studies are encouraging from the view point of subseasonal prediction, and are not entirely inconsistent with the sorts of complimentary studies mentioned above, there are a number of issues to consider that might impact the limit of predictability estimate they provide. First, the model has been

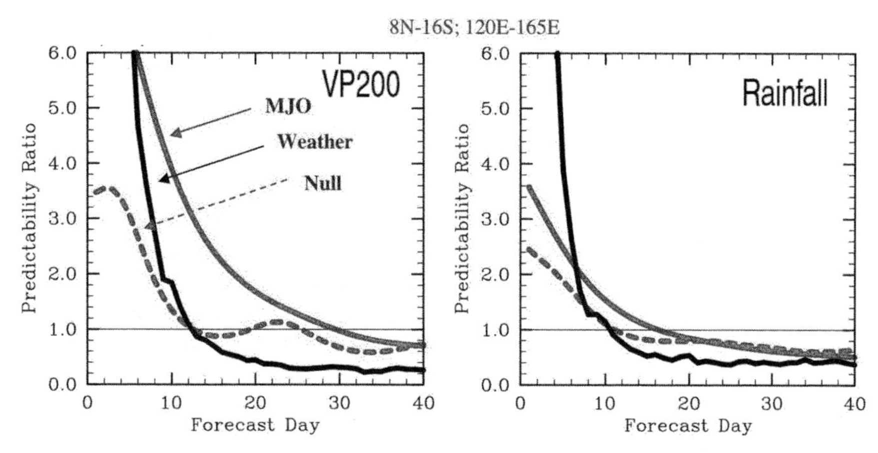

Figure 12.12. Predictability measure (defined as the ratio of the MJO "signal" and the MJO forecast error; see Waliser *et al.*, 2003b) vs. lead time based on 120 northern hemisphere winter MJO twin-predictability forecast cases for VP200 (*left*) and rainfall (*right*) from the NASA GLA model for 120 active/strong MJO cases (solid black), 30 weak/null MJO cases (dashed gray), and the unfiltered "weather" variations (using the 120 active MJO cases; solid gray) for the region 8°N–16°S and 120°E–165°E.

shown to have too much high-frequency low-wave-number activity (Slingo *et al.*, 1996). Relative to the MJO, this variability would be considered to be disorganized, errant convective activity that may erode the relatively smooth evolution of the MJO and thus diminish its predictability. Second, these simulations were carried out with fixed climatological SST values. A previous study with this model showed that coupled SSTs tend to have an enhancing and organizing influence on the MJO, making it stronger and more coherent (Waliser *et al.*, 1999a). Thus the exclusion of SST coupling may lead to an underestimate of the predictability as well. Both of these issues would appear to have a direct relation to the methods and results associated with the Krishnamurti *et al.* studies discussed above.

There are also a number of aspects associated with the model and/or analysis to suggest that the above results might overestimate the predictability of the MJO. The first is that the model's coarse resolution and inherent reduced degrees of freedom relative to the true atmosphere may limit the amount of small-scale variability that would typically erode large time and space scale variability. However, it is important to note in this regard that the low-order EOFs of intraseasonally filtered model output typically do not capture as much variability as analogous EOFs of observed quantities. Thus the model's MJO itself still has room to be more robust and coherent which would tend to enhance predictability. In addition to model shortcomings, the simple manner in which perturbations were added to the initial conditions may also lead to an overestimate of the predictability. The perturbation structure and the size of the perturbations may be too conservative and not adequately represent the type of initial condition error that would be found in an operational context. However, even if that is the case, it would seem that adequate size "initial" errors would occur in the forecast in a matter of a day or two and thus one would expect this aspect to overestimate the predictability by only a couple of days, if at all.

In order to address some of the uncertainties mentioned above, an analogous study for boreal summer conditions using the European Centre for Medium-range Weather Forecast–Hamburg atmospheric model (ECHAM) AGCM has recently been undertaken (Liess *et al.*, 2004). The modeling and analysis framework is similar to that described above with two important exceptions. First, rather than select a large number of events (i.e., ~15–20) for each of the 4 phases of the boreal summer ISO (i.e., convection in Indian, maritime continent, western Pacific, central Pacific) and performing only a few (i.e., 2) perturbation experiments with each, this study has selected 3 relatively strong events and performs a larger ensemble of forecasts for each of the 4 phases (i.e., 15). In addition, rather than use simply determined perturbations, this study uses the breeding method (Toth and Kalnay, 1993; Cai *et al.*, 2003). The left panels of Figure 12.13 show precipitation (upper) and 200-hPa velocity potential (right) from the individual members of one of the 15-member ensembles (i.e., one phase of one event). Evident is the expected spread of the forecasts with lead time. The right panels of Figure 12.13 quantify this spread in terms of a "signal-to-noise" ratio, defined as in the Waliser *et al.* (2003) study described above. These results suggest that the boreal summer ISO exhibits dynamical predictability with lead times potentially up to and beyond 30 days. These lead times are at least as large, if not larger, than those found in the Waliser *et al.* studies highlighted above. However, it should be noted that the events analyzed here are the strongest 3 events in a 10-year model simulation record, and those above were based on both strong and moderate size events which could account for the difference. In any case, even though the above results do not take into account systematic model bias relative to the observations, they, along with many of the other studies discussed above, indicate that a promising avenue and timescale of operational prediction lies ahead.

12.5 REAL-TIME FORECASTS

Based on the qualified success of some of the MJO prediction efforts discussed above, namely those associated with empirical models, a number of forecast schemes have been implemented in real time. The first of these was associated with the Wheeler and Weickmann (2001) scheme described in Section 12.2. This scheme has been operational for about 3 years and provides forecasts out to about 2–3 weeks lead time for not only the MJO but other coherent modes of tropical variability (e.g., Kelvin waves and mixed Rossby–gravity waves). A second, somewhat related effort, that has been developed more recently builds on the study by Lo and Hendon (2000) and utilizes what is referred to as an all-season Real-time Multivariate MJO (RMM) index (Wheeler and Hendon, 2003). The index results from projecting daily data onto the first two modes of a combined EOF of tropical (15°N–15°S) OLR, and zonal winds at 850 and 200 hPa. This projection

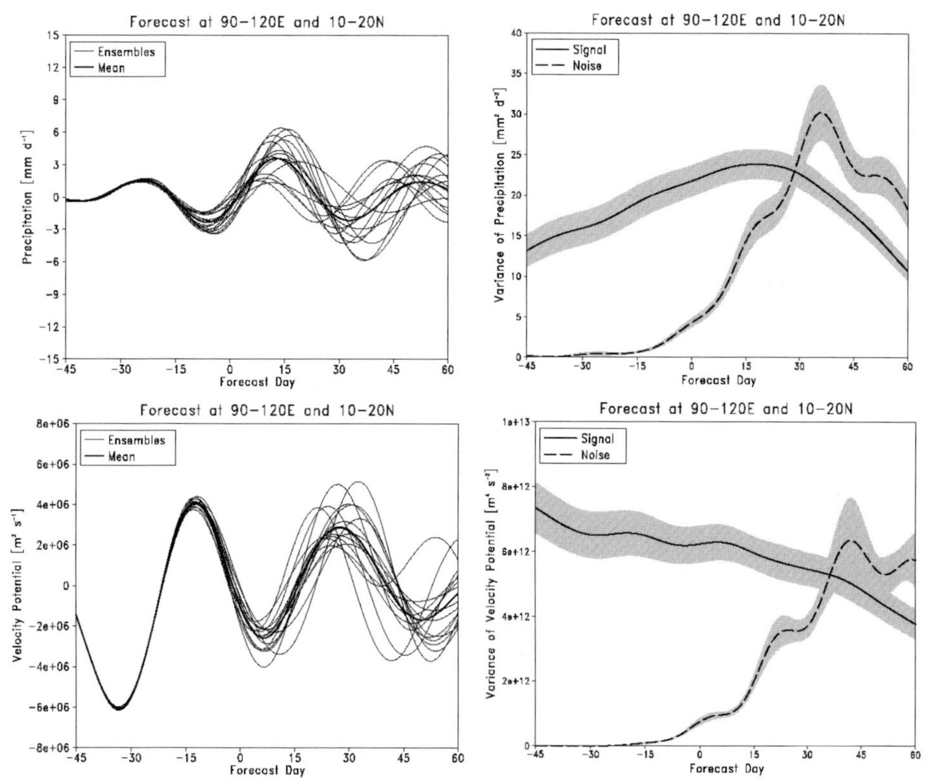

Figure 12.13. (*left*) Fifteen-member ensemble forecast of a boreal summer ISO event, in a given phase of the event (out of four defined phases), using the ECHAM5 AGCM. Data are taken from 90°E–120°E and 10°N–20°N and are 30–90-day filtered precipitation (*upper*) and 200-hPa velocity potential (*lower*) anomalies. (*right*) The signal-to-noise ratios (using the methods in Waliser *et al.*, 2003, see discussion in Section 12.4) for precipitation (*upper*) and 200-hPa velocity potential (*lower*) anomalies when combining 15-member ensembles from 3 different model ISO events and including all 4 phases of each event (i.e., $N = 168$).
From Liess *et al.* (2004).

onto the EOF pair, along with the prior removal of an estimate of the data's very low-frequency components (e.g., ENSO) via their relationship to interannual SST variability, remove the need to perform time filtering to identify the MJO. The values of the index (actually two indices, one amplitude time series for mode 1 (RMM1) and one for mode 2 (RMM2)) at any given time can be used for monitoring. In addition, seasonally and time-lag dependent regression can be used to forecast the evolution of these indices or any associated field, using as predictors RMM1 and RMM2 at the initial day. Skill scores in terms of spatial correlation are about 0.6 for a 12-day forecast and 0.5 for a 15-day forecast. The advantages of the method are that it has a seasonal dependence built in and it can be easily adapted for forecasting nearly any field related to the MJO (see Chapter 5 for more details).

Charles Jones and his colleagues have produced real-time predictions of the MJO for about two years. The scheme utilized has evolved over this time period and the most recent version is described in Jones *et al.* (2003b). The model is based on band-passed (20–90 days) OLR, and zonal winds at 850 and 200 hPa. Upon filtering, a combined EOF of the three fields is computed and then the PCs are separated into summer and winter. A seasonally dependent regression model is then formed at every given lead between 1 and 10 pentads. The model utilizes the first five PCs from the EOF analysis and the five most recent values of the PCs. The model is found to exhibit winter and summer skills comparable to the other empirical models described in Section 12.2.

In quite a different approach, stemming from a somewhat different and/or more comprehensive objective, Matt Newman and his colleagues have developed and implemented a real-time forecasting scheme that has applicability to the MJO based on what is often referred to as the Linear Inverse Model (LIM; Winkler *et al.*, 2001; Newman *et al.*, 2003). The LIM is based on NCEP–NCAR Reanalysis data (Kalnay *et al.*, 1996) that has had the annual cycle removed, been smoothed with a 7-day running mean filter, gridded to T21 spatial resolution, and been reduced by EOF decomposition. The specific fields used include global 250 and 750-hPa streamfunction and tropical column-integrated diabatic heating. For the northern hemisphere winter (summer) model, the first 30 (30) streamfunction and 7 (20) diabatic heating EOFs are used. In this model, historical data are used to define the relationship between a given state (i.e., a weekly average) and conditions one week later, with the process being iterated to produce multi-week forecasts. The advantage of the model is that it includes both tropical (in terms of diabatic heating – hence a prediction of the MJO) and extra-tropical (in terms of streamfunction) forecasts. In this way, the interaction between can be more readily examined and diagnosed. For tropical forecasts of diabatic heating, the LIM slightly outperforms a research version of the NCEP MRF model at lead times of 2 weeks for both northern hemisphere summer and winter, particularly in regions where the MJO is most strongly affecting the diabatic heating field (Figure 12.14).

Van den Dool and Qin (1996) developed a generalized wave propagating forecasting technique that they refer to as "empirical wave propagation" (EWP). EWP is a "phase-shifting" technique that allows one, in the diagnostic step, to determine the amplitude-weighted-average climatological phase speed of anomaly waves (e.g., equatorial MJO), where the waves are represented as either zonal or spherical harmonics. The diagnostic step results in a table of phase speed (or one-day displacement) for waves in the anomaly field as a function of zonal wave number, calendar month, and latitude, based on a specified (model or observed) data set. Its first application was to mid-latitude Rossby waves as diagnosed from 500-hPa geopotential height fields (Qin and van den Dool, 1996). More recently, it has been applied to the MJO (van den Dool and Saha, 2002) and implemented in real time. Figure 12.15 shows the results from the diagnostic step based on analysis of five years of 200-hPa velocity potential analysis data for all seasons. In this case, the wave number 1 disturbance propagates at about $5\,\mathrm{m\,s^{-1}}$ and has an amplitude of about $5\times10^{6}\,\mathrm{m^2\,s^{-1}}$. In the forecast step, given an initial anomaly field, one projects

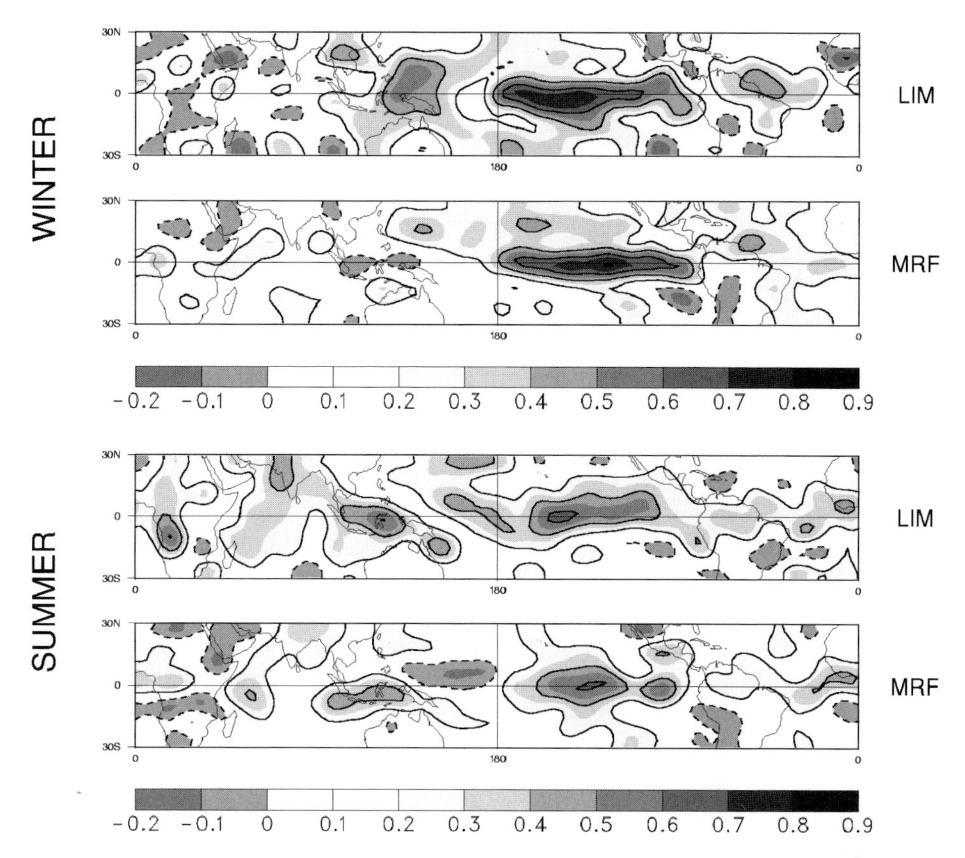

Figure 12.14. Anomaly correlations between forecast and verifiation column-integrated diabatic heating using the LIM forecast model (Winkler *et al.*, 2001; Newman *et al.*, 2003) and a research version of the NCEP MRF model (i.e., MRF98) for both the northern hemisphere winter (*top*) and summer (*bottom*). Forecasts were made for June–August periods for the years 1979–2000. Solid (dashed) contours indicate positive (negative) values.

the initial condition onto sines and cosines or spherical harmonics, then propagates each wave over the longitude displacement provided by the table, and transforms the field back to physical space. This technique is particularly well suited for empirically forecasting the large-scale upper-level anomalies associated with the MJO.

It is almost a certainty that the MJO's greatest impact based on sheer numbers of people and the severity of losses in agriculture and economics is associated with the Asian summer monsoon. Motivated by this, Webster and Hoyos (2003) have developed an empirical model for predicting Indian district rainfall and the Brahmaputra and Ganges River discharge into Bangladesh on 20–25-day timescales. The empirical model is physically based with predictors drawn from the composite structure of the monsoon ISV (e.g., Indian Ocean SST, precipitation over India, upper level easterly jet, surface winds over the Arabian Sea). In essence, the model

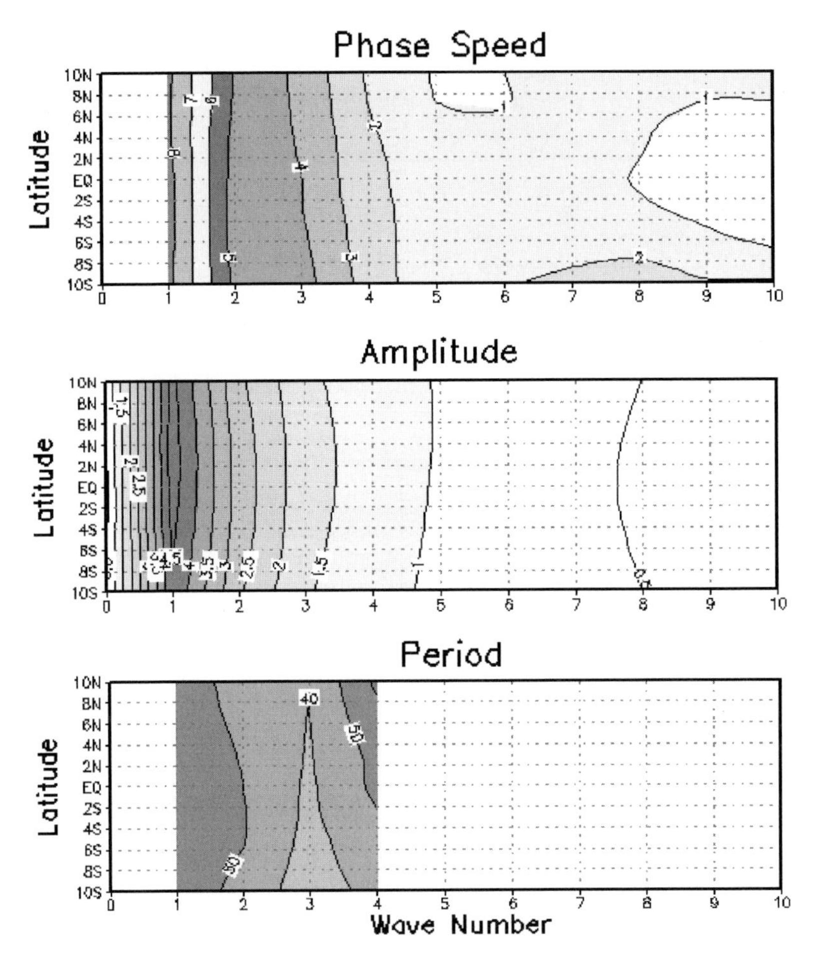

Figure 12.15. Diagnostic information provided by the EWP technique for observed 200-hPa velocity potential anomalies for all seasons combined for the years 1979–1983. The horizontal axis in each plot is global wave number. The units are $\mathrm{m\,s^{-1}}$, $1 \times 10^6\,\mathrm{m^2\,s^{-1}}$, and days, for the upper, middle, and lower plots, respectively.
Based on van den Dool and Qin (1996) and van den Dool (2002).

is Bayesian and uses a wavelet technique to separate significant spectral bands. The model has been used successfully to predict rainfall in hindcast mode. For example, Figure 12.16 shows observed and 20-day forecasts of 5-day average rainfall amounts over the Ganges Valley for the summers of 1999–2002. From these hindcasts, it appears the model is well adept at capturing the seasonal, interannual, and subseasonal rainfall variability. The model was also used for the first time during the summer of 2003 in a real-time operational mode in the Climate Forecast Application in Bangladesh project as part of a 3-tier forecasting system wherein seasonal outlooks are given every month for the ensuing 6 months, a 20–25-day forecast is

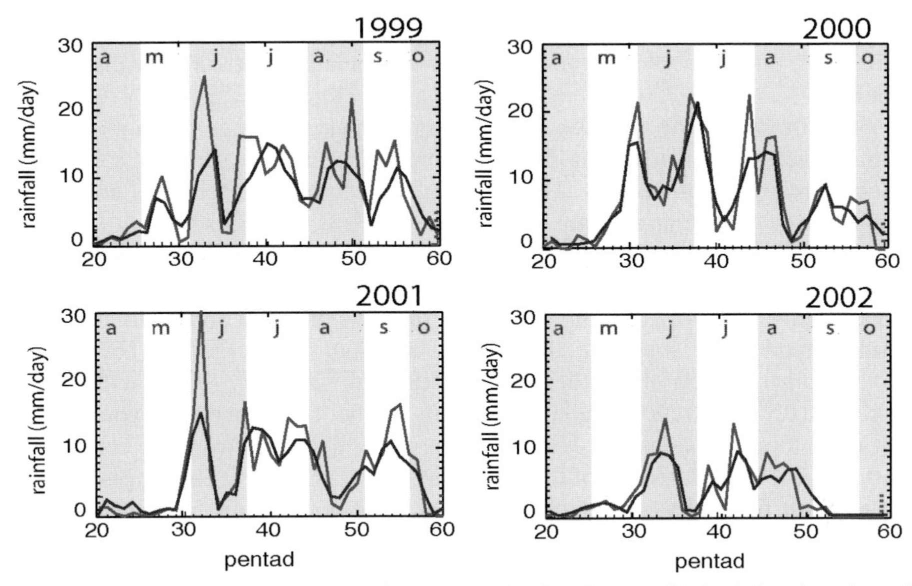

Figure 12.16. Gray lines indicate the forecasts at 20-day (4 pentads) lead time based on the empirical model of Webster and Hoyas (2003) of precipitation averaged over the Ganges Valley for the summers of 1999–2002. Black lines indicate observations. Gray and white background denotes months.

prepared every 5 days, and a 1–5-day forecast is prepared daily. At the time of writing, these forecasts of precipitation and river discharge have been integrated into the Bangladesh forecasting system on an experimental basis.

Based on the sorts of activities and preliminary successes described above, along with the needs to take a more systematic approach to diagnosing problems in dynamical forecasts of the MJO, an experimental MJO prediction program has recently been formulated and is in the process of being implemented. The formal components of this program arose from two parallel streams of activity. The first was the occurrence of the sub-seasonal workshop mentioned in the introduction of this chapter (Schubert *et al.*, 2002) and the recognition of the importance of the MJO in regards to the potential skill to be had from sub-seasonal predictions. The second stream of activity ensued from the priorities and recommendations of the US CLIVAR Asian–Australian Monsoon Working Group (AAMWG). In their 2001 research prospectus (AAMWG *et al.*, 2001) as well as in subsequent deliberations with the US CLIVAR Scientific Steering Committee, recommendations were made to develop an experimental prediction program due to the significant influence that the MJO has on the character and evolution of Asian–Australian monsoons. These streams of activity led to an informal (E-mail) discussion among a number of MJO forecast enthusiasts during the summer and fall of 2002 that helped to formulate the framework for such a program and to the identification of a sponsor that could provide scientific and technical support as well as serve as the data host/server (i.e., NOAA's Climate Diagnostics Center). Invitations to participate in the program were subsequently sent out to a number of empirical modelers and international forecasting agencies, and an implementation meeting was held in June 2003 (Waliser *et al.*, 2003a).

The motivation for the above experimental program involves not only the obvious objective of forecasting MJO variability but also to serve as a basis for model intercomparison studies. The latter includes using the forecasts and biases in model error growth as a means to learn more about, and possibly rectify, model shortcomings, and also includes using the empirical models to provide some measure of the expectations that should be attributed to the dynamical models in terms of MJO predictive skill. In addition, it is hoped that this program and its forecasts will provide a modeling resource to those trying to diagnose interactions between the MJO and other aspects of weather and sub-seasonal variability (e.g., Pacific–North American (PNA), Arctic Oscillation (AO)). While the immediate goal of the program is to assemble and provide what is readily available from the community in terms of 2–4-week forecasts of the MJO, there are a number of challenges faced by such an effort that are worth highlighting. The most notable involve 1) how to deal with forecast models that have yet or routinely do not have a lead-dependent forecast climatology which is necessary to remove a model's systematic biases, 2) the degree that coupled models and ensembles need to be or can be incorporated into the project, 3) the manner the MJO signal(s) are to be extracted from the heterogeneous set of models (e.g., empirical and numerical), and 4) the general logistical problems of dealing with assembling a very non-uniform set of forecast products from different agencies and researchers in near real time and streamlining them for the purpose of this project.

12.6 DISCUSSION

The review of the studies examined in this chapter was meant to summarize the historical developments associated with MJO forecasting and provide a brief description of the current state of affairs in regards to MJO prediction capability and what is known of its inherent limits of predictability. Overall, there is enough evidence to suggest that MJO prediction can be approached with considerable optimism due to the facts that our capabilities as yet seem far from saturating their potential, and once exploited in an operational sense, will provide a unique and important link between our more mature areas of forecasting, namely weather and ENSO. At present, one of our greatest challenges is still to develop robust and realistic representations of the MJO in our weather and climate forecast models (Chapter 11). Once we have such a capability, we not only have a means to improve predictions of low-frequency weather variations in the tropics that are directly impacted by the MJO, including the onsets and breaks of the Asian and Australian summer monsoons (e.g., Yasunari, 1980; Lau and Chan, 1986; Hendon and Liebmann, 1990a, b; see also Chapters 2, 3, and 5), but we will also likely improve forecasts associated with a number of processes remote to the MJO (see Chapter 4). These include wintertime mid-latitude circulation anomalies (e.g., Weickmann, 1983;

Liebmann and Hartmann, 1984; Weickmann *et al.*, 1985; Lau and Phillips, 1986; Ferranti *et al.*, 1990; Higgins and Schubert, 1996; Higgins and Mo, 1997; Jones *et al.*, 2003a), summertime precipitation variability over Mexico and South America as well as to austral wintertime circulation anomalies over the Pacific–South American sector (e.g., Nogues-Paegle and Mo, 1997; Mo and Higgins, 1998b; Jones and Schemm, 2000; Mo, 2000b; Paegle *et al.*, 2000), extreme events in rainfall variability along the western U.S.A. (e.g., Mo and Higgins, 1998a, c; Higgins *et al.*, 2000; Jones, 2000; Whitaker and Weickmann, 2001), and the development of tropical storms/hurricanes in both the Atlantic and Pacific sectors (Maloney and Hartmann, 2000a, b; Mo, 2000a; Higgins and Shi, 2001).

From the discussion in this chapter, as well as those in Chapters 7, 10, and 11, a number of areas of research and development present themselves. These include a more complete understanding of the role that coupling to the ocean plays in maintaining, and in particular forecasting, the MJO (e.g., Flatau *et al.*, 1997; Wang and Xie, 1998; Waliser *et al.*, 1999a; Kemball-Cook *et al.*, 2002; Fu *et al.*, 2003). For example, a number of recent studies have indicated an error in the phase relation between the convection and SSTAs associated with the MJO in GCM simulations using specified SSTs, whereas coupled simulations tend to reproduce the observed phase relationship (Wu *et al.*, 2002; Fu and Wang, 2003; Zheng *et al.*, 2003). These sorts of studies not only imply the importance of incorporating MJO-related SSTAs but also necessitate they be coupled (i.e., forecast) as well. In addition, there has been virtually no research (albeit the somewhat related work by Krishnamurti *et al.* discussed in Section 12.3) done on model initialization/data assimilation issues in terms of what are the critical criteria to meet in order to adequately initialize the state of the MJO. Related to this are issues regarding the importance of the basic state of the forecast model and how an incorrect basic state might negatively impact the maintenance and propagation of the MJO (Inness *et al.*, 2003; Liess and Bengtsson, 2003; Liess *et al.*, 2003; Sperber *et al.*, 2003). Additional avenues of research include exploring the methods proposed by Krishnamurti *et al.* (1990) with other present-day forecast systems and on more MJO cases as well as exploring the possibility of assimilating empirically-derived forecasts of the MJO into extended-range weather forecasts in order to improve their forecasts of the MJO as well as the remote processes and secondary circulations they interact with. In addition to the above, there is clearly a need for additional dynamical predictability studies of the MJO using other GCMs as well as sensitivity studies to test the effects of SST coupling and ENSO state, the impacts from/on mid-latitude variability, and the influence of the size and type of initial condition perturbations and definition of predictability.

12.7 APPENDIX

Excerpt from John von Neumann (1955):

> It seems quite plausible from general experience that in any mathematical problem it is easiest to determine the solution for shorter periods, over which the extrapolation

parameter is small. The next most difficult problem to solve is that of determining the asymptotic conditions – that is, the conditions that exist over periods for which the extrapolation parameter is very large, say near infinity. Finally, the most difficult is the intermediate range problem, for which the extrapolation parameter is neither very small nor very large. In this case the neglect of either extreme is forbidden. On the basis of these considerations, it follows that there is a perfectly logical approach to any computational treatment of the problem of weather prediction. The approach is to try first short-range forecasts, then long-range forecasts of those properties of the circulation that can perpetuate themselves over arbitrarily long periods of time (other things being equal), and only finally to attempt forecast for medium–long time periods which are too long to treat by simple hydrodynamic theory and too short to treat by the general principles of equilibrium theory.

12.8 ACKNOWLEDGMENTS

This work was supported by the National Science Foundation (ATM-0094416), the National Oceanographic and Atmospheric Administration (NA16GP2021), and the National Atmospheric and Aeronautics Administration (NAG5-11033).

12.9 REFERENCES

AAMWG, US CLIVAR, W. K. M. Lau, S. Hastenrath, B. Kirtman, T. N. Krishnamurti, R. Lukas, J. McCreary, J. Shukla, J. Shuttleworth, D. Waliser, *et al.* (2001) An Asian–Australian Monsoon Research Prospectus by the Asian-Australian Monsoon Working Group, 46 pp.

Barnston, A. G., M. H. Glantz, and Y. X. He (1999) Predictive skill of statistical and dynamical climate models in SST forecasts during the 1997–98 El Niño episode and the 1998 La Niña onset. *Bull. Amer. Meteorol. Soc.*, **80**, 217–243.

Cai, M., E. Kalnay, and Z. Toth (2003) Bred vectors of the Zebiak-Cane model and their potential application to ENSO predictions. *J. Climate*, **16**, 40–56.

Cane, M. A., S. E. Zebiak, and S. C. Dolan (1986) Experimental Forecasts of El-Niño. *Nature*, **321**, 827–832.

Chen, T. C. and J. C. Alpert (1990) Systematic errors in the annual and intraseasonal variations of the planetary-scale divergent circulation in NMC medium-range forecasts. *J. Atmos. Sci.*, **118**, 2607–2623.

Ferranti, L., T. N. Palmer, F. Molteni, and K. Klinker (1990) Tropical–extratropical interaction associated with the 30–60-day oscillation and its impact on medium and extended range prediction. *J. Atmos. Sci.*, **47**, 2177–2199.

Flatau, M., P. J. Flatau, P. Phoebus, and P. P. Niller (1997) The feedback between equatorial convection and local radiative and evaporative processes: The implications for intraseasonal oscillations. *J. Atmos. Sci.*, **54**, 2373–2386.

Fu, X. and B. Wang (2003) Different solutions of intraseasonal oscillation exist in atmosphere–ocean coupled model and atmosphere-only model. *Journal of Climate*, submitted.

Fu, X., B. Wang, T. Li, and J. McCreary (2003) Coupling between northward-propagating boreal summer ISO and Indian Ocean SST: Revealed in an atmosphere–ocean coupled model. *J. Atmos. Sci.*, **60**, 1733–1753.

Goswami, B. N. and P. Xavier (2003) Potential predictability and extended range prediction of Indian summer monsoon breaks. *Geophys. Res. Lett.*, **30**, 1966, doi:10.1029/2003GL017,810.

Graham, N. E. and T. P. Barnett (1995) ENSO and ENSO-related predictability. Part 2: northern-hemisphere 700-Mb height predictions based on a hybrid coupled ENSO model. *J. Climate*, **8**, 544–549.

Hendon, H. H. and B. Liebmann (1990a) The intraseasonal (30–50 Day) oscillation of the Australian summer monsoon. *J. Atmos. Sci.*, **47**, 2909–2923.

Hendon, H. H. and B. Liebmann (1990b) A composite study of onset of the Australian summer monsoon. *J. Atmos. Sci.*, **47**, 2227–2240.

Hendon, H. H., C. D. Zhang, and J. D. Glick (1999) Interannual variation of the Madden–Julian Oscillation during austral summer. *J. Climate*, **12**, 2538–2550.

Hendon, H. H., B. Liebmann, M. Newman, J. D. Glick, and J. K. E. Schemm (2000) Medium-range forecast errors associated with active episodes of the Madden–Julian Oscillation. *Mon. Weather Rev.*, **128**, 69–86.

Higgins, R. W. and S. D. Schubert (1996) Simulations of persistent North Pacific circulation anomalies and interhemispheric teleconnections. *J. Atmos. Sci.*, **53**, 188–207.

Higgins, R. W. and K. C. Mo (1997) Persistent North Pacific circulation anomalies and the tropical intraseasonal oscillation. *J. Climate*, **10**, 223–244.

Higgins, R. W. and W. Shi (2001) Intercomparison of the principal modes of interannual and intraseasonal variability of the North American Monsoon System. *J. Climate*, **14**, 403–417.

Higgins, R. W., J. K. E. Schemm, W. Shi, and A. Leetmaa (2000) Extreme precipitation events in the western United States related to tropical forcing. *J. Climate*, **13**, 793–820.

Inness, P. M., J. M. Slingo, E. Guilyardi, and J. Cole (2003) Simulation of the Madden–Julian Oscillation in a coupled general circulation model. Part II: The role of the basic state. *J. Climate*, **16**, 365–382.

Jones, C. (2000) Occurrence of extreme precipitation events in California and relationships with the Madden–Julian Oscillation. *J. Climate*, **13**, 3576–3587.

Jones, C. and J. K. E. Schemm (2000) The influence of intraseasonal variations on medium- to extended-range weather forecasts over South America. *Mon. Wea. Rev.*, **128**, 486–494.

Jones, C., D. E. Waliser, J. K. E. Schemm, and W. K. M. Lau (2000) Prediction skill of the Madden and Julian Oscillation in dynamical extended range forecasts. *Clim. Dyn.*, **16**, 273–289.

Jones, C., D. E. Waliser, K. M. Lau, and W. Stern (2003a) The Madden–Julian Oscillation and its impact on Northern Hemisphere weather predictability. *Mon. Wea. Rev.*, submitted.

Jones, C., L. M. V. Carvalho, R. W. Higgins, D. E. Waliser, and J. K. E. Schemm (2003b) Statistical forecast skill of tropical intraseasonal convective anomalies. *J. Climate*, submitted.

Kalnay, E., M. Kanamitsu, R. Kistler, W. Collins, D. Deaven, L. Gandin, M. Iredell, S. Saha, G. White, J. Woollen, *et al.* (1996) The NCEP/NCAR 40-year reanalysis project. *Bull. Amer. Meteorol. Soc.*, **77**, 437–471.

Kalnay, E. R., W. Balgovind, D. Chao, J. Edelmann, L. Pfaendtner, L. Takacs, and K. Takano (1983) *Documentation of the GLAS fourth order general circulation model* (Volume 1, NASA Tech. Memo. No. 86064). NASA Goddard Space Flight Center, Greenbelt, M.D.

Kemball-Cook, S., B. Wang, and X. Fu (2002) Simulation of the ISO in the ECHAM4 model: The impact of coupling with an ocean model. *J. Atmos. Sci.*, **59**, 1433–1453.

Keppenne, C. L. and M. Ghil (1992) Adaptive filtering and prediction of the Southern Oscillation index. *J. Geophys. Res. Atmos.*, **97**, 20449–20454.

Kirtman, B. P., J. Shukla, B. H. Huang, Z. X. Zhu, and E. K. Schneider (1997) Multiseasonal predictions with a coupled tropical ocean–global atmosphere system. *Mon. Wea. Rev.*, **125**, 789–808.

Kousky, V. E. and M. T. Kayano (1993) Real-time monitoring of intraseasonal oscillations. 18th Annual Climate Diagnostics Workshop, 1–5 November, 1993, Boulder, CO.

Krishnamurti, T. N., D. K. Oosterhof, and A. V. Mehta (1988) Air sea interaction on the time scale of 30 to 50 days. *J. Atmos. Sci.*, **45**, 1304–1322.

Krishnamurti, T. N., S. O. Han, and V. Misra (1995) Prediction of the dry and wet spell of the Australian monsoon. *Int. J. Climatol.*, **15**, 753–771.

Krishnamurti, T. N., M. Subramaniam, D. K. Oosterhof, and G. Daughenbaugh (1990) Predictability of low-frequency modes. *Meteorol. Atmos. Phys.*, **44**, 63–83.

Krishnamurti, T. N., M. Subramaniam, G. Daughenbaugh, D. Oosterhof, and J. H. Xue (1992) One-month forecasts of wet and dry spells of the monsoon. *Mon. Wea. Rev.*, **120**, 1191–1223.

Lau, K. M. and P. H. Chan (1986) Aspects of the 40–50 day oscillation during the northern summer as inferred from outgoing longwave radiation. *Mon. Wea. Rev.*, **114**, 1354–1367.

Lau, K. M. and T. J. Phillips (1986) Coherent fluctuations of extratropical geopotential height and tropical convection in intraseasonal time scales. *J. Atmos. Sci.*, **43**, 1164–1181.

Lau, K. M. and F. C. Chang (1992) Tropical intraseasonal oscillation and its prediction by the NMC operational model. *J. Climate*, **5**, 1365–1378.

Liebmann, B. and D. L. Hartmann (1984) An observational study of tropical midlatitude interaction on intraseasonal time scales during winter. *J. Atmos. Sci.*, **41**, 3333–3350.

Liess, S. and L. Bengtsson (2003) The intraseasonal oscillation in ECHAM4. Part II: Sensitivity studies. *Clim. Dyn.*, **22**, 671–688.

Liess, S., D. E. Waliser, and S. Schubert (2004a) Predictability studies of the intraseasonal oscillation with the ECHAM5 GCM. *J. Atmos. Sci.*, submitted.

Liess, S., L. Bengtsson, and K. Arpe (2004b) The Intraseasonal Oscillation in ECHAM4. Part I: Coupled to a comprehensive ocean model. *Clim. Dyn.*, **22**, 653–688.

Lo, F. and H. H. Hendon (2000) Empirical extended-range prediction of the Madden–Julian Oscillation. *Mon. Wea. Rev.*, **128**, 2528–2543.

Lorenz, E. N. (1965) A study of the predictability of a 28-variable atmospheric model. *Tellus*, **17**, 321–333.

Lorenz, E. N. (1982) Atmospheric predictability experiments with a large numerical-model. *Tellus*, **34**, 505–513.

Maloney, E. D. and D. L. Hartmann (2000a) Modulation of eastern North Pacific hurricanes by the Madden–Julian Oscillation. *J. Climate*, **13**, 1451–1460.

Maloney, E. D. and D. L. Hartmann (2000b) Modulation of hurricane activity in the Gulf of Mexico by the Madden–Julian Oscillation. *Science*, **287**, 2002–2004.

Matsuno, T. (1966) Quasi-geostrophic motions in the equatorial area. *J. Met. Soc. Jap.*, **44**, 25–43.

Mo, K. C. (2000a) The association between intraseasonal oscillations and tropical storms in the Atlantic basin. *Mon. Wea. Rev.*, **128**, 4097–4107.

Mo, K. C. (2000b) Intraseasonal modulation of summer precipitation over North America. *Mon. Wea. Rev.*, **128**, 1490–1505.

Mo, K. C. (2001) Adaptive filtering and prediction of intraseasonal oscillations. *Mon. Wea. Rev.*, **129**, 802–817.

Mo, K. C. and R. W. Higgins (1998a) Tropical convection and precipitation regimes in the western United States. *J. Climate*, **11**, 2404–2423.

Mo, K. C. and R. W. Higgins (1998b) The Pacific–South American modes and tropical convection during the Southern Hemisphere winter. *Mon. Wea. Rev.*, **126**, 1581–1596.

Mo, K. C. and R. W. Higgins (1998c) Tropical influences on California precipitation. *J. Climate*, **11**, 412–430.

Neuman, J. V. (1955) Some remarks on the problem of forecasting climate fluctuations. *"Dynamics of Climate": The Proceedings of a Conference on the Application of Numerical Integration Techniques to the Problem of the General Circulation*, Pergamon Press, 137.

Newman, M., P. D. Sardeshmukh, C. R. Winkler, and J. S. Whitaker (2003) A study of subseasonal predictability. *Mon. Wea. Rev.*, **131**, 1715–1732.

Nogues-Paegle, J. and K. C. Mo (1997) Alternating wet and dry conditions over South America during summer. *Mon. Wea. Review*, **125**, 279–291.

Paegle, J. N., L. A. Byerle, and K. C. Mo (2000) Intraseasonal modulation of South American summer precipitation. *Mon. Wea. Rev.*, **128**, 837–850.

Palmer, T. N. (1993) Extended-range atmospheric prediction and the Lorenz model. *Bull. Amer. Meteorol. Soc.*, **74**, 49–65.

Qin, J. and H. M. van den Dool (1996) Simple extensions of an NWP model. *Mon. Wea. Rev.*, **124**, 277–287.

Salby, M. L. and H. H. Hendon (1994) Intraseasonal behavior of clouds, temperature, and motion in the tropics. *J. Atmos. Sci.*, **51**, 2207–2224.

Schemm, J. E., H. M. van den Dool, and S. Saha (1996) A multi-year DERF experiment at NCEP. *11th Conference on Numerical Weather Prediction, August 19–13*, Norfolk, Virginia, 47–49.

Schubert, S., R. Dole, H. M. van den Dool, M. Suarez, and D. Waliser (2002) Proceedings from a workshop on "Prospects for improved forecasts of weather and short-term climate variability on subseasonal (2 week to 2 month) time scales", 16–18 April 2002, Mitchellville, MD, NASA/TM 2002-104606, vol. 23, 171 pp.

Shukla, J. (1985) Predictability. *Adv. Geophys.*, **28B**, 87–122.

Slingo, J. M., K. R. Sperber, J. S. Boyle, J. P. Ceron, M. Dix, B. Dugas, W. Ebisuzaki, J. Fyfe, D. Gregory, J. F. Gueremy, *et al.* (1996) Intraseasonal oscillations in 15 atmospheric general circulation models: Results from an AMIP diagnostic subproject. *Clim. Dyn.*, **12**, 325–357.

Sperber, K. R., J. M. Slingo, P. M. Inness, and W. K. M. Lau (1997) On the maintenance and initiation of the intraseasonal oscillation in the NCEP/NCAR reanalysis and in the GLA and UKMO AMIP simulations. *Clim. Dyn.*, **13**, 769–795.

Sperber, K. R., J. M. Slingo, P. M. Inness (2003) The Madden–Julian Oscillation in GCMs in Research Activities in Atmospheric and Oceanic Modelling, Report No. 33, WMO/TD-No. 1161, p. 09-01.

Sud, Y. C. and G. K. Walker (1992) A review of recent research on improvement of physical parameterizations in the GLA GCM. In: D. R. Sikka and S. S. Singh (eds), *Physical Processes in Atmospheric Models*. Wiley Eastern Ltd., pp. 422–479.

Thompson, P. D. (1957) Uncertainty of initial state as a factor in the predictability of large scale atmospheric flow patterns. *Tellus*, **9**, 275–295.

Toth, Z. and E. Kalnay (1993) Ensemble forecasting at NMC: The generation of perturbations. *Bull. Amer. Meteorol. Soc.*, **74**, 2330–2371.

van den Dool, H. M. (1994) Long-range weather forecasts through numerical and empirical-methods. *Dyn. Atmos. Oceans*, **20**, 247–270.

van den Dool, H. M. and J. Qin (1996) An efficient and accurate method of continuous time interpolation of large-scale atmospheric fields. *Mon. Weather Rev.*, **124**, 964–971.

van den Dool, H. M. and S. Saha (2002) Analysis of propagating modes in the tropics in short AMIP runs. *AMIPII workshop. WMO. Toulouse, November 12–15, 2002.*

Vautard, R. and M. Ghil (1989) Singular spectrum analysis in nonlinear dynamics, with applications to paleoclimatic time-series. *Physica D*, **35**, 395–424.

von Neumann, J. (1955) Some remarks on the problem of forecasting climate fluctuations. *Dynamics of Climate. The proceedings of a conference on the application of numerical integration techniques to the problem of the general circulation*, Pergammon Press, 137.

von Storch, H. and J. Xu (1990) Principal oscillation pattern analysis of the 30- to 60-day oscillation in the tropical troposphere. *Clim. Dyn.*, **4**, 175–190.

Waliser, D., Z. Zhang, K. M. Lau, and J. H. Kim (2001) Interannual sea surface temperature variability and the predictability of tropical intraseasonal variability. *J. Atmos. Sci.*, **58**, 2595–2614.

Waliser, D., S. Schubert, A. Kumar, K. Weickmann, and R. Dole (2003a) Proceedings from a workshop on "Modeling, Simulation and Forecasting of Subseasonal Variability", 4–5 June 2003, University of Maryland, College Park, Maryland, NASA/TM 2003-104606, vol. 25, 67 pp.

Waliser, D. E., K. M. Lau, and J. H. Kim (1999a) The influence of coupled sea surface temperatures on the Madden–Julian Oscillation: A model perturbation experiment. *J. Atmos. Sci.*, **56**, 333–358.

Waliser, D. E., C. Jones, J. K. E. Schemm, and N. E. Graham (1999b) A statistical extended-range tropical forecast model based on the slow evolution of the Madden–Julian Oscillation. *J. Climate*, **12**, 1918–1939.

Waliser, D. E., K. M. Lau, W. Stern, and C. Jones (2003b) Potential predictability of the Madden–Julian Oscillation. *Bull. Amer. Meteor. Soc.*, **84**, 33–50.

Waliser, D. E., W. Stern, S. Schubert, and K. M. Lau (2003c) Dynamic predictability of intraseasonal variability associated with the Asian summer monsoon. *Q. J. R. Meteor. Soc.*, **129**, 2897–2925.

Waliser, D. E., K. Jin, I. S. Kang, W. F. Stern, S. D. Schubert, M. L. Wu, K. M. Lau, M. I. Lee, J. Shukla, V. Krishnamurthy *et al.* (2003d) AGCM simulations of intraseasonal variability associated with the Asian summer monsoon. *Clim. Dyn.*, **21**, 423–446.

Wang, B. and X. S. Xie (1998) Coupled modes of the warm pool climate system. Part 1: The role of air–sea interaction in maintaining Madden–Julian Oscillation. *J. Climate*, **11**, 2116–2135.

Webster, P. J. and C. Hoyos (2003) Forecasting monsoon rainfall and river discharge variability on 20–25 day time scales. *Bull. Amer. Meteorol. Soc.*, in press.

Weickmann, K. M. (1983) Intraseasonal circulation and outgoing longwave radiation modes during northern hemisphere winter. *Mon. Wea. Rev.*, **111**, 1838–1858.

Weickmann, K. M., G. R. Lussky, and J. E. Kutzbach (1985) Intraseasonal (30–60 day) fluctuations of outgoing longwave radiation and 250-Mb stream-function during northern winter. *Mon. Wea. Rev.*, **113**, 941–961.

Wheeler, M. and G. N. Kiladis (1999) Convectively coupled equatorial waves: Analysis of clouds and temperature in the wavenumber–frequency domain. *J. Atmos. Sci.*, **56**, 374–399.

Wheeler, M. and K. M. Weickmann (2001) Real-time monitoring and prediction of modes of coherent synoptic to intraseasonal tropical variability. *Mon. Wea. Rev.*, **129**, 2677–2694.

Wheeler, M. and H. Hendon (2003) An all-season real-time multivariate MJO index: Development of an index for monitoring and prediction. *Monthly Weather Review*, submitted.

Whitaker, J. S. and K. M. Weickmann (2001) Subseasonal variations of tropical convection and week-2 prediction of wintertime western north American rainfall. *J. Climate*, **14**, 3279–3288.

Winkler, C. R., M. Newman, and P. D. Sardeshmukh (2001) A linear model of wintertime low-frequency variability. Part I: Formulation and forecast skill. *J. Climate*, **14**, 4474–4494.

Wu, M. L. C., S. Schubert, I. S. Kang, and D. E. Waliser (2002) Forced and free intra-seasonal variability over the South Asian monsoon region simulated by 10 AGCMs. *J. Climate*, **15**, 2862–2880.

Yasunari, T. (1980) A quasi-stationary appearance of the 30–40 day period in the cloudiness fluctuations during the summer monsoon over India. *J. Met. Soc. Jap.*, **59**, 336–354.

Zheng, Y., D. E. Waliser, W. F. Stern, and C. Jones (2003) The role of coupled sea surface temperatures in the simulation of the tropical intraseasonal oscillation. *J. Climate*, **17**, 4109–4134.

Index

Note: page numbers in italics refer to figures.